Noninvasive Prenatal Testing (NIPT)

D1354621

Noninvasive Prenatal Testing (NIPT)

Applied Genomics in Prenatal Screening and Diagnosis

Edited by

Lieve Page-Christiaens

Hanns-Georg Klein

ACADEMIC PRESS

An imprint of Elsevier

Academic Press is an imprint of Elsevier
125 London Wall, London EC2Y 5AS, United Kingdom
525 B Street, Suite 1650, San Diego, CA 92101, United States
50 Hampshire Street, 5th Floor, Cambridge, MA 02139, United States
The Boulevard, Langford Lane, Kidlington, Oxford OX5 1GB, United Kingdom

Notices
Knowledge and best practice in this field are constantly changing. As new research and experience broaden our understanding,
changes in research methods, professional practices, or medical treatment may become necessary.

Practitioners and researchers must always rely on their own experience and knowledge in evaluating and using any information,
methods, compounds, or experiments described herein. In using such information or methods they should be mindful of their own
safety and the safety of others, including parties for whom they have a professional responsibility.

To the fullest extent of the law, neither the Publisher nor the authors, contributors, or editors, assume any liability for any injury and/
or damage to persons or property as a matter of products liability, negligence or otherwise, or from any use or operation of any
methods, products, instructions, or ideas contained in the material herein.

Library of Congress Cataloging-in-Publication Data
A catalog record for this book is available from the Library of Congress

British Library Cataloguing-in-Publication Data
A catalogue record for this book is available from the British Library

ISBN 978-0-12-814189-2

For information on all Academic Press publications
visit our website at https://www.elsevier.com/books-and-journals

www.elsevier.com • www.bookaid.org

Publisher: John Fedor
Acquisition Editor: Peter B. Linsley
Editorial Project Manager: Carlos Rodriguez
Production Project Manager: Sreejith Viswanathan
Cover Designer: Christian Bilbow

Typeset by SPi Global, India

We dedicate this book to all the authors who, with no other incentive
but sharing their knowledge and expertise, have invested their precious time
in creating this book.

Lieve Page-Christiaens
Hanns-Georg Klein

Contents

vii

SECTION 3   CLINICAL INTEGRATION

CHAPTER 15 Ethics of Cell-Free DNA-Based Prenatal Testing for Sex Chromosome Aneuploidies and Sex Determination.. 251

Wybo Dondorp, Angus Clarke, Guido de Wert

CHAPTER 16 Cost-Effectiveness of Cell-Free DNA-BASED Noninvasive Prenatal Testing: Summary of Evidence and Challenges.............................. 269

Stephen Morris, Caroline S. Clarke, Emma Hudson

SECTION 4 THE FUTURE

Contributors

Stephanie Allen
Birmingham Women's and Children's NHS Foundation Trust, Edgbaston; West Midlands Regional Genetics Laboratory, Birmingham Women's and Children's NHS Foundation Trust, Birmingham, United Kingdom

Arthur L. Beaudet
Department of Molecular and Human Genetics, Baylor College of Medicine, Houston, TX, United States

D. Stephen Charnock-Jones
Department of Obstetrics and Gynaecology, University of Cambridge, NIHR Cambridge Comprehensive Biomedical Research Centre; Centre for Trophoblast Research (CTR), Department of Physiology, Development and Neuroscience, University of Cambridge, Cambridge, United Kingdom

S.H. Cheng
Li Ka Shing Institute of Health Sciences; Department of Chemical Pathology, The Chinese University of Hong Kong, Prince of Wales Hospital, Shatin, New Territories, Hong Kong SAR, China

Angus Clarke
Institute of Cancer and Genetics, Cardiff University, United Kingdom

Caroline S. Clarke
Research Department of Primary Care & Population Health, University College London, London, United Kingdom

Frederik B. Clausen
Department of Clinical Immunology, Rigshospitalet, Copenhagen University Hospital, Copenhagen, Denmark

Guido de Wert
Department of Health, Ethics & Society, Research Schools CAPHRI & GROW, Maastricht University, Maastricht, The Netherlands

Zandra C. Deans
UK NEQAS for Molecular Genetics, GenQA Edinburgh Office, The Royal Infirmary of Edinburgh, Edinburgh, United Kingdom

Y.M. Dennis Lo
Li Ka Shing Institute of Health Sciences; Department of Chemical Pathology, The Chinese University of Hong Kong, Prince of Wales Hospital, Shatin, New Territories, Hong Kong SAR, China

Wybo Dondorp
Department of Health, Ethics & Society, Research Schools CAPHRI & GROW, Maastricht University, Maastricht, The Netherlands

Jane Fisher
Antenatal Results and Choices (ARC), London, United Kingdom

Francesca Gaccioli
Department of Obstetrics and Gynaecology, University of Cambridge, NIHR Cambridge Comprehensive Biomedical Research Centre; Centre for Trophoblast Research (CTR), Department of Physiology, Development and Neuroscience, University of Cambridge, Cambridge, United Kingdom

Amy Gerrish
Birmingham Women's and Children's NHS Foundation Trust, Edgbaston, Birmingham, United Kingdom

T. Harasim
Center for Human Genetics and Laboratory Diagnostics, Dr. Klein, Dr. Rost and Colleagues, Martinsried, Germany

Verena Haselmann
Institute for Clinical Chemistry, Medical Faculty of the University of Heidelberg, University Hospital Mannheim, Mannheim, Germany

Ros J. Hastings
Cytogenomics External Quality Assessment (CEQAS), GenQA Oxford Office, Women's Centre, Oxford University Hospitals NHS Trust, Oxford, United Kingdom

Lidewij Henneman
VU University Medical Center, Amsterdam, The Netherlands

Emma Hudson
Department of Applied Health Research, University College London, London, United Kingdom

Mark D. Kilby
Centre for Women's & Newborn Health, Institute of Metabolism & Systems Research, University of Birmingham, Birmingham Health Partners; West Midlands Fetal Medicine Centre, Birmingham Women's and Children's NHS Foundation Trust, Birmingham, United Kingdom

Fiona L. Mackie
Institute of Metabolism and Systems Research, University of Birmingham, Birmingham, United Kingdom

Stephen Morris
Department of Applied Health Research, University College London, London, United Kingdom

Dale Muzzey
Counsyl Inc., South San Francisco, CA, United States

Maria Neofytou
Center for Human Genetics, KU Leuven, Leuven, Belgium

D. Oepkes
Department of Obstetrics, Leiden University Medical Center, Leiden, The Netherlands

Pranav Pandya
Fetal Medicine Department, University College London Hospital, London, United Kingdom

Mark D. Pertile
Victorian Clinical Genetics Services, Murdoch Children's Research Institute; Department of Paediatrics, University of Melbourne, Melbourne, VIC, Australia

Elizabeth Quinlan-Jones
Department of Clinical Genetics, Birmingham Women's & Children's NHS Foundation Trust; Institute of Metabolism and Systems Research, University of Birmingham, Birmingham, United Kingdom

Adalina Sacco
Fetal Medicine Department, University College London Hospital, London, United Kingdom

Peter W. Schenk
Center for Human Genetics and Laboratory Diagnostics (AHC), Martinsried, Germany

Gordon C.S. Smith
Department of Obstetrics and Gynaecology, University of Cambridge, NIHR Cambridge Comprehensive Biomedical Research Centre; Centre for Trophoblast Research (CTR), Department of Physiology, Development and Neuroscience, University of Cambridge, Cambridge, United Kingdom

C. Ellen van der Schoot
Department of Experimental Immunohematology, Sanquin Research; Landsteiner Laboratory, Academic Medical Centre, University of Amsterdam, Amsterdam, The Netherlands

Carla van El
VU University Medical Center, Amsterdam, The Netherlands

Joris Robert Vermeesch
Center for Human Genetics, KU Leuven, Leuven, Belgium

E.J.T. Verweij
Department of Obstetrics & Gynaecology, Erasmus MC, Rotterdam, The Netherlands

Liesbeth Vossaert
Department of Molecular and Human Genetics, Baylor College of Medicine, Houston, TX, United States

A. Wagner
Center for Human Genetics and Laboratory Diagnostics, Dr. Klein, Dr. Rost and Colleagues, Martinsried, Germany

Dian Winkelhorst
Department of Experimental Immunohematology, Sanquin Research; Department of Obstetrics and Gynaecology, Leiden University Medical Center, Amsterdam, The Netherlands

Nicola Wolstenholme
UK NEQAS for Molecular Genetics, GenQA Edinburgh Office, The Royal Infirmary of Edinburgh, Edinburgh; Cytogenomics External Quality Assessment (CEQAS), GenQA Oxford Office, Women's Centre, Oxford University Hospitals NHS Trust, Oxford, United Kingdom

Elizabeth Young
Birmingham Women's and Children's NHS Foundation Trust, Edgbaston, Birmingham, United Kingdom

Acronyms and Abbreviations

95%CI	95% confidence interval
AC	amniocentesis
aCGH	array comparative genomic hybridization
ACOG	American College of Obstetricians and Gynecologists
ADAM12	ADAM metallopeptidase domain 12
ADM	adrenomedullin
AFP	alpha-fetoprotein
ART	assisted reproductive technologies
BMI	body mass index
BW	birthweight
C14MC	chromosome 14 miRNA cluster
C19MC	chromosome 19 miRNA cluster
CBS	circular binary segmentation
CD	cluster of differentiation
CEQAS	Cytogenomic External Quality Assessment Scheme
CER	cost-effectiveness ratio
CF	cystic fibrosis
cfDNA NIPT	cell-free DNA-based noninvasive prenatal testing
cfDNA	cell-free deoxyribose nucleic acid
cffDNA	cell-free fetal DNA
cfpDNA	cell-free placental DNA
cfRNA	cell free RNA
cfRNA-seq	cell-free RNA-seq
CFS	charge flow separation
CFTS	combined first trimester screening
CI	confidence interval
CMA	chromosomal microarray analysis
CNV	copy number variant
CPM	confined placental mosaicism
CRH	corticotropin releasing hormone
cSMART	circulating single-molecule amplification and resequencing technology
CTC	circulating tumor cell
CV	chorion villus
CVS	chorionic villus sampling
DA	diamniotic
DC	dichorionic
ddPCR	droplet digital PCR
DGC	density gradient centrifugation
DIA	dimeric inhibin A
DMD	Duchenne muscular dystrophy
DNA	deoxyribonucleic acid
DOP-PCR	degenerate oligonucleotide primed-PCR
DR	detection rate
DZ	dizygotic

EFNA3	ephrin-A3
EFW	estimated fetal weight
ELISA	enzyme-linked immunosorbent assay
EMQN	European Molecular Quality Network
ENG	endoglin
EpCAM	epithelial cadherin adhesive molecule
EQA	external quality assessment
ERVWE1	syncytin
EVs	extracellular vesicles
FACS	fluorescence-activated cell sorting
FASTQ	text file format name for storing sequencing data
FBS	fetal blood sampling
FcRn	neonatal Fc-receptor
FCT	first trimester combined test
FF	fetal fraction
FGR	fetal growth restriction
FISH	fluorescent in situ hybridization
FLT1	fms-like tyrosine kinase
FMH	fetomaternal hemorrhage
FN	false negative
FNAIT	fetal and neonatal alloimmune thrombocytopenia
fnRBC	fetal nucleated red blood cell
FP	false positive
FPR	false positive rate
FSTL3	follistatin like 3
GA	gestational age
GDM	gestational diabetes mellitus
GPA	glycophorin A
HbF	fetal hemoglobin
hCG	human chorionic gonadotropin
HDFN	hemolytic disease of the fetus and newborn
HIF2α	hypoxia-inducible factor 2α
HLA	human leukocyte antigen
HMM	hidden Markov model
HO-1	haem oxygenase-1
HOXA9	homeobox-A9
HPA	human platelet antigen
HTRA1	HtrA serine peptidase 1
ICER	incremental cost-effectiveness ratio
ICH	intracranial hemorrhage
ICSI	intracytoplasmic sperm injection
IGF1 and IGF2	insulin-like growth factors 1 and 2
IGF1R	IGF1 receptor
IGFBP-1	insulin like growth factor binding protein 1
IGFBP2	IGF binding protein 2
INHA	inhibin α subunit

INHB	inhibin β subunit
IQR	interquartile range
ISET	isolation by size of epithelial tumor/trophoblastic cells
ISPD	International Society of Prenatal Diagnosis
IUFD	intrauterine fetal death
IUGR	intrauterine growth restriction
IUT	intrauterine transfusion
IVF	in vitro fertilization
LCM	laser capture microdissection
LDHA	lactate dehydrogenase A
LEP	leptin
LIANTI	linear amplification by transposon insertion
lncRNA	long noncoding RNA
LTC	long-term culture
MA	monoamniotic
MACS	magnetic-activated cell sorting
MALBAC	multiple annealing and looping-based amplification
MB	megabase
MC	monochorionic
MCA-PSV	middle cerebral artery peak systolic velocity
MDA	multiple displacement amplification
miRNA	microRNA
MPS	massively parallel sequencing
MPSS	massively parallel shotgun sequencing
mRNA	messenger RNA
MX	monosomy X
MZ	monozygotic
NGS	next generation sequencing
NHS	National Health Service
NICE	National Institute for Health and Care Excellence
NICHD	National Institute of Child Health and Human Development
NIFTY	NICHD Fetal Cell Isolation Study
NIPD	noninvasive prenatal diagnosis
NIPS	noninvasive prenatal screening
NIPT	noninvasive prenatal testing
NPV	negative predictive value
nRBC	nucleated RBC
NTDs	neural tube defects
PAPP-A	pregnancy-associated plasma protein A
PCR	polymerase chain reaction
PEP-PCR	primer extension preamplification-PCR
PGH or GH2	placental growth hormone
PI	propidium iodide
piRNA	piwi-interacting RNA
PLAC1, PLAC3, and PLAC4	placenta-specific 1, 3, and 4
PLAT	plasminogen activator

PlGF	placenta growth factor
PPV	positive predictive value
PRL	prolactin
PSG1	pregnancy specific beta-1-glycoprotein
PT	proficiency testing
QALY	quality-adjusted life year
qPCR	quantitative PCR
RAPID study	Reliable, Accurate Prenatal, non-Invasive Diagnosis study
RBC	red blood cell
Rc	ceiling ratio
RCOG	Royal College of Obstetrics and Gynaecologists
RHD	*RHD* gene, assessed by genotyping
RhD	RhD polypeptide, assessed by serology
RHD	Rhesus D blood group determined by genotyping
RhD	Rhesus D blood group determined by serologic assays
RHD	Rhesus D by genotyping
RhD	Rhesus D by serotyping
RHDO	relative haplotype dosage analysis
RMD	relative mutation dosage
RNA	ribonucleic acid
RNA-seq	RNA sequencing
ROS	reactive oxygen species
SELP	selectin P
SERPINE1	plasminogen activator inhibitor-1
SGA	small for gestational age
SGD	single gene disorder
SMA	spinal muscular atrophy
SNP	single nucleotide polymorphism
SNV	single nucleotide variant
SOGC	Society of Obstetricians and Gynaecologists of Canada
STC	short-term culture
STR	short tandem repeat
T2, T13, T18, T21	trisomy 2, trisomy 13, trisomy 18, trisomy 21
TFM	true fetal mosaicism
TN	true negative
TP	true positive
TRIC	trophoblast retrieval and isolation from the cervix
TRIDENT study	Trial by Dutch laboratories for Evaluation of Non-Invasive Prenatal Testing
TUNEL	terminal deoxynucleotidyl transferase-mediated dUTP nick end labeling
uE3	unconjugated estriol
UK NEQAS	UK National External Quality Assessment Service
UK	United Kingdom
UPD	uniparental disomy
US	United States
VEGFA	vascular endothelial growth factor A
VEGFR	vascular endothelial growth factor receptor
WBC	white blood cell

WGA	whole genome amplification
WGS	whole genome sequencing
wkGA	weeks of gestational age
ZFY	Y-chromosome-specific zinc finger protein
Z-score	measure for the difference between expected and observed chromosome read density

CELL-FREE DNA (cfDNA): OVERVIEW AND TECHNOLOGY

FETAL DNA IN MATERNAL PLASMA: AN AMAZING TWO DECADES

1

Y.M. Dennis Lo*,†, S.H. Cheng*,†
Li Ka Shing Institute of Health Sciences, The Chinese University of Hong Kong, Prince of Wales Hospital, Shatin, New Territories, Hong Kong SAR, China* Department of Chemical Pathology, The Chinese University of Hong Kong, Prince of Wales Hospital, Shatin, New Territories, Hong Kong SAR, China†

In 1997, Lo et al. demonstrated that during pregnancy, fetal DNA could be seen in the plasma and serum of pregnant women [1]. Subsequent work has demonstrated the gestational variations [2] and rapid clearance of circulating fetal DNA following delivery [3]. Circulating fetal DNA has been found to consist of short fragments of DNA, which have a size distribution shorter than that of circulating maternally derived DNA in maternal plasma [4].

Diagnostically, we have witnessed a rapid evolution of this young field in the last two decades. This book provides a summary of the important developments during this period. Early work had focused on the detection of DNA sequences that the fetus had inherited from its father, and which were absent in the pregnant mother's genome, such as the Y chromosome of a male fetus [5], the *RHD* gene of a RhD-positive fetus carried by a RhD-negative mother [6,7], and a mutation inherited by the fetus from its father, but which is absent in its mother's genome [8].

These early applications have been more recently joined by those using newer technologies such as microfluidics digital PCR, droplet digital PCR, and massively parallel sequencing. Hence, noninvasive prenatal testing of single gene disorders has now been extended to elucidation of the paternal and maternal inheritances of the fetus in autosomal recessive disorders [9–11] and sex-linked disorders [12,13].

The area for noninvasive prenatal testing that has received the most attention over the last few years is its use for the detection of fetal chromosomal aneuploidies using massively parallel sequencing [14,15]. Since the first large-scale clinical trials demonstrating the robustness of this technology in 2011 [16,17], this technology has been quickly introduced into clinical practice in dozens of countries around the world. The detection of chromosomal aneuploidies has been rapidly followed by the demonstration that such an approach can also be used to detect subchromosomal aberrations [18,19].

Perhaps the ultimate illustration of the diagnostic potential of fetal DNA in maternal plasma is the demonstration that the entire fetal genome could be determined from maternal plasma [20–22]. These earlier works have recently been followed by the elucidation of a so-called second generation noninvasive fetal genome using newer sequencing and bioinformatics approaches [23]. By means of such methods, fetal de novo mutations could be examined in a genome-wide manner from maternal plasma and the maternal inheritance of the fetus could also be determined with a resolution of approximately two orders of magnitude higher than from previous attempts [23].

Noninvasive Prenatal Testing (NIPT). https://doi.org/10.1016/B978-0-12-814189-2.00001-3

Hence, one can see from the above that developments of the diagnostic applications of fetal DNA in maternal plasma have been most remarkable over the last two decades. However, there is still much to be learnt. For example, the biological characteristics of circulating fetal DNA still remain to be completely elucidated. Emerging areas of investigation include the relationship between circulating DNA and nucleosomal structure [24,25], the existence of preferred plasma DNA fragment endpoints [23], and the tissues of origin of circulating DNA [26,27]. Finally, a thought-provoking and unresolved problem is whether circulating fetal DNA has any biological or pathogenic functions. Hence, the next two decades will certainly be very exciting.

REFERENCES

[1] Lo YM, Corbetta N, Chamberlain PF, Rai V, Sargent IL, Redman CW, et al. Presence of fetal DNA in maternal plasma and serum. Lancet 1997;350(9076):485–7.

[2] Lo YM, Tein MS, Lau TK, Haines CJ, Leung TN, Poon PM, et al. Quantitative analysis of fetal DNA in maternal plasma and serum: implications for noninvasive prenatal diagnosis. Am J Hum Genet 1998;62(4):768–75.

[3] Lo YM, Zhang J, Leung TN, Lau TK, Chang AM, Hjelm NM. Rapid clearance of fetal DNA from maternal plasma. Am J Hum Genet 1999;64(1):218–24.

[4] Chan KC, Zhang J, Hui AB, Wong N, Lau TK, Leung TN, et al. Size distributions of maternal and fetal DNA in maternal plasma. Clin Chem 2004;50(1):88–92.

[5] Costa JM, Benachi A, Gautier E. New strategy for prenatal diagnosis of X-linked disorders. N Engl J Med 2002;346(19):1502.

[6] Faas BH, Beuling EA, Christiaens GC, von dem Borne AE, van der Schoot CE. Detection of fetal *RHD*-specific sequences in maternal plasma. Lancet 1998;352(9135):1196.

[7] Lo YM, Hjelm NM, Fidler C, Sargent IL, Murphy MF, Chamberlain PF, et al. Prenatal diagnosis of fetal RhD status by molecular analysis of maternal plasma. N Engl J Med 1998;339(24):1734–8.

[8] Chiu RW, Lau TK, Leung TN, Chow KC, Chui DH, Lo YM. Prenatal exclusion of beta thalassaemia major by examination of maternal plasma. Lancet 2002;360(9338):998–1000.

[9] New MI, Tong YK, Yuen T, Jiang P, Pina C, Chan KC, et al. Noninvasive prenatal diagnosis of congenital adrenal hyperplasia using cell-free fetal DNA in maternal plasma. J Clin Endocrinol Metab 2014;99(6):E1022–30.

[10] Hui WW, Jiang P, Tong YK, Lee WS, Cheng YK, New MI, et al. Universal haplotype-based noninvasive prenatal testing for single gene diseases. Clin Chem 2017;63(2):513–24.

[11] Barrett AN, McDonnell TC, Chan KC, Chitty LS. Digital PCR analysis of maternal plasma for noninvasive detection of sickle cell anemia. Clin Chem 2012;58(6):1026–32.

[12] Tsui NB, Kadir RA, Chan KC, Chi C, Mellars G, Tuddenham EG, et al. Noninvasive prenatal diagnosis of hemophilia by microfluidics digital PCR analysis of maternal plasma DNA. Blood 2011;117(13):3684–91.

[13] Hudecova I, Jiang P, Davies J, Lo YMD, Kadir RA, Chiu RWK. Noninvasive detection of F8 int22h-related inversions and sequence variants in maternal plasma of hemophilia carriers. Blood 2017;130(3):340–7.

[14] Chiu RW, Chan KC, Gao Y, Lau VY, Zheng W, Leung TY, et al. Noninvasive prenatal diagnosis of fetal chromosomal aneuploidy by massively parallel genomic sequencing of DNA in maternal plasma. Proc Natl Acad Sci USA 2008;105(51):20458–63.

[15] Fan HC, Blumenfeld YJ, Chitkara U, Hudgins L, Quake SR. Noninvasive diagnosis of fetal aneuploidy by shotgun sequencing DNA from maternal blood. Proc Natl Acad Sci USA 2008;105(42):16266–71.

[16] Chiu RW, Akolekar R, Zheng YW, Leung TY, Sun H, Chan KC, et al. Non-invasive prenatal assessment of trisomy 21 by multiplexed maternal plasma DNA sequencing: large scale validity study. BMJ 2011;342: c7401.

[17] Palomaki GE, Kloza EM, Lambert-Messerlian GM, Haddow JE, Neveux LM, Ehrich M, et al. DNA sequencing of maternal plasma to detect Down syndrome: an international clinical validation study. Genet Med 2011;13(11):913–20.

[18] Srinivasan A, Bianchi DW, Huang H, Sehnert AJ, Rava RP. Noninvasive detection of fetal subchromosome abnormalities via deep sequencing of maternal plasma. Am J Hum Genet 2013;92(2):167–76.

[19] Yu SC, Jiang P, Choy KW, Chan KC, Won HS, Leung WC, et al. Noninvasive prenatal molecular karyotyping from maternal plasma. PLoS One 2013;8(4):e60968.

[20] Lo YMD, Chan KCA, Sun H, Chen EZ, Jiang P, Lun FMF, et al. Maternal plasma DNA sequencing reveals the genome-wide Genetic and mutational profile of the fetus. Sci Transl Med 2010;2(61):61ra91.

[21] Kitzman JO, Snyder MW, Ventura M, Lewis AP, Qiu R, Simmons LE, et al. Noninvasive whole-genome sequencing of a human fetus. Sci Transl Med 2012;4(137):137ra176.

[22] Fan HC, Gu W, Wang J, Blumenfeld YJ, El-Sayed YY, Quake SR. Non-invasive prenatal measurement of the fetal genome. Nature 2012;487(7407):320–4.

[23] Chan KC, Jiang P, Sun K, Cheng YK, Tong YK, Cheng SH, et al. Second generation noninvasive fetal genome analysis reveals de novo mutations, single-base parental inheritance, and preferred DNA ends. Proc Natl Acad Sci USA 2016;113(50):E8159–68.

[24] Snyder MW, Kircher M, Hill AJ, Daza RM, Shendure J. Cell-free DNA comprises an in vivo nucleosome footprint that informs its tissues-of-origin. Cell 2016;164(1-2):57–68.

[25] Straver R, Oudejans CB, Sistermans EA, Reinders MJ. Calculating the fetal fraction for noninvasive prenatal testing based on genome-wide nucleosome profiles. Prenat Diagn 2016;36(7):614–21.

[26] Sun K, Jiang P, Chan KC, Wong J, Cheng YK, Liang RH, et al. Plasma DNA tissue mapping by genome-wide methylation sequencing for noninvasive prenatal, cancer, and transplantation assessments. Proc Natl Acad Sci USA 2015;112(40):E5503–12.

[27] Guo S, Diep D, Plongthongkum N, Fung HL, Zhang K. Identification of methylation haplotype blocks aids in deconvolution of heterogeneous tissue samples and tumor tissue-of-origin mapping from plasma DNA. Nat Genet 2017;49(4):635–42.

UNDERSTANDING THE BASICS OF NGS IN THE CONTEXT OF NIPT

2

Dale Muzzey

Counsyl Inc., South San Francisco, CA, United States

INTRODUCTION: EVOLUTION OF SEQUENCING TECHNOLOGY

The term "next-generation sequencing" (NGS) begs the questions of what "first-generation sequencing" was and how NGS is both similar to and different from its predecessor. Sanger developed the first generation of DNA sequencing in the 1970s [1,2]. His eponymous sequencing approach is an in vitro adaptation of the cellular replication machinery that cleverly leverages unextendable DNA bases. These modified bases are introduced at low concentration in a reaction minimally containing (1) a high concentration of extendable bases, (2) the single-stranded DNA template to be sequenced, (3) a short oligonucleotide primer complementary to the template onto which new bases could be synthesized, and (4) DNA polymerase enzymes that execute the extension reaction. Early Sanger sequencing experiments involved four independent reactions, each containing a single type of unextendable base (A, T, G, or C). Whenever a polymerase randomly incorporates one of the unextendable bases into the nascent DNA molecule (e.g., an unextendable G in the nascent strand incorporated opposite a C in the template), further synthesis would terminate, yielding a truncated copy of the template. Critically, since all nascent strands anchor from the same oligonucleotide primer, the position of extension termination—and hence the length of the nascent DNA strand—is a direct proxy for the base at the $3'$ end of the molecule. By using gel electrophoresis to resolve the respective lengths of terminated molecules in each of the four reactions, it is possible to infer the sequence of the entire template.

Sanger sequencing became slightly more scalable with the introduction of unextendable bases that were uniquely dyed (Fig. 1). Rather than achieving base-specific information by partitioning into four reactions, a capillary electrophoresis instrument coupled with a fluorescent dye detector could resolve both the relative sizes of fragments and the identity of their terminating bases [3–5]. To criticize these machines as unscalable neglects one of their unmitigated triumphs: they were the workhorses that sequenced the very first human genome in the 1990s [6–9]. However, with a cost in the billions of dollars and with a timeline on the order of years, genome sequencing would remain prohibitive in a clinical setting without a major technological leap.

NGS revolutionized genome sequencing by overcoming many of the limitations of the Sanger technique [10], yet the most pervasive NGS methodology shares much in common with its predecessor. As described in further detail later, NGS also leverages extension termination and fluorescent bases, and it relies upon DNA polymerases that append a single base at a time to a nascent DNA molecule. Indeed, in many respects, an NGS experiment is comparable to performing millions or

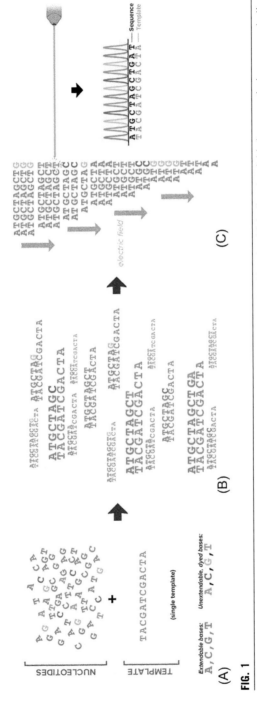

FIG. 1

Overview of Sanger sequencing. (A) The Sanger synthesis reaction contains a nucleotide mix with both extendable bases and a low concentration of unextendable, dyed bases (a DNA primer, DNA polymerase, and buffer are also required an not shown). (B) Extended molecules terminate at different locations with the molecule having the color of the last incorporated base. (C) Arrangement of the terminated molecules in an electric field—DNA is negatively charged—and passage through a capillary equipped with a dye detector identifies fluorescence peaks that reveal the original template's sequence.

billions of Sanger reactions in parallel (hence the NGS moniker "massively parallel sequencing"). This explosive increase in throughput shattered some of the barriers (e.g., cost and turnaround time) that had largely prevented the use of genomics in routine clinical care, and it paved the way for cfDNA-based prenatal testing.

HOW NGS WORKS

The role of an NGS device is to distill a specially prepared library of DNA molecules into a long text file of sequences, one line for each sequenced molecule. NGS sequencers perform this molecule-to-text mapping across many research and clinical contexts, spanning everything from RNA sequencing in broccoli [11] to ribosome profiling in bacteria [12] to DNA sequencing for NIPT in pregnant women [13]. These applications of NGS are distinguished primarily by the steps upstream of DNA being injected into the sequencer, termed "library preparation." Mirroring the diversity of upstream sample preparation methods is a comparably vast range of downstream analyses, one of which is the analysis method for NIPT, covered in detail in Chapter 3. In addition to describing how an NGS machine sequences DNA, this section discusses NIPT-specific workflows upstream and downstream of sequencing.

UPSTREAM OF THE SEQUENCER

DNA extraction

As the name implies, cell-free DNA is not found in the blood cells, but rather must be extracted from the plasma. cfDNA fragments are relics of dead cells [14]: when a cell undergoes programmed cell death ("apoptosis"), a suite of enzymes is synthesized that digest the genomic DNA [15]. These enzymes can only access DNA that is not packaged into nucleosomes [16], the octamers of histone proteins that regulate both gene expression and genome topology in the cell [17]. The lack of accessibility to nucleosomal DNA means that the ~150 nucleotide (nt) fragments of DNA encircling each nucleosome survive the apoptosis process, and these fragments ejected from the dying cell constitute the cfDNA that is sequenced to give so-called NGS reads, described later in more detail.

To extract cfDNA from plasma, the blood must first be spun in a centrifuge to separate plasma, buffy coat (which contains the white blood cells), and erythrocytes.[1] Approximately 55% of whole blood is plasma. When aspirating the plasma from centrifuged blood, care must be taken to avoid the buffy coat because the high concentration of maternal DNA in the white blood cells would dilute the scarce placenta-derived cfDNA and reduce or altogether eliminate sensitivity for fetal aneuploidy detection.

Standard and commercially available DNA extraction techniques can purify sufficient cfDNA from an aspirated plasma sample to power the analysis [18,19]. The typical concentration of cfDNA fragments in plasma is only 5–50 ng/mL, and this low level of cfDNA in plasma is noteworthy because it imposes a lower bound on the volume of blood required for cfDNA-based prenatal testing. If blood volume is too low—or if the extraction is inefficient—then the number of copies of the genome in the extracted sample is so low that the subtle changes in fetal chromosomal abundance may not be

---

[1]Centrifugation speed must be high enough to separate the blood components but low enough to avoid hemolysis, which could dilute placenta-derived cfDNA. Hemolysis could also result from other mishandling of the sample prior to extraction, for example, by storage at extreme temperature.

detectable. For instance, if only 10 copies of the genome are present in the extracted sample, it is likely not possible to detect a 2% change in abundance of chr21. Conversely, an efficient extraction should yield enough genome equivalents to empower detection of fetal chromosomal aneuploidies even at low fetal fraction. The number of genome equivalents required from extraction depends on the subsequent NIPT analysis to be performed. Whole-genome sequencing (WGS) NIPT requires very few cfDNA fragments at any given site, so the volume of blood drawn from the patient can be quite low [13], and multiple extraction attempts can be made from a single blood collection tube. By contrast, targeted techniques like the single-nucleotide polymorphism (SNP) method require hundreds of genome equivalents at each interrogated site, such that allele balance can be measured with high precision (more information on WGS and SNP methods will be given in Chapter 3) [20]. Therefore blood collection volumes for SNP method NIPT testing are typically higher than for WGS.

Since cfDNA concentrations are so low, it is not trivial to measure whether enough cfDNA was extracted to power NIPT. Extracted DNA is typically amplified by PCR prior to NGS; therefore even an inefficient extraction can yield plenty of DNA for sequencing, meaning that an inefficient extraction is not reflected by the depth of NGS sequencing. Fortunately, it is possible to detect inefficient extraction via the "complexity" of the sequencing data. For instance, in the context of WGS, it is expected that, with efficient extraction, genomic positions will have either 0 or 1 aligned sequenced fragment (with the majority having 0) because the sequencing will be Poisson sampling from a very rich initial pool of genomic material [21]. Following an inefficient extraction, however, the pool of original genomic material is very sparse, meaning that sites will tend to have 0 or $\gg 1$ mapped fragments, resulting in data with low "complexity." If extraction efficiency is very high, then PCR may not be required to yield sufficient DNA for sequencing; such "PCR-free" library preparation is expected to have high library complexity. It is important to monitor the sequencing complexity of NGS data to ensure that claims of fetal ploidy have adequate statistical power.

Library preparation

An NGS machine can only sequence an ensemble of DNA molecules—called a "library"—that have been properly prepared. Specifically, for the Illumina devices that predominate in the clinical genomics setting, each DNA molecule in the library must have a common set of flanking adapters (sequences typically ~50 nt and specified by the manufacturer), with all 5′ ends sharing one sequence that differs from the one sequence shared by all 3′ ends[2] [22]. Flanking all molecules with common adapters permits efficient amplification and extension of the entire library using only a single pair of primers. Such amplification and extension occur (1) upstream of the sequencer to attain a sufficient input concentration (this step is optional), (2) inside the NGS machine immediately prior to sequencing during the "cluster amplification" procedure (described later), and (3) during the sequencing reaction itself (also described later).

The most common process by which all 5′ ends have one adapter and all 3′ ends have a different common adapter involves clever molecular biology [23]. First, all double-stranded cfDNA molecules—which may have short single-stranded overhangs at the termini—are incubated with a polymerase enzyme that both trims back 3′ overhangs, fills in 5′ overhangs, and appends an adenine (A) base to the 3′ of all molecules, that is, creating an A overhang (Fig. 2). Finally, the DNA fragments

---

[2]DNA is a directional polymer with a backbone that is a series of bonded sugar molecules. Each sugar has five carbons (referred to as 1′, 2′, 3′, 4′, and 5′, with each pronounced "1 prime," "2 prime," etc.), and the bonding of one sugar to the next occurs via a phosphodiester bond between the 5′ and 3′ carbons.

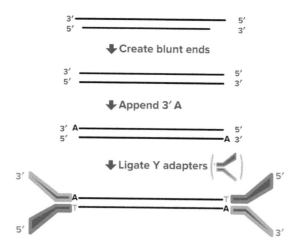

FIG. 2

NGS library preparation for cfDNA. Blunt-ended cfDNA results from trimming overhanging 3′ ends (*top left*) and filling in overhanging 5′ ends (*top right*). Appending A bases to the 3′ ends of double-stranded cfDNA enables ligation of Y adapters.

are mixed with "Y adapters" and a ligase. The Y adapters contain two single strands of DNA that are complementary at one end (the trunk of the "Y") but not at the other (the branches of the "Y"). The double-stranded portion of the Y adapter has a T overhang, which means that it will hybridize to the cfDNA fragments' A overhangs. The asymmetrical design of the two strands in the Y adapter ensures that, subsequent to ligation of Y adapters at each end of the cfDNA molecule, each of the resulting single strands has a common 5′-specific adapter and a common 3′-specific adapter.

Sequencing biases are minimized when the fragments undergoing NGS are of similar length [24]. In most NGS applications, an in vitro fragmentation reaction—followed possibly by a size-selection step—produces fragments of comparable and acceptable size. For NIPT, however, no such steps are needed, as the in vivo DNA fragmentation that occurs during apoptosis yields fragments with highly uniform length near 150 nt [25,26]. In fact, the length of cfDNA-fragment processing in vivo is so precise and reproducible that even subtle differences between placental nucleosomes and those from other tissues cause cfDNA-length disparities that can be analyzed to give an estimate of the fetal fraction (placental fragments are systematically shorter than nonplacental) [27].

cfDNA fragment length can be captured via WGS-based NIPT but not via SNP-based NIPT due to differences in their library preparation approaches. WGS-based NIPT simply appends Y-adapters to cfDNA molecules that are unmodified (other than the blunt ending and A-tailing described previously). This simple library preparation workflow is preferable for WGS-based NIPT because the goal is to have an unbiased sampling of all cfDNA extracted from the plasma. For SNP-based NIPT, however, only the cfDNA fragments that overlap informative SNP sites provide insight into possible fetal aneuploidy. Therefore SNP-based NIPT requires a molecular enrichment of fragments of interest, which can be achieved via a multiplex PCR reaction [20]. In multiplex PCR, hundreds to thousands of different primer sets can be mixed together in a single PCR tube with the cfDNA extracted from a given sample; with proper primer design and reaction conditions, fragments from each of the targeted locations can be massively enriched in preparation for sequencing. Adapter sequences can either be appended to the multiplex primers themselves—in which case multiplex PCR alone yields an NGS-competent library—or Y-adapter ligation can occur subsequent to multiplex PCR. The reason length-based information is lost in such a reaction is that the amplicon length is dictated by the designed primers, not by the template cfDNA fragment off of which they amplify.

The form factor of NGS machines necessitates one more critical step during NIPT library preparation, the barcoding of samples [28]. Illumina sequencing data is sold in units of flowcells, where each flowcell yields hundreds of millions or billions of reads. The number of reads per flowcell exceeds what is needed for a single sample. Therefore it is economically advantageous to load many samples onto a single flowcell, a process called "multiplexing" (Fig. 3).

However, unlike other lab devices that maintain physical separation of samples throughout the assay workflow—for example, qPCR machines, ELISA devices, and capillary sequencers—an NGS flowcell eliminates physical separation between samples during sequencing. Thus a "demultiplexing" mechanism is needed by which NGS data can be parsed into sample-specific cohorts again after sequencing. Demultiplexing is enabled by sample-specific barcodes, short (typically ~6–8 nt) DNA sequences included in the set of Y adapters used to create a library for a given sample. Critically, the barcode will differ from sample to sample, but all molecules from a given sample will have the same barcode. The NGS machine emits one text file for barcode sequences and a separate text file of cfDNA fragment sequences, where rows in each file correspond to the same molecule (i.e., the sequence in the first line of the barcode file is the barcode of the cfDNA molecule whose sequence is the first to appear in the cfDNA-fragment sequence file). Using these files, it is possible to parse the whole-flowcell sequencing files into sample-specific files, thereby recapitulating in silico the physical separation of samples that existed when the samples were being processed in multiwell plates upstream of sequencing.

THE ACT OF NGS: FROM MOLECULAR LIBRARY TO TEXT FILE

Though multiple post-Sanger sequencing technologies exist that could each claim the title of "NGS", in the clinical genomics setting of NIPT, the term "NGS" effectively implies Illumina-style sequencing, as it is the predominant platform. Therefore, below we describe the "sequencing-by-synthesis" process by which Illumina sequencers execute NGS [22].

Cluster generation

To gain intuition into the Illumina-style NGS workflow, recall from the earlier comparison of Sanger sequencing and NGS that both methods involve the determination of DNA sequence by measuring a fluorescence signal one base at a time. Therefore at the most fundamental level, an NGS machine must be able to resolve single molecules and capture the series of fluorescent signals that correspond to the molecules' respective DNA sequences. The process of cluster generation ensures that single molecules are resolvable and that the fluorescence signal for a single molecule can be adequately captured.

The first step of cluster generation [22] is the loading of a DNA library (chemically denatured into single strands) into a glass chamber called a flowcell. The surfaces of the flowcell are coated with oligonucleotides homologous to the adapter sequences appended to cfDNA fragments during library preparation. Single-stranded fragments anneal to the surface of the flowcell at random positions. The position to which a fragment anneals is critical, as that fragment will remain in the same position of the

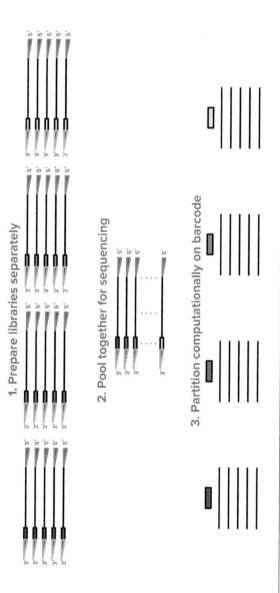

FIG. 3

Multiplexing and demultiplexing with molecular barcodes. Sample-specific molecular barcode sequences (*colored rectangles*) are appended during library preparation. Samples are pooled for multiplexed sequencing and then demultiplexed computationally based on the identity of the barcode.

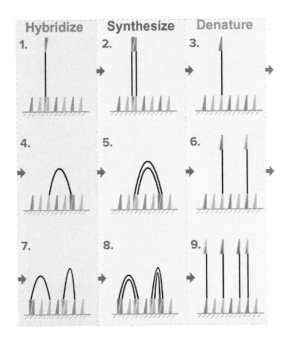

FIG. 4

Bridge amplification overview. PCR using flowcell-bound primers that are complementary to NGS library fragments clonally amplify the initially hybridized fragment in close proximity, yielding a cluster of duplicated molecules.

flowcell throughout the entire NGS procedure. The concentration of the loaded library must be carefully calibrated: if the concentration is too high, multiple library fragments could occupy the same location in the flowcell, obscuring the ability to detect fragment-specific fluorescence; if the loaded concentration is too low, then the sequencing capacity of the flowcell is underutilized, which could cause the sequencing depth per sample to be too low for confident aneuploidy detection.

The second step of cluster generation is called bridge amplification and is a PCR reaction that occurs on the surface of the flowcell. This localized amplification of DNA hybridized to the flowcell surface is needed because the intensity of a single fluorescent base incorporated into a single molecule undergoing sequencing is too weak for the NGS machine's camera to capture. To boost the fluorescence signal to a detectable level, the original library fragment is proximally copied hundreds to thousands of times to yield a dense clonal population of fragments called a "cluster." Bridge amplification—schematized in Fig. 4—utilizes the oligonucleotides affixed to the flowcell surface. These oligos serve as primers for each successive round of amplification. Because they are bonded to the glass slide, each single-stranded molecule becomes primed for a subsequent round of extension by bending over to make a bridge with a flowcell-bonded oligo. The final step of cluster generation is to use enzymatic cleavage and chemical denaturation to remove the same single strand in each duplex (e.g., remove the one with a pink primer bound to the flowcell), which leaves single-stranded DNA molecules that all have an identical sequence (see top panel of Fig. 5, where all single-stranded molecules have pink primer on top and blue primer bonded to the flowcell).

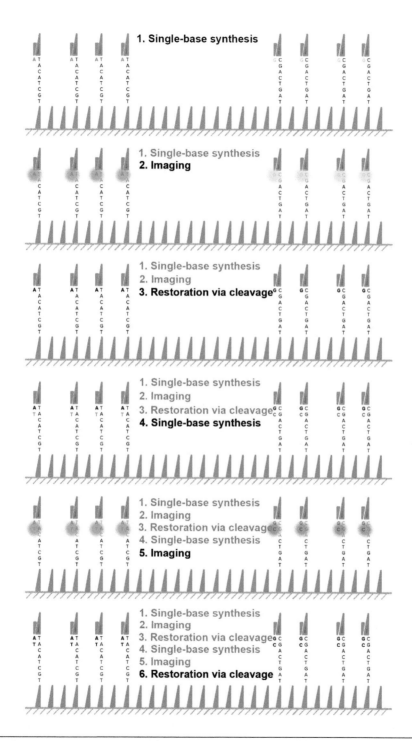

FIG. 5

Two rounds of sequencing-by-synthesis NGS. For two different template molecules and their corresponding clusters (*left* and *right*), NGS proceeds by cycling through the indicated steps of synthesis, imaging, and restoration via cleavage.

Sequencing cycle

Once clusters have been amplified, the sequencing reaction can commence [22]. The first step is to introduce a sequencing primer into the flowcell. The sequencing primer anneals to a common sequence embedded in the adapter molecule of each fragment. The primer's 3′ end is immediately adjacent to the cfDNA insert; sequencing, therefore, will decode the sequence beginning at one of the ends of the cfDNA fragment. Next, the NGS machine floods the flowcell with a reaction mix containing DNA polymerases and fluorescently labeled, unextendable nucleotides (Fig. 5). The polymerases extend off of the sequencing primer, incorporating whichever fluorescently labeled base is complementary to the template molecule. Because these bases are unextendable, extension only proceeds for a single base before terminating. At this point, the unincorporated nucleotides and remaining extension mix are flushed from the flowcell chamber, and imaging begins. A camera scans over the entire surface of the flowcell (for newer Illumina devices, both the top and bottom of the flowcell are scanned), capturing and storing images of the clusters. The color of the cluster will correspond to the base that was just incorporated, and the cluster was only visible due to the bridge amplification process. After image capture, a chemical cocktail enters the flowcell that removes the fluorescence moiety from the recently incorporated bases and restores their ability to be extended. This restoration reaction is critical, as it primes each molecule to undergo another round of extension and image capture. Indeed, the entire cycle of extension, imaging, and restoration may be repeated several hundreds of times (based on user preference), with each iteration of the cycle deciphering an additional base of the sequence for all clusters on the flowcell surface. The number of cycles determines the ultimate length of the reads that will be available for mapping.

The number of sequencing cycles is typically short (25–36) in NIPT applications [13,29]. Unlike other types of genetic testing, where sequenced molecules are analyzed to discover novel genomic variants (thereby conferring value to longer reads), current cfDNA NIPT applications screening for chromosomal abnormalities are not trying to reveal novel variation at the level of single bases. For SNP-based NIPT, multiplex primers can be designed such that the SNPs they interrogate are relatively close to the sequencing primer. For WGS-based NIPT, sequencing need only proceed until reads can be uniquely mapped. These shorter read lengths are attractive for two purposes relevant to NIPS: (1) short reads enable fast turnaround time, as the duration of NGS scales linearly with sequencing length, and (2) short reads cost less than longer reads, supporting the affordability of the screen.

Paired-end sequencing

NGS machines can only resolve the sequence at the terminus of a DNA fragment because the oligonucleotides that prime the sequencing reaction anneal to the adapter DNA flanking the fragment [22]. Further, since nucleotide extension can only proceed from 5′ to 3′, only one end can be sequenced from the flowcell-anchored single-stranded fragments generated during cluster amplification. However, via a process called "paired-end" sequencing, the sequence originating at both ends of a fragment can be resolved. As the name implies, paired-end sequencing executes two rounds of the single-end sequencing process described previously, with the rounds primed by different sequencing primers and separated by an intervening strand-switching mechanism. In the first step of this strand-switching process (Fig. 6), double-stranded DNA in the flowcell is denatured; this separates the original single-stranded molecules in the cluster from the nascent single strands synthesized off of the first-round sequencing primer. These unanchored nascent strands are washed from the flowcell, effectively returning the flowcell to its original status after cluster amplification. To capture sequence from the

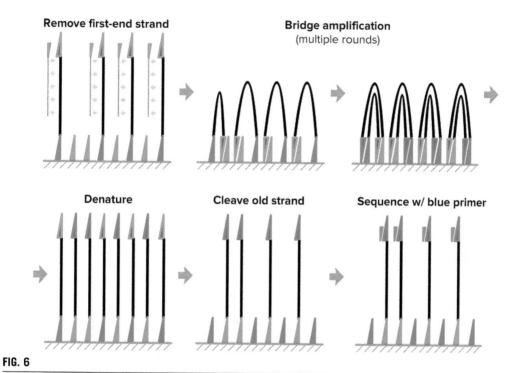

Remove first-end strand

Bridge amplification
(multiple rounds)

Denature

Cleave old strand

Sequence w/ blue primer

FIG. 6

Strand-switching mechanism to enable paired-end sequencing. Following sequencing from the template's first end with the *pink primer*, denaturation removes the synthesized strand. Multiple rounds of bridge amplification followed by denaturation and cleavage of the already sequenced single-strand enables sequencing from the opposite end of the molecule from the *blue primer*.

other end of the fragment, however, the reverse complement of strands in the original cluster must be synthesized so that the extension reaction proceeds from 5′ to 3′. Single-stranded fragments in a cluster are reverse complemented by first undergoing one round of bridge amplification, which creates a cluster in which both strands are present. Next, the original strand is removed by introducing a molecule that specifically cleaves only one of the flowcell-anchored oligos. This process prepares the cluster to undergo extension with the opposite sequencing primer and yields sequencing information from the fragment's other end.

Image analysis and sequencing metrics

The steps described previously illustrate how an NGS machine converts the molecular information encoded in a DNA sequence into a stack of images, yet the intended output is a text file of sequence information, not a stack of high-resolution images. "Base calling" software integrated into the sequencing device performs this last transformation (Fig. 7) [22]. The software aims to detect the location of each cluster in each image and tracks both the position and color throughout the stack. The processes of cluster identification and color coding have evolved on the Illumina platform. In early sequencers, clusters could form in random positions across the slide and nucleotides had four colors; therefore the

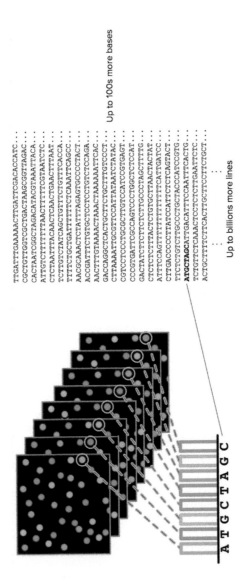

FIG. 7

Image analysis and FASTQ generation. By tracking the location and color of each single cluster through every cycle of base extension, NGS software can infer the sequence of each template molecule, writing it to a large FASTQ file containing the sequence and quality metrics of every fragment in the flowcell.

machine captured and analyzed four images at every cycle. Newer machines, however, employ a "patterned" flowcell, where the cluster amplification chemistry has been changed such that a single template molecule fills one of many etched wells on the flowcell that collectively form a honeycomb pattern. This pattern simplifies—and, consequently, quickens—image analysis. Also speeding the image analysis process is the use of only two colors in base encoding. By modifying the adenine base to bear both green and red fluorophores, all four bases can be resolved simply by looking at the red and green channels: A is red and green, C is red but not green, T is green but not red, and G has no color.

For each cluster the software analyzes, it emits both a "base call" (i.e., A, T, C, G, or N, in the case the base is indecipherable) and a quality score. On a scale of 1–40, the quality score describes how confident the software is about the base call, thereby adding an analog-like component to an otherwise digital assertion about the base. Quality scores are important, as they enable bioinformatic filtering of the sequencing data. Reads composed entirely of low-quality bases, for instance, should likely be discarded. Similarly, with a technique like SNP-based NIPT, where the identity of single bases at a SNP is important as it impacts the observed allele balance (see Chapter 3 for more on the SNP-method algorithm), accounting for quality scores could boost aneuploidy detection performance.

Once the image analysis determines the sequence of bases and quality scores for a given cluster, it writes that information to a FASTQ file. The cluster is also assigned a name, which typically includes the positional coordinates of the cluster within the flowcell. Taken together, this information constitutes an NGS "read." Current NGS machines can measure billions of reads per flowcell, so FASTQ files are very large text files. Importantly, generation of the FASTQ file marks completion of the NGS machine's original mandate: converting a molecular library of DNA fragments into a text-based list of sequences.

DOWNSTREAM OF THE SEQUENCER: DEMULTIPLEXING AND ALIGNMENT

Following the completion of sequencing and base calling are the steps of demultiplexing and alignment, which are common across NGS-based NIPT approaches and precede the platform-specific analyses described in Chapter 3. Demultiplexing, covered earlier, is the process by which NGS reads are assigned to their sample of origin based on the sequence of their corresponding molecular barcode. Barcode reads and cfDNA reads are written to different FASTQ files (in the case of paired-end sequencing, there are two cfDNA FASTQ files and one barcode file), but their respective rows correspond to each other. Therefore the demultiplexing process is fairly straightforward (and is typically included with the sequencing device's software). The user first loads a mapping of barcodes to sample names into the software. Next, a simple script plows through the FASTQ files for the barcodes and cfDNA reads, copying the cfDNA-read information to sample-specific FASTQ files based on the barcode. To minimize the discarding of reads based on minor NGS mistakes during barcode sequencing (e.g., an incorrect base call at one of the barcode's bases), barcodes are typically chosen such that they are sufficiently dissimilar to other barcodes that they can be clearly identified despite one or more mismatches [30].

The fundamental premise of cfDNA NIPT—that is, determining if the fetal genome has anomalous dosage of a particular region—presumes that cfDNA molecules have been mapped to their respective regions of origin. This mapping occurs during the alignment phase. The basic idea of alignment is simple: for a given read's sequence of tens to hundreds of letters, find the position where that series of letters appears in a sequence of roughly 3 billion letters (i.e., the human reference genome). Though this concept is simple, it is not trivial to execute in an efficient manner. As NIPT analyses typically consider only uniquely mapping reads—foregoing those with redundant sequence that could come from multiple locations—it is necessary to consider the entire genome as a source for the read. It would be sufficient but woefully inefficient to scan across the entire genome for every read and inspect whether the read matches the reference at each offset. This naive approach would require 3 billion comparisons per read (one for each offset in the reference), and doing this procedure for the billions of reads from a single flowcell would be $\sim 10^{18}$ computations. To complicate matters further, NGS reads frequently differ from the reference at specific bases (e.g., SNPs). Therefore determining whether a read maps to a particular reference position is not as easy as finding a perfect match; instead, the mapping algorithm should capture whether the read is a close match at a particular position.

The viability of NGS as a tractable technology—both in general and for NIPT in particular—stems from development of fast algorithms capable of aligning millions of reads to the human genome in minutes [31,32]. One of the key insights of the developers of these algorithms is that, unlike the experimental data itself, the reference genome is static. Therefore any preprocessing of the genome that could facilitate searching would have a large payoff in performance. Indeed, the dominant alignment software packages share a step upstream of the alignment itself during which they create a set of index files of the reference genome. The primary index file is an exquisitely permuted version of the genome that both retains all information about the original genome sequence and reorders it in a way to facilitate quick searching. This special ordering enables reads to be mapped one base at time: with each successive base in a read (e.g., a G base), nearly 75% of the transformed genome can be disregarded for all future searching (e.g., those with A, C, or T). Iterating over successive bases allows rapid convergence to the true origin(s) of the read. Indeed, armed with this transformed genome, a read can sometimes be uniquely mapped in just 10–20 computations (i.e., after 10–20 bases), in terrifically stark contrast to the naive algorithm considered previously that required 3 billion computations to compare a read to the sequence at each genomic position. These alignment algorithms implement nuanced modifications to be robust to gaps and mismatches, but support for these features adds minimal overhead. In the end, the preprocessed genome index is what makes alignment plausible on the timescale required for clinical cell-free-based NIPT.

ALTERNATIVE NGS AND NON-NGS TECHNOLOGIES FOR NIPT

Although nearly all sequencing-based NIPT offerings utilize Illumina machines, any technology—sequencing based or otherwise—that can characterize the original genomic location of a large number of cfDNA molecules could perform cell-free DNA-based NIPT if it were to operate with sufficiently high speed and throughput. Illumina's particular sequencing-by-synthesis technology has supplanted other related approaches (e.g., pyrosequencing) and sequencing-by-ligation platforms (e.g., SOLiD) because Illumina machines' cost per base is low and the time per base is amenable to the demanded clinical turnaround time [10]. However, a competing technology with faster and cheaper sequencing could quickly transform the NIPT sequencing landscape. For instance, a still-nascent technology that could soon compete with the sequencing-by-synthesis approach is nanopore sequencing [33]. Nanopores resolve DNA sequence by measuring voltage signatures as a long DNA molecule traverses a protein pore embedded in an otherwise charge-blocking membrane. Each group of nucleotides (e.g., GCGTA) within the pore has as a characteristic voltage level that a base calling algorithm can deconvolve into a particular DNA sequence when considering a molecule's entire voltage trajectory. The speed and scalability of nanopores would be attractive for cell-free DNA-based NIPT applications, and even the primary limitation that nanopore developers have struggled to overcome—a high error rate that complicates variant identification in patients' genomes—may have minimal downside for depth-based cfDNA NIPT, which can tolerate errors as long as alignment algorithms can still map reads to their original location in the genome (see Chapter 3). Because nanopores are optimized for sequencing extremely long DNA molecules approaching hundreds of thousands of bases—vastly higher than the ~150 in a single cfDNA fragment—optimal library preparation for nanopore sequencing could involve extensive concatemerization of cfDNA, such that hundreds of cfDNA molecules are stitched together into a single molecule. Importantly, this discussion of nanopore use in cfDNA NIPT is speculative and does not imply they have particularly great promise for cfDNA NIPT; indeed, they are currently better suited to other genomics applications that motivate their current development. Rather, nanopores underscore the idea that sequencing technology continues to evolve and that new techniques may emerge at any moment: sequencing-by-synthesis approaches are ubiquitous today in NIPT sequencing-based assays not because they are intrinsically superior, but because their cost and time per base is currently the best in a rapidly evolving field.

As noted previously, an NIPT-suitable DNA technology need only be able to map cfDNA fragments to their genomic locations quickly and affordably, but microarrays demonstrate that this technology does not strictly need to sequence the DNA. Microarray-based cfDNA NIPT offerings measure the abundance of hundreds of thousands of cfDNA fragments via specific hybridization probes that tile genomic regions of interest [34]. Microarrays (described further in Chapter 3) can even identify the abundance of particular alleles; such sampling at highly polymorphic SNP sites provides information about fetal fraction. The idea of using homologous DNA (e.g., the hybridization probes on microarrays) to quantify sequences of interest—rather than sequencing them directly—also underlies recent attempts to perform cfDNA NIPT with quantitative PCR [35], where a properly chosen set of primers can measure cfDNA abundance in NIPT-relevant regions.

CONCLUSION

For both trivially obvious and more subtle reasons, NGS is particularly well suited to NIPT applications. On the obvious side, NGS provides the digital, nucleotide-level data capable of powering both the depth- and allele-based NIPT workflows that require measurement of cfDNA fragment identity and abundance (discussed further in Chapter 3). Critically, an NGS machine generates these data affordably and quickly, such that NIPT via NGS satisfies the constraints of providers, payers, and patients.

On the subtle side, NGS also captures signals that correlate with placenta-derived cfDNA. For instance, paired-end NGS provides a measure of fragment length, and placental-derived fragments tend to be shorter than their maternally derived counterparts [26]. Placental fragments also have a unique DNA methylation profile that NGS can detect subsequent to bisulfite treatment [36] (bisulfite converts unmethylated C bases to uracil—which sequences like T bases—whereas methylated C bases remain unchanged). Finally, NGS reports the precise position of the end of a cfDNA fragment at single-nucleotide resolution, and this end-position information contains valuable placental signal because there is a systematic difference in the maternal and placental nucleosome positions that determine cfDNA termini locations [37]. Analysis algorithms that extract and amplify these and other placental signals can increase sensitivity for fetal chromosomal abnormalities.

Cell-free DNA-based NIPT is quickly becoming routine in prenatal clinical care, in large part due to the maturation of NGS as a means to read and count cfDNA. This widespread clinical adoption both incentivizes cost-cutting technological advancements and generates the extremely large data sets that facilitate discovery of sensitivity-boosting placenta-specific signals. As such, the ability to quantify and interpret cfDNA will surely continue to evolve at a rapid rate, and these improvements will increase both the performance and accessibility of cell-free DNA-based NIPT.

ACKNOWLEDGMENTS

Clement Chu and Mark Theilmann provided helpful comments on the manuscript.

REFERENCES

[1] Sanger F, Nicklen S, Coulson AR. DNA sequencing with chain-terminating inhibitors. Proc Natl Acad Sci USA 1977;74:5463–7.

[2] Dovichi NJ, Zhang J. How capillary electrophoresis sequenced the human genome. Angew Chem Int Ed 2000;.

[3] Smith LM, Sanders JZ, Kaiser RJ, Hughes P, Dodd C, Connell CR, et al. Fluorescence detection in automated DNA sequence analysis. Nature 1986;321:674–9.

[4] Prober JM, Trainor GL, Dam RJ, Hobbs FW, Robertson CW, Zagursky RJ, et al. A system for rapid DNA sequencing with fluorescent chain-terminating dideoxynucleotides. Science 1987;238:336–41.

[5] Swerdlow H, Gesteland R. Capillary gel electrophoresis for rapid, high resolution DNA sequencing. Nucleic Acids Res 1990;18:1415–9.

[6] National Research Council (U.S.). Committee on Mapping and Sequencing the Human Genome. Alberts B. Report of the committee on mapping and sequencing the human genome. National Academies; 1988.

[7] Lander ES, Linton LM, Birren B, Nusbaum C, Zody MC, Baldwin J, et al. Initial sequencing and analysis of the human genome. Nature 2001;409:860–921.

[8] International Human Genome Sequencing Consortium. Finishing the euchromatic sequence of the human genome. Nature 2004;431:931–45.

[9] Collins FS, Morgan M, Patrinos A. The human genome project: lessons from large-scale biology. Science 2003;300:286–90.

[10] Goodwin S, McPherson JD, McCombie WR. Coming of age: ten years of next-generation sequencing technologies. Nat Rev Genet 2016;17:333–51.

[11] Gao J, Yu X, Ma F, Li J. RNA-seq analysis of transcriptome and glucosinolate metabolism in seeds and sprouts of broccoli (Brassica oleracea var. italic). PLoS One 2014;9:e88804.

[12] Li G-W, Oh E, Weissman JS. The anti-Shine-Dalgarno sequence drives translational pausing and codon choice in bacteria. Nature 2012;484:538–41.

[13] Fan HC, Blumenfeld YJ, Chitkara U, Hudgins L, Quake SR. Noninvasive diagnosis of fetal aneuploidy by shotgun sequencing DNA from maternal blood. Proc Natl Acad Sci USA 2008;105:16266–71.

[14] Taglauer ES, Wilkins-Haug L, Bianchi DW. Review: cell-free fetal DNA in the maternal circulation as an indication of placental health and disease. Placenta 2014;35(Suppl):S64–8.

[15] Enari M, Sakahira H, Yokoyama H, Okawa K, Iwamatsu A, Nagata S. A caspase-activated DNase that degrades DNA during apoptosis, and its inhibitor ICAD. Nature 1998;391:43–50.

[16] Sakahira H, Enari M, Nagata S. Cleavage of CAD inhibitor in CAD activation and DNA degradation during apoptosis. Nature 1998;391:96–9.

[17] Kouzarides T. Chromatin modifications and their function. Cell 2007;128:693–705.

[18] Sorber L, Zwaenepoel K, Deschoolmeester V, Roeyen G, Lardon F, Rolfo C, et al. A comparison of cell-free DNA isolation kits: isolation and quantification of cell-free DNA in plasma. J Mol Diagn 2017;19:162–8.

[19] Pérez-Barrios C, Nieto-Alcolado I, Torrente M, Jiménez-Sánchez C, Calvo V, Gutierrez-Sanz L, et al. Comparison of methods for circulating cell-free DNA isolation using blood from cancer patients: impact on biomarker testing. Transl Lung Cancer Res 2016;5:665–72.

[20] Zimmermann B, Hill M, Gemelos G, Demko Z, Banjevic M, Baner J, et al. Noninvasive prenatal aneuploidy testing of chromosomes 13, 18, 21, X, and Y, using targeted sequencing of polymorphic loci. Prenat Diagn 2012;32:1233–41.

[21] Fan HC, Quake SR. Sensitivity of noninvasive prenatal detection of fetal aneuploidy from maternal plasma using shotgun sequencing is limited only by counting statistics. PLoS One 2010;5:e10439.

[22] Bentley DR, Balasubramanian S, Swerdlow HP, Smith GP, Milton J, Brown CG, et al. Accurate whole human genome sequencing using reversible terminator chemistry. Nature 2008;456:53–9.

[23] Quail MA, Kozarewa I, Smith F, Scally A, Stephens PJ, Durbin R, et al. A large genome center's improvements to the Illumina sequencing system. Nat Methods 2008;5:1005–10.

[24] Minarik G, Repiska G, Hyblova M, Nagyova E, Soltys K, Budis J, et al. Utilization of Benchtop next generation sequencing platforms ion torrent PGM and MiSeq in noninvasive prenatal testing for chromosome 21 trisomy and testing of impact of in silico and physical size selection on its analytical performance. PLoS One 2015;10:e0144811.

[25] Fan HC, Blumenfeld YJ, Chitkara U, Hudgins L, Quake SR. Analysis of the size distributions of fetal and maternal cell-free DNA by paired-end sequencing. Clin Chem 2010;56:1279–86.

[26] Jiang P, Lo YMD. The long and short of circulating cell-free DNA and the ins and outs of molecular diagnostics. Trends Genet 2016;32:360–71.

[27] Yu SCY, Chan KCA, Zheng YWL, Jiang P, Liao GJW, Sun H, et al. Size-based molecular diagnostics using plasma DNA for noninvasive prenatal testing. Proc Natl Acad Sci USA 2014;111:8583–8.

[28] Wong KH, Jin Y, Moqtaderi Z. Multiplex Illumina sequencing using DNA barcoding. Curr Protoc Mol Biol 2013;Chapter 7:Unit 7.11.

[29] Chiu RWK, Chan KCA, Gao Y, Lau VYM, Zheng W, Leung TY, et al. Noninvasive prenatal diagnosis of fetal chromosomal aneuploidy by massively parallel genomic sequencing of DNA in maternal plasma. Proc Natl Acad Sci USA 2008;105:20458–63.

[30] Mir K, Neuhaus K, Bossert M, Schober S. Short barcodes for next generation sequencing. PLoS One 2013;8:e82933.

[31] Li H, Durbin R. Fast and accurate short read alignment with Burrows-Wheeler transform. Bioinformatics 2009;25:1754–60.

[32] Langmead B, Trapnell C, Pop M, Salzberg SL. Ultrafast and memory-efficient alignment of short DNA sequences to the human genome. Genome Biol 2009;10:R25.

[33] Feng Y, Zhang Y, Ying C, Wang D, Du C. Nanopore-based fourth-generation DNA sequencing technology. Genomics Proteomics Bioinformatics 2015;13:4–16.

[34] Juneau K, Bogard PE, Huang S, Mohseni M, Wang ET, Ryvkin P, et al. Microarray-based cell-free DNA analysis improves noninvasive prenatal testing. Fetal Diagn Ther 2014;36:282–6.

[35] Kazemi M, Salehi M, Kheirollahi M. MeDIP real-time qPCR has the potential for noninvasive prenatal screening of fetal trisomy 21. Int J Mol Cell Med 2017;6:13–21.

[36] Sun K, Jiang P, Chan KCA, Wong J, Cheng YKY, Liang RHS, et al. Plasma DNA tissue mapping by genome-wide methylation sequencing for noninvasive prenatal, cancer, and transplantation assessments. Proc Natl Acad Sci USA 2015;112:E5503–12.

[37] Chan KCA, Jiang P, Sun K, Cheng YKY, Tong YK, Cheng SH, et al. Second generation noninvasive fetal genome analysis reveals de novo mutations, single-base parental inheritance, and preferred DNA ends. Proc Natl Acad Sci USA 2016;113:E8159–68.

GLOSSARY

Cluster generation and bridge amplification In an Illumina NGS device, single DNA molecules are locally duplicated many times via bridge amplification to generate discrete clusters of clonal DNA. The fluorescence yielded by single-base extension at each cluster is sufficiently bright for the device to detect and later decode into a sequence of DNA bases.

DNA barcode A short DNA sequence (typically <10 bases) appended to each fragment in a sequencing library that can trace the NGS read back to its upstream source, for example, its sample or molecule of origin.

Flowcell A glass or plastic chamber equipped with fluidic machinery that enables flooding of the chamber with contents of interest. For NGS, molecules to be sequenced affix to the inner walls of the chamber, and the fluidics introduce chemicals that drive the many cycles of a sequencing reaction.

Multiplexing and demultiplexing Sample-specific barcodes enable the loading of many different samples into a single NGS flowcell (multiplexing). Computational parsing of barcode sequences allows reads from the flowcell to be mapped to files specific to particular samples (demultiplexing).

NGS Next-generation sequencing is a method that adapts its predecessor technology (Sanger sequencing) to be far more scalable, enabling the simultaneous sequencing of an ensemble of billions of DNA fragments.

Sanger sequencing A method developed in the 1970s and subsequently used to sequence the first human genome assembly. In one reaction, Sanger sequencing resolves the sequence of ~400 bases of a single template molecule.

Sequencing library An ensemble of DNA molecules prepared in such a way to render them capable of being sequenced. For Illumina machines, a library has appropriate adapter sequences ligated to the termini of each DNA fragment, allowing hybridization to the flowcell.

Y adapter Double-stranded molecules that are complementary on one side (the base of the Y) and noncomplementary on the other side (the branches of the Y). Y adapters are used when preparing an NGS library such that one sequence is appended to all 5′ ends and a different sequence to all 3′ ends. They often contain a molecular barcode used for multiplexing.

THE TECHNOLOGY AND BIOINFORMATICS OF CELL-FREE DNA-BASED NIPT

3

Dale Muzzey

Counsyl Inc., South San Francisco, CA, United States

INTRODUCTION

The primary challenge of cell-free DNA ("cfDNA") based prenatal testing (commonly called noninvasive prenatal testing or NIPT) is to identify fetal chromosomal anomalies from maternal plasma samples, where maternal cfDNA molecules far outnumber their fetal counterparts. Further complicating this task is the fact that the relative amounts of fetal and maternal fragments (i.e.; the fetal fraction) varies across pregnancies and gestational ages, reaching as high as 40% and as low as 1%. Therefore, to achieve maximal sensitivity and specificity, a cfDNA test's analysis pipeline must determine both the fetal fraction ("FF") of a sample and the likelihood of an aneuploid fetus. Several elegant molecular and bioinformatic strategies have emerged to infer FF and ploidy status reliably and at low cost, yielding a range of cfDNA test offerings that hold in common a remarkably high clinical sensitivity that has driven the rapid adoption of this screening modality. In this chapter, the analysis fundamentals are described for three cfDNA test platforms: whole-genome sequencing (WGS), single-nucleotide polymorphism (SNP), and microarray. The methodologies of aneuploidy identification, FF inference, and microdeletion detection are discussed in turn. These topics provide a basic understanding of how cell-free DNA based prenatal testing works in routine pregnancies and lay the groundwork for a discussion at the end of the chapter about edge cases that will themselves become common as use of the screening grows.

ANEUPLOIDY IDENTIFICATION
WGS-BASED NIPT

To help guide understanding of the algorithm underlying whole-genome sequencing (WGS)-based NIPT analysis [1–3], it is useful first to consider the ultimate goal of NIPT: robust identification of fetal copy-number anomalies in large genomic regions at low FF. The "copy-number anomalies" include trisomies and monosomies in the fetus; "large genomic regions" include whole chromosomes and relatively smaller variants like microdeletions, and we stipulate "low FF" because if the test is sensitive at low FF, then it will surely work well at high FF. This goal provides a framework to evaluate the challenges the analysis algorithm must overcome. For instance, in a pregnancy with 2% FF in which

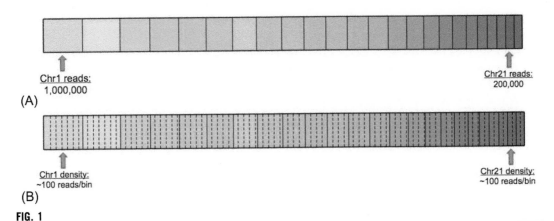

FIG. 1

Read density is preferable to total reads when comparing chromosome dosage. (A) Each box depicts a human chromosome with the width of the box proportional to the chromosome's size. Even in the case of disomy of chr1 and chr21, the number of reads mapped to each chromosome can differ substantially. (B) Schematic of equal-size tiled bins across the genome. Bins are typically tens of kilobases in length, much smaller than shown. Comparisons of chromosome dosage are more straightforward after calculating bin density.

the fetus has trisomy 21—chr21 comprises 1.7% of the genome—only 1 in 3000 cfDNA fragments actually derives from the fetal copies of chr21 $(1.7\% * 2\%)^{-1}$. Therefore to detect a 50% increase in the number of fetal copies of chr21 (i.e., from disomic to trisomic) with high statistical confidence, it is not sufficient to sequence thousands or even hundreds of thousands of random cfDNA fragments. Indeed, millions of sequenced cfDNA fragments are required.

The first step of the WGS analysis pipeline is to align the millions of sequenced cfDNA fragments ("reads") to the human reference genome. Off-the-shelf software packages can map millions of reads to the 3-billion-base human reference genome in the order of minutes [4,5]. Considering the goal of detecting a low-FF T21 sample, the key question is whether the number of reads mapping to chr21 is higher than expected. A naive approach would be to compare the level of chr21 reads to the reads mapped to a known disomic chromosome: for example, assume that there are two fetal copies of chr1 and use the number of reads mapped to chr1 as an estimate for the number of mapped reads to a disomic chr21 (Fig. 1A). However, for an ordinary euploid sample, chr1 may have 1,000,000 reads while chr21 may have only 200,000, illustrating the folly of comparing read totals across chromosomes: chr1 trivially has ∼10× more than chr21 because it is ∼10× longer. Instead, for the comparisons of cfDNA abundance that underlie WGS-based analyses, it is best to use a length-agnostic measurement, such as the density of reads (Fig. 1B).

Read density in WGS-based NIPT algorithms is typically assessed by tiling each chromosome with nonoverlapping bins of equal size, counting the number of reads per bin, and averaging the reads per bin over a region of interest (e.g., a microdeletion or whole chromosome) [1,2,6]. This process of averaging over bins requires two important choices when implementing the algorithm: the bin size to use and the method for calculating an average.

Bin size choice

There is no strictly optimal bin size, yet noteworthy factors inform the choice. One key factor in bin size choice is the total number of reads the sample receives, which was asserted above to be in the tens of millions. In general, total reads and bin size are inversely proportional, with deeper sequencing

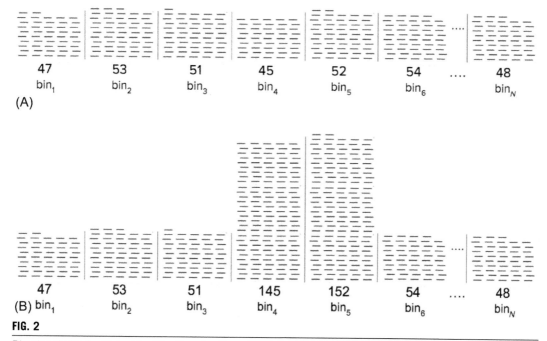

FIG. 2

Bin-size choice facilitates identification of outliers. (A) There will naturally be bin-to-bin variability in the number of reads observed per bin, and this variability in observed reads scales as the square root of the average. With an average of 50, the noise is ~7 reads per bin, making it straightforward—even at the level of a single bin—to observe and remove large outliers (B), e.g., caused by maternal CNVs.

permitting smaller bins. Ideally, bins should be small enough to allow clear differentiation of aneuploidy-scale deviations in signal from large, spurious deviations that should be omitted from the dataset (e.g., biological events like maternal CNVs or analytical artifacts like alignment mistakes; Fig. 2). For instance, if bins are so small that the average bin count is only two reads, the discreteness of NGS reads means that it is not straightforward to interpret a bin with four reads as being from a normal, aneuploid, or maternal-CNV-harboring chromosome. However, with larger bins that have 50 reads on average, it is much simpler to distinguish a deflection consistent with aneuploidy (e.g., 55 reads) from a maternal-CNV-caused deviation (e.g., 100 reads).

While the discretization of individual reads described previously provides an upward pressure on bin size, the existence of localized, anomalous genomic regions exerts a downward pressure that supports the use of smaller bins. These anomalous regions could be bins or portions of bins that have stochastic, idiosyncratic, and/or highly skewed levels of mapped reads; even very high-quality samples contain such regions. Fortunately, such regions are rare across the entire genome, yet they deviate greatly from one of the key underlying assumption of WGS, that is, that cfDNA fragments are uniformly sampled. In practice, the few anomalous bins are discarded from the analysis, demonstrating one reason why a large number of relatively small bins are beneficial. To see another reason, consider a 5-kb segment that is anomalous and rightfully caused its encompassing bin to be discarded. If bins were 20 kb, only 15 kb worth of nonanomalous signal would effectively be lost due to the anomaly. However, if bins were 100 kb, then 95 kb of valid sequence would be lost, illustrating the potentially substantial loss of signal resulting from using large bins. To see this argument from another perspective, suppose small anomalous regions occur every ~100 kb. If bin size were set to 200 kb, then nearly every bin would harbor signal-compromising anomalies that jeopardize the ability to accurately identify aneuploidy.

Consistent with the two arguments previously—which exert upward and downward forces on bin size—most WGS-based NIPT algorithms described in the literature use 50 kb bins.

Calculating an average bin value for a region and sources of error

Though it sounds straightforward to calculate an average of reads per bin across the many bins that tile a genomic region of interest, a lot of algorithmic nuance is required to maximize the screen's sensitivity and specificity given a sample's sequenced cfDNA fragments. To appreciate the performance gains from special algorithmic care, recall that WGS-based NIPT is a sampling problem: the sequenced reads in each bin are simply a sampling of the fragments present in the maternal plasma, which are themselves only samples from apoptotic cells in the placenta and other parts of the body. As in any sampling problem (e.g., polling prospective voters before an election), in WGS-based NIPT there is a margin of error by which the observed average reads per bin differ from its underlying and typically unobservable true value. Note that if the error in the calculated average reads per bin is too large, it causes false ploidy calls, thereby undercutting the goal of NIPT. The error can be represented as the standard error, $\sigma/\mathrm{sqrt}(N)$, where σ is the standard deviation of reads per bin, and N is the number of bins. As N is an immutable property of a chromosome and the selected bin size, it is the standard deviation of reads per bin (σ) that drives the error of the observed bin count average. The aim, then, of a well-crafted WGS-based NIPT algorithm is to remove sources of bias that inflate σ such that it can descend to its theoretical minimum (which, by Poisson statistics, is $\sigma \sim \mathrm{sqrt}(\text{reads per bin})$) [6].

There are multiple sources of bias in WGS data, but methods exist to remove them or mitigate their effects. As described later, some biases (e.g., GC bias, nonunique sequence, etc.) have known molecular underpinnings and can be corrected with approaches tailored to their origin, while other biases may not have a known source but can be diminished through general robust analysis methods.

Sample-to-sample differences in the propensity with which fragments of particular GC content get sequenced can inflate σ above its theoretical minimum [7,8]. This so-called GC bias arises due to the different thermodynamics of G:C base pairs, which have three hydrogen bonds, and A:T base pairs, which have two. GC bias can be introduced at any level of the NIPT workflow, including DNA extraction, NGS library preparation, and sequencing itself. To appreciate the manifestation of GC bias in WGS data, recall that the expectation is for WGS to yield a uniform sampling of reads across the genome. However, the empirical distribution of reads is not strictly uniform and instead correlates in part with GC content: cfDNA fragments that have high (~80%) or low (~20%) GC content often generate fewer reads than expected relative to fragments with average GC content (40%–60%). To correct for GC bias, the algorithm should first assess the extent of bias on a sample-specific basis by calculating the ratio between the observed number of reads with a given GC content and the number of such fragments in the whole genome (Fig. 3). Next, observed fragments can be scaled by their particular GC content and known GC bias: for example, if fragments with 20% GC content were observed to be sequenced with only 50% relative efficiency, then each read observed from a fragment with 20% GC content should be scaled by $1/0.5 = 2$.

Redundant regions of the genome can also cause nonuniformity in WGS coverage [9] and reduce NIPT accuracy if unaddressed [10]. In redundant regions, it is impossible for the NGS read alignment software to infer the true genomic origin of a fragment. As such, even if a region occurs thousands of

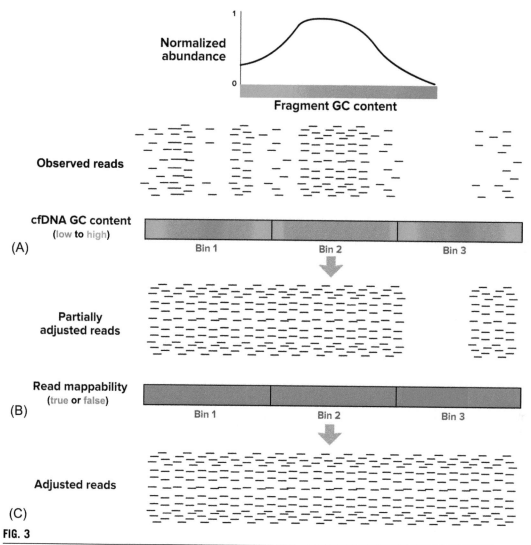

FIG. 3

Correcting for GC and mappability biases. (A) Despite an expectation of uniform coverage, sequenced cfDNA fragments map with nonuniform density (*bottom*), driven in part by GC bias (*top*). (B) Correction for GC bias involves scaling the observed reads by a correction factor derived from the aggregate GC bias plot (show in A, *top*); e.g., in *red-shaded regions* with low GC content and normalized abundance near 50%, each observed read is scaled by 2 (i.e., 1/0.5). (C) Some residual coverage gaps after GC-bias correction stem from mappability, where "mappable" positions are those where fragments align uniquely. Scaling a bin's reads by the reciprocal of mappability increase the uniformity of coverage.

times in the genome (e.g., Alu repeats), the aligner may pile reads derived from all of those regions into just a single region; this skewed parsing of reads could create a single bin with a tremendously high number of reads, and many other bins with depressed read values. One way to address degenerate read mappability is to flag redundant regions in advance, mask them altogether to prohibit mapped reads from counting toward the bin total, and then scale the observed reads in a region by the reciprocal of its share of mappable bases (e.g., a bin with 60 reads that is 75% mappable would be scaled to have 80 reads) (Fig. 3).

Maternal copy-number variants ("CNVs") are another source of potential bias in WGS NIPT data [11–13], and their clear biological origin enables specific handling during analysis. In particular, maternal CNVs have a deflection in reads per bin that is very large but also predictable, 50% more or less than the disomic baseline. If such a deflection were of fetal origin, it would be consistent with 100% FF, which is clearly impossible. Maternal CNVs' predictable signal makes it possible to computationally scan across the genome and identify contiguous spans of bins where the bin counts are increased or decreased by 50%. These bins can then be omitted from the calculation of the region average. Some of the early WGS-based NIPT algorithms did not attempt to identify maternal CNVs, meaning their constituent bins were not omitted from subsequent analysis. These strongly skewed bins increased the mean bin value on CNV-harboring chromosomes to the point that false-positive aneuploidy calls were yielded in several cases. These false-positive results revealed not only a shortcoming in maternal CNV handling, but also a suboptimal, nonrobust method of calculating the average (i.e., the mean), described further later.

Even if a source of bias has unknown molecular origin, its impact can be mitigated if the bias is systematic. For instance, if a bin has elevated reads relative to other bins reproducibly across all samples and with no obvious explanation, it can nevertheless be included in the analysis after a normalization step: for example, by removing signal from the first several principal components [14] or, more simply, by calculating the median number of reads for the bin across a large sample cohort (where the median sample can be assumed to be disomic) and then subtracting this average from the bin in all samples. Applying this normalization procedure across all bins—whether or not they are outliers—reduces the systematic variance in the data and ultimately serves to reduce σ.

Finally, because any outlying bins that were not filtered out using the previous strategies could yield false results, it is important to reject such bins [6,15] and/or use robust measures for both the average and the dispersion of bin density (Fig. 4). In particular, for the average, it is best to use the median rather than the mean, as the median will not be skewed by a strong outlier, caused by biological phenomena (e.g., a maternal CNV) or analytical artifacts (e.g., alignment mistakes). For the dispersion, it is best to avoid the standard deviation, as it is susceptible to outliers; the interquartile range (IQR) is a more robust estimator of dispersion. To see the power of using the median and IQR relative to the mean and standard deviation, consider an analysis pipeline that was unaware of the existence of maternal CNVs (mCNVs). With the mean and standard deviation, an mCNV spanning ~2% of a chromosome would be enough to strongly risk a false-positive result, whereas using the median and IQR does not risk a false positive until an mCNV spans ~30% of the chromosome.

Taken together, the processes of aligning reads to bins, correcting for GC bias and mappability, filtering outlier bins, and averaging in a robust manner yield an observed read density for a region of interest (μ_{obs}), but deducing an aneuploidy call from μ_{obs} requires a couple other pieces of information. In particular, to say whether μ_{obs} is statistically higher than expected, it is critical to know both the expected density under a disomic hypothesis (μ_{exp}) and the average deviation between μ_{obs} and μ_{exp}, such that assessment of a statistically significant difference is possible.

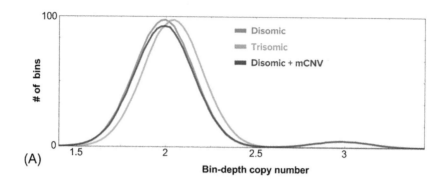

	Mean	STDEV	Median	IQR
Disomic	0.00	0.33	0.00	0.12
Trisomic	0.10	0.33	0.10	0.12
Disomic + mCNV	0.10	0.55	0.02	0.13

(B)

FIG. 4

Outlier-insensitive measures are important for NIPT analysis robustness. (A) Distributions of bin depths (see Figs. 2 and 3 for depiction of bin depth) are shown for three simulated scenarios, each centered at a copy number of two, which assumes the maternal background is disomic. Fetal trisomy shifts the distribution rightward in proportion to the sample's FF. A disomic sample harboring an mCNV (*blue*) has a minority of bins at highly elevated depth corresponding to CN~3. (B) Troublingly, the disomic+mCNV case has the same mean as the trisomic case, and there is a large increase in the standard deviation relative to the disomic and trisomic cases. By contrast, the median appropriately has a large deflection for the trisomic sample but not the disomic+mCNV sample. Additionally, the relative increase in IQR is far less than the relative change in standard deviation.

Expected read density

There are multiple ways to calculate μ_{exp}, with some methods being sample specific, others being region specific, and still others leveraging data across both samples and regions. In a sample-specific approach from the early days of WGS-based NIPT, a so-called reference chromosome was assigned to each chromosome of interest [16], where the pair of chromosomes had similar features, for example, GC content. On the assumption that the reference chromosome k was disomic, then the expected read density for the region of interest i in sample j ($\mu_{exp,i,j}$) was assumed to be equal to the reference chromosome density ($\mu_{obs,k,j}$). However, this approach has the shortcoming of putting all the eggs in one basket, so to speak: an uncorrected anomaly on the reference chromosome could corrupt aneuploidy calling for the chromosome of interest. A further downside of the reference chromosome approach is that it only leverages a small subset of the data gathered from sequencing the whole genome, as it only considers a minority of 22 chromosomes (exempting the chromosome of interest itself). This one limitation reveals yet another: with a sample-specific approach, there is only a limited number of potential reference chromosomes, and there may not be a suitable reference for a particular chromosome of interest.

A pan-sample, single-region approach obviates the concern about comparing to a reference region disparate from the chromosome of interest, because all comparisons are relative to the lone chromosome of interest [15]. Specifically, $\mu_{exp,i,j}$ for region i in sample j is the average of $\mu_{obs,i,k}$ for region i in all samples $k \neq j$. This approach must be exercised with caution for two main reasons. First, recall that the chromosome of interest is itself rendered interesting because it has elevated incidence of aneuploidy; thus there may be aneuploid samples in the background set (i.e., the samples $k \neq j$) that could skew a calculation of the average of $\mu_{\text{obs},i,k}$. Ideally, the algorithm should identify and exclude such samples, but at the very least, using a robust estimator like the median to calculate the average of $\mu_{\text{obs},i,k}$ is important. Second, this approach does not account for sample-specific deviations from the background cohort. For instance, even with seamless adherence to laboratory best practices, there are sometimes rare samples whose distribution of sequenced fragments across the genome differs from that of the majority of samples. If the background majority of samples has relatively homogeneous read density in the region of interest (i.e., uniform $\mu_{\text{obs},i}$), but sample j systematically deviates from the background, sample j could spuriously appear aneuploid. Therefore like the sample-specific approach described earlier, the single-region method has shortcomings.

In principle, the most powerful and robust way to calculate $\mu_{\text{exp},i,j}$ is to leverage all regions of interest in all samples. Such an approach would harness the virtue of region-specific calculations (e.g., using chr21 read density in background samples to infer chr21 in the sample of interest) while also accounting for sample-specific effects (e.g., if the sample of interest deviates from the background cohort). A machine-learning model (e.g., linear regression) can serve this dual purpose [8]. The training set for such a model is a matrix of $\mu_{\text{obs},i,j}$ for many regions i and samples j. For a given region of interest i, the model determines how best to weight all other regions $k \neq i$ in order to best predict region i across background samples. In a way, the regression model can be considered an expansion of the reference chromosome method. The latter effectively used a weight of 1.0 for the reference chromosome, whereas the former might use weights of 0.12 for chr1, -0.05 for chr2, 0.2 for chr3, and so on, where weights are derived to yield the best prediction for the chromosome of interest across samples. Once the model learns this optimal weighting, it can predict the expected read density for the region i in the sample of interest j by calculating a weighted sum based on the other regions in sample j (e.g., $0.12 * \mu_{\text{obs},\text{chr1},j} + (-0.05) * \mu_{\text{obs},\text{chr2},j} + 0.2 * \mu_{\text{obs},\text{chr3},j} + \cdots$).

Calculating and interpreting the z-score

Beyond determining the observed bin density and the expected disomic density as described previously, one final and relatively straightforward calculation is needed to assess aneuploidy: the average difference between the expected and observed densities across a large cohort of samples. Like most other calculations in NIPT, this expected deviation ($\sigma_{\text{O-E}}$) should also be robust to outliers—for example, by using the IQR or median absolute deviation rather than the standard deviation—because high-FF aneuploid samples will manifest as outliers, where observed density far exceeds disomic expectation. Once μ_{obs}, μ_{exp}, and $\sigma_{\text{O-E}}$ have been determined, it is possible to calculate the z-score:

$$z = \left(\mu_{\text{obs}} - \mu_{\text{exp}}\right) / \sigma_{\text{O-E}}$$

z-Scores near zero suggest that the sample does not deviate from the disomic hypothesis. By contrast, samples with high-magnitude z-scores are consistent with aneuploidy: trisomy if the z-score is highly positive and monosomy if the z-score is strongly negative.

The chosen threshold, or cutoff, between euploidy and aneuploidy has a large impact on the sensitivity and specificity of the test. In general, low thresholds sacrifice specificity to achieve high sensitivity, whereas high thresholds prioritize specificity over sensitivity.

Assuming that z-scores of truly disomic samples are normally distributed, a z-score threshold of three equates to a minimum expected specificity of ~99.9% (i.e., one false positive per 1000 samples).

It is possible to achieve even higher levels of specificity by making the euploid-to-aneuploid cutoff aware of sample-specific FF [14]. To see this, recall that μ_{obs} in an aneuploid sample scales with FF (e.g., a high FF trisomic sample has very large deviation between μ_{obs} and μ_{exp} and, thus should yield a very high z-score). Thus if the sample has a known FF value that should give $z \sim 20$ when trisomic, then it is no longer prudent to have a z-score cutoff of three. The threshold could be raised to, say, five, which (1) boosts specificity by almost eliminating the possibility of a disomic sample being a false positive by chance, and (2) has virtually no downward impact on sensitivity, since the z-score of an aneuploid sample still vastly exceeds the cutoff.

Sex chromosome aneuploidies

Whereas detecting autosomal aneuploidies is a one-dimensional problem that primarily focuses on the change in depth of a single chromosome, identifying sex-chromosome aneuploidies requires a two-dimensional analysis with simultaneous consideration of chrX and chrY [17]. For instance, the difference between an MX female and a euploid XY male cannot be deciphered from chrX alone, as both samples are effectively monosomy on chrX. But, they are distinguished by chrY: the MX sample has effectively zero depth from chrY, whereas the XY sample has an increase in chrY signal comparable to the deficit in chrX signal. Determining ploidy status on females—that is, MX, XX, or XXX—is effectively an autosome-like, one-dimensional analysis once it is established that chrY is absent.

For males, however, who may be XXY, XYY, or XY, ploidy-status resolution is more complicated, as samples' positions in the chrX-vs-chrY 2D space become relevant. Assigning genotypes depends on how closely a given sample's position in the plot corresponds to each respective genotype's hypothetical region. XXY samples have no depletion in chrX relative to the maternal background, though they have significant presence of chrY. As such, they are expected to occupy the territory on a vertical line emanating from the origin. To determine where XYY samples are expected to be in the chrX-vs-chrY plot, the scaling relationship between chrX and chrY signal (i.e., the slope of the blue region in Fig. 5) must be determined for XY males, which are in abundance. Next, the

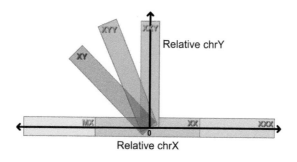

FIG. 5

Sex-chromosome ploidy status determined by position in 2D plot of chrX vs chrY. The normalized bin depths calculated from chrX and chrY for each sample determine the sample's position in the previous 2D plot, where schematized regions indicate fetal genotypes. Being near the origin means that the fetal genotype matches the maternal background, i.e., XX. The chrY values in all males and the chrX values in aneuploid females (i.e., MX or XXX) emanate away from the origin in proportion to FF. Each NIPT algorithm must determine how to assign genotypes in both the overlapping regions near the origin and the white space between regions.

expected chrX-vs-chrY relationship for XYY can be calculated, for example, by doubling the slope for XY males. Finally, each sample must be assigned to a ploidy status based on its position relative to each hypothesis line. Distinguishing the three male genotypes is challenging at very low fetal fraction (i.e., near the origin in Fig. 5) because the hypotheses converge. As XY samples far outnumber XXY and XYY in the general population, it is expected that samples near the origin are XY, and they are, therefore typically reported as such. Though this approach ensures high specificity for sex calling, it necessarily reduces the sensitivity for XXY and XYY at low fetal fraction, a limitation that applies to all forms of NIPT.

SINGLE-NUCLEOTIDE POLYMORPHISM-BASED NIPT

Trisomy causes an overrepresentation of a particular genomic region, and this excess DNA can create different signals in NGS data that can be measured during NIPT. The most obvious is an increase in the number of cfDNA fragments deriving from the aneuploid region; indeed, this NGS depth signal is the one that underlies WGS-based NIPT. But, the excess fetal cfDNA fragments can also alter the allele balance at particular sites. For instance, whereas a G/T SNP (single-nucleotide polymorphism) in the maternal genome has a ~50% allele balance, the contribution of the fetal genome in a pregnant mother may skew the allele balance away from 50%, for example, if the fetus is G/G at the same site. Such deflections in allele balance can be analyzed to infer ploidy via the SNP-based method of NIPT (Fig. 6) [18–20].

SNP-based NIPT requires allele balance measurements across thousands of sites that tile regions of interest. These requirements pose two important challenges: (1) how to measure allele balance from cfDNA and (2) which sites should be interrogated.

The need to measure allele balance means that a fundamentally different technique of NGS library preparation is required for SNP-based NIPT [19] relative to WGS-based NIPT. With WGS library preparation and at the level of sequencing required for WGS-based NIPT, only ~20% of genomic sites have even a single read covering them, making it virtually meaningless to assess the fraction of alleles at a given site: no fraction can be measured at the 80% of sites with no coverage, and the allele balance is either 0% or 100% at sites with only a single NGS read. Allele balance could be coarsely measured if WGS depth were increased 100-fold (i.e., giving ~20× everywhere), but the cost of the test would also increase to a prohibitive level ~100-fold higher. The solution for getting high depth at particular sites affordably is to perform a multiplex PCR enrichment of targeted sites. A single multiplex PCR reaction—containing many primer pairs that flank particular bases of interest—can amplify thousands of unique genomic locations, yielding an NGS library from which the depth at each site can affordably be in the hundreds. With hundreds of reads at a given site, measuring the allele balance is both meaningful and easy.

The targeted nature of SNP-based NIPT requires judicious choice of which sites to amplify. As described in more detail later, the most informative sites are highly polymorphic in the population, as the allele balance deflections associated with aneuploidy can only be measured where the mother and/or the fetus is heterozygous. To see why only these highly diverse sites are useful, consider the example in Fig. 6. For the indicated site, the mother is homozygous for the reference allele (by convention, reference alleles at all sites are simply called "A" and nonreference are called "B"), and the disomic fetus is heterozygous ("A/B"). In a 20% FF sample, the allele balance (fraction of A bases) of maternal and fetal cfDNA together is 90% (where the 10% drop in A is due to the single B allele in the

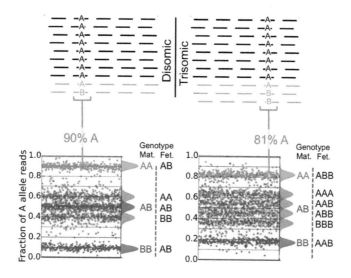

FIG. 6

The SNP method detects aneuploidy via deflections in allele balances. The *left panel* shows a disomic fetus, and the *right panel* shows a trisomic fetus. In a 20% fetal-fraction pregnancy, the majority of cfDNA fragments (*lines at top*) are maternally derived (*black*), with the minority coming from the placenta (*blue*). Only a subset of cfDNA fragments interrogate SNP sites, i.e., those positions depicted as having A and B alleles. Via multiplex PCR, the SNP method amplifies only the cfDNA fragments containing SNPs and sequences them to yield an allele balance measurement, e.g., the fraction of A bases at a site. Each spot in the *bottom panels* represents a different SNP, with shading corresponding to the maternal genotype and distributions of allele balances shown at the right of each plot. In the case of a paternally inherited trisomy (*right side of figure*; note extra B allele at a site where the mother is homozygous for A), the allele-balance distributions shift significantly relative to the disomic case (*left side of figure*). The SNP quantifies the magnitude of such signal shifts to detect aneuploidy.

Adapted from Artieri CG, Haverty C, Evans EA, Goldberg JD, Haque IS, Yaron Y, et al. Noninvasive prenatal screening at low fetal fraction: comparing whole-genome sequencing and single-nucleotide polymorphism methods. Prenat Diagn 2017;37(5):482–90.

fetus). If the fetus has a paternally inherited trisomy where the father contributes two copies of the B-harboring chromosome, however, then the allele balance shifts to 81%. Measuring such shifts across many sites can assess the likelihood of euploidy vs aneuploidy, discussed further later. This example highlights why polymorphism is important for the SNP method: if the mother and fetus were both homozygous A/A, there is no difference in the allele balance between fetal disomy and trisomy, as both would be 100% A.

The need for heterozygosity has several important implications for SNP-based NIPT. First, unlike WGS-based NIPT, where the number of informative observations (i.e., the bins discussed earlier) is a constant function of chromosome size, the number of informative SNPs will differ with each pregnancy based on the genotype of mother and father. In particular, useful SNP data becomes sparse as relatedness between the parents increases; at the extreme, SNP-based NIPT is poorly suited to consanguineous couples, since fewer SNP positions are divergent between the mother and fetus than in nonconsanguineous couples. Second, the ethnicity of the mother and father can influence the number of informative SNPs, as a highly polymorphic site in patients of European descent may not be polymorphic in

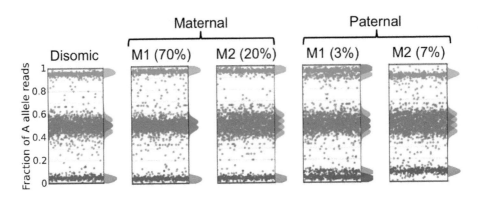

FIG. 7

SNP method analysis depends on the parental and meiotic origins of trisomy. All five panels depict a modeled pregnancy with 10% fetal fraction. For the four aneuploidy scenarios at right, the parent of origin (maternal or paternal) and meiotic stage of origin (M1 or M2) are accompanied by the frequency of such events in parentheses. Allele balance distributions in the paternal M2 trisomy differs conspicuously from the disomic case, consistent with very high sensitivity for such trisomies. The more common maternal M1 trisomies, however, are more challenging to detect by eye and, consequently, by the algorithm as well.

Adapted from Artieri CG, Haverty C, Evans EA, Goldberg JD, Haque IS, Yaron Y, et al. Noninvasive prenatal screening at low fetal fraction: comparing whole-genome sequencing and single-nucleotide polymorphism methods. Prenat Diagn 2017;37(5):482–90.

people of Asian descent. In sum, due to the variable number of SNPs per pregnancy, the sensitivity of SNP-based NIPT is intimately tied not only to fetal fraction (as is the case for WGS-based NIPT), but also to the ethnicity and relatedness of the parents. In light of these factors, careful quality control is required to ensure confident results.

Finally, the deflections in allele balance at the core of SNP-based NIPT analysis are highly dependent on both the parental and meiotic origin of an aneuploidy [21]. Fig. 7 depicts SNP-based NIPT data for five pregnancies, each with 10% FF. The pattern of SNP allele balances in paternally inherited trisomies—which account for ~10% of all T21 cases—is conspicuously different from the disomic case (e.g., note downward shift of red SNPs, upward shift of blue SNPs, and dilation of the green region), and such trisomies are handily detected with very high confidence. The far more common maternally inherited trisomies, however, deviate less from the disomic case (especially trisomies originating in the M1 phase of meiosis, which account for 70% of T21 cases), and these trisomies are more challenging for SNP-based NIPT to detect, especially at low FF [21].

Quantifying aneuploidy likelihood in SNP-based NIPT

SNP-based NIPT detects aneuploidy by enumerating various ploidy hypotheses and evaluating which is the most likely given the observed data [19,20,22]. The root hypotheses include disomy, maternally inherited M1 trisomy (one copy of each maternal chromosomes inherited), maternally inherited M2 trisomy (two copies of a single maternal chromosome inherited), paternally inherited M1 trisomy, and paternally inherited M2 trisomy. Due to the possibility of recombination during meiosis 1, however, the algorithm must entertain a combinatorial exploration of hypotheses: for example, a paternally inherited trisomic chromosome could actually switch between having M1-like spans (both paternal alleles present) and M2-like spans (two copies of a single paternal allele present). The likelihood of each hypothesis is evaluated across the many SNPs in the dataset, either at the single SNP level or as small groups of frequently coinherited SNPs called haplotypes [22]. Following is a description of how the single-SNP math works. The exact method for analyzing haplotypes has not been described in the literature and is therefore omitted here. But, the key principle is highly similar between analyzing single SNPs and haplotype blocks. The key difference is that haplotypes simply allow particular hypotheses under evaluation to be better defined, which could boost signal if the parental genomes indeed have particular haplotypes and could, conversely, have minimal impact on sensitivity if the haplotype is absent or interrupted.

Consider a single site at which a 94% allele balance (fraction of A alleles) is observed among 50 NGS reads (47 A alleles, and 3 B alleles) using SNP-based NIPT in a pregnancy with 10% FF. Further assume that the population frequency of the A and B alleles in the population is 50%. The goal is to evaluate which ploidy hypothesis best applies. Mathematically, this can be expressed as:

$$p(94\%|\ gt_M, gt_F, 10\%FF, ploidy\_hypothesis, 50\times)$$

where gt_M and gt_F are the maternal and fetal genotypes, respectively. Because the maternal genome is predominant in cfDNA, the observation of an allele balance near 100% strongly suggests that gt_M is AA. As the fetal genotype is unknown, the probability calculation must take the weighted sum of probabilities across potential fetal genotypes, altering the earlier equation to yield:

$$\sum_{gt_F \in (AA, AB, BB)} p(94\%|\ AA_M, gt_F, 10\%FF, ploidy\_hypothesis, 50\times)*p(gt_F|\ AA_M, gt_P, ploidy\_hypothesis)$$

where gt_P is the paternal genotype. As gt_P itself may be unknown, the earlier equation may need further expansion to take a weighted sum over the paternal genotypes AA, AB, and BB (not shown). For the sake of seeing a number produced by the earlier equation, let us assume that gt_P is known to be BB and then evaluate one step of the summation in the earlier equation where gt_F and the ploidy hypothesis under consideration are AB and disomy, respectively. The rightmost term of the equation—$p(AB_F|\ AA_M, BB_P, disomy)$—is simply 100%, because a disomic fetus will certainly be AB if the mother is AA and the father is BB. The leftmost term—$p(94\%\ |\ AA_M, AB_F, 10\%FF, disomy, 50\times)$—can be thought of as a coin-flipping problem. Having 10%FF coupled with AA_M and AB_F means that there are effectively 19 A alleles for every 1 B allele; put differently, the coin being flipped is highly biased and has a 95% chance of being A and a 5% chance of being B. Having $50\times$ NGS depth at a site with 94% allele balance means that the observed coin flipping yielded 47 A alleles and 3 B alleles. The binomial distribution describes probabilities in cases like this, and it would give the following:

$$p(94\%|\ AA_M, AB_F, 10\%FF, disomy, 50\times) = \binom{50}{47}*0.95^{47}*0.05^{(50-47)} \sim 0.2199$$

The previous calculation is only a partial evaluation of the probability of observing 94% allele balance at the SNP site; the full summation would require comparable calculations where gt_F equals AA and BB. Further, analogous calculations would need to be performed for each hypothesis under consideration at this single SNP, just not the disomic case evaluated previously. Next, because all of the math previously is only for a single SNP, similar calculations must be performed across all such SNPs for each region of interest. Finally, to compare hypotheses, the likelihood of each hypothesis is calculated as the product of probabilities across SNPs. The algorithm returns an aneuploid call if the likelihood score of a nondisomic hypothesis exceeds that of the disomic hypothesis by an amount sufficient to ensure high overall specificity of the test.

MICROARRAY-BASED NIPT

At both the molecular and algorithmic levels, NIPT using microarray [23] combines features of both the WGS and SNP methods. At the molecular level, no cfDNA molecules are sequenced by a microarray. Rather, the abundance of fluorescently labeled cfDNA is measured in a sequence-specific manner via hybridization to synthetic oligonucleotide probes affixed to known positions on the microarray slide. The fluorescence intensity at each probe—an analog signal distinct from the digital signals returned from NGS employed by the WGS and SNP methods—reports the level of a particular cfDNA in a single sample. Because microarrays only gather signal at locations with custom probes, they must be specially designed for NIPT, for example, with probes that tile the chromosomes of interest. This targeting of certain cfDNA molecules with probes is analogous to the multiplex PCR of the SNP method and very different from the untargeted ascertainment of WGS.

At the algorithmic level, microarray-based NIPT typically resembles WGS more than SNP. Each probe position can be treated like a bin from WGS-based analysis, where the abundance of cfDNA measured at the probe corresponds to the depth of NGS reads in a given bin. As with WGS analysis, minimization of error per observation (bin or probe) increases sensitivity; therefore it is customary for microarrays to have probe redundancy, where each probe may be spotted on the array at multiple different locations. The form factor and lack of multiplexing (i.e., the ability to test more than one sample per array; see Chapter 2) of microarrays poses a notable constraint: because each microarray can only accommodate a fixed number of probes, expanding the content of this targeted NIPT test could require a large jump in cost (e.g., doubling the number of arrays needed per sample). The NGS-based offerings are not similarly constrained, since a single NGS flowcell yields far more reads than a single sample could need; therefore panel expansions are more easily accommodated by multiplexing fewer samples on the flowcell.

FETAL FRACTION ESTIMATION

As the name suggests, fetal fraction (FF) describes the portion of a cfDNA sample that is derived from placental tissue. As cfDNA-based prenatal testing aims to detect genomic features in the fetus, FF is a key determinant of the screen's sensitivity, with increasing FF yielding higher sensitivity when all else is equal. This section details how FF can be measured with each platform.

WGS-BASED NIPT

There are two basic features of a cfDNA molecule detectable by NGS: the sequence of a fragment (which dictates where it aligns in the genome) and its length (paired-end sequencing required). Interestingly, both pieces of data—when aggregated across wide swaths of the genome—contain extractable information about FF for WGS-based NIPT.[1]

As described earlier, the aneuploidy-calling algorithm from WGS data leverages binned read counts, and FF determination does the same. In the most basic case, consider an aneuploid pregnancy,

---

[1]Note that the allele balance of multiple reads at a single site—which will play a fundamental role in SNP-based FF inference—is not a plausible signal for WGS-based FF determination: whether the read harbors a SNP is mostly uninformative, as the depth is insufficient to measure a useful allele-balance signal.

for example, T21. Bins on chr21 will, on average, have higher depth. Critically, the extent of their greater depth depends linearly on FF, as chr21 fragments are increasingly likely to be sequenced as the share of placenta-derived cfDNA grows. This linear correspondence means that the gain in chr21 average bin depth in an aneuploid fetus is itself a measure of FF. Fortunately, the approach of using an aneuploid chromosome to report FF is not restricted to fetuses at risk for disease. Indeed, the concept can be recast: direct FF measurement is possible when the placental signal differs from the maternal background. By this logic, FF measurement in male-fetus pregnancies is very straightforward: the observed average bin count value on chrX is lower than in the maternal background and the signal on chrY exceeds maternal background. By measuring the average bin depth on chrX relative to the average depth on disomic autosomes, FF can be derived based on the following equations:

$$chrX_{obs} = (\text{Maternal contribution}) + (\text{Fetal contribution})$$

A male-fetus pregnancy is effectively monosomic for chrX, so the previous can be reexpressed as:

$$chrX_{obs} = (1 - FF)*disomic_{chrX} + FF*monosomic_{chrX}$$

Assuming depth normalization such that $disomic_{chrX} = 1$ and $monosomic_{chrX} = 0.5 * disomic_{chrX}$:

$$chrX_{obs} = (1 - FF)*1 + FF*0.5*1$$

Solving for FF gives:

$$FF = 2*(1 - chrX_{obs})$$

Therefore if the average bin depth on chrX is 95% of the disomic background, then $FF = 10\%$.

The critical shortcoming of the previous approach to FF measurement is that it only applies to a fraction of pregnancies: aneuploid fetuses and euploid male fetuses. A more nuanced analysis is needed to estimate FF for all pregnancies, but it has important parallels with the prior method. In the same way the prior method applied a mathematical operation (the median) to a collection of bins with depth deflected away from normal (those on chrX), FF can also be inferred using a slightly more complicated operation—a weighted sum—on bin counts across the genome. Specifically, using a large cohort of male-fetus pregnancies in which bin counts are measured and FF is known, a linear regression model can be trained that determines how much to weight each bin such that a weighted sum of autosomal bin counts yields an accurate FF value [24]. Mathematically, the trained regression model infers FF as follows:

$$FF = \sum_{bin \in autosomes} weight_{bin}*depth_{bin}$$

Training of the regression model requires thousands of male-fetus samples with known FF, and error in the regression's predictions drop as more training samples are added, though there are decreasing marginal gains as cohort size grows.

It is not immediately obvious why a weighted sum of autosomal bin counts should reflect FF, but there is a plausible explanation rooted in molecular biology (Fig. 8). Genomic DNA is wrapped around nucleosomes, eight-subunit complexes of histone proteins that play an important role in regulating gene expression and managing genome topology [25]. DNA approximately 150 nt in length encircles each nucleosome, and linker DNA between nucleosomes ranges in length, though is typically <1000 nt.

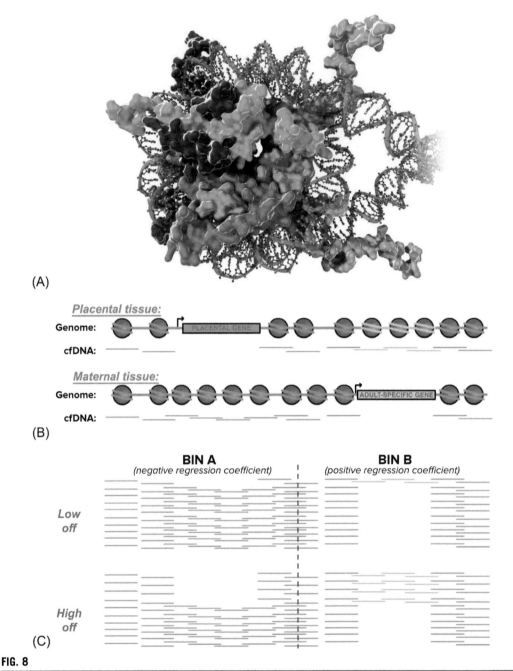

(A)

(B)

(C)

FIG. 8

Intuition underlying fetal fraction prediction via regression of bin-count depth. (A) Cellular DNA comprising the human genome wraps around nucleosomes, octameric protein complexes shown here in pink. The ~150 nt encircling the nucleosome persists as cfDNA when the cell dies. (B) Tissue-specific cfDNA abundance results from the differential positioning of nucleosomes, which itself stems from the tissue-specific expression of particular genes. (C) Different FF levels—i.e., different mixing of the placental and maternal tissue from (B)—will have subtle differences in the number of sequenceable fragments reads per bin. A regression model for FF can utilize these disparities, e.g., by applying a negative coefficient to Bin A (depth is anticorrelated with FF) and a positive coefficient to Bin B (depth is correlated with FF).

Expression of genes whose DNA is packaged in nucleosomes tends to be repressed because it is less accessible to regulatory and transcription factors; by contrast, genes with lower nucleosome occupancy tend to be more highly expressed due to higher accessibility (Fig. 8B).

When cells die and undergo apoptosis, nucleases digest the linker DNA between nucleosomes, but the 150-nt nucleosomal fragments persist and are purged from the cell [26,27]. These ejected fragments comprise the cfDNA tested via NIPT. FF inference would be simple if it were possible to attribute reads to placental tissue. But, it is not possible to infer the precise tissue of origin for a given fragment once it enters the bloodstream, as it mixes with fragments from other types of cells. Fortunately, each tissue type has a unique nucleosome occupancy profile [28], due in part to its tissue-specific gene expression pattern. In placental tissue, for example, a region with low gene expression (e.g., a postnatally expressed gene) and high nucleosome occupancy would contribute many cfDNA fragments, and vice versa. Critically, the region's cfDNA abundance should correlate with FF, because FF effectively reports the placental contribution to total cfDNA (Fig. 8C). This example illustrates why cfDNA abundance levels in a region may scale with FF, and that is precisely the signal the FF regression model uses. Bins with counts that positively correlate with FF have strongly positive weights; those with negative correlation have negative weights, and bins whose counts do not correspond with FF receive near-zero weights. Promising recent work shows that cfDNA positional correlation with FF exists down to the level of single bases (i.e., a "bin" can range from 1 base to tens of thousands of bases), meaning that predictive models of FF can operate over many length scales [29].

Aside from determining its positional origin in the genome, the sequence of a cfDNA fragment may contain other information about FF. For instance, fetal genomes systematically differ from their maternal counterparts in their patterns of DNA methylation [30]. Bisulfite treatment enables such methylation to be captured by NGS because unmethylated cytosine bases—but not methylated cytosines—convert to uracil bases and, consequently, manifest as a mismatch to the reference genome. Measuring the extent of methylation per fragment determines the probability that the fragment is placenta derived, and aggregating these signals across all sequenced reads gives a prediction of FF for the entire sample.

In addition to placental nucleosome positioning providing a signal for FF estimation, FF can be inferred from the length of reads in WGS-based NIPT (even better FF estimation could result from combining positional and length signals as in Ref. [29]). Fragment length is determined via paired-end sequencing, where the two ends of a cfDNA molecule are mapped to the genome, and the difference between mapping coordinates is calculated. The tissue of origin influences fragment length, as the number of DNA bases remaining after nuclease digestion is slightly smaller for nucleosomes in placental tissue than in maternal tissue. A sample-rich training set mapping out the relationship between the cfDNA length distribution and FF enables prediction of FF for a new sample where only its fragment length distribution has been measured.

SNP-BASED NIPT

Differences in signal between the fetal and maternal contributions to cfDNA are the substrate for inferring FF. In WGS, this signal is weak in any individual bin but strong when integrated across the genome. With the SNP method, by contrast, even a single site where the fetal genotype differs from the maternal genotype can provide a low-error reflection of FF. When averaged across merely tens of sites, FF can be resolved with low error [22].

FF measurement in SNP-based NIPT involves only a minor modification to the aneuploidy identification algorithm described earlier. Aneuploidy with the SNP method is assessed by taking FF as given and then considering the likelihood of the aggregate allele-balance data across many SNPs by cycling through different ploidy hypotheses, for example, disomic, M1-maternal trisomic, and so on. To infer FF, this process is inverted: disomy is assumed, and then the likelihood of the data is calculated by cycling through a range of FF values, ultimately picking the most likely FF. As aneuploid regions will manifest as having different FF than disomic regions, this analysis must be performed in a robust manner, for example, by considering each region separately and rejecting outliers.

MICROARRAY-BASED NIPT

The process of inferring FF with microarrays and the SNP method is highly similar [23]. Certain microarrays can measure SNP-specific signal. By amplifying regions known to have high diversity in the population and then measuring the extent to which the observed allelic balance differs from the maternal genotype (e.g., a site with 3% allele balance, after noise correction, suggests that the maternal signal would be 0%, and the DNA contributing the 3% signal is fetal-derived), FF can be directly detected.

MICRODELETION IDENTIFICATION

The sensitivity of aneuploidy detection algorithms—irrespective of platform—depends mostly on FF and the number of observations measured from a chromosome of interest. Previous sections described how observations are combined (e.g., via calculation of a median for WGS or evaluation of maximum likelihood model for SNP) into an assertion of ploidy. These observations are bins in the WGS method, polymorphic sites where maternal and fetal genotypes differ in the SNP method, and probes in the microarray method. As the number of observations grows, confidence in ploidy calls also improves, all else being equal. Conversely, sensitivity is reduced in regions, such as microdeletions, that only support a limited number of observations. Commonly screened microdeletions range in size from 2 to 25 MB, which is approximately 4% to 50% of chr21. Their relatively small size means that a limited number of bins span the region (WGS method) [14], and there are relatively fewer polymorphic sites available for the SNP method [31]. The microarray method can, in principle, target an almost limitless number of sites within a microdeletion, but the method becomes constrained by cost and assay form factor as each microarray can only accommodate a finite number of probes.

Diversity in the boundaries of microdeletions further complicates their identification during NIPT. To appreciate the algorithmic challenge posed by varied breakpoint locations, consider two different 22q11.2 deletions, one with the most common length of 2.5 MB and a longer one that spans 5 MB that encompass the 2.5 MB deletion. If the algorithm only considers fixed boundaries, there are two potential problems. First, an algorithm that only assays the genomic region most frequently spanned by empirically observed microdeletions (i.e., the 2.5 MB region) forgoes signal in the 5 MB deletion that would boost sensitivity of the larger deletion. Second, by contrast, if the algorithm uses the 5 MB boundary, it will have diminished sensitivity for the 2.5 MB deletion, since half of the 5 MB window will be euploid and weaken the deletion signal.

Implementing an algorithm to identify dynamic breakpoints can overcome the shortcomings of applying fixed boundaries to a dynamic boundary problem. Algorithms that can identify such breakpoints include hidden Markov models (HMMs) [32,33] and circular binary segmentation (CBS) [14,34]. HMM use is routine in bioinformatics pipelines on genomic data and are described in detail elsewhere. In brief, HMMs operate by traversing the observed data points (bins, SNPs, or probes) spanning a region (e.g., the superset of data points that could be included in 22q11.2), entertaining a copy-number change at each data point, and choosing the copy-number trajectory through the data that has highest likelihood. In the presence of a detectable microdeletion, the most likely trajectory should be disomic up- and downstream of the deletion and monosomic within it. The HMM must account for FF, as the change in signal due to a copy-number change will be proportional to FF. CBS operates by partitioning the data points spanning a region into groups and then testing whether there is a statistically significant difference in the signal between groups. Every set of plausible partitions is considered, and the most likely partition defines the boundaries of a possible microdeletion. Importantly, neither HMM nor CBS analysis alone identifies a microdeletion; rather, they define a region that may contain a microdeletion, and the aneuploidy algorithms described earlier must be applied to these putative deletion regions to assess whether they are significant and truly consistent with aneuploidy.

ALGORITHM PERFORMANCE ON BIOLOGICAL EDGE CASES

NIPT platforms are optimized to distinguish euploidy from aneuploidy in the majority of pregnancies, yet they have partially overlapping virtues and limitations in less common cases that are worthy of consideration.

A shared limitation of all techniques is the difficulty of handling mosaicism [35,36]. In addition to the challenges of interpreting the clinical impact of mosaicism, its analytical detection is complicated. For instance, in cases where the placenta is mosaic for an aneuploidy, the FF associated with the mosaic aberration will necessarily be less than the total FF of the pregnancy: the lowered FF of the mosaic tissue reduces the sensitivity of identifying an abnormality.

Nonsingleton pregnancies [37,38] pose a related challenge in that each fetus will only contribute a fraction of the total FF of the pregnancy. With each NIPT platform, it is relatively straightforward to process monozygotic twins because, at the assay level, the pregnancy will effectively behave as if it were a singleton. Dizygotic twins, however, are more difficult to assess: as with mosaicism, an aneuploid dizygotic twin will only contribute a share of total FF, and both twins will not necessarily have equal contributions.

The NIPT platforms diverge in their analysis of pregnancies conceived with a donor ovum and in cases of triploidy. For the WGS and microarray methods, no special handling is required for donor pregnancies, as they do not interrogate or rely upon genotypic differences between the genomes of the fetus and the pregnant woman. The math underlying the SNP method, however, must be altered to account for the fact that the fetus will not have inherited chromosomes from the pregnant woman in case of egg donation. As such, more genotype hypotheses must be considered in the likelihood model, which could lessen sensitivity at low FF. Triploid fetuses are challenging to detect for the WGS and microarray methods, as the uniform fetal ploidy across the genome means that depth does not vary between chromosomes. It is possible that male-fetus triploids could be detected by comparing the total FF of the pregnancy—inferred from autosomal bin counts as described earlier—to the FF measured from chrX and chrY (where total FF might be 15%, FF(chrX) is 5%, and FF(chrY) is 10%). The SNP method's ability to detect allele balance at many SNPs facilitates identification of "diandric" (paternally inherited) triploidy, but "digynic" (maternally inherited) triploidy detection via SNP-based NIPT has not been reported [39].

The fact that the microarray and SNP methods are targeted assays, whereas WGS surveys the whole genome, means that WGS is better suited to identify rare, novel variants in unexpected locations. Such variants include rare autosomal aneuploidies [40], such as T7, T9, and T16, which in nonmosaic form are typically associated with miscarriage or stillbirth. In mosaic form or when seemingly restricted to the placenta they can be associated with congenital anomalies, placental dysfunction, and uniparental disomy (see Chapter 7). When occurring in nonmosaic form in a singleton pregnancy the $FF_{rare\_trisomy}$ will equal the $FF_{whole\_pregnancy}$. When occurring in mosaic form or due to a vanishing twin the $FF_{rare\_trisomy}$ will be lower than the $FF_{whole\_pregnancy}$. Maternal malignancy also often manifests as copy-number aberrations distributed throughout the genome [41,42], which is trivially conducive to WGS NIPT and would only be detected by targeted assays if the targeted regions were mutated. Finally, genome-wide detection of unascertained, novel microdeletions is only easily possible via a WGS technique.

CONCLUSION

The various cfDNA-based NIPT platforms have much in common and yet also differ in fundamental ways. Most importantly, they are united in having high clinical sensitivity for common aneuploidies, which is the goal of current cfDNA-based prenatal testing programs and lends perspective to the discussion of their nuanced differences. Their other commonality is that each uses a plethora of independent observations tiled across a region of interest to provide signal of a fetal deviation from the maternal background. Many observations are needed to combat the fact that the noise associated with each single measurement swamps out the occasionally miniscule signal of a low-FF chromosomal anomaly. The algorithms differ greatly, however, in both the way that cfDNA is prepared for measurement and the way that both ploidy status and FF are mathematically inferred. As clinical adoption becomes more widespread, further development of the methods is expected to continue at a torrid pace. This progress should give higher sensitivity at lower cost, thereby both increasing the clinical performance of cfDNA based NIPT and broadening access to this important screening.

ACKNOWLEDGMENTS

Carrie Haverty provided helpful comments on the manuscript.

REFERENCES

[1] Fan HC, Blumenfeld YJ, Chitkara U, Hudgins L, Quake SR. Noninvasive diagnosis of fetal aneuploidy by shotgun sequencing DNA from maternal blood. Proc Natl Acad Sci USA 2008;105:16266–71.

[2] Chiu RWK, Chan KCA, Gao Y, Lau VYM, Zheng W, Leung TY, et al. Noninvasive prenatal diagnosis of fetal chromosomal aneuploidy by massively parallel genomic sequencing of DNA in maternal plasma. Proc Natl Acad Sci USA 2008;105:20458–63.

[3] Straver R, Sistermans EA, Holstege H, Visser A, Oudejans CBM, Reinders MJT. WISECONDOR: detection of fetal aberrations from shallow sequencing maternal plasma based on a within-sample comparison scheme. Nucleic Acids Res 2014;42:e31.

[4] Langmead B, Trapnell C, Pop M, Salzberg SL. Ultrafast and memory-efficient alignment of short DNA sequences to the human genome. Genome Biol 2009;10:R25.

[5] Li H, Durbin R. Fast and accurate short read alignment with Burrows-Wheeler transform. Bioinformatics 2009;25:1754–60.

[6] Fan HC, Quake SR. Sensitivity of noninvasive prenatal detection of fetal aneuploidy from maternal plasma using shotgun sequencing is limited only by counting statistics. PLoS One 2010;5:e10439.

[7] Benjamini Y, Speed TP. Summarizing and correcting the GC content bias in high-throughput sequencing. Nucleic Acids Res 2012;40:e72.

[8] Johansson LF, de Boer EN, de Weerd HA, van Dijk F, Elferink MG, Schuring-Blom GH, et al. Novel algorithms for improved sensitivity in non-invasive prenatal testing. Sci Rep 2017;7:1838.

[9] Derrien T, Estellé J, Marco Sola S, Knowles DG, Raineri E, Guigó R, et al. Fast computation and applications of genome mappability. PLoS One 2012;e30377:7.

[10] Thorne CD, NP GDBDL, Benjamini Y, Speed TP, et al. Investigating and correcting plasma DNA sequencing coverage bias to enhance aneuploidy discovery. PLoS One 2014;9:e86993.

[11] Snyder MW, Simmons LE, Kitzman JO, Coe BP, Henson JM, Daza RM, et al. Copy-number variation and false positive prenatal aneuploidy screening results. N Engl J Med 2015;372:1639–45.

[12] Strom CM, Maxwell MD, Owen R. Improving the accuracy of prenatal screening with DNA copy-number analysis. N Engl J Med 2017;376:188–9.

[13] Chudova DI, Sehnert AJ, Bianchi DW. Copy-number variation and false positive prenatal screening results. N Engl J Med 2016;375:97–8.

[14] Zhao C, Tynan J, Ehrich M, Hannum G, McCullough R, Saldivar J-S, et al. Detection of fetal subchromosomal abnormalities by sequencing circulating cell-free DNA from maternal plasma. Clin Chem 2015; 61:608–16.

[15] Jensen TJ, Zwiefelhofer T, Tim RC, Džakula Ž, Kim SK, Mazloom AR, et al. High-throughput massively parallel sequencing for fetal aneuploidy detection from maternal plasma. PLoS One 2013;8:e57381.

[16] Sehnert AJ, Rhees B, Comstock D, de Feo E, Heilek G, Burke J, et al. Optimal detection of fetal chromosomal abnormalities by massively parallel DNA sequencing of cell-free fetal DNA from maternal blood. Clin Chem 2011;57:1042–9.

[17] Mazloom AR, Džakula Ž, Oeth P, Wang H, Jensen T, Tynan J, et al. Noninvasive prenatal detection of sex chromosomal aneuploidies by sequencing circulating cell-free DNA from maternal plasma. Prenat Diagn 2013;33:591–7.

[18] Zimmermann B, Hill M, Gemelos G, Demko Z, Banjevic M, Baner J, et al. Noninvasive prenatal aneuploidy testing of chromosomes 13, 18, 21, X, and Y, using targeted sequencing of polymorphic loci. Prenat Diagn 2012;32:1233–41.

[19] Samango-Sprouse C, Banjevic M, Ryan A, Sigurjonsson S, Zimmermann B, Hill M, et al. SNP-based non-invasive prenatal testing detects sex chromosome aneuploidies with high accuracy. Prenat Diagn 2013;33:643–9.

[20] Hall MP, Hill M, Zimmermann B, Sigurjonsson S, Westemeyer M, Saucier J, et al. Non-invasive prenatal detection of trisomy 13 using a single nucleotide polymorphism- and informatics-based approach. PLoS One 2014;9:e96677.

[21] Artieri CG, Haverty C, Evans EA, Goldberg JD, Haque IS, Yaron Y, et al. Noninvasive prenatal screening at low fetal fraction: comparing whole-genome sequencing and single-nucleotide polymorphism methods. Prenat Diagn 2017;37(5):482–90.

[22] Rabinowitz M, Gemelos G, Banjevic M, Ryan A, Demko Z, Hill M, Zimmerman B, Baner J. Methods for non-invasive prenatal ploidy calling. 20140162269.

[23] Juneau K, Bogard PE, Huang S, Mohseni M, Wang ET, Ryvkin P, et al. Microarray-based cell-free DNA analysis improves noninvasive prenatal testing. Fetal Diagn Ther 2014;36:282–6.

[24] Kim SK, Hannum G, Geis J, Tynan J, Hogg G, Zhao C, et al. Determination of fetal DNA fraction from the plasma of pregnant women using sequence read counts. Prenat Diagn 2015;35:810–5.

[25] Kouzarides T. Chromatin modifications and their function. Cell 2007;128:693–705.

[26] Enari M, Sakahira H, Yokoyama H, Okawa K, Iwamatsu A, Nagata S. A caspase-activated DNase that degrades DNA during apoptosis, and its inhibitor ICAD. Nature 1998;391:43–50.

[27] Sakahira H, Enari M, Nagata S. Cleavage of CAD inhibitor in CAD activation and DNA degradation during apoptosis. Nature 1998;391:96–9.

[28] Ernst J, Kheradpour P, Mikkelsen TS, Shoresh N, Ward LD, Epstein CB, et al. Mapping and analysis of chromatin state dynamics in nine human cell types. Nature 2011;473:43–9.

[29] Chan KCA, Jiang P, Sun K, Cheng YKY, Tong YK, Cheng SH, et al. Second generation noninvasive fetal genome analysis reveals de novo mutations, single-base parental inheritance, and preferred DNA ends. Proc Natl Acad Sci USA 2016;113:E8159–68.

[30] Sun K, Jiang P, Chan KCA, Wong J, Cheng YKY, Liang RHS, et al. Plasma DNA tissue mapping by genome-wide methylation sequencing for noninvasive prenatal, cancer, and transplantation assessments. Proc Natl Acad Sci USA 2015;112:E5503–12.

[31] Wapner RJ, Babiarz JE, Levy B, Stosic M, Zimmermann B, Sigurjonsson S, et al. Expanding the scope of noninvasive prenatal testing: detection of fetal microdeletion syndromes. Am J Obstet Gynecol 2015;212: 332.e1–9.

[32] McCallum KJ, Wang J-P. Quantifying copy number variations using a hidden Markov model with inhomogeneous emission distributions. Biostatistics 2013;14:600–11.

[33] Wang H, Nettleton D, Ying K. Copy number variation detection using next generation sequencing read counts. BMC Bioinformatics 2014;15:109.

[34] Talevich E, Shain AH, Botton T, Bastian BC. CNVkit: genome-wide copy number detection and visualization from targeted DNA sequencing. PLoS Comput Biol 2016;12:e1004873.

[35] Grati FR, Malvestiti F, Ferreira JCPB, Bajaj K, Gaetani E, Agrati C, et al. Fetoplacental mosaicism: potential implications for false-positive and false-negative noninvasive prenatal screening results. Genet Med 2014;16:620–4.

[36] Wang Y, Chen Y, Tian F, Zhang J, Song Z, Wu Y, et al. Maternal mosaicism is a significant contributor to discordant sex chromosomal aneuploidies associated with noninvasive prenatal testing. Clin Chem 2014;60:251–9.

[37] Tan Y, Gao Y, Lin G, Fu M, Li X, Yin X, et al. Noninvasive prenatal testing (NIPT) in twin pregnancies with treatment of assisted reproductive techniques (ART) in a single center. Prenat Diagn 2016;36:672–9.

[38] Fosler L, Winters P, Jones KW, Curnow KJ, Sehnert AJ, Bhatt S, et al. Aneuploidy screening using noninvasive prenatal testing in twin pregnancies. Ultrasound Obstet Gynecol 2017;49(4):470–7.

[39] Curnow KJ, Wilkins-Haug L, Ryan A, Kırkızlar E, Stosic M, Hall MP, et al. Detection of triploid, molar, and vanishing twin pregnancies by a single-nucleotide polymorphism–based noninvasive prenatal test. Am J Obstet Gynecol 2015;212:79.e1–9.

[40] Pertile MD, Halks-Miller M, Flowers N, Barbacioru C, Kinnings SL, Vavrek D, et al. Rare autosomal trisomies, revealed by maternal plasma DNA sequencing, suggest increased risk of feto-placental disease. Sci Transl Med 2017;9, pii: eaan 1240.

[41] Bianchi DW, Chudova D, Sehnert AJ, Bhatt S, Murray K, Prosen TL, et al. Noninvasive prenatal testing and incidental detection of occult maternal malignancies. JAMA 2015;314:162–9.

[42] Lo YMD. Noninvasive prenatal testing complicated by maternal malignancy: new tools for a complex problem. NPJ Genom Med 2016;1:15002.

GLOSSARY

Aneuploidy Abnormality in the number of chromosomes.

Fetal fraction The share of cfDNA derived from placental tissue.

GC bias Reactions involving the amplification of DNA—such as PCR and NGS library preparation—may be biased toward amplification of fragments with a particular GC content (i.e., the fraction of G and C bases in the sequence).

Hidden Markov model and circular binary segmentation When applied in the context of identifying genomic deletions and duplications, these methods partition a sequence region of different copy number.

Maternal CNV A deletion or duplication in the maternal genome that can complicate identification of fetal aneuploidies in NIPT.

Nucleosome A complex between an octamer of histone proteins and $\sim$150 bases of genomic DNA. Nucleosomes regulate genome structure and gene expression.

CELL-FREE DNA IN CLINICAL PRACTICE

PRENATAL SCREENING FOR COMMON ANEUPLOIDIES BEFORE AND AFTER THE INTRODUCTION OF CELL-FREE DNA-BASED PRENATAL TESTING

Adalina Sacco*, Pranav Pandya[†]

Fetal Medicine Department, University College London Hospital, London, United Kingdom Fetal Medicine Department, University College London Hospital, London, United Kingdom[†]

PRENATAL SCREENING BEFORE CELL-FREE DNA-BASED NIPT

ANEUPLOIDIES

Aneuploidies, the presence of an abnormal number of chromosomes, are conditions associated with significant morbidity and mortality. Aneuploidies affect approximately 1 in 160 live births [1], although the overall incidence is higher due to natural pregnancy losses and termination of affected pregnancies. The commonest aneuploidies affecting live births are trisomies 21, 18, and 13 and monosomy X. Trisomies describe an extra copy of a particular chromosome and usually occur due to maternal meiotic nondysjunction. Monosomy X describes a missing copy of the X chromosome in females (45X0, Turner syndrome) which also usually occurs due to maternal meiotic nondysjunction. Sex chromosome aneuploidies such as 45XO and 47XXY (Klinefelter syndrome) are not commonly screened for because in the absence of structural anomalies they would not satisfy the criteria for antenatal screening, and there is insufficient data on the accuracy of cfDNA-based testing for this indication [2]. Sex chromosome aneuploidies will therefore not be discussed further in this chapter, but are extensively addressed in Chapter 15.

Trisomy 21 (T21), Down's syndrome, is the commonest trisomy compatible with life. It affects 1.08 per 1000 live births in the United Kingdom [3] and is associated with a number of defects which vary between people in their presence and severity. Possible features include cardiac anomalies, duodenal atresia, mild-to-moderate learning difficulties, hypothyroidism, hearing and vision disorders, seizures, increased risks of certain cancers, particularly leukemia, and Alzheimer's disease.

Trisomy 18 (T18), Edwards' syndrome, affects 1 in 6000 to 1 in 8000 live births [4]. The prevalence at the time of screening, 12 weeks, is 1 in 600 for a 35-year-old woman; between 12 weeks and 40 weeks

Noninvasive Prenatal Testing (NIPT). https://doi.org/10.1016/B978-0-12-814189-2.00004-9

the fetal death rate is around 80%. Possible features include cardiac defects, renal anomalies, severe learning disabilities, omphalocele, central nervous system defects, breathing and feeding difficulties, and physical deformities such as "rocker-bottom" feet, "clenched" hands, micrognathia, and low-set ears. The majority of live-born babies die in the first few days or weeks of life and less than 10% survive to 1 year of age [5].

Trisomy 13 (T13), Patau's syndrome, affects approximately 1 in 8000 to 1 in 12,000 live births. The prevalence at the time of screening, 12 weeks, is 1 in 1800 for a 35-year-old woman; again, there is a fetal death rate between 12 and 40 weeks of around 80%. Possible features include cardiac defects, renal anomalies, severe learning disabilities, omphalocele, microcephaly, holoprosencephaly, deafness, seizures, and cleft lip and palate. As with trisomy 18, the vast majority of fetuses die in utero, the majority of live-born babies die in the first few days or weeks of life and less than 10% survive to 1 year of age [5].

SCREENING BY MATERNAL AGE

The rate of autosomal aneuploidies increases significantly with advancing maternal age; the risk of trisomy 21 at 12 weeks of gestation rises from 1 in 1000 for a woman aged 20 years to 1 in 250 for a woman aged 35 years [6]. Similarly, the risk of trisomy 18 increases from 1 in 2500 to 1 in 600 and the risk of trisomy 13 increases from 1 in 8000 to 1 in 1800, over the same time period. Prior to the advent of detailed antenatal ultrasonography and maternal serum biochemistry, maternal age was used as a screening tool for the detection of trisomies. However, maternal age is a poor screening test for trisomies as the majority of babies affected are born to women under the age of 35 years. During the 1970s and 1980s, the estimated detection rate with maternal age as a screening tool was 30% [6] (Fig. 1).

Once identified as "high risk" for trisomies by maternal age, the diagnostic option available to the mother was invasive testing, which less than half of women at the time opted for. Amniocentesis in the 1980s was estimated to carry a 1.0% additional risk of spontaneous pregnancy loss [7]. As the vast majority of pregnancies conceived by women over the age of 35 years are not affected by trisomies,

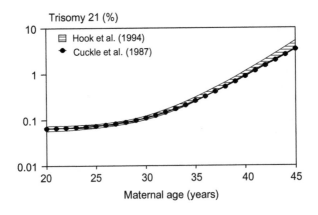

FIG. 1

Maternal age and increasing aneuploidy risk [6a,6b].

From thesis P. Pandya with permission.

using maternal age alone as a screening tool to offer amniocentesis would therefore lead to a high loss of nonaffected fetuses with a low detection rate, therefore it is no longer recommended.

SCREENING BY MATERNAL BIOCHEMISTRY

Measurement of maternal serum biochemistry was initially intended to screen for neural tube defects; however, it was shown that alpha-fetoprotein (AFP) was low in fetuses with trisomy 21, leading to further research in this area. Serum biochemistry was initially suggested as a screening option for women under 35 years of age but, as it was increasingly shown to be more sensitive than maternal age alone, it also became an option for women over 35 years of age [8]. As several maternal biochemical blood markers are affected by trisomic pregnancies, these can be measured and converted to a gestational age-specific multiple of the median (MoM), which is compared to a level at which one would expect an unaffected pregnancy. Markers investigated during the 1980s and 1990s for second trimester screening included AFP, human chorionic gonadotropin (hCG), unconjugated estriol and inhibin A.

Trisomy 21 is typically associated with low levels of AFP and estriol, and high levels of hCG and inhibin A. Adding the results of these biomarkers to maternal age increases the sensitivity of second trimester screening in a stepwise fashion, as follows [9,10]:

- Double test (AFP and hCG): 55%–60% detection, 5% false positive rate.
- Triple test (AFP, hCG and E3): 60%–65% detection, 5% false positive rate.
- Quadruple test (AFP, hCG, inhibin A, unconjugated estriol): 65%–70% detection, 5% false positive.

Using free instead of total hCG increases the maximum detection rate by about 5% in all the earlier tests.

In the first trimester, hCG and PAPP-A can be used for serum screening. Trisomy 21 is typically associated with high levels of hCG, as in the second trimester, but low levels of PAPP-A. Combining maternal age with hCG and PAPP-A levels in the first trimester gives a detection rate of approximately 65% with a false positive rate of 5% [6]. This detection rate decreases as the first trimester progresses, as PAPP-A is a more powerful marker at earlier gestations.

SCREENING BY ULTRASOUND

Maternal serum biochemistry alters with gestational age and calculation of the MoM requires accurate pregnancy dating. Ultrasound was initially introduced to date the pregnancy prior to measurement of AFP. In addition, ultrasound scanning equipment, image resolution, and technique were also advancing. In the 1990s, evidence accumulated that increased fluid at the back of the fetal neck (nuchal translucency, NT) (Fig. 2) in the first trimester was associated with fetal trisomies [11], as well as other chromosomal and structural abnormalities. Screening for trisomy 21 using maternal age and NT together gives a detection rate of 75%, with a false positive rate of 5% [12]. Adding in first trimester maternal serum biochemistry increases the detection rate further, and so the combined test (maternal age, NT, hCG, and PAPP-A) has a detection rate for trisomy 21 of 85%–90% with a 5% false positive rate (Table 1).

Comparisons of first trimester screening by the combined test and second trimester screening by the quadruple test have shown the combined test to be superior. However, performing both sets of screening in a sequential fashion, that is, maternal age, nuchal translucency, and maternal serum biochemistry

Table 1 Performance of Serum Screening and Combined Testing for Trisomy 21		
Screening Method	**Detection Rate (%)**	**False Positive Rate (%)**
Maternal age (MA) alone	30	5
MA + serum biochemistry		
Double test (AFP + HCG)	55–60	5
Triple test (AFP + HCG + inhibin A)	60–65	5
Quadruple test (AFP + HCG + inhibin A + unconjugated estriol)	65–70	5
MA + nuchal translucency (NT)	75–80	5
Combined test (MA + NT + HCG + PAPPA)	85–90	5

in both first and second trimesters, may have a higher detection rate (90%–95%) for trisomy 21 [13]. However, sequential screening introduces diagnostic delay and requires a second blood test and has not been widely accepted by women or healthcare professionals.

It is important to note that all the rates discussed previously are for the detection of trisomy 21. Trisomies 18 and 13 are also associated with an increased NT and low PAPP-A in the first trimester but are instead associated with low levels of hCG. The use of the previous combined screening protocol will identify approximately 75% of fetuses with trisomy 18 or 13 when using algorithms for trisomy 21 but over 90% when using trisomy 18/13 algorithms [5]. As in over two-thirds of fetuses with these trisomies, there are associated anatomical defects that can be detected early in pregnancy (e.g. holoprosencephaly, exomphalos), their overall detection rate is likely to be significantly increased with the use of routine first trimester ultrasound scanning.

In the United Kingdom and in many parts of the world, the current trisomy screening recommendations [14] are to use the combined test in the first trimester and, if that results in a low risk (less than 1

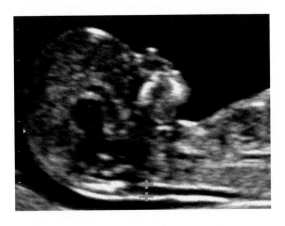

FIG. 2

Ultrasound image of normal and increased nuchal translucency.

From "Fetal Medicine: Basic Science and Clinical Practice." Chapter on First Trimester Anomalies; courtesy Mr. Fred Ushakov.
Publisher Elsevier, with permission.

in 150 at term), no further invasive testing is performed. For women presenting in the second trimester (between 14+2 and 20+0 weeks' gestation), the quadruple test is performed with the same cutoff of 1 in 150. All women are subsequently offered a detailed anomaly scan at 18–20+6 weeks of gestation, at which time the detection of fetal anomalies will result in the offer of invasive testing.

UPTAKE OF SCREENING

Screening for fetal aneuploidy is parental choice and uptake should not be used as a marker of screening performance. Data regarding the uptake of fetal trisomy screening in the United Kingdom is therefore not collected, and so estimates are calculated based on the total number of tests and on the total number of births [15] (60%–74%) or retrospective questionnaire studies [16] (65%). In the United States, screening uptake appears similar [17] (70%), but increases in areas where state law requires all women to be offered screening [18] (80%). A retrospective Canadian study [19] showed similar rates of screening uptake (62.2%) with wide regional variation (27.8%–80.3%), and an Australian study [20] showed a slightly lower uptake of screening (45%). Within Europe, rates of screening uptake also vary widely, from under 30% in the Netherlands to over 90% in Denmark [21]. The availability, cost, and societal perspective of screening is probably responsible for this wide variation in uptake.

DIAGNOSTIC TESTS

Invasive testing for trisomies by amniocentesis or chorionic villus sampling (CVS) is diagnostic and not a screening test. It is offered in place of screening in many countries either if the risk is already judged to be high (e.g., ultrasound abnormalities, balanced parental translocations, previously affected pregnancies) or for maternal choice [22]. As discussed previously, invasive testing is traditionally quoted to carry a 1.0% risk of additional spontaneous pregnancy loss, based on a large randomized trial of amniocentesis in 1986, in which the actual pregnancy loss rate was 1.7% in the study group and 0.7% in controls [7]. Amniocentesis prior to 15 weeks' gestation has been associated with a higher rate of fetal loss and talipes [23] and is therefore not recommended. CVS prior to 10 weeks' gestation has been associated with limb hypoplasia [24] and is also not recommended.

Observational studies in the early 2000s suggested the fetal loss rate following second trimester amniocentesis to be lower than previously thought [25,26]. Rates of fetal loss following CVS were quoted as 1.9% in 2009 [27]. More recent publications have suggested that fetal loss following amniocentesis and CVS to be lower than previously described. A 2015 study [28] found a procedure-related risk for amniocentesis of 0.11%, with an actual pregnancy loss rate of 0.81% in the study group and 0.67% in controls, and a procedure-related risk for CVS of 0.22%, with an actual pregnancy loss rate of 2.18% in the study group and 1.79% in controls. In 2016, a population-based study [29] of nearly 150,000 women found no increased risk of fetal loss with either amniocentesis or CVS.

Table 2 Performance of cfDNA NIPT vs Combined Test

Aneuploidy	Screening Method	Detection Rate (%)	False Positive Rate (%)
Trisomy 21	Combined test	85–90	5
	NIPT	99.7	0.04
Trisomy 18	Combined test	75	5
	NIPT	97.9	0.04
Trisomy 13	Combined test	75	5
	NIPT	99.0	0.04

USE OF CELL-FREE DNA-BASED NIPT IN HIGH- AND LOW-RISK POPULATIONS
CELL-FREE DNA NIPT USE IN ANEUPLOIDY SCREENING

Since 2010, the use of NIPT for aneuploidy screening has been explored extensively. The reader is referred to Chapter 3 of this book for discussion on methodology of cfDNA NIPT use in this situation and to Chapter 6 for discussion of cfDNA NIPT use in twin pregnancies.

The most recent meta-analysis [30] included 35 studies on cfDNA NIPT use in aneuploidy screening and only included studies in which data on pregnancy outcome were provided for more than 85% of the study population. It found a 99.7% detection rate for trisomy 21, with a false positive rate of 0.04%. This compares favorably with the combined screening detection rate of 85%–90% and false positive rate of 5%, as described previously. In this same meta-analysis, trisomy 18 was found to have a detection rate of 97.9%, with a false positive rate of 0.04% and trisomy 13 was found to have a detection rate of 99.0%, with a false positive rate of 0.04%. This also compares favorably with standard combined screening (Table 2).

Despite these high detection rates and low false positive rates, it is important to emphasize that cfDNA NIPT is still a screening test and that women with a high risk result require invasive testing to confirm the findings. cfDNA NIPT is not considered diagnostic as the "fetal DNA" detected is actually placental in origin; it originates from apoptosis of placental cytotrophoblast and syncytiotrophoblast cells [31]. In cases of placental mosaicism, the cytotrophoblast may contain an abnormal cell line which is not present in the fetus. Although CVS also obtains placental tissue, this is considered diagnostic when it analyses both the cytotrophoblast and the mesenchyme, which is more likely to represent the fetal karyotype [32]. Additionally, a "vanishing twin," maternal chromosome abnormalities, and maternal (malignant) disease may all affect the cell-free DNA in maternal plasma and suggest an abnormality which is not fetal in origin. For more information, the reader is referred to Chapter 5 of this book.

cfDNA NIPT used for aneuploidy screening may fail to provide a result in some cases; this is most commonly due to a low fetal fraction, which itself may be due to a number of reasons. In the aforementioned meta-analysis [30], the reported failure rates ranged from 0% to 12.2%. A study [33] investigating failed results found that fetal fraction was reduced with rising maternal body mass index, increased maternal age, South Asian racial origin compared to Caucasian origin, and in assisted conceptions. They found a failure rate of 2.9% in unaffected pregnancies, 1.9% in trisomy 21, 8.0% in trisomy 18, and 6.3% in trisomy 13. In trisomies 18 and 13 there is a relatively small placenta, reflected by low PAPP-A levels, and this may be one of the reasons for cfDNA NIPT failure. This reinforces the central role of a detailed first trimester scan assessing fetal anatomy and specifically looking for major anomalies that may be associated with these trisomies.

NIPT USE IN HIGH-RISK POPULATIONS

High-risk populations have been described as women with an increased risk following conventional screening (combined or quadruple test). Different studies have used different cutoffs to offer cfDNA NIPT in a contingent model of screening. In the meta-analysis [30] described previously, 30 of the 35 studies were in high-risk pregnancies. The authors found no difference in screening performance between high-risk and routine populations in subgroup analysis.

Two major studies have investigated the use of NIPT in high-risk populations, with the offer of NIPT contingent on results from standard screening.

In the first [34], over 11,000 pregnancies in two UK hospitals were included. Women underwent routine combined screening in the first trimester; those with a risk greater than 1 in 100 at 12 weeks (equivalent to 1 in 150 live birth risk) (high risk) were offered invasive testing, NIPT or no further testing, and those with a risk between 1 in 101 and 1 in 2500 (intermediate risk) were offered cfDNA NIPT or no further testing. There were 47 fetuses with trisomy 21 in this study; 41 (87%) of them were in the high-risk group, 5 cases were in the intermediate-risk group, and 1 case was in the low-risk group. Of the 24 fetuses with trisomy 18, 22 (92%) were in the high-risk group and 2 were in the intermediate-risk group. All 4 (100%) of the fetuses with trisomy 13 were in the high-risk group. NIPT failed to provide a result after first sampling in 2.7% of cases. In 63.0% of those retested, the second test provided a result. cfDNA NIPT correctly identified 97.7% of cases of trisomy 21 tested for (43/44), with one false negative case. cfDNA NIPT correctly identified 87.5% (21/24) of trisomy 18 cases tested for and 50% (2/4) of trisomy 13 cases tested for (Table 3).

Not all patients opted for further investigations after initial screening. In the high-risk group, 37.6% (173/460) opted for invasive testing, 60.0% (276/460) for cfDNA NIPT, and 2.4% (11/460) declined further investigation. In the intermediate-risk group, 91.5% (3249/3552) opted for cfDNA NIPT and 8.5% (303/3552) declined further investigations. Overall, taking into account test performance and

Table 3 Follow-Up Tests and Detected Cases in 460 Patients With High (Greater Than 1:100 at 12 Weeks) and 3552 Patients With Intermediate (Between 1:101 and 1:2500 Risk) Risk After First Trimester Combined Test

Outcome	n	High Risk (n = 460)				Intermediate Risk (n = 3552)			Low Risk (n = 7680)
		Total	CVS	cfDNA	No test	Total	cfDNA	No Test	No Test
Trisomy 21	47	41	27	13	1	5	4	1	1
Trisomy 18	24	22	17	5	0	2	2	0	0
Trisomy 13	4	4	3	1	0	0	0	0	0
Nontrisomy	11,617	393	126	257	10	3545	3243	302	7679
Total	11,692	460	173	276	11	3552	3249	303	7680

From Gil MM, Revello R, Poon LC, Akolekar R, Nicolaides KH. Clinical implementation of routine screening for fetal trisomies in the UK NHS: cell-free DNA test contingent on results from first-trimester combined test. Ultrasound Obstet Gynecol 2016;47:45–52.

patient choice, contingent screening detected 91.5% (43/47) cases of trisomy 21, and 100% (28/28) cases of trisomy 18 or 13. NIPT was associated with a 43% reduction in invasive testing in the high-risk group, from 65.6% performed the previous year in high-risk patients to 37.6% in this study. Patients were more likely to opt for invasive testing with increasing risk of trisomies and increasing nuchal translucency and were more likely to opt against invasive testing if they were of Afro-Caribbean origin or given the combined screening results on a different visit to having the NT ultrasound scan (i.e., their unit did not offer a "one-stop" service). In the intermediate-risk group, 91.5% (3249/3552) opted for cfDNA NIPT and 8.5% (303/3553) had no further investigations. The option of cfDNA NIPT therefore increased the number of women who accepted any form of further testing (cfDNA NIPT or Invasive Prenatal Diagnosis). Therefore while one may argue that offering cfDNA NIPT in a contingent model may reduce detection rate, in reality because more women will have testing the overall detection rate will increase. Termination of pregnancy following prenatal diagnosis of trisomy 21 was 74% overall. For women in the high-risk group, the rate of termination was 92.6% in those who chose invasive testing which was subsequently positive, and 35.7% in those who chose cfDNA NIPT and subsequently received a high-risk result (9/13 of these women proceeded to invasive testing, 5/13 had a termination of pregnancy, and 4/13 had no further testing). For women in the intermediate-risk group there were five cases of trisomy 21. In four of these cases the parents had opted for cfDNA NIPT, which gave a high risk result for three of them. Two of these three opted for termination of pregnancy. These figures suggest that women were using cfDNA NIPT for information and preparedness but not necessarily to proceed to termination of pregnancy.

The RAPID (Reliable, Accurate, Prenatal non-Invasive Diagnosis) study [35], conducted from our institution, investigated the use of cfDNA NIPT in high-risk populations and included over 30,000 patients in eight UK maternity units. Women underwent routine combined or quadruple testing depending on gestational age; those with a risk greater than 1 in 150 were offered invasive testing, NIPT or no further testing, and those with a risk between 1 in 151 and 1 in 1000 were offered NIPT or no further testing. There were 934 women with a risk of greater than 1 in 150, 17.8% (166/934) opted for invasive testing, 74.4% (695/934) opted for NIPT, and 7.8% (73/934) declined further investigations. Of the women with a risk between 1 in 151 and 1 in 1000, the proportion opting for NIPT was 80.3% (1799/2241). The authors concluded that in England with an estimated 698,500 births per annum, offering NIPT as a contingent test to women with a Down's syndrome screening risk of at least 1/150 would increase detection by 195 cases (95% uncertainty interval 34–480) with 3368 (95% uncertainty interval 2279–4027) fewer invasive tests and 17 (95% uncertainty interval 7–30) fewer procedure-related miscarriages for a nonsignificant difference in total costs.

NIPT USE IN LOW-RISK POPULATIONS

The studies described previously evaluated cfDNA NIPT in a clinical setting which was contingent on routine screening. The use of cfDNA NIPT as primary screening, without prior routine screening, has been reviewed in several studies, shown in Table 4.

The data show that cell-free DNA-based NIPT can be used as a primary screening method with improved detection rates, much higher positive predictive value, and reduced false positive rates for trisomy 21 compared to standard screening.

At the moment, the main limiting factors for implementation of cfDNA as primary screening are the cost of testing, and management of cases where NIPT fails to provide a result.

Table 4 Studies of cfDNA NIPT for Trisomy 21 Detection in Low-Risk Populations

Study	Number of Pregnancies	Detection Rate (%)	False Positive Rate (%)	Positive Predictive Value
Nicolaides et al. [36]	2049	100	0.1	
Dan et al. [37]	11,105	100	0.03	
Bianchi et al. [38]	2052	100	0.3	45.5 (vs 4.2 using standard screening)
Norton et al. [39]	15,841	100	0.06	80.9 (vs 3.4 using standard screening)
Zhang et al. [40]	147,314	99.17	0.05	92.19

FUTURE OF CONVENTIONAL SCREENING
CONSEQUENCES FOR EXISTING SERVICES

Cell-free DNA-based NIPT is clearly a major advance for prenatal screening, and its use has rapidly progressed during this decade. However, it seems unlikely that cfDNA NIPT will entirely replace existing services, such as ultrasound scanning and invasive testing, for several reasons. Firstly, cfDNA NIPT requires an ultrasound scan to be performed for demonstrating viability, dating, and to assess for multiple pregnancy or "vanishing twin," all of which has an effect on test performance. First trimester scanning also offers an early chance to assess for major structural abnormalities such as anencephaly, holoprosencephaly, and ventral wall defects which cannot be assessed by cfDNA NIPT.

Secondly, the presence of a fetal anomaly detected on ultrasound scanning should prompt the offer of invasive testing directly rather than cfDNA NIPT which, as previously discussed, remains a screening test and is not diagnostic. Trisomies 21, 18, and 13 are not the only abnormalities which are clinically useful to detect; chromosomal microarray analysis has been shown to identify additional, clinically significant information in 1.7% of pregnancies when compared to standard karyotyping when the indication for testing was maternal age or high-risk screening, and in 6.0% of pregnancies when compared to standard karyotyping in the presence of a structural anomaly [41]. A positive cfDNA NIPT will always need to be confirmed with invasive testing, even if the indication is an ultrasound anomaly. Choosing cfDNA NIPT in case of ultrasound anomalies will therefore cause delay in diagnosis and incur extra cost versus going straight to invasive testing.

Thirdly, there will be a small number of women, also discussed previously, who will not receive a result with NIPT ("test failure"). In this situation, the options would be to repeat the test, perform standard combined screening, or proceed straight to invasive testing. This latter option is recommended by the ACOG [42,43], due to the increased risk of aneuploidy with a "test failure." Also, given the ability of invasive testing to assess for chromosomal anomalies other than trisomies, including clinically significant chromosomal copy number variations, and the most recent data quoted earlier in this chapter suggesting that the risk of associated pregnancy loss may be much less in modern practice than previously suggested, it has been argued that all women should be offered diagnostic testing directly without prior screening [44].

Serum biochemistry may also retain a role in prenatal screening outside of aneuploidy risk assessment; it is well documented that low PAPP-A levels are associated with growth restriction, preterm delivery, hypertensive disorders of pregnancy, and stillbirth. Many first trimester prediction models for preeclampsia or stillbirth include PAPP-A values, and models such as that used in the ASPRE trial [45] have shown that treating women at high risk for preeclampsia with aspirin reduces the incidence of the disease. Several angiogenic markers, such as placental growth factor (PlGF) and soluble fms-like tyrosine kinase 1 (sFlt-1) have also been shown to be associated with an increased risk of developing preeclampsia [46] and may have a role in routine first trimester screening in the future.

However, despite the continued clinical value of ultrasound scanning, invasive testing, and serum biochemistry, it seems likely that increasing use of cfDNA NIPT will lead to a reduction in biochemical testing, and cytogenetic/molecular assessment of fetal cells or tissue. As noted in a recent RCOG Scientific Impact Paper [47], overall there will be a significant reduction in the number of invasive procedures and this will impact future training of specialists in fetal medicine. It seems likely that invasive procedures will be undertaken in fewer, larger centers than at present.

CELL-FREE DNA-BASED NIPT IMPLEMENTATION MODELS, COSTS, AND TYPES OF TEST

The two main methods of implementing cfDNA NIPT screening which have been used to date are as follows:

- High-risk or contingency screening: offering NIPT to a selected group of women, based on their results from standard screening.
- Low-risk or population screening: offering NIPT to all women, in place of standard screening.

The evidence for both these models has been discussed previously in this chapter. It is clear that the detection rate for cfDNA NIPT in contingency screening will depend on the risk level at which cfDNA NIPT is offered and the uptake of testing. It is also very likely that cost will play a role in which of these models is implemented and which cutoff is used in a contingent model. Although the cost of cfDNA NIPT is currently high in many countries, wider introduction and advances in sequencing technology would be expected to decrease expenditure in invasive testing. In the United Kingdom, the RAPID trial [35] previously discussed found that offering cfDNA NIPT to women with a combined screening result of more than 1 in 150 would not have a significant effect on cost. It is likely that costs of cfDNA NIPT will continue to decrease as technology advances and volume increases; it may be that, as this happens, it becomes cost effective to offer at lower risks or on a population basis.

Several commercial companies currently offer cfDNA NIPT screening—this financial drive may well affect implementation via consumer advertising and medical education. It may be the case that the private sector involvement has led to a publication bias in cfDNA NIPT studies. It is difficult to compare the performance of cfDNA NIPT tests from different companies. Two main methods of cfDNA NIPT are currently used: massively parallel shotgun sequencing (MPSS) and targeted sequencing (TS). Single-nucleotide polymorphism (SNP) is a form of targeted sequencing. A recent Cochrane review [48] found no difference in clinical sensitivity or specificity for either method. This review concluded that "NIPT appears sensitive and highly specific for detection of fetal trisomies 21, 18 and 13 in high-risk populations. There is a paucity of data on the accuracy of NIPT as a first-tier screening test in a population of unselected pregnant women." The Cochrane review included four of the five publications in Table 4. Cell-free DNA testing at present is not well validated for microdeletions or duplications but this is likely to change in the future with improving technology and increasing the depth of sequencing.

ETHICAL ISSUES

When considering the implementation of cfDNA NIPT for aneuploidy screening, it is obvious that a number of ethical and moral issues are raised, which it is important to be cognizant of. The Nuffield Council on Bioethics report [49] explores many of these in depth; we will highlight a few of the main ones here.

- Harm reduction: as widespread cfDNA NIPT implementation is likely to reduce the rate of invasive testing, it will therefore reduce the number of procedure-related miscarriages (which may be small).
- Justice/fairness: many women are already accessing cfDNA NIPT in many countries around the world in the private sector. In some cases, such as the United Kingdom, a few hospitals currently offer cfDNA NIPT in a contingent model in a public health setting whereas other publicly funded hospitals do not. One could argue that making a highly accurate test only available to women who can afford it, or on the basis of postcode, is unfair to others.
- Reproductive autonomy: by offering more accurate screening than the current standard, cfDNA NIPT furthers the ability of women and their partners to choose the circumstances of their pregnancy and either make plans for giving birth to a child affected by aneuploidy or make plans for termination of the pregnancy.
- Information giving/understanding: as cfDNA NIPT is not associated with a miscarriage risk and is quick and simple to do, it may be that the implications of test outcomes are not fully explained or considered. As the Nuffield Council report discusses: "some healthcare professionals may be focusing on medical problems when imparting information about Down's syndrome, without describing more fully what it can be like to have a child with Down's syndrome. The provision of accurate, balanced information that supports all screening choices equally, and the need for sufficient time to discuss any concerns are essential requirements for the introduction of NIPT in the NHS."
- Sex determination: although not the primary aim of aneuploidy screening, cfDNA NIPT (as with invasive karyotyping) is able to determine the sex of the fetus at an early gestation. This may increase the risk of sex selective terminations taking place. We recommend that unless clinically indicated (e.g., suspected congenital adrenal hyperplasia), routine fetal sex determination should not be offered by cfDNA NIPT.
- Termination of pregnancy: as the Nuffield Council report discusses, "Introducing cfDNA NIPT in the NHS could lead to an increase in the number of terminations following a diagnosis of Down's, Edwards' or Patau's syndrome. Some believe this amounts to eugenics. If this leads to a significant reduction in the number of people born and living with these syndromes, it is possible that the quality of health and social care they receive and the importance attributed to research into these syndromes will be affected. Making cfDNA NIPT available in the NHS could be perceived as sending negative and hurtful messages about the value of people with the syndromes being tested for." This issue has been raised as a concern of many forms of screening prior to NIPT including the current combined screening which, as discussed previously, is widely used across the world.

INTERNATIONAL IMPLEMENTATION OF CELL-FREE DNA-BASED NIPT

In writing this section, we are aware that the international use of cfDNA NIPT is rapidly evolving and likely to change very quickly from what is described as follows.

It has been shown that in the majority of high- and middle-income countries, cfDNA NIPT is available [50]. In the majority of these countries this is through the private sector. In a number of countries, such as Belgium, The Netherlands, Singapore, Australia, Canada, and the United States, cfDNA NIPT can also be available through the public sector. Prices have been quoted as ranging from $350 (Australia) to $2900 (United States), with an average cost of $874 worldwide.

In the United Kingdom, the National Screening Committee recommendation [51] has been that women who are at higher risk (defined as greater than 1 in 150) should be offered cfDNA NIPT as an additional step in the screening pathway along with the options of invasive testing or no further testing. This will come into effect in the public sector in Autumn 2018.

The ACOG Committee Opinion [42] has also recommended a contingency model of NIPT implementation rather than population-based screening.

Early 2017 the Netherlands has implemented primary or population-based screening. As will be the case in the UK, fetal sex is not communicated. Part of the costs are paid by the patient and the costs are equal for first trimester combination test and cfDNA NIPT. Initial experience suggests an increase in uptake from approximately 30% for combined screening to 40% for cfDNA NIPT. Termination of pregnancy rate in women who have a trisomy confirmed remains unchanged compared to combined screening. The Dutch laboratories use massively parallel sequencing as a technology, and perform a 3-year study into the impact of analysis filters on the quality of the cfDNA NIPT test, into the number and type of additional findings and their clinical utility, and their impact on the participants. Participants can elect whether to have additional findings communicated to them or not.

The International Society for Prenatal Diagnosis [52] have stated that the following options are all currently considered appropriate:

- NIPT as a primary test offered to all pregnant women
- NIPT secondary to a high-risk assessment based on serum and ultrasound screening protocols
- NIPT contingently offered to a broader group of women ascertained as having high or intermediate risks by conventional screening.
- Standard combined (ultrasound nuchal translucency at 11–13 completed weeks combined with serum markers at 9–13 weeks) or quadruple screening, or both (sequential screening).

The International Society of Ultrasound in Obstetrics and Gynaecology (ISUOG) Updated Consensus Statement from 2017 [53] stresses the importance of a first trimester ultrasound scan between 11 and 13 +6 weeks, regardless of a woman's intention to undergo cfDNA testing. The statement advises that following a normal early pregnancy scan, as defined by ISUOG guidelines, three options regarding screening or testing for trisomy 21 and, to a lesser extent, trisomies 18 and 13 should be explained, and might be considered for women who wish to have further risk assessment: (1) screening strategies based on individual risk calculated from maternal age and nuchal translucency measurement and/or maternal serum markers and/or other ultrasound markers in the first trimester, if needed and requested followed by cfDNA NIPT or invasive testing. Expert opinion suggests that in women considered at very high (>1:10) risk after combined screening cfDNA testing should not routinely replace invasive testing (2) cfDNA NIPT as a first-line screening test (3) invasive testing based on a woman's preference or background risk without risk calculation. ISUOG confirms that experience in low-risk women confirms the high detection rates published for high-risk populations and states that using cfDNA-based NIPT for intermediate or low-risk patients might be endorsed as a widely available option when new data emerge and costs decrease.

The American College of Medical Genetics and Genomics [54] states that "New evidence strongly suggests that NIPS can replace conventional screening for Patau's, Edwards and Down syndromes across the maternal age spectrum, for a continuum of gestational age beginning at 9 and 10 weeks, and for patients who are not significantly obese." They further recommend allowing patients to select diagnostic or screening approaches for the detection of genomic changes that are consistent with their personal goals and preferences.

SUMMARY

Cell-free DNA-based NIPT is a major advancement in prenatal screening for common aneuploidies. However, we do not believe it will ever entirely replace ultrasound and/or maternal serum biochemistry because of their benefits above and beyond screening for aneuploidy and are mindful of the ethical issues that come with such new advances. The implementation of this technology into common practice is already happening at a rapid pace and will require robust standards and monitoring as it continues.

REFERENCES

[1] Nussbaum RL, McInnes RR, Willard HF. Thompson & Thompson genetics in medicine. 7th ed. Philadelphia: Saunders/Elsevier; 2007.
[2] Non-invasive Prenatal Testing: Ethical Issues. Nuffield Council on Bioethics; 2017.
[3] Irving C, Basu A, Richmond S, Burn J, Wren C. Twenty-year trends in prevalence and survival of Down syndrome. Eur J Hum Genet 2008;16:1336–40.
[4] Cereda A, Carey J. The trisomy 18 syndrome. Orphanet J Rare Dis 2012;7:81.
[5] Rasmussen SA, Wong LY, Yang Q, May KM, Friedman JM. Population-based analyses of mortality in trisomy 13 and trisomy 18. Pediatrics 2003 April;111(4 Pt 1):777–84.
[6] Nicolaides KH. Screening for fetal aneuploidies at 11 to 13 weeks. Prenat Diagn 2011;31:7–15.
[6a] Cuckle HS, Wald NJ, Thompson SG. Estimating a woman's risk of having a pregnancy associated with Down's syndrome using her age and serum alpha-fetoprotein level. Br J Obstet Gynaecol 1987;94:387–402.
[6b] Hook EB, Cross PK, Regal RR. The frequency of 47,+21, 47,+18, and 47,+13 at the uppermost extremes of maternal ages: results on 56,094 fetuses studied prenatally and comparisons with data on live births. Hum Genet 1984;211–20.
[7] Tabor A, Philip J, Madsen M, Bang J, Obel EB, Nørgaard-Pedersen B. Randomised controlled trial of genetic amniocentesis in 4606 low-risk women. Lancet 1986;1(8493):1287–93.
[8] Driscoll D, Gross S. Prenatal screening for aneuploidy. N Engl J Med 2009;360:2556–62.
[9] Wald N, Huttly W, Hackshaw A. Antenatal screening for Down's syndrome with the quadruple test. Lancet 2003;361:835–6.
[10] Cuckle H, Benn P, Wright D. Down syndrome screening in the first and/or second trimester: model predicted performance using meta-analysis parameters. Semin Perinatol 2005;29(4):252–7.
[11] Nicolaides KH, Azar G, Byne D, Mansur C, Marks K. Fetal nuchal translucency: ultrasound screening for chromosomal defects in first trimester of pregnancy. BMJ 1992;304:86789.
[12] Nicolaides KH. Nuchal translucency and other first-trimester sonographic markers of chromosomal abnormalities. Am J Obstet Gynecol 2004;191:45–67.

[13] Malone FD, Canick JA, Ball RH, Nyberg DA, Comstock CH, Bukowski R, et al. First- and Second-Trimester Evaluation of Risk (FASTER) Research Consortium. First-trimester or second-trimester screening, or both, for Down's syndrome. N Engl J Med 2005;353:2001–11.

[14] NHS Fetal Anomaly Screening Programme (FASP): programme handbook June 2015. http://www.gov.uk/government/publications/fetal-anomaly-screening-programme-handbook.

[15] Morris JK, Springett A. The National Down Syndrome Cytogenetic Register for England and Wales: 2013 annual report. London, United Kingdom: Queen Mary University of London, Barts and The London School of Medicine and Dentistry; 2014.

[16] Rowe R, Puddicombe D, Hockley C, Redshaw M. Offer and uptake of prenatal screening for Down syndrome in women from different social and ethnic backgrounds. Prenat Diagn 2008;28:1245–50.

[17] Palomaki GE, Knight GJ, McCarthy JE, Haddow JE, Donhowe JM. Maternal serum screening for Down syndrome in the United States: a 1995 survey. Am J Obstet Gynecol 1997;176(5):1046–51.

[18] Kuppermann M, Learman LA, Gates E, Gregorich SE, Nease RF Jr, Lewis J. Beyond race or ethnicity and socioeconomic status: predictors of prenatal testing for Down syndrome. Obstet. Gynecol. 2006;107(5):1087.

[19] Hayeems RZ, Campitelli M, Ma X, Huang T, Walker M, Guttmann A. Rates of prenatal screening across health care regions in Ontario, Canada: a retrospective cohort study. CMAJ Open June 11, 2015 vol. 3 no. 2E236-2E243.

[20] Jaques AM, Collins VR, Muggli EE, Amor DJ, Francis I, Sheffield LJ, et al. Uptake of prenatal diagnostic testing and the effectiveness of prenatal screening for Down syndrome. Prenat Diagn 2010;30:522–30.

[21] Crombag N, Vellinga Y, Kluijfhout S, Bryant L, Ward P, Iedema-Kuiper R, et al. Explaining variation in Down's syndrome screening uptake: comparing the Netherlands with England and Denmark using documentary analysis and expert stakeholder interviews. BMC Health Serv Res 2014;14:437.

[22] Screening for Fetal Aneuploidy. American College of Obstetricians and Gynaecologists Practice Bulletin Number 163; 2016.

[23] Sundberg K, Bang J, Smidt-Jensen S, Brocks V, Lundsteen C, Parner J, et al. Randomised study of risk of fetal loss related to early amniocentesis versus chorionic villus sampling. Lancet 1997;350:697–703.

[24] Firth HV, Boyd PA, Chamberlain P, MacKenzie IZ, Lindenbaum RH, Huson SM. Severe limb abnormalities after chorionic villus sampling at 56-66 days' gestation. Lancet 1991;337:762–3.

[25] Nassar A, Martin D, Gonzalez-Quintero VH, Gomez-Marin O, Salman F, Gutierrez A, et al. Genetic amniocentesis complications: is the incidence overrated? Gynecol Obstet Invest 2004;58(2):100–4.

[26] Odibo AO, Gray DL, Dicke JM, Stamilio DM, Macones GA, Crane JP. Revisiting the fetal loss rate after second-trimester genetic amniocentesis: a single center's 16-year experience. Obstet Gynecol 2008 Mar;111(3):589–95.

[27] Tabor A, Vestergaard CH, Lidegaard Ø. Fetal loss rate after chorionic villus sampling and amniocentesis: an 11-year national registry study. Ultrasound Obstet Gynecol 2009;34(1):19–24.

[28] Akolekar R, Beta J, Picciarelli G, Ogilvie C, D'Antonio F. Procedure-related risk of miscarriage following amniocentesis and chorionic villus sampling: a systematic review and meta-analysis. Ultrasound Obstet Gynecol 2015;45:16–26.

[29] Wulff CB, Gerds TA, Rode L, Ekelund CK, Petersen OB, Tabor A and the Danish Fetal Medicine Study Group. Risk of fetal loss associated with invasive testing following combined first-trimester screening for Down syndrome: a national cohort of 147 987 singleton pregnancies. Ultrasound Obstet Gynecol 2016;47:38–44.

[30] Gil MM, Accurti V, Santacruz B, Plana MN, Nicolaides KH. Analysis of cell-free DNA in maternal blood in screening for aneuploidies: updated meta-analysis. Ultrasound Obstet Gynecol 2017;50:302–14.

[31] Flori E, Doray B, Gautier E, Kohler M, Ernault P, et al. Circulating cell-free fetal DNA in maternal serum appears to originate from cyto- and syncytio-trophoblastic cells. Case report. Hum Rerpod 2004 Mar;19(3):723–4.

[32] Grati FR, Malvestiti F, Ferreira JCPB, Bajaj K, Gaetani E, et al. Fetoplacental mosaicism: potential implications for false-positive and false-negative noninvasive prenatal screening results. Genetics in Medicine 2014;16:620–4.

[33] Revello R, Sarno L, Ispas A, Akolekar R, Nicolaides KH. Screening for trisomies by cell-free DNA testing of maternal blood: consequences of a failed result. Ultrasound Obstet Gynecol 2016;47:698–704.

[34] Gil MM, Revello R, Poon LC, Akolekar R, Nicolaides KH. Clinical implementation of routine screening for fetal trisomies in the UK NHS: cell-free DNA test contingent on results from first-trimester combined test. Ultrasound Obstet Gynecol 2016;47:45–52.

[35] Chitty LS, Wright D, Hill M, Verhoef TI, Daley R, Lewis C, et al. Uptake, outcomes, and costs of implementing non-invasive prenatal testing for Down's syndrome into NHS maternity care: prospective cohort study in eight diverse maternity units. BMJ 2016;354:i3426.

[36] Nicolaides KH, Syngelaki A, Ashoor G, Birdir C, Touzet G. Noninvasive prenatal testing for fetal trisomies in a routinely screened first-trimester population. Am J Obstet Gynecol 2012;207:374.e1–6.

[37] Dan S, Wang W, Ren J, Li Y, Hu H, Zu Z, et al. Clinical application of massively parallel sequencing-based prenatal noninvasive fetal trisomy test for trisomies 21 and 18 in 11105 pregnancies with mixed risk factors. Prenat Diagn 2012;32:1225–32.

[38] Bianchi DW, Lamar Parker R, Wentworth J, Madankumar R, Saffer C, Das AF, et al. DNA sequencing versus standard prenatal aneuploidy screening. N Engl J Med 2014;370:799–808.

[39] Norton ME, Jacobsson B, Swamy GK, Laurent LC, Ranzini AC, Brar H, et al. Cell-free DNA analysis for noninvasive examination of trisomy. N Engl J Med 2015;372:1589–97.

[40] Zhang H, Gao Y, Jiang F, Fu M, Yuan Y, Guo Y, et al. Non-invasive prenatal testing for trisomies 21, 18 and 13: clinical experience from 146958 pregnancies. Ultrasound Obstet Gynecol 2015;45:530–8.

[41] Wapner RJ, Martin CL, Levy B, Ballif BC, Eng CM, et al. Chromosomal microarray versus karyotyping for prenatal diagnosis. N Engl J Med 2012;367(23):2175–84.

[42] The American College of Obstetricians and Gynecologists. Screening for fetal aneuploidy. Practice Bulletin Number 163, 2016.

[43] The American College of Obstetricians and Gynecologists. Cell-free DNA screening for fetal aneuploidy. Committee Opinion Number 640, 2015.

[44] Evans MI, Wapner RJ, Berkowitz RL. Noninvasive prenatal screening or advanced diagnostic testing: caveat emptor. Am J Obstet Gynecol 2016;215(3):298–305.

[45] Rolnik DL, Wright D, Poon LC, O'Gorman N, Syngelaki A, et al. Aspirin versus placebo in pregnancies at high risk for preterm preeclampsia. N Engl J Med 2017;377:613–22.

[46] Monte S. Biochemical markers for prediction of preeclampsia: review of the literature. J Prenat Med 2011;5(3):69–77.

[47] Royal College of Obstetricians and Gynaecologists. Non-invasive Prenatal Testing for Chromosomal Abnormality using Maternal Plasma DNA. Scientific Impact Paper No. 15, 2014.

[48] Badeau M, Lindsay C, Blais J, Nshimyumukiza L, Takwoingi Y, et al. Genomics-based non-invasive prenatal testing for detection of fetal chromosomal aneuploidy in pregnant women. Cochrane Database Syst Rev 2017;11:CD011767.

[49] Nuffield Council on Bioethics. Non-invasive prenatal testing: ethical issues. March 2017.

[50] Minear MA, Lewis C, Pradhan S, Chandrasekharan S. Global perspectives on clinical adoption of NIPT. Prenat Diagn 2015;35:959–67.

[51] UK National Screening Committee Non-Invasive Prenatal Testing (NIPT) recommendation. 2016.

[52] Benn P, Borrell A, Chiu RWK, Cuckle H, Dugoff L, et al. Position statement from the Chromosome Abnormality Screening Committee on behalf of the Board of the International Society for Prenatal Diagnosis. Prenat Diagn 2015;35:725–34.

[53] Salomon LJ, Alfirevic Z, Audibert F, et al. ISUOG updated consensus statement on the impact of non-invasive prenatal testing (NIPT) on prenatal ultrasound practice. Ultrasound Obstet Gynecol 2017;49:815–6.

[54] Gregg AR, Skotko BG, Benkendord JL, Monaghan KG, Bajaj K, Best RG, et al. Noninvasive prenatal screening for fetal aneuploidy, 2016 update: a position statement of the American College of Medical Genetics and Genomics. Genet Med 2016; https://doi.org/10.1038/gim.2016.97.

WHY CELL-FREE DNA-BASED NONINVASIVE PRENATAL TESTING FOR FETAL CHROMOSOME ANOMALIES IS NOT DIAGNOSTIC

T. Harasim, A. Wagner

Center for Human Genetics and Laboratory Diagnostics, Dr. Klein, Dr. Rost and Colleagues, Martinsried, Germany

INTRODUCTION

Since the commercial introduction of cfDNA-based prenatal tests in 2011, a plethora of studies with designs of proof of concept, validation, case reports, large cohort studies, and meta-analyses have been published. Although cfDNA-based prenatal screening for trisomy 21, 18, and 13 has made big steps forward from the very early stages of its use, it soon became evident that false-positive and false-negative results might occur and that many of these have biological causes. Knowledge and understanding of the causes of a false result is essential to enable clinicians and genetic counselors to counsel the patients comprehensively and appropriately, both prior to a test as well as after having received the test result. This will ensure responsible application of an innovative prenatal testing method and allow evidence-based decision making for a mother and subsequently her fetus.

DIAGNOSTIC TESTS AND SCREENING TESTS

The performance of a new test method is usually described by four characteristics: sensitivity, specificity, and positive and negative predictive value. The latter two depend on the prevalence of the disease in a given population. The combination of sensitivity and false-positive rate $(1 - \text{specificity})$ allows a test method to be classified as a screening- or a diagnostic test for a given prevalence and cutoff. Sensitivity is defined as the ability of a test to correctly identify patients with a disease. In contrast, specificity describes the ability of a test to identify persons without the disease. These two test variables are fundamentally different between a diagnostic test and a screening test: diagnostic tests require both sensitivity and specificity to be as close as possible to 100%, whereas screening tests are usually characterized by either a high sensitivity or a high specificity. cfDNA-based prenatal testing

Noninvasive Prenatal Testing (NIPT). https://doi.org/10.1016/B978-0-12-814189-2.00005-0

for trisomy 21 has both, a high sensitivity and a high specificity [1–3]. The occurrence of false results though prevents it from being a diagnostic test. Most of these false results find their origin in the placenta or in the mother. They may have an irreversible impact on pregnancy management and decisions about pregnancy termination. cfDNA-based prenatal testing is still considered as a screening test implicating confirmation of all positive test results by an independent test method.

SENSITIVITY AND SPECIFICITY OF CELL-FREE DNA-BASED PRENATAL TESTING AND THEIR CONSEQUENCES FOR PREDICTIVE VALUES

Several meta-analyses on cfDNA-based prenatal testing have been published to date [1–5]. To illustrate the current performance of cfDNA-based prenatal tests, the results of Mackie et al. are shown here [2].

Table 1 shows the sensitivity and specificity of cfDNA-based prenatal tests for trisomy 13, 18, 21 and monosomy X, which for trisomy 21 reach levels close to 100%. The sensitivity is highest for trisomy 21 and decreases for trisomy 18, monosomy X, and trisomy 13. Neither the cfDNA NIPT method (whole genome or targeted sequencing; qPCR or microarray data were not included) nor the test population's a priori risk (high risk vs low risk) impacted test characteristics. Reasons for false results mentioned in individual studies were as follows: fetal fraction below a predefined cutoff, confined placental mosaicism (CPM), and maternal copy number variations (CNVs) were the most common explanations provided. Gil et al. [4], who only included studies reporting clinical outcome in more than 85% of cases, reported slightly higher detection rates: 99.7% for trisomy 21, 97.9% for trisomy 18, 99% for trisomy 13, and 95.8% for monosomy X.

To further illustrate the performance metrics, we assume a hypothetical general obstetric population of 100,000 patients with a trisomy 21 prevalence of 0.5%. This translates into 500 pregnant women carrying a fetus with trisomy 21. cfDNA-based prenatal testing, with its meta-analytical sensitivity of 99.4% identifies 497 of them, while three cases will be missed. From the 99,500 pregnancies without trisomy 21, ~100 will receive a false-positive result given the specificity of 99.9%. A positive predictive value (PPV) can be calculated with these values and shows how likely it is that a pregnant woman with a trisomy 21 positive cfDNA NIPT report actually carries a fetus with trisomy 21: given a disease prevalence of 0.5%, the PPV will be 83.2%. By comparison, in a higher risk population with a trisomy 21 prevalence of 3%, 3000 out of 100,000 patients carry a fetus with trisomy 21: 2982 trisomy 21 cases will be detected but 18 will be missed; 97,000 pregnant women are trisomy 21 negative, but 97 will receive a false-positive result. The PPV will be 96.8%. These hypothetical calculations demonstrate that PPVs depend on disease prevalence and that the personal a priori risk influences the significance of an abnormal cfDNA NIPT result for the individual patient. As trisomy negative pregnancies are always in excess in a pregnant population (representing a high nondisease prevalence) and

Table 1 Sensitivity and Specificity of cfDNA-Based NIPT [2]				
Performance	**Trisomy 21**	**Trisomy 18**	**Trisomy 13**	**Monosomy X**
n tested	148,344	146,940	134,691	6712
Sensitivity (%;95%CI)	99.4 (98.3–99.8)	97.7 (95.2–98.8)	90.6 (82.3–95.8)	92.9 (74.1–98.4)
Specificity (%;95%CI)	99.9 (99.9–100.0)	99.9 (99.8–100.0)	100.0 (99.9–100.0)	99.9 (99.5–99.9)

Table 2 Definitions of Sensitivity, Specificity, and Positive and Negative Predictive Value

		Outcome		Predictive Value of the Test
		Trisomy	**Disomy**	
NIPT result	Abnormal	True positive (TP)	False positive (FP)	Positive predictive value (PPV) TP/TP+FP
	Normal	False negative (FN)	True negative (TN)	Negative predictive value (NPV) TN/TN+FN
	Test performance	Sensitivity TP/TP+FN False-negative rate (FNR) = 1 − sensitivity	Specificity TN/TN+FP False-positive rate (FPR) = 1 − specificity	

meta-analytical specificity values of cfDNA-based NIPT reach almost 100% independent of the chromosome tested, the negative predictive value (NPV) is always close to 100% both in high- and low-risk populations (Table 2).

FALSE-POSITIVE RESULTS OF CELL-FREE DNA-BASED PRENATAL TESTING

The main reason why cfDNA-based prenatal testing is currently classified as a screening test is the occurrence of false-positive results. These have important implications for pregnancy management and parental decisions.

BIOLOGICAL REASONS FOR FALSE-POSITIVE RESULTS OF CELL-FREE DNA-BASED PRENATAL TESTING

Confined placental mosaicism

CPM is the existence of two (or more) cell lines with different chromosomal complements in a feto-placental unit derived from a single zygote [6]. Genetic mosaicism can be a result of two types of cell division errors (Figs. 1 and 2). First, after the development of a euploid zygote, mitotic errors can occur in part of the cells, and three different cell lines may be generated: a normal diploid cell line, a trisomic cell line, and a monosomic one; in general, the latter cannot proliferate further and undergoes apoptosis. Depending on the developmental stage of the embryo at the time of this mitotic error, a generalized- or a confined placental mosaicism evolves.

A second origin of CPM are meiotic errors during germ cell production. These result in a trisomic zygote; in later developmental stages this aneuploid zygote can be "rescued" by an additional mitotic error. The karyotype in the embryo depends on its developmental stage at the time of rescue: rescue in early stages implies a generalized mosaicism; a late stage trisomy rescue can confine the trisomy to the extra-embryonic compartment, the future placenta. The postzygotic correction can also lead to (mosaic) uniparental disomy (UPD), a disomy with a pair of homologous chromosomes from one parent only.

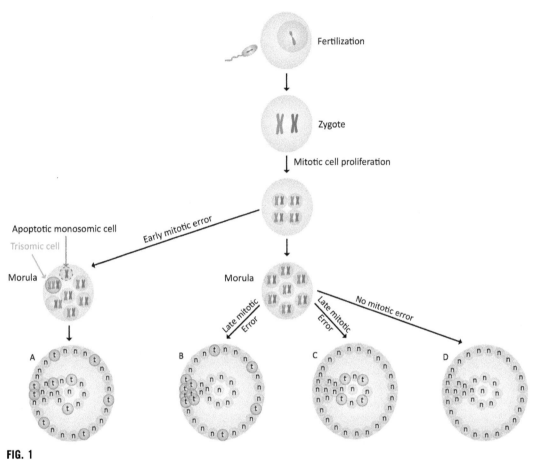

FIG. 1

Mosaicism due to mitotic nondysjunction. After a haploid oocyte is fertilized by a haploid sperm cell, the corresponding diploid zygote starts mitotic cell division. A mitotic error at the early stages of embryogenesis results in three cell types: trisomic cells, disomic cells, and monosomic cells. The latter usually undergo apoptosis. Depending on the time of mitotic error the following situations are possible: Embryo A: early mitotic error at the morula stage; generalized mosaicism of normal (n) and trisomic (t) cells in extra-embryonic tissue (future placenta) as well as ICM (future embryo). Embryo B: late mitotic error, after compartmentalization of extraembryonic tissue and ICM, CPM with trisomic cells in the extraembryonic compartment only. Embryo C: late mitotic error, after compartmentalization of extraembryonic tissue and the ICM, fetoplacental discordance with normal placenta and mosaic trisomic embryo, or (not shown) complete trisomic embryo. Percentages of abnormal cells in mosaic compartments range from only a few abnormal cells to close to 100% abnormal cells. Since only samples of these compartments are analyzed, these percentages always are a proxy based on how representative a sample is for the whole. Embryo D: no mitotic error, euploid embryo and an euploid placenta. *CPM*, confined placental mosaicism; *ICM*, inner cell mass (future embryo); *n*, normal/euploid; *t* trisomic.

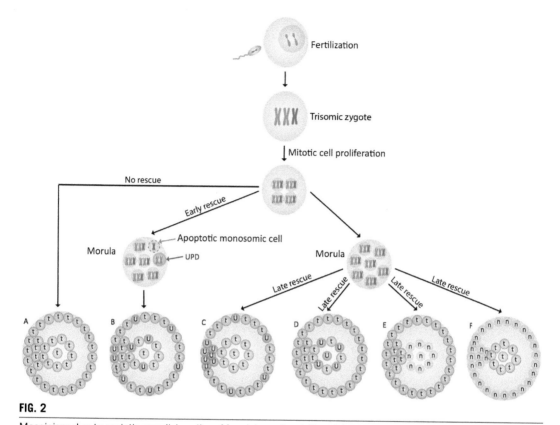

FIG. 2

Mosaicism due to meiotic nondisjunction. Mosaicism of meiotic origin can evolve from a trisomic zygote that is "rescued" in later developmental stages by a mitotic error. Depending on which chromosome is discarded, the supernumerary one, or one of the original euploid set, this mitotic error can result in uniparental or biparental disomic cells. The former is the case in the embryos B, C, and D, and the latter in embryos E and F. Depending on the timing of rescue the following situations are possible: Embryo A: no mitotic rescue, nonmosaic trisomic future embryo and future placenta. Embryo B: early rescue at the morula stage discards the only chromosome that is present as a single copy, generalized mosaicism of embryo and mosaicism with euploid cells with uniparental disomy and trisomic cells in the future placenta. Not shown: early rescue at the morula stage discarding one of the two chromosomes that is present as a double copy, generalized mosaicism of embryo and mosaicism with euploid cells with biparental disomy and trisomic cells in the future placenta. Embryo C: late mitotic rescue, after compartmentalization of ICM and extraembryonic tissue, trisomy in the future embryo and mosaicism with euploid cells with uniparental disomy (U) and trisomic cells in the future placenta. Embryo D: a late mitotic rescue, after compartmentalization of ICM and extraembryonic tissue, mosaicism with cells with uniparental disomy and trisomic cells in the future embryo and trisomy in the future placenta. Embryo E: a late mitotic rescue, after compartmentalization of ICM and extraembryonic tissue, biparental euploid future embryo (full or mosaic, mosaic not shown) and trisomy in the future placenta. Embryo F: a late mitotic rescue, after compartmentalization of ICM and extraembryonic tissue, trisomy in the future embryo and a biparental euploid cells (full or mosaic, mosaic not shown) in the future placenta. Percentages of abnormal cells in mosaic compartments range from only a few abnormal cells to close to 100% abnormal cells. Since only samples of these compartments are analyzed, these percentages always are a proxy based on how representative a sample is for the whole. *n*, normal; *ICM*, inner cell mass (future embryo); *t*, trisomic; *UPD*, uniparental disomy.

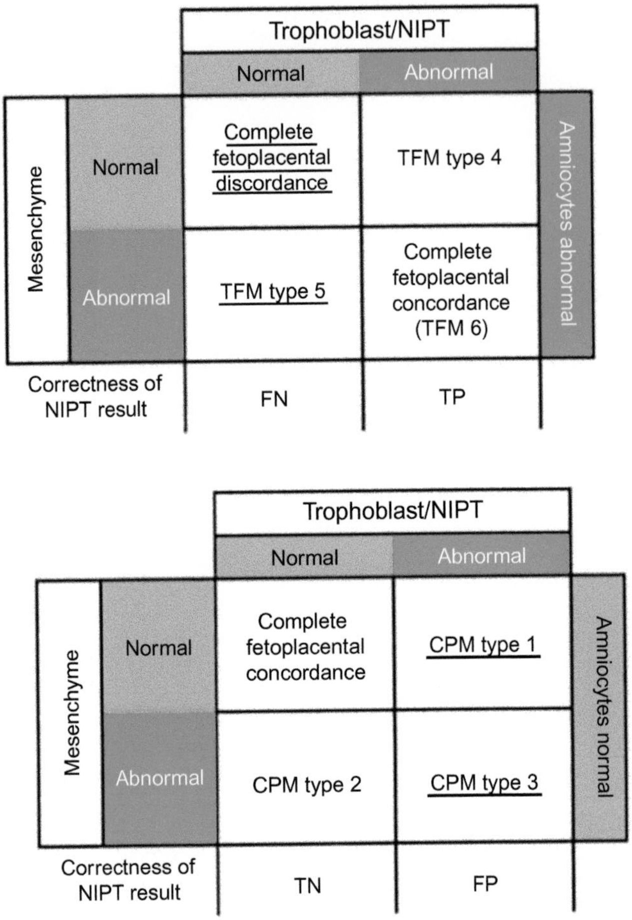

FIG. 3

Distribution of placental cell lines in confined placental mosaicism (CPM) and true fetal mosaicism (TFM). *Red background* represents aneuploid cell lines and green background represents euploid cell lines. (A) Types of TFM (amniocytes with abnormal karyotype). (B) Types of CPM (amniocytes with normal karyotype). *FP*, false positive; *TN*, true negative; *TP*, true positive.

Depending on the affected tissue types, one can categorize fetoplacental discrepancies in three subtypes of CPM when the karyotype in amniocytes is normal and three subtypes of true fetal mosaicism (TFM) when the karyotype in amniocytes is abnormal [7]. Differences between each category and expected cfDNA NIPT results are outlined in Fig. 3.

Cell-free "fetal" DNA originates from the cytotrophoblast and syncytiotrophoblast. Aneuploidy in this external layer of the chorionic villus leads to CPM types 1 and 3 and causes false-positive cfDNA NIPT results. In TFM type 5 and complete fetoplacental discordance, either the cytotrophoblast or both cytotrophoblast and mesenchyme are normal, while the fetus is not, generating false-negative cfDNA NIPT results [8]. Experience with chorion villus (CV) analysis shows an overall prevalence of mosaicism of 2%. CPM type 1 occurs more frequently than CPM type 3. TFM type 5 is the least probable scenario [9,10]. Trisomy 18 and trisomy 21 are less frequently involved (4% and 2% of mosaic cases, respectively) than trisomy 13 and monosomy X (22% and 59% of all cases) [11,12]. Also, the likelihood that a mosaic trisomy in CV is confirmed in the fetus differs per chromosome. Fetal confirmation rates were highest for trisomy 21 (34%) followed by monosomy X (26%), trisomy 18 (17%), and trisomy 13 (2.4%) [9]. Similar figures were reported in another publication by the same group [11]. Taken together, these data explain why cfDNA-based NIPT has better PPVs for trisomy 21 than for trisomy 18 and 13: trisomy 21 rarely presents as a mosaic in chorionic villi and if so, is more likely to be confirmed in the fetus.

Even if not confirmed in the fetus, CPM is not harmless. A (mosaic) trisomic placenta is not a healthy placenta as has already been demonstrated many years ago [13]. Spontaneous miscarriage, intrauterine fetal growth restriction, preterm delivery, and fetal demise can all be a result of altered placental function due to CPM [13–15]. This will be discussed in more detail in Chapter 7. Imprinting effects due to UPD after trisomy rescue are responsible for several clinical syndromes, for example, transient neonatal diabetes (paternal UPD 6), Silver-Russell syndrome (maternal UPD 7), and Prader-Willi and Angelman syndrome (maternal and paternal UPD15) [16]. Apart from imprinting effects, UPD also increases the risk for recessive disorders due to loss of heterozygosity.

Any CPM with a sufficient percentage of trisomic cells in the cytotrophoblast can be identified by cfDNA NIPT. Mosaicism is likely if the trisomic fraction is lower than the fetal fraction [17]. Another point of consideration is that in case of a small placenta as a result of CPM the fetal fraction may be lower, and this might translate in a cfDNA NIPT test failure. This correlates with the observation that there is a 2.5–6.7 times higher prevalence of aneuploidy among failed cfDNA NIPT cases analyzed with targeted enrichment-based NIPT methods requiring a minimum fetal fraction of 4% [18,19] as compared to MPS-based cfDNA NIPT tests [20–22]. Additionally, CPM can negatively influence the effective trisomic fraction. The fetal fraction, by definition, is the sum of all locations of the placental cytotrophoblast releasing cfDNA into the maternal blood stream. CPM might not be distributed equally throughout the placenta: some regions can have alternate ratios of normal to trisomic cells compared to the rest of the placental regions. This phenomenon is illustrated in a case report, where a mosaic double trisomy 18+21 (T18+T21) resulted in a NIPT result positive for T21, but negative for T18 [23]. By multilocus placenta biopsy, regions in the placenta could be identified, which had significantly more T21 positive cells (61%) than T18 positive cells (22%). These regions were responsible for a decreased T18 trisomic fraction, also called the "effective" fetal fraction for T18. The term effective fetal fraction in the context of CPM is also illustrated well by a case report of low-level CPM of trisomy 18 where the effective fetal fraction was calculated [24]: in multiple placental biopsies, the mosaic level for T18 was on average 30% for all cells. The overall fetal fraction was 7.4%, and the mosaicism hence resulted in an effective fetal fraction of $\sim 2.2\%$, which was too low for the NIPT test method used to detect the trisomy 18.

In summary, in this section we have described the types, frequency, origins, and other clinical effects of CPM including its influence on cfDNA NIPT accuracy. In the context of cfDNA NIPT, CPM has to be considered one of the main reasons for "false" NIPT results. Strictly spoken, these results are not false, since cfDNA NIPT is a liquid biopsy of the placenta and accurately reflects the karyotype of the cytotrophoblastic layer and the proportion of abnormal cells. But due to the phenomenon of fetoplacental discrepancies, cfDNA NIPT does not always reflect the fetal genome. The term "cell-free fetal DNA" is therefore incorrect and should be replaced by "cell-free placental DNA" or "cell-free DNA" since maternal plasma will always contain a pool of both, fetal and maternal cfDNA.

Vanishing twins

A second cause of false-positive cfDNA NIPT results is vanishing twins. A vanishing twin is defined as a fetus of a multigestational pregnancy that died in utero. A large cohort study reanalyzed biobanked mother-child samples of cases of false-positive prenatal fetal rhesus D (*RHD*) assignment and estimated the frequency of vanishing twins to be 0.6% [25]. The reported incidence of vanishing twins after IVF/ICSI was ~13% of all twin pregnancies [26]. The frequency of twin pregnancies increases with maternal age and with the use of assisted reproduction. In a study including 32 vanishing twins, median maternal age was significantly higher than in viable twins [27].

In spontaneous abortions both in cases with a normal and an abnormal karyotypes placental cfDNA and total cfDNA increases [28,29]. A dichorionic twin placenta releases two individual fractions of cfDNA into the maternal circulation and these are analyzed in concert during cfDNA-based NIPT. An aneuploid vanishing twin might therefore cause a false-positive cfDNA NIPT result for the fetus that is alive [30]. Theoretically, an "overrepresented" euploid vanishing twin, releasing excess amounts of cfDNA in the maternal circulation, could mask aneuploidy in the fetus that is alive and lead to a false-negative cfDNA NIPT result. This has not been described in literature. Along the same lines an unnoticed male vanishing twin can be responsible for a discrepancy between fetal sex according to cfDNA NIPT results and ultrasound imaging. Fetal sex discordances are discussed further in this chapter.

In cases of vanishing twins with discordant sex, comparing sex-dependent with sex-independent methods of fetal fraction estimation allows to deduce the fetal subfraction of each twin [31–34]. Likewise, if the vanishing twin is aneuploid and the living twin is not, an aneuploid fraction that is significantly lower than the fetal fraction can point toward a vanishing twin, or placental mosaicism [35].

The effect of vanishing twins on NIPT results depends on their ability to release cfDNA into the maternal blood stream. There is little data about how long after demise the placenta of the demised fetus still contributes to the "fetal" DNA pool in maternal plasma. SNP technologies allow to identify an extra haplotype. With this technology the presence of a third haplotype has been demonstrated in known vanishing twin cases up to 8 weeks after the estimated spontaneous demise of the cotwin. [27]. Thurik et al. demonstrated that a vanishing twin was the most likely cause of genotyping a 27-week fetus as *RHD* positive, while the newborn was serotyped RhD negative, meaning that in this case, cfDNA of a vanishing twin had persisted into the second trimester of pregnancy [25].

Early detection of a vanishing twin can be achieved by ultrasound examination. In the context of first trimester screening, early ultrasound with attention for the eventual presence of a vanishing twin is indispensable for correct test result interpretation. Likewise, an abnormal cfDNA NIPT result should be followed by an ultrasound explicitly looking for a vanishing twin.

The use of cfDNA NIPT in intact twin pregnancies is discussed in more detail in Chapter 6.

Rare autosomal trisomies and large segmental aberrations

Rare autosomal trisomies are trisomies other than those involving the chromosomes 13, 18, 21, X and Y. In nonmosaic form, these trisomies are usually not compatible with life and they have therefore rarely been seen as results of invasive prenatal diagnosis in amniotic fluid cells. In mosaic form, rare autosomal trisomies have been associated with intrauterine growth restriction, fetal death, true fetal mosaicism, and UPD. All cell-free DNA NIPT methods that are based on comparison of targeted chromosomes with reference chromosomes will be influenced by a trisomy or monosomy or large CNV in the reference chromosomes. Whole genome sequencing and analysis offers an opportunity to fully understand the underlying biological causes of these unusual results and decrease the false-positive rate [36–38]. Bioinformatic causes of false-positive results are discussed in Chapter 3 and Rare autosomal trisomies in Chapter 7.

Fetal sex discordances

Noninvasive fetal sex determination can reliably be done by quantitative PCR (qPCR). Amplification of the sex determining region of the Y-chromosome (*SRY*) gene, either exclusively or in combination with the *DYS14* marker, is sufficient to achieve high sensitivity and specificity, provided the presence of enough cell-free placental DNA is controlled for [39–41]. A meta-analysis reviewing 90 studies published between 1997 and 2010 evaluated the performance of noninvasive cfDNA-based fetal sex determination and calculated a median sensitivity of 96.6% and a median specificity of 98.9% [42]. For fetal sex determination for medical reasons in the context of noninvasive prenatal diagnosis (NIPD), as for women who are carriers of sex-linked disorders, the specificity should be 100%. For NIPD stringent diagnostic algorithms and additional tests to control for the presence of fetal DNA have therefore been put in place [38].

There are biological reasons for discordancies in sex assignment between cfDNA NIPT, ultrasonography, and fetal or newborn karyotyping [43] (Fig. 4).

If ultrasound and karyotyping results indicate a male fetus, but cfDNA-based testing reports female sex, a low fetal fraction, placental 46,XX/46,XY or 45,X/46,XY mosaicism and (theoretically) demise of a female cotwin can be an explanation.

If the karyotype and the cfDNA result indicate a female fetus, but ultrasound shows male genitalia, an "XX male" due to a translocation of part of the short arm of the Y-chromosome, including the *SRY* gene, onto the X-chromosome is a first possibility [44]. Other possible causes are congenital adrenal hyperplasia [45], an androgen producing tumor or exogenous androgens, all causes of virilization of a female fetus.

The third situation is when ultrasound and karyotyping both indicate a female fetus, but cfDNA testing detects Y-chromosomal DNA: placental 46,XY/46,XX mosaicism, demise of a male cotwin, history of transplantation with male donor [46], and recent blood transfusion from a male donor are possible explanations.

Finally, if a karyotypical male fetus with Y sequences detected at cfDNA testing shows female genitalia during ultrasound examination, feminization can be the result of Smith-Lemli-Opitz syndrome or by mutations in several sex developmental genes, for example, androgen receptor [47], 17ß-Hydroxysteroid dehydrogenase, steroid 5-alpha reductase, or *SRY*.

Maternal factors

Maternal acquired or constitutional genomic alterations can be responsible for false prenatal cfDNA-based testing results. An age-related mosaic loss of a maternal X-chromosome can cause a false-positive NIPT result for monosomy X [17,48]. The chance of an age-related mosaic loss is not equal for both X-chromosomes: preferentially, the inactivated version of the chromosome is lost [49,50].

Additionally, maternal malignancies can result in multiple whole chromosome or segmental gains or losses, which are absent in the fetus [51]. Although maternal cancer during pregnancy is rare (0.1%), hematologic malignancies or solid organ tumors releasing cell-free tumor DNA into the maternal circulation have been described as a cause of false cfDNA testing results [52–54]. These can be detected

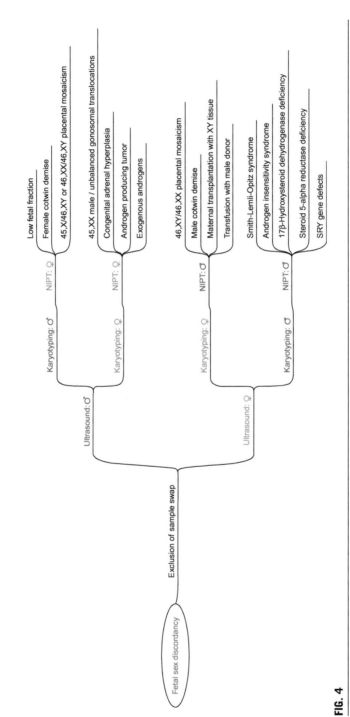

FIG. 4

Causes of discordances in fetal sex between cell-free DNA-based noninvasive testing, karyotyping and imaging.

by whole genome-based NIPT methods and diagnostic follow-up algorithms have been proposed [55]. Also, nonmalignant maternal diseases such as systemic lupus erythematodes with anti-double-strand DNA antibodies and severe maternal vitamin B12 deficiency have been described in association with abnormal cfDNA profiles [56,57].

Benign maternal de novo or inherited duplications or deletions represent another reason for false-positive and false-negative NIPT results. These variants translate to more or less cfDNA reads of the corresponding chromosome in the cfDNA assay [58]. The sensitivity for false results due to maternal CNVs is highest for shorter test chromosomes and decreases with higher fetal fractions [59].

Confirmation of abnormal cell-free DNA results

The overview shows why abnormal cfDNA-based NIPT results have to be confirmed by a diagnostic test, a genetic analysis in chorionic villi, or amniotic fluid cells.

Analysis of amniocytes is considered superior to analysis of chorionic villi, since part of the false-positive results are due to discrepancies between trophoblast and fetus. Amniocytes are strongly representative of the fetus and mosaicism in amniotic fluid cells is much rarer than in chorionic villi. The drawback of amniocentesis (AC) is that it is considered safe from 15 weeks of gestation only and not earlier [60]. Given that cfDNA-based prenatal testing is offered from the 10th week onward, confirming abnormal cfDNA results with an amniocentesis will cause a delay in obtaining the final results. Most women will perceive this waiting period as stressful. In the worst case, it might lead to pregnancy termination without confirming cfDNA-based NIPT results. This scenario can be avoided by choosing CVS, which is usually done between weeks 11 and 14 of gestation. The drawback of CVS, however, is the possible detection of type 3 mosaicism confined to the placenta (CPM type 3), leading to a false-positive result, and true fetal mosaicism type 5 (TFM type 5), leading to a false-negative result if only cytotrophoblast is analyzed [61].

Also, mosaicism between cytotrophoblast and mesenchyme can be detected (CPM types 1 and 2 and TFM types 4 and 5) making a second invasive procedure, amniocentesis, necessary to ascertain the fetal karyotype. As reviewed by Grati et al. [11] and described in the section on "Confined placental mosaicism", mosaicism for trisomy 21, trisomy 18, trisomy 13, and monosomy X was found in, respectively, 2%, 4%, 22%, and 59% of abnormal CVS cases. The anomaly was confirmed by amniocentesis in, respectively, 44%, 14%, 4%, and 26% only. These figures can be used as a proxy for the risk that amniocentesis will be needed as a second invasive test when CVS is chosen to confirm an abnormal cfDNA NIPT result. In view of the figures, one can defend that NIPT results indicating a trisomy 13 or a monosomy X should preferentially be confirmed by AC, since the risk for obtaining a mosaic result after CVS is too high, and especially trisomy 13 is rarely confirmed in amniocentesis. Monosomy X, if present in the fetus, almost universally ends in miscarriage.

To allow accurate interpretation of CVS results, cytotrophoblast cells (STC) as well as mesenchymal cells (LTC) should always be analyzed. If results of either or both are mosaic or if there is a discrepancy between the 2 cell layers, amniocentesis and FISH analysis of a large number of amniocytes is indicated, to differentiate CPM from TPM (true fetal mosaicism). Van Opstal and Srebniak advised along these lines in a recent special report [62].

FALSE-NEGATIVE RESULTS OF CELL-FREE DNA-BASED PRENATAL TESTING

In theory, all phenomena responsible for false-positive cfDNA-based prenatal test results can also cause false-negative cfDNA test results [63]. The most important cause of false-negative results is, however, a fetal fraction that is too low, or too low for a given sequencing depth. Causes of low fetal fraction are high maternal BMI, early gestational age, hemolysis in the sample due to storage, or sending conditions or maternal therapy such as anticoagulation medication [64–66]. An increased BMI is associated with a higher number of adipocytes releasing maternal cfDNA into the blood and diluting the placental cfDNA [67]. Most cfDNA NIPT tests currently offered estimate fetal fraction and do not give a test result if the fetal fraction is inadequate for the given technology, as such reducing the risk for false-negative results.

The second most important cause for false-negative results is TFM type 5, where a normal cytotrophoblast will generate false-negative cfDNA test while mesenchyme and fetus are abnormal. In early pregnancy cytotrophoblast and syncytiotrophoblast are the sole sources of cell-free placental DNA (cfDNA) in maternal plasma and mesenchymal cells do not shed cfDNA in the mothers' blood. Therefore cell-free DNA-based NIPT will never outperform STC chorionic villus analysis. At chorionic villus sampling (CVS) not only cytotrophoblast but also mesenchymal cells are prepared for karyotyping, the so-called short- and long-term cultured villi (STC and LTC). The mesenchyme is more representative of the fetus than the cytotrophoblast and combined analysis of STC- and LTC-cultured villi is considered diagnostic. In a study on 5967 consecutive cases of CVS, 2% and 7.3% of, respectively, trisomy 21 and trisomy 18 cases would potentially have generated false-negative results if only STC villi had been analyzed, due to the biological phenomenon of absence of the aberration in the cytotrophoblast [63]. In another series describing 60,347 consecutive CVS analyses, 1317 (2.18%) mosaic cases were detected and in 5.7% of the ones confirmed by amniocentesis (1001) TFM type 5 with normal cytotrophoblast and abnormal mesenchyme and fetus was seen [9].

Although never described in literature, in theory a normal vanishing twin can cover up the presence of an affected twin by its temporarily overrepresented fetal fraction. Likewise, theoretically maternal deletions on the target chromosomes can result in a false-negative result when they cooccur with a trisomic pregnancy.

CONCLUSION

In summary, there is a plethora of phenomena and reasons for false results of cfDNA-based tests. It is therefore critical, that before cfDNA-based prenatal testing is performed a full patient history is taken, ultrasound excludes vanishing twins as well as confirming vitality and checking number of fetuses, and that patients receive comprehensive genetic counseling. Explaining the placental origin of cfDNA allows the patient a clear view of the abilities and limitations of cfDNA-based prenatal screening. If cfDNA-based tests indicate a trisomy, confirmatory testing is needed before drawing conclusions.

Continuing method optimization may allow to further decrease false-positive and false-negative results among others by improving identification of anomalies of placental and maternal origin.

REFERENCES

[1] Gil M, Accurti V, Santacruz B, Plana M, Nicolaides K. Analysis of cell-free DNA in maternal blood in screening for aneuploidies: updated meta-analysis. Ultrasound Obstet Gynecol 2017;50(3):302–14.

[2] Mackie F, Hemming K, Allen S, Morris R, Kilby M. The accuracy of cell-free fetal DNA-based non-invasive prenatal testing in singleton pregnancies: a systematic review and bivariate meta-analysis. BJOG 2017;124:32–46.

[3] Badeau M, Lindsay C, Blais J, Nshimyumukiza L, Takwoingi Y, Langlois S, et al. Genomics-based non-invasive prenatal testing for detection of fetal chromosomal aneuploidy in pregnant women. Cochrane Database Syst Rev 2017.

[4] Gil MM, Akolekar R, Quezada MS, Bregant B, Nicolaides KH. Analysis of cell-free DNA in maternal blood in screening for aneuploidies: meta-analysis. Fetal Diagn Ther 2014;35:156–73.

[5] Taylor-Phillips S, Freeman K, Geppert J, Agbebiyi A, Uthman OA, Madan J, et al. Accuracy of non-invasive prenatal testing using cell-free DNA for detection of Down, Edwards and Patau syndromes: a systematic review and meta-analysis. BMJ Open 2016;6:e010002.

[6] Lestou VS, Kalousek DK. Confined placental mosaicism and intrauterine fetal growth. Arch Child Fetal Neonatal Ed 1998;79:F223–6.

[7] Grati FR, Malvestiti F, Ferreira JCPB, Bajaj K, Gaetani E, Agrati C, et al. Fetoplacental mosaicism: potential implications for false-positive and false-negative noninvasive prenatal screening results. Genet Med 2014.

[8] Hochstenbach R, Nikkels PGJ, Elferink MG, Oudijk MA, van Oppen C, van Zon P, et al. Cell-free fetal DNA in the maternal circulation originates from the cytotrophoblast: proof from an unique case. Clin Case Rep 2015;3:489–91.

[9] Malvestiti F, Agrati C, Grimi B, Pompilii E, Izzi C, Martinoni L, et al. Interpreting mosaicism in chorionic villi: results of a monocentric series of 1001 mosaics in chorionic villi with follow-up amniocentesis: interpreting mosaicism in chorionic villi. Prenat Diagn 2015;35:1117–27.

[10] Kalousek DK, Vekemans M. Confined placental mosaicism. J Med Genet 1996;33.

[11] Grati FR, Bajaj K, Malvestiti F, Agrati C, Grimi B, Malvestiti B, et al. The type of feto-placental aneuploidy detected by cfDNA testing may influence the choice of confirmatory diagnostic procedure: confirmatory procedure after cfDNA testing. Prenat Diagn 2015;35:994–8.

[12] Hook EB, Warburton D. Turner syndrome revisited: review of new data supports the hypothesis that all viable 45,X cases are cryptic mosaics with a rescue cell line, implying an origin by mitotic loss. Hum Genet 2014;133:417–24.

[13] Lestou VS, Kalousek DK. Confined placental mosaicism and intrauterine fetal growth. Arch Dis Child-Fetal Neonatal Ed 1998;79:F223–6.

[14] Mardy A, Wapner RJ. Confined placental mosaicism and its impact on confirmation of NIPT results. Am J Med Genet C Semin Med Genet 2016;172:118–22.

[15] Wilkins-Haug L, Quade B, Morton CC. Confined placental mosaicism as a risk factor among newborns with fetal growth restriction. Prenat Diagn 2006;26:428–32.

[16] Robinson WP, Peñaherrera MS, Jiang R, Avila L, Sloan J, McFadden DE, et al. Assessing the role of placental trisomy in preeclampsia and intrauterine growth restriction. Prenat Diagn 2010;30:1–8.

[17] Russell LM, Strike P, Browne CE, Jacobs PA. X chromosome loss and ageing. Cytogenet Genome Res 2007;116:181–5.

[18] Norton ME, Jacobsson B, Swamy GK, Laurent LC, Ranzini AC, Brar H, et al. Cell-free DNA analysis for non-invasive examination of trisomy. N Engl J Med 2015;372:1589–97.

[19] Pergament E, Cuckle H, Zimmermann B, Banjevic M, Sigurjonsson S, Ryan A, et al. Single-nucleotide polymorphism–based noninvasive prenatal screening in a high-risk and low-risk cohort. Obstet Gynecol 2014;124:210–8.

[20] Yaron Y. The implications of non-invasive prenatal testing failures: a review of an under-discussed phenomenon: clinical implications of test failures in NIPT, Prenat Diagn 2016;36(5):391–6.

[21] Palomaki GE, Deciu C, Kloza EM, Lambert-Messerlian GM, Haddow JE, Neveux LM, et al. DNA sequencing of maternal plasma reliably identifies trisomy 18 and trisomy 13 as well as Down syndrome: an international collaborative study. Genet Med 2012;14:296–305.

[22] Taneja PA, Snyder HL, de Feo E, Kruglyak KM, Halks-Miller M, Curnow KJ, et al. Noninvasive prenatal testing in the general obstetric population: clinical performance and counseling considerations in over 85 000 cases. Prenat Diagn 2016;36:237–43.

[23] Mao J, Wang T, Wang B-J, Liu Y-H, Li H, Zhang J, et al. Confined placental origin of the circulating cell free fetal DNA revealed by a discordant non-invasive prenatal test result in a trisomy 18 pregnancy. Clin Chim Acta 2014;433:190–3.

[24] Gao Y, Stejskal D, Jiang F, Wang W. False-negative trisomy 18 non-invasive prenatal test result due to 48, XXX,+18 placental mosaicism. Ultrasound Obstet Gynecol 2014;43:477–8.

[25] Thurik FF, Ait Soussan A, Bossers B, Woortmeijer H, Veldhuisen B, Page-Christiaens GCML, et al. Analysis of false-positive results of fetal *RHD* typing in a national screening program reveals vanishing twins as potential cause for discrepancy: discordant fetal *RHD* screening results and vanishing twins. Prenat Diagn 2015;35:754–60.

[26] Márton V, Zádori J, Kozinszky Z, Keresztúri A. Prevalences and pregnancy outcome of vanishing twin pregnancies achieved by in vitro fertilization versus natural conception. Fertil Steril 2016;106:1399–406.

[27] Curnow KJ, Wilkins-Haug L, Ryan A, Kırkızlar E, Stosic M, Hall MP, et al. Detection of triploid, molar, and vanishing twin pregnancies by a single-nucleotide polymorphism–based noninvasive prenatal test. Am J Obstet Gynecol 2015;212:79.e1–9.

[28] Yin A, Ng EHY, Zhang X, He Y, Wu J, Leung KY. Correlation of maternal plasma total cell-free DNA and fetal DNA levels with short term outcome of first-trimester vaginal bleeding. Hum Reprod 2007;22:1736–43.

[29] Lim JH, Kim MH, Han YJ, Lee DE, Park SY, Han JY, et al. Cell-free fetal DNA and cell-free total DNA levels in spontaneous abortion with fetal chromosomal aneuploidy. PLoS One 2013;8:e56787.

[30] Futch T, Spinosa J, Bhatt S, de Feo E, Rava RP, Sehnert AJ. Initial clinical laboratory experience in noninvasive prenatal testing for fetal aneuploidy from maternal plasma DNA samples. Prenat Diagn 2013;33:569–74.

[31] Kim SK, Hannum G, Geis J, Tynan J, Hogg G, Zhao C, et al. Determination of fetal DNA fraction from the plasma of pregnant women using sequence read counts: determination of fetal DNA fraction from the plasma of pregnant women using sequence read counts. Prenat Diagn 2015;35:810–5.

[32] Bruno DL, Ganesamoorthy D, Thorne NP, Ling L, Bahlo M, Forrest S, et al. Use of copy number deletion polymorphisms to assess DNA chimerism. Clin Chem 2014. .

[33] Jiang P, Chan KCA, Liao GJW, Zheng YWL, Leung TY, Chiu RWK, et al. FetalQuant: deducing fractional fetal DNA concentration from massively parallel sequencing of DNA in maternal plasma. Bioinformatics 2012;28:2883–90.

[34] Tong YK, Chiu RWK, Leung TY, Ding C, Lau TK, Leung TN, et al. Detection of restriction enzyme digested target DNA by PCR amplification using a stem-loop primer: application to the detection of hypomethylated fetal DNA in maternal plasma. Clin Chem 2007;53:1906–14.

[35] Grömminger S, Yagmur E, Erkan S, Nagy S, Schöck U, Bonnet J, et al. Fetal aneuploidy detection by cell-free DNA sequencing for multiple pregnancies and quality issues with vanishing twins. J Clin Med 2014;3:679–92.

[36] Pertile MD, Halks-Miller M, Flowers N, Barbacioru C, Kinnings SL, Vavrek D, et al. Rare autosomal trisomies, revealed by maternal plasma DNA sequencing, suggest increased risk of feto-placental disease. Sci Transl Med 2017;9:eaan1240.

[37] Van Opstal D, van Maarle MC, Lichtenbelt K, Weiss MM, Schuring-Blom H, Bhola SL, et al. Origin and clinical relevance of chromosomal aberrations other than the common trisomies detected by genome-wide NIPS: results of the TRIDENT study. Genet Med 2017.

[38] Bianchi DW. Should we 'open the kimono' to release the results of rare autosomal aneuploidies following non-invasive prenatal whole genome sequencing? Prenat Diagn 2017;37:123–5.

[39] Scheffer PG, van der Schoot CE, Page-Christiaens GC, Bossers B, van Erp F, de Haas M. Reliability of fetal sex determination using maternal plasma. Obstet Gynecol 2010;115:117–26.

[40] Galbiati S, Smid M, Gambini D, Ferrari A, Restagno G, Viora E, et al. Fetal DNA detection in maternal plasma throughout gestation. Hum Genet 2005;117:243–8.

[41] Fernández-Martínez FJ, Galindo A, Garcia-Burguillo A, Vargas-Gallego C, Nogués N, Moreno-García M, et al. Noninvasive fetal sex determination in maternal plasma: a prospective feasibility study. Genet Med 2012;14:101–6.

[42] Wright CF, Wei Y, Higgins JP, Sagoo GS. Non-invasive prenatal diagnostic test accuracy for fetal sex using cell-free DNA a review and meta-analysis. BMC Res Notes 2012;5:476.

[43] Grati FR. Implications of fetoplacental mosaicism on cell-free DNA testing: a review of a common biological phenomenon. Ultrasound Obstet Gynecol 2016;48:415–23.

[44] Mansfield N, Boogert T, McLennan A. Prenatal diagnosis of a 46,XX male following noninvasive prenatal testing. Clin Case Rep 2015;3:849–53.

[45] Colmant C, Morin-Surroca M, Fuchs F, Fernandez H, Senat M-V. Non-invasive prenatal testing for fetal sex determination: is ultrasound still relevant? Eur J Obstet Gynecol Reprod Biol 2013;171:197–204.

[46] Bianchi DW, Parsa S, Bhatt S, Halks-Miller M, Kurtzman K, Sehnert AJ, et al. Fetal sex chromosome testing by maternal plasma DNA sequencing: clinical laboratory experience and biology. Obstet Gynecol 2015;125:375–82.

[47] Franasiak JM, Yao X, Ashkinadze E, Rosen T, Scott RT. Discordant embryonic aneuploidy testing and prenatal ultrasonography prompting androgen insensitivity syndrome diagnosis. Obstet Gynecol 2015;125:383–6.

[48] Guttenbach M, Koschorz B, Bernthaler U, Grimm T, Schmid M. Sex chromosome loss and aging: in situ hybridization studies on human interphase nuclei. Am J Hum Genet 1995;57:1143.

[49] Hando JC, Tucker JD, Davenport M, Tepperberg J, Nath J. X chromosome inactivation and micronuclei in normal and Turner individuals. Hum Genet 1997;100:624–8.

[50] Tucker JD, Nath J, Hando JC. Activation status of the X chromosome in human micronucleated lymphocytes. Hum Genet 1996;97:471–5.

[51] Bianchi DW, Chudova D, Sehnert AJ, Bhatt S, Murray K, Prosen TL, et al. Noninvasive prenatal testing and incidental detection of occult maternal malignancies. JAMA 2015;314:162.

[52] Pavlidis NA. Coexistence of pregnancy and malignancy. Oncologist 2002;7:279–87.

[53] Amant F, Verheecke M, Wlodarska I, Dehaspe L, Brady P, Brison N, et al. Presymptomatic identification of cancers in pregnant women during noninvasive prenatal testing. JAMA Oncol 2015;1:814.

[54] Vandenberghe P, Wlodarska I, Tousseyn T, Dehaspe L, Dierickx D, Verheecke M, et al. Non-invasive detection of genomic imbalances in Hodgkin/Reed-Sternberg cells in early and advanced stage Hodgkin's lymphoma by sequencing of circulating cell-free DNA: a technical proof-of-principle study. Lancet Haematol 2015;2:e55–65.

[55] Bianchi DW. Unusual prenatal genomic results provide proof-of-principle of the liquid biopsy for cancer screening. Clin Chem 2018;64(2):254–6.

[56] Chan RWY, Jiang P, Peng X, Tam L-S, Liao GJW, Li EKM, et al. Plasma DNA aberrations in systemic lupus erythematosus revealed by genomic and methylomic sequencing. Proc Natl Acad Sci USA 2014;111:E5302–11.

[57] Schuring-Blom H, Lichtenbelt K, van Galen K, Elferink M, Weiss M, Vermeesch JR, et al. Maternal vitamin B12 deficiency and abnormal cell-free DNA results in pregnancy: maternal vitamin B12 deficiency and abnormal NIPT outcome. Prenat Diagn 2016;36:790–3.

[58] Snyder MW, Simmons LE, Kitzman JO, Coe BP, Henson JM, Daza RM, et al. Copy-number variation and false positive prenatal aneuploidy screening results. N Engl J Med 2015;372:1639–45.

[59] Zhang H, Zhao Y-Y, Song J, Zhu Q-Y, Yang H, Zheng M-L, et al. Statistical approach to decreasing the error rate of noninvasive prenatal aneuploid detection caused by maternal copy number variation. Sci Rep 2015;5.

[60] Alfirevic Z, Mujezinovic F, Sundberg K. Amniocentesis and chorionic villus sampling for prenatal diagnosis. In: The Cochrane Collaboration, editor. Cochrane Database of Systematic Reviews, John Wiley & Sons, Ltd; Chichester, UK, 2003.

[61] Uquillas K, Chan Y, King JR, Randolph LM, Incerpi M. Chorionic villus sampling fails to confirm mosaic trisomy 21 fetus after positive cell-free DNA: CVS after cell-free DNA. Prenat Diagn 2017;37:296–8.

[62] Van Opstal D, Srebniak MI. Cytogenetic confirmation of a positive NIPT result: evidence-based choice between chorionic villus sampling and amniocentesis depending on chromosome aberration. Expert Rev Mol Diagn 2016;16:513–20.

[63] Van Opstal D, Srebniak MI, Polak J, de Vries F, Govaerts LCP, Joosten M, et al. False negative NIPT results: risk figures for chromosomes 13, 18 and 21 based on chorionic villi results in 5967 cases and literature review. PLoS One 2016;11:e0146794.

[64] Barrett AN, Zimmermann BG, Wang D, Holloway A, Chitty LS. Implementing prenatal diagnosis based on cell-free fetal DNA: Accurate identification of factors affecting fetal DNA yield. PLoS One 2011;6:e25202...

[65] Burns W, Koelper N, Barberio A, Deagostino-Kelly M, Mennuti M, Sammel MD, et al. The association between anticoagulation therapy, maternal characteristics, and a failed cfDNA test due to a low fetal fraction. Prenat Diagn 2017;37:1125–9.

[66] Ma G-C, Wu W-J, Lee M-H, Lin Y-S, Chen M. The use of low molecular weight heparin reduced the fetal fraction and rendered the cell-free DNA testing for trisomy 21 false negative: false negative cfDNA for fetal trisomy 21 due to heparin use. Ultrasound Obstet Gynecol 2018;51(2):274–7.

[67] Haghiac M, Vora NL, Basu S, Johnson KL, Presley L, Bianchi DW, et al. Increased death of adipose cells, a path to release cell-free DNA into systemic circulation of obese women. Obesity 2012;20:2213–9.

THE ROLE OF CELL-FREE DNA-BASED PRENATAL TESTING IN TWIN PREGNANCY

6

Fiona L. Mackie*, Mark D. Kilby[†,‡]

Institute of Metabolism and Systems Research, University of Birmingham, Birmingham, United Kingdom* Centre for Women's & Newborn Health, Institute of Metabolism & Systems Research, University of Birmingham, Birmingham Health Partners, Birmingham, United Kingdom[†] *West Midlands Fetal Medicine Centre, Birmingham Women's and Children's NHS Foundation Trust, Birmingham, United Kingdom*[‡]

INTRODUCTION

Due to an increasing trend of advancing maternal age at conception, and the increased use of assisted reproductive technologies (ART), the number of twin and higher order multiple pregnancies is increasing in many countries, despite the introduction of the single embryo transfer policy [1].

Compared to singletons, twin pregnancies are at increased relative risk of many obstetric complications, including aneuploidy [2]. This is probably the indirect consequence of the fact that older women not only have a higher risk of aneuploidy pregnancies, but also a higher risk of naturally conceiving twin pregnancies. Additionally, older women are more likely to need ART to conceive, and ART is a risk factor for twin pregnancies as well as for fetal aneuploidy [3].

Risks of prenatal invasive testing are higher in twins than in singletons [4] and accurate noninvasive prenatal testing (NIPT) would therefore be welcomed by mothers of twin pregnancies, particularly by those women who had difficulties conceiving. Additionally, in twin pregnancies, there is also a greater potential for sampling errors at chorionic villous sampling (CVS) as well as at amniocentesis [5]. The currently used conventional prenatal screening strategies: the first trimester combined test and the second trimester serum screening test have a high false positive rate that surpasses that in singleton pregnancies [6–9] and may be even higher in pregnancies conceived by ART [10,11].

Data from published studies informing the use of cell-free DNA-based noninvasive prenatal testing (cfDNA NIPT) in twins are relatively small compared to that of singleton pregnancies. There is debate as to whether the diagnostic accuracy of cfDNA NIPT in twin pregnancy is equivalent to that of singletons, and what role cfDNA NIPT should have in routine clinical care. This is against the backdrop of the increasing use in clinical practice of cfDNA testing both in state-run and private healthcare systems.

In this chapter, we will outline the evidence and the current opinions of the professional bodies, review the differences in cfDNA NIPT accuracy between twins and singletons, and stratify the available evidence on impact of chorionicity.

Noninvasive Prenatal Testing (NIPT). https://doi.org/10.1016/B978-0-12-814189-2.00006-2

ZYGOSITY, CHORIONICITY, AND AMNIONICITY

Twin pregnancies can be classified according to zygosity, chorionicity, and amnionicity, all of which have implications on prenatal diagnosis and subsequent management. Zygosity refers to the number of zygotes. Monozygotic (MZ) twins originate from one zygote (one sperm and one oocyte), and dizygotic (DZ) twins originate from two zygotes (two sperm and two oocytes). Consequently, MZ twins are genetically "identical" although there are rare exceptions reported of heterokaryotypic MZ twins [12–19]. It is assumed that a third of twins are monozygous. DZ twins are genetically nonidentical. In MZ twins, it can be assumed that the cfDNA NIPT result will reflect the genetic makeup of both twins, whereas in DZ twins it is more complicated as, in case of an anomaly, the twins are most likely to be discordant, that is, one twin will be affected but the other twin will not.

However, in clinical practice, the prenatal risk is allocated by the chorionicity (placental type) and amnionicity. Chorionicity refers to the number of placentas: monochorionic (MC) twins share one placenta and have interfetal vascular placental anastomoses, whereas dichorionic (DC) twins have two individual placental masses. Up to 15% of twins are MC and 85%–90% are DC. Amnionicity refers to the number of amniotic fluid sacs: monoamniotic (MA) twins have one sac; diamniotic (DA) twins have two sacs. The chorionicity and amnionicity should be determined in all twin pregnancies at the first trimester ultrasound scan [20,21], as each combination (DCDA, MCDA, MCMA) is managed differently antenatally due to different perinatal/fetal risks and complications including: twin-twin transfusion syndrome, selective intrauterine growth restriction, intrauterine death, and cord entanglement [22,23]. The sensitivity and specificity of "correct" first trimester ultrasound chorionicity allocation is >99%. Clinically, zygosity cannot be assumed from chorionicity and/or amnionicity. Although all DZ twins are DC, and all MC twins are MZ, not all DC twins are DZ, and not all MZ twins are MC. Fifteen percent of DC twins are MZ; of the MZ twins 33% are DCDA, 66% MCDA, and 1% MCMA. Consequently, this adds a layer of complexity to cfDNA NIPT in twins, particularly in cases of DC twins which although most likely to be DZ and thus discordant for aneuploidy, may be MZ. Table 1 outlines the theoretical genetic risks in twin pregnancies.

VANISHING TWINS

An increasingly recognized problem of twinning is the "vanishing twin," whereby single twin demise is noted on ultrasound prior to 14 weeks' gestation. In such cases, the "vanishing twin" may be "reabsorbed" and difficult to visualize on ultrasound later in pregnancy. A vanishing twin has little consequence to the surviving cotwin. It may though have consequences for false positive rates in cfDNA NIPT (of a presumed singleton pregnancy).

HIGHER ORDER MULTIPLE PREGNANCIES

There is little data from cohort studies on cfDNA NIPT in higher order multiples (triplets, quadruplets, etc.), including the effect that selective reduction has on cfDNA NIPT results of the surviving fetuses [24]. Therefore, conclusions on test accuracy in higher order multiple pregnancies cannot be confidently drawn and are not covered in this chapter.

Table 1 Genetic Risks in Twin Pregnancies, Based on One-Third of Twins Being Monozygous

	Chromosome Abnormality	X-Linked (Fetal Sexing Only)	X-Linked (Specific Diagnostic Test)	Autosomal Recessive
Risk for singleton pregnancy	Y[a]	1/2	1/4	1/4
Risk of at least one twin being affected	5/3 Y	2/3	3/8	3/8
Risk of both twins being affected	~1/3 Y	1/3	1/8	1/8

[a]*Risk Y will vary with maternal age, family history, etc.*
Table amended from Harper P. Twins and prenatal diagnosis. In: Practical Genetic Counselling. 7th ed: CRC Press; 2010. p. 132.

ROLE OF CELL-FREE DNA NIPT IN TWIN PREGNANCIES

The majority of research on cfDNA NIPT in twin pregnancies has investigated detecting aneuploidy, with few studies looking at fetal Rhesus status, or fetal sexing. The focus of this section will be on autosomal and sex chromosome aneuploidy testing, but we will briefly cover the other areas as well. Again, it is essential to emphasize that the clinical efficacy has to be assessed separately for both DC and MC twin pregnancies.

CELL-FREE DNA NIPT FOR FETAL ANEUPLOIDY IN TWIN PREGNANCIES
TRISOMY 21 TEST ACCURACY

A systematic review by Gil et al., performed in 2015, on cfDNA NIPT for detecting aneuploidy demonstrated a lower detection rate (DR) and higher false positive rate (FPR) for detecting trisomy 21 in twins overall (DR: 93.7% [95%CI 83.6–99.2], FPR: 0.23% [95%CI 0.00–0.92], 5 studies, 430 tests, $I^2 = 0.0\%$) compared to singletons (DR: 99.2% [95%CI 98.5–99.6], FPR: 0.09% [95%CI 0.05–0.14], 24 studies, 22,659 tests, $I^2 = 0.0\%$) [25]. The same authors updated their systematic review 12 months later and revealed a higher DR and lower FPR in twins when testing for trisomy 21 (DR: 100% [95% CI 95.2–100], FPR: 0% [95%CI 0–0.003], 5 studies, 1135 tests, $I^2 = 0\%$) compared to their previous systematic review results, despite the short time period between reviews [26]. All studies performed cfDNA NIPT after 10 weeks gestation, some were performed in "high-risk populations" [5,27,28], some in mixed-risk populations [29,30]. The updated review published in 2017 included 3 new studies, an additional 934 pregnancies [28–30]. The improved accuracy over the 12 months may be due to the increased number of tests included in the updated meta-analysis (1135 vs 430 tests), or be the result of technological advances and improved sequencing platforms. Another important difference in the 2017 review is that the authors changed their study inclusion criteria to exclude the 3 case-control studies, equating to 231 tests [31–33]. Case-control studies are less informative than cohort studies because they do not reflect the prevalence of the condition. Finally, the updated review also demonstrated an improved test accuracy in singleton pregnancies over the same time period: DR: 99.7% [95%CI 99.1–99.9], FPR: 0.04% [95%CI

0.02–0.08], 30 studies, 226,995 tests, $I^2 = 1.2\%$. This suggests improvements in cfDNA NIPT technology or bioinformatics most likely play an important role.

Since the publication of Gil's updated review, three additional cohort studies by Du (92 tests), Fosler (115 tests), and Le Conte (420 tests) [18,34,35] and one meta-analysis by Liao [36] have been published. The three new studies demonstrated high test accuracy in twin pregnancies, comparable to that of singleton pregnancies, with a sensitivity of 1.00 (95%CI 0.63, 1.00) and specificity of 1.00 (95% CI 0.99, 1.00) which equated to 1 false positive result in 627 tests. Fig. 1 displays the improvement in test accuracy seen across the updated systematic reviews by Gil, and the individual results of the three additional studies which we have included in "(3) Gil meta-analysis 2017 updated." Seven tests reported by Benachi [28] and included in Gil's updated analysis were also included in the 492 twin pregnancies described by Le Conte, therefore we have not included Benachi in the "(3) Gil meta-analysis 2017 updated" to avoid double counting. This study included 2 true positive results.

The meta-analysis by Liao included only one of the new studies [34] due to the search strategy being curtailed at July 1, 2016. For trisomy 21 it reported a sensitivity of 0.99 (95%CI 0.92, 1.00) and a specificity of 1.00 (95%CI 0.99, 1.00) based on 10 studies, equating to 2093 tests. However, it included case-control studies, and we cannot exclude the possibility of overlapping cohorts in 2 studies, potentially leading to double counting 323 participants, hence the results should be interpreted with caution.

The reviews by Gil and Liao demonstrate that the number of tests evaluated in twins is still substantially lower than in singletons (1135 and 2093 vs 226,995 tests, respectively). Even if the new 627 tests are added to results by Gil, the total number of trisomy 21 affected cases evaluated is relatively small (30/1762) and therefore the results from the twin cohort studies need to be interpreted with caution.

TRISOMY 18 AND TRISOMY 13 TEST ACCURACY (TABLES 2 AND 3)

Studies using cfDNA NIPT to screen for trisomy 18 and trisomy 13 involved even smaller cohorts. Only 3 studies in the review by Gil 2017 evaluated trisomy 18 [27–29], and 2 studies assessed trisomy 13 [28,29], therefore, a meta-analysis was not performed. Unfortunately, placental tissue was not available for analysis in the 2 false-negative trisomy 18 cases [27,29]. In the study by Huang [27] the false negative result occurred in a discordant MZ twin whereby karyotyping confirmed 1 normal twin and 1 with trisomy 18, whereas NIPT reported a normal pregnancy. The authors state that discordant MZ twins are rare, and that this may be due to "trisomic rescue" or a postzygotic event which will be discussed in the section on "Technical Issues With cfDNA NIPT for Aneuploidy Screening in Twins." The false negative result in Sarno [29] was also in a discordant twin pregnancy, although the authors state the pregnancy was DC, they do not report the zygosity. When Canick et al. [31] examined fetal fraction, they found that the lowest fetal fraction of a twin cohort ($n = 24$ tests) was in a DZ pregnancy discordant for trisomy 13. The three studies published later [18,34,35] reported lower test accuracy for detecting trisomy 18 and trisomy 13, as compared to trisomy 21, but the results are difficult to interpret due to the low number of cases examined. Two of the three studies each reported 1 case of trisomy 18 [34,35] both of which were accurately detected by cfDNA NIPT. The study by Du [18] reported 1 false positive result for trisomy 13 and had no trisomy 13 cases in the cohort, and Le Conte [35], who also included the samples from Benachi [28] reported one true positive result for trisomy 13 and no false positive or false negative results. The review by Liao finally also reported lower test accuracy for trisomy 18 with a sensitivity of 0.85 (95%CI 0.55, 0.98) and specificity of 1.00 (95%CI

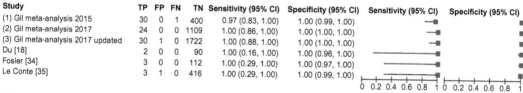

Study	TP	FP	FN	TN	Sensitivity (95% CI)	Specificity (95% CI)	Sensitivity (95% CI)	Specificity (95% CI)
(1) Gil meta-analysis 2015	30	0	1	400	0.97 (0.83, 1.00)	1.00 (0.99, 1.00)		
(2) Gil meta-analysis 2017	24	0	0	1109	1.00 (0.86, 1.00)	1.00 (1.00, 1.00)		
(3) Gil meta-analysis 2017 updated	30	1	0	1722	1.00 (0.88, 1.00)	1.00 (1.00, 1.00)		
Du [18]	2	0	0	90	1.00 (0.16, 1.00)	1.00 (0.96, 1.00)		
Fosler [34]	3	0	0	112	1.00 (0.29, 1.00)	1.00 (0.97, 1.00)		
Le Conte [35]	3	1	0	416	1.00 (0.29, 1.00)	1.00 (0.99, 1.00)		

FIG. 1

Forest plot of cfDNA NIPT for trisomy 21 in twin pregnancies.

0.99, 1.00) based on 5 studies, equating to 1167 tests, 13 cases of trisomy 18. The results of their trisomy 13 meta-analysis reported a sensitivity of 1.00 (95%CI 0.30, 1.00) and specificity of 1.00 (95%CI 0.99, 1.00) based on 3 studies, equating to 605 tests, 3 cases of trisomy 13.

In singletons cfDNA-based prenatal screening for trisomy 18 and trisomy 13 is less accurate than for trisomy 21. The additional complexities of twin pregnancies add to this challenge.

SEX CHROMOSOME ANEUPLOIDY

As far as the authors are aware, no studies have specifically addressed sex chromosome aneuploidy in twin pregnancies, nor do commercial companies offer this in DC twin pregnancies.

DETERMINING ZYGOSITY BY CELL-FREE DNA NIPT

In 2013 a small proof-of-principle study by Qu et al. demonstrated that it may be possible to determine zygosity using cfDNA NIPT [37]. The authors studied maternal blood samples from eight twin pregnancies (4 DCDA, 3 MCDA, 1 MCMA) at 19–30 weeks' gestation. They estimated the separate fetal fractions of each twin by genotyping 900,000 single nucleotide polymorphisms (SNPs) using massively parallel shotgun sequencing (MPSS) and calculated the concordance rates between the blocks of SNPs. Comparison with cord blood samples taken from each twin after delivery confirmed that the MCDA and MCMA twins were MZ with a genotype concordance rate of >99.6%, and that all the

Author	Total Tests in Cohort	T18 Cases	Sensitivity (95%CI)	Specificity (95%CI)
Table 2 cfDNA NIPT Accuracy for Trisomy 18 in Twin Pregnancies in Published Cohort Studies (Case-Control Studies Excluded)				
Huang [27]	187	2	0.50 (0.01, 0.99)	1.00 (0.98, 1.00)
Benachi [28][a]	7	0	Not estimable	1.00 (0.59, 1.00)
Sarno [29]	417	4	0.75 (0.19, 0.99)	1.00 (0.99, 1.00)
Fosler [34]	115	1	1.00 (0.03, 1.00)	1.00 (0.97, 1.00)
Le Conte [35]	420	1	1.00 (0.03, 1.00)	1.00 (0.99, 1.00)
Du [18]	92	0	Not estimable	1.00 (0.96, 1.00)

[a]These seven tests are also included in Le Conte [35].

Table 3 cfDNA NIPT Accuracy for Trisomy 13 in Twin Pregnancies in Published Cohort Studies (Case-Control Studies Excluded)

Author	Total Tests in Cohort	T13 Cases	Sensitivity (95%CI)	Specificity (95%CI)
Benachi [28][a]	7	0	Not estimable	1.00 (0.59, 1.00)
Sarno [29]	417	1	0.00 (0.00, 0.97)	1.00 (0.99, 1.00)
Fosler [34]	115	0	Not estimable	1.00 (0.97, 1.00)
Le Conte [35]	420	1	1.00 (0.03, 1.00)	1.00 (0.99, 1.00)
Du [18]	92	0	Not estimable	0.99 (0.94, 1.00)

[a]These seven tests are also included in Le Conte [35].

DCDA twins were DZ as the genotype concordance rate was 75.6%–80%. Unfortunately, the study did not include any DCDA twins which were MZ. In October 2017, Natera announced to have developed the first antenatal cfDNA NIPT able to identify zygosity; however, the associated test accuracy and validation data for the test is yet to be published [38].

TECHNICAL ISSUES WITH CELL-FREE DNA NIPT FOR ANEUPLOIDY SCREENING IN TWIN PREGNANCIES

The total fraction of cfDNA in twin pregnancies is higher than in singletons [31]. As twin pregnancies have a higher placental mass, and cfDNA originates from trophoblasts this has biological plausibility. The same applies to the theory that there is an increased feto-maternal interface in twin pregnancies that consequently increases the transfer of cfDNA into maternal circulation [39]. It was initially thought that each twin contributes an equal quantity of cfDNA, but this has since been demonstrated not to be true. The contribution of cfDNA from DZ twins can differ almost twofold [37,40]. This means that although the total fetal fraction may appear normal, the fetal fraction of the affected fetus may be below the threshold for detection at a given sequencing depth, and consequently the affected twin may be missed. Bevilacqua et al. [41] sent cfDNA samples of two women to two laboratories in Belgium and two outside of Belgium. One sample was from a pregnant woman with a DC twin pregnancy at 17 weeks of gestation where 1 fetus had trisomy 21. The trisomy 21 was not detected by any of the laboratories, despite total fetal fractions of 10% and 16.4% being reported, although the breakdown from each twin was not determined. The authors hypothesize that there may have been a significant technical challenge with this sample or that the trisomic fetus may have been underrepresented. Therefore, some advocate the use of the lower fetal fraction in the twin pregnancy to be used for aneuploidy assessment [33,42], thereby reducing the rate of false negative results, but subsequently increasing the rate of failed/no-result tests [43]. However, the systematic review by Liao et al. found no difference in test accuracy between studies that did and did not have a fetal fraction cutoff [36].

The rate of failed tests for cfDNA NIPT in twin pregnancies for aneuploidy varies from 0% [27] to 9.4% [29]. The main reason studies cite for test failure is a low fetal fraction and the variation in test failure may be a reflection of whether or not the lowest twin's fetal fraction is used [33,42]. Another significant predictor of test failure in twin as well as singleton pregnancies is high maternal weight [29,35]. This is due to accelerated turnover of maternal adipocytes which release increased amounts of *maternal* cell-free DNA into the circulation and dilute the placental DNA [43]. Finally, in vitro fertilization (IVF) conception is also associated with an increased test failure rate due to low fetal fraction [29]. This may be related to lower placental mass and function in IVF pregnancies, and thus lower levels of cfDNA, as reflected by lower pregnancy-associated plasma protein A (PAPP-A) levels in IVF conceived singleton pregnancies. PAPP-A levels are an indirect marker of placental mass and function [29,44,45].

As explained previously, zygosity has important implications: DZ fetuses are more likely to be discordant for a chromosome anomaly than MZ twins. If fetuses are concordant for a chromosome anomaly, they both will contribute to the trisomic fetal fraction and this will improve the detection rate. If fetuses are discordant for a given anomaly there is an elevated risk that the fetal fraction of the abnormal fetus will fall below the threshold for detection. In twin pregnancies, we therefore theoretically expect a higher detection rate in MZ twins as opposed to DZ twins. Interestingly, no studies have looked at the effect of zygosity on the accuracy of cfDNA NIPT for aneuploidy. The following inaccurate cfDNA NIPT results in twin pregnancies have been published: of the four false negative cases occurring in twins discordant for aneuploidy, three were DC [29,33] and one was MC [27]. Of the three false positive results, one was a DC unlike sex and hence DZ twin pregnancy [18], one was MC with no placental tissue available for analysis [35], and one inaccurate result was not discussed in detail [29]. Even when MZ twins are viewed as identical, this may be considered an oversimplification as there are various causes of discordance in MZ twins, including: unequal blastomere allocation, anaphase lagging, post-zygotic chromosomal nondisjunction, point mutations in a single gene, trisomy rescue, as well as epigenetic changes [14].

Studies stratifying the effect of chorionicity are important and informative. The study by Lau et al. on cfDNA NIPT with massively parallel sequencing (MPS) showed higher median cfDNA concentrations in MC than in DC twins, with MC twins having a concentration similar to that of singleton pregnancies [29]. In theory this should make cfDNA NIPT in MC twins as efficient, if not better than in singletons [5]. However, it is difficult to assess the accuracy of cfDNA NIPT in MC twin pregnancies specifically, because there have been very few aneuploidy cases tested in MC twins [36].

An added layer of complexity in cfDNA NIPT for aneuploidy for twin pregnancy is that even if it would be possible to determine from the cfDNA NIPT result that the fetuses are discordant for the anomaly, unless there are ultrasound markers, it would not be possible to know which fetus is affected [40]. And finally, reasons for inaccurate cfDNA NIPT results in singleton pregnancies are also applicable to twin pregnancies: confined placental mosaicism (CPM), maternal chromosomal abnormalities, maternal malignancy, vanishing twins [41]. It is important to state though that despite these technical issues and the dearth of research, the accuracy of cfDNA NIPT in twin pregnancies is still higher than the accuracy of conventional aneuploidy screening. A systematic review found that the first trimester combined test in twins, with a cutoff risk of 1:100–1:300, had a pooled sensitivity of 0.893 (95%CI 0.797, 0.947), and a pooled specificity of 0.946 (95%CI 0.933, 0.957) based on 5 studies, which equated to 6397 tests [7]. The DR of trisomy 21 with a 1:150 risk cutoff using the second trimester "quad" test in MC twin pregnancies is 80% which is comparable to singletons, but in DC twin pregnancies the detection rate of the "quad" test is only 30%–50% [46]. Both types of twins having a 3% screen positive rate using the "quad" test, this means that invasive testing would be recommended in 3% of patients tested.

MANAGEMENT FOLLOWING SUSPECTED/HIGH RISK FOR ANEUPLOIDY CELL-FREE DNA NIPT RESULT IN TWIN PREGNANCIES

Currently, following a suspected/high risk for aneuploidy cfDNA NIPT result, parents are offered invasive diagnostic testing either CVS or amniocentesis, depending amongst others on gestational age. The call for a first trimester scan in patients prior to cfDNA NIPT testing [47] has particular relevance for twin pregnancies as ultrasonographic determination of chorionicity is vital for antenatal care [21,23] and important for cfDNA NIPT result interpretation. In addition, the first trimester ultrasound scan may be used to exclude structural anomalies in the twin pregnancy, although the sensitivity for detection is relatively low [48]. If the twins have been categorized as MC on first trimester ultrasound scan, the high risk for aneuploidy is considered to apply to both fetuses because the twins will be MZ, although there are exceptions. If the twins are DC, the risk is *likely* to only apply to one of the fetuses but here as well there are exceptions. Without knowledge of the zygosity, it is not possible to know whether a set of same sex DC twins are DZ or MZ. Fetal sex can only be accurately determined by ultrasound from 17 weeks' gestation. It is important that chorionicity and amnionicity are accurately documented in the first trimester because in MC twins concordant for growth and anatomy, one might choose to sample only one amniotic fluid sac at amniocentesis, while in DC twins or those with uncertain chorionicity, both amniotic fluid sacs need to be sampled [22]. If the results of invasive testing show that both twins are affected, the couple may decide to terminate the whole pregnancy. If the results of invasive testing demonstrate that one twin is affected, the couple may decide to have a selective termination of pregnancy. Selective termination of pregnancy in DC twins is performed by intracardiac injection of potassium chloride into the affected twin and carries a risk of up to 9.6% of demise of the nonaffected cotwin, depending on gestational age [49,50]. Selective termination of pregnancy in the rare discordant MC pregnancy cannot be performed by intracardiac or intravascular injections because of intertwin vascular anastomoses in the placenta. In this case the procedure is performed by ultrasound-directed radiofrequency ablation of the umbilical cord in the first and early second trimester, and bipolar cord diathermy for more advanced gestational ages. This procedure has a risk of preterm labor, and a 10.6%–14.7% risk of demise of the nonaffected cotwin [51]. Neurologic damage in the nonaffected surviving cotwin has been described in 4.5%–8% [51,52].

COUNSELING IN TWIN PREGNANCIES

Parents should be aware that although accuracy of cfDNA NIPT in twin pregnancy is improving, the numbers are still low and there is less evidence for accuracy than in singleton pregnancies. As with singleton pregnancies it should be emphasized that cfDNA NIPT is not a diagnostic test and confirmation via invasive testing is required before definitive conclusions can be drawn. The risk of miscarriage following invasive testing in twins is higher than in singletons and this may potentially involve one or two unaffected fetuses. Parents also need to understand the risks associated with selective feticide if that is a choice they are considering. The role of cfDNA NIPT for aneuploidy in singleton pregnancies in the UK is currently being assessed by evaluative implementation as a contingent screening test for those with a risk from conventional screening higher than 1:150. Twin pregnancies are not included in this evaluation [53]. Some argue NIPT should be used as a first-line screening test, given that its accuracy is higher than conventional "combined" or "quad" testing screening.

NIPT FOR RHESUS D STATUS IN TWIN PREGNANCIES

To the authors' knowledge, there have been no studies looking specifically at fetal Rhesus D status and NIPT in twin pregnancies. The use of cfDNA to determine fetal Rhesus D (*RHD*) status or other blood group types depends on country and healthcare service (see Chapter 8 for more information on Fetal Blood Group Typing). In current UK clinical practice, cfDNA-based determination of fetal Rhesus D status is considered a diagnostic test and is available on the National Health Service (NHS) for women with a history of Rhesus disease [54]. The method used is real-time quantitative polymerase chain reaction (PCR) amplification. Another application of cfDNA-based testing is fetal *RHD* screening in RhD negative women to allow targeted administration of anti-D prophylaxis to Rhesus D negative women who are pregnant with a *RHD* positive fetus. While there have been several large studies in singletons demonstrating high test accuracy, there have been relatively few twin pregnancies tested. In theory, the test accuracy of fetal RhD status in twin pregnancies should be the same as in singletons because of the way the test works. If cfDNA testing is performed in women known to be Rhesus D negative, the presence of *RHD* positive genes in the cfDNA sample is considered diagnostic as to the presence of at least one *RHD* positive fetus. MC twins are MZ and have concordant Rhesus D statuses. In the case of a positive cfDNA result in DC twins, it is not known whether both fetuses are *RHD* positive, or just one. Irrespective, the treatment is the same: a RhD negative woman will need antenatal prophylactic anti-D even if only one twin is *RHD* positive. Studies that have reported test accuracy in twin pregnancies have reported high test accuracies of 100% sensitivity and 100% specificity for diagnosing at least one fetus as Rhesus D positive, or both fetuses as Rhesus D negative ($n=56$ tests) [55–58], including in the first trimester of pregnancy ($n=62$ tests) [59]. Indeed, cfDNA may be considered superior to invasive diagnostic testing as demonstrated by a case in which the cfDNA NIPT demonstrated the presence of at least one *RHD* positive fetus in a RhD negative mother who then underwent technically difficult amniocentesis in which the results indicated that both fetuses were RhD negative; however, neonatal serology confirmed that one twin was RhD positive, leading the authors to believe that the same amniotic sac had been sampled twice [54].

CELL-FREE DNA NIPT FOR FETAL SEXING FOR MEDICAL REASONS IN TWIN PREGNANCIES

To the authors' knowledge, there have been no studies looking specifically at fetal sexing for medical reasons using cfDNA NIPT in twin pregnancies. In the same way cfDNA NIPT works for fetal Rhesus D status, the diagnosis of fetal sex is made based on the presence of non-maternal genes in this case from the Y chromosome. Therefore, it is possible to tell if there is at least one male fetus or two female fetuses [55]. In DC twins, it is not possible to tell if it is one male or two male fetuses. A small study ($n=55$ twin pregnancies) by Smid [60] using real-time quantitative PCR reported that the signal was significantly stronger in DC twin pregnancies with two male fetuses, as compared to singleton male fetus pregnancies. However, they reported no differences between MC twins with two male fetuses,

DC twins with two male fetuses, or DC twins with one male and one female fetus. Similar findings have also been echoed by Attilakos who evaluated 65 twin pregnancies [39].

A SUMMARY OF PROFESSIONAL BODIES' RECOMMENDATIONS ON CELL-FREE DNA TESTING IN TWIN PREGNANCIES

There are no specific guidelines for cfDNA NIPT in twins for fetal Rhesus status or fetal sexing. With regards to aneuploidy, the professional societies generally advise caution (Table 4). The UK's Royal College of Obstetricians and Gynaecologists (RCOG) published a scientific opinion paper in 2014 and stated that there are several unanswered questions and additional complexities when cfDNA NIPT is performed in multiple pregnancies [61]. In 2015 the American College of Obstetricians and Gynecologists (ACOG) advised that "cell-free DNA screening is not recommended for women with multiple gestations" due to the paucity of data [62]. The Society of Obstetricians and Gynecologists of Canada (SOGC) updated their guidance in September 2017 and stated that 'there is less validation data for using cfDNA for aneuploidy screening in twin pregnancies compared to singletons, therefore it should be undertaken with caution' [63]. The International Society for Prenatal Diagnosis (ISPD) stated in 2015 that the accuracy of cfDNA NIPT for aneuploidy in twin pregnancies is similar to that in singletons [64].

The majority of research papers assessing cfDNA-based prenatal screening for aneuploidy testing in twin pregnancies state that more data are needed before it can be offered as a routine clinical service [18,31,41,65]. The authors of the recent study by Le Conte state it can be used in the second trimester, but more data are needed to support its use in the first trimester [35].

Table 4 Opinions of Professional Bodies on the Use of cfDNA NIPT for Aneuploidy Screening in Twin Pregnancies

Professional Body	Year of Latest Publication	Recommendation/Opinion
Royal College of Obstetricians and Gynaecologists (RCOG)	2016	There are several unanswered questions and additional complexities when performed in multiple pregnancies. The RCOG Green top Guideline 2016 indicates that NIPT shows promise in MC twins but data based on very small number.
American College of Obstetricians and Gynecologists (ACOG)	2015	"Cell-free DNA screening is not recommended for women with multiple gestations" due to the paucity of data
Society of Obstetricians and Gynecologists of Canada (SOGC)	2017	"There is less validation data for using cfDNA for aneuploidy screening in twin pregnancies compared to singletons, therefore it should be undertaken as caution"
International Society for Prenatal Diagnosis (ISPD)	2015	The accuracy of cfDNA NIPT for aneuploidy in twin pregnancies is similar to that in singletons

CONCLUSION

The accuracy of cfDNA-based prenatal screening for aneuploidy, fetal Rhesus D status, and fetal sex appears comparable in twin pregnancies and singleton pregnancies but substantially fewer tests have been performed in twin pregnancies to underpin this claim, especially in DC twin pregnancies. This and other issues complicating any prenatal testing in twin pregnancies should be discussed with prospective parents by a professional familiar with the specific issues.

REFERENCES

[1] ONS. Birth Characteristics in England and Wales: 2015, London: ONS; 2016. Available from: https://www.ons.gov.uk/peoplepopulationandcommunity/birthsdeathsandmarriages/livebirths/bulletins/birthcharacteristicsinenglandandwales/2015.

[2] Glinianaia S, Rankin J, Sturgiss S, Ward Platt M, Crowder D, Bell R. The north of England survey of twin and multiple pregnancy. Twin Res Hum Genet 2013;16(1):112–6.

[3] Hansen M, Kurinczuk JJ, Milne E, de Klerk N, Bower C. Assisted reproductive technology and birth defects: a systematic review and meta-analysis. Hum Reprod Update 2013;19(4):330–53.

[4] Agarwal K, Alfirevic Z. Pregnancy loss after chorionic villus sampling and genetic amniocentesis in twin pregnancies: a systematic review. Ultrasound Obstet Gynecol 2012;40(2):128–34.

[5] Lau TK, Jiang F, Chan MK, Zhang H, Salome Lo PS, Wang W. Non-invasive prenatal screening of fetal down syndrome by maternal plasma DNA sequencing in twin pregnancies. J Matern Fetal Neonatal Med 2013;26(4):434–7.

[6] Spencer K. Screening for trisomy 21 in twin pregnancies in the first trimester using free β-hCG and PAPP-A, combined with fetal nuchal translucency thickness. Prenat Diagn 2000;20(2):91–5.

[7] Prats P, Rodríguez I, Comas C, Puerto B. Systematic review of screening for trisomy 21 in twin pregnancies in first trimester combining nuchal translucency and biochemical markers: a meta-analysis. Prenat Diagn 2014;34(11):1077–83.

[8] Vink J, Wapner R, D'Alton ME. Prenatal diagnosis in twin gestations. Semin Perinatol 2012;36(3):169–74.

[9] Cleary-Goldman J, Berkowitz RL. First trimester screening for down syndrome in multiple pregnancy. Semin Perinatol 2005;29(6):395–400.

[10] Gjerris AC, Tabor A, Loft A, Christiansen M, Pinborg A. First trimester prenatal screening among women pregnant after IVF/ICSI. Hum Reprod Update 2012;18(4):350–9.

[11] Bellver J, Casanova C, Garrido N, Lara C, Remohí J, Pellicer A, et al. Additive effect of factors related to assisted conception on the reduction of maternal serum pregnancy-associated plasma protein A concentrations and the increased false-positive rates in first-trimester Down syndrome screening. Fertil Steril 2013;100(5):1314–20.

[12] Egan E, Reidy K, O'Brien L, Erwin R, Umstad M. The outcome of twin pregnancies discordant for trisomy 21. Twin Res Hum Genet 2013;17(1):38–44.

[13] Chang Y, Yi W, Chao A, Chen K, Cheng P, Wang T, et al. Monozygotic twins discordant for trisomy 21: discussion of etiological events involved. Taiwan J Obstet Gynecol 2017;56:681–5.

[14] Machin G. Non-identical monozygotic twins, intermediate twin types, zygosity testing, and the non-random nature of monozygotic twinning: a review. Am J Med Genet C Semin Med Genet 2009;151C(2):110–27.

[15] Choi SA, Ko JM, Shin CH, Yang SW, Choi JS, Oh SK. Monozygotic twin discordant for down syndrome: mos 47,XX,+21/46,XX and 46,XX. Eur J Pediatr 2013;172(8):1117–20.

[16] Ramsey KW, Slavin TP, Graham G, Hirata GI, Balaraman V, Seaver LH. Monozygotic twins discordant for trisomy 13. J Perinatol 2012;32(4):306–8.

[17] Reuss A, Gerlach H, Bedow W, Landt S, Kuhn U, Stein A, et al. Monozygotic twins discordant for trisomy 18. Ultrasound Obstet Gynecol 2011;38(6):727–8.

[18] Du E, Feng C, Cao Y, Yao Y, Lu J, Zhang Y. Massively parallel sequencing (MPS) of cell-free fetal DNA (cffDNA) for trisomies 21, 18, and 13 in twin pregnancies. Twin Res Hum Genet 2017;20(3):242–9.

[19] McFadden P, Smithson S, Massaro R, Huang J, Prado G, Shertz W. Monozygotic twins discordant for trisomy 13. Pediatr Dev Pathol 2017;20(4):340–7.

[20] Sepulveda W, Sebire N, Hughes K, Odibo A, KH N. The lambda sign at 10–14 weeks of gestation as a predictor of chorionicity in twin pregnancies. Ultrasound Obstet Gynecol 1996;7:421–3.

[21] NICE. Multiple pregnancy. The management of twin and triplet pregnancies in the antenatal period. NICE clinical guideline 129. Excellence NIfHaC. Manchester: NICE; 2011.

[22] Kilby M, Bricker L. RCOG green-top guideline No. 51: management of monochorionic twin pregnancy. BJOG 2016;124:e1–e45.

[23] Khalil A, Rodgers M, Baschat A, Bhide A, Gratacos E, Hecher K, et al. ISUOG practice guidelines: role of ultrasound in twin pregnancy. Ultrasound Obstet Gynecol 2016;47(2):247–63.

[24] Xanthopoulou L, Shaw A, Wyatt M, Kotzadamis A, Golubeva A, Pissaridou S, et al. Fetal reduction after multiple pregnancies: implications and considerations for noninvasive prenatal testing. Prenat Diagn 2016;36(Supp 1):75.

[25] Gil M, Quezada M, Revello R, Akolekar R, Nicolaides K. Analysis of cell-free DNA in maternal blood in screening for fetal aneuploidies: updated meta-analysis. Ultrasound Obstet Gynecol 2015;45:249–66.

[26] Gil MM, Accurti V, Santacruz B, Plana MN, Nicolaides KH. Analysis of cell-free DNA in maternal blood in screening for aneuploidies: updated meta-analysis. Ultrasound Obstet Gynecol 2017;50(3):302–14.

[27] Huang X, Zheng J, Chen M, Zhao Y, Zhang C, Liu L, et al. Noninvasive prenatal testing of trisomies 21 and 18 by massively parallel sequencing of maternal plasma DNA in twin pregnancies. Prenat Diagn 2014;34(4):335–40.

[28] Benachi A, Letourneau A, Kleinfinger P, Senat M-V, Gautier E, Favre R, et al. Cell-free DNA analysis in maternal plasma in cases of fetal abnormalities detected on ultrasound examination. Obstet Gynecol 2015;125(6):1330–7.

[29] Sarno L, Revello R, Hanson E, Akolekar R, Nicolaides KH. Prospective first-trimester screening for trisomies by cell-free DNA testing of maternal blood in twin pregnancy. Ultrasound Obstet Gynecol 2016;47 (6):705–11.

[30] Tan Y, Gao Y, Lin G, Fu M, Li X, Yin X, et al. Noninvasive prenatal testing (NIPT) in twin pregnancies with treatment of assisted reproductive techniques (ART) in a single center. Prenat Diagn 2016;36(7):672–9.

[31] Canick J, Kloza E, GM L-M, Haddow J, Ehrich J, Ehrich M, et al. DNA sequencing of maternal plasma to identify down syndrome and other trisomies in multiple gestations. Prenat Diagn 2012;32:730–4.

[32] Grömminger S, Yagmur E, Erkan S, Nagy S, Schöck U, Bonnet J, et al. Fetal aneuploidy detection by cell-free DNA sequencing for multiple pregnancies and quality issues with vanishing twins. J Clin Forensic Med 2014;3(3):679.

[33] Gil MM, Quezada MS, Bregant B, Syngelaki A, Nicolaides KH. Cell-free DNA analysis for trisomy risk assessment in first-trimester twin pregnancies. Fetal Diagn Ther 2014;35(3):204–11.

[34] Fosler L, Winters P, Jones KW, Curnow KJ, Sehnert AJ, Bhatt S, et al. Aneuploidy screening by non-invasive prenatal testing in twin pregnancy. Ultrasound Obstet Gynecol 2017;49(4):470–7.

[35] Le Conte G, Letourneau A, Jani J, Kleinfinger P, Lohmann L, Costa J-M, et al. Cell-free fetal DNA analysis in maternal plasma as a screening test for trisomy 21, 18 and 13 in twin pregnancies. Ultrasound Obstet Gynecol 2017; [epub ahead of print].

[36] Liao H, Liu S, Wang H. Performance of non-invasive prenatal screening for fetal aneuploidy in twin pregnancies: a meta-analysis. Prenat Diagn 2017;37:874–82.

[37] Qu J, Leung T, Jiang P, Liao G, Cheng Y, Sun H, et al. Noninvasive prenatal determination of twin zygosity by maternal plasma DNA analysis. Clin Chem 2013;59:427–35.

[38] Natera. Natera's Panorama non-invasive prenatal test now available for screening twin pregnancies. California: Natera; 2017 [press release].

[39] Attilakos G, Maddocks D, Davies T, Hunt L, Avent N, Soothill P, et al. Quantification of free fetal DNA in multiple pregnancies and relationship with chorionicity. Prenat Diagn 2011;31:967–72.

[40] Leung TY, Qu JZZ, Liao GJW, Jiang P, Cheng YKY, Chan KCA, et al. Noninvasive twin zygosity assessment and aneuploidy detection by maternal plasma DNA sequencing. Prenat Diagn 2013;33(7):675–81.

[41] Bevilacqua E, Guizani M, Cos Sanchez T, Jani JC. Concerns with performance of screening for aneuploidy by cell-free DNA analysis of maternal blood in twin pregnancy. Ultrasound Obstet Gynecol 2016;47(1):124–5.

[42] Struble C, Syngelaki A, Oliphant A, Song K, Nicolaides K. Fetal fraction estimate in twin pregnancies using directed cell-free DNA analysis. Fetal Diagn Ther 2014;35:199–203.

[43] Bevilacqua E, Gil MM, Nicolaides KH, Ordoñez E, Cirigliano V, Dierickx H, et al. Performance of screening for aneuploidies by cell-free DNA analysis of maternal blood in twin pregnancies. Ultrasound Obstet Gynecol 2015;45(1):61–6.

[44] Amor DJ, Xu JX, Halliday JL, Francis I, Healy DL, Breheny S, et al. Pregnancies conceived using assisted reproductive technologies (ART) have low levels of pregnancy-associated plasma protein-a (PAPP-A) leading to a high rate of false-positive results in first trimester screening for down syndrome. Hum Reprod 2009;24(6):1330–8.

[45] Kagan KO, Wright D, Spencer K, Molina FS, Nicolaides KH. First-trimester screening for trisomy 21 by free beta-human chorionic gonadotropin and pregnancy-associated plasma protein-A: impact of maternal and pregnancy characteristics. Ultrasound Obstet Gynecol 2008;31(5):493–502.

[46] Public Health England. Fetal anomaly screening programme: standards. Available from: https://www.gov.uk/government/publications/fetal-anomaly-screening-programme-standards; 2015.

[47] Alfirevic Z, Bilardo C, Salomon L, Tabor A. Women who choose cell-free DNA testing should not be denied first-trimester anatomy scan. BJOG 2017;124(8):1159–61.

[48] D'Antonio F, Familiari A, Thilaganathan B, Papageorghiou A, Manzoli L, Khalil A, et al. Sensitivity of first-trimester ultrasound in the detection of congenital anomalies in twin pregnancies: population study and systematic review. Acta Obstet Gynecol Scand 2016;95(12):1359–67.

[49] Bigelow CA, Factor SH, Moshier E, Bianco A, Eddleman KA, Stone JL. Timing of and outcomes after selective termination of anomalous fetuses in dichorionic twin pregnancies. Prenat Diagn 2014;34(13):1320–5.

[50] Dural O, Yasa C, Kalelioglu I, Can S, Yılmaz G, Corbacioglu A, et al. Comparison of perinatal outcomes of selective termination in dichorionic twin pregnancies performed at different gestational ages. J Matern Fetal Neonatal Med 2017;30(12):1388–92.

[51] Gaerty K, Greer RM, Kumar S. Systematic review and meta-analysis of perinatal outcomes after radiofrequency ablation and bipolar cord occlusion in monochorionic pregnancies. Am J Obstet Gynecol 2015;213(5):637–43.

[52] van Klink J, Koopman H, Middeldorp J, Klumper F, Rijken M, Oepkes D, et al. Long-term neurodevelopmental outcome after selective feticide in monochorionic pregnancies. BJOG 2015;1517–24.

[53] UKNSC. UK NSC non-invasive prenatal testing (NIPT) recommendation. London: UKNSC; 2016. Available from: http://legacy.screening.nhs.uk/screening-recommendations.php [press release].

[54] Finning K, Martin P, Soothill P, Avent N. Prediction of fetal D status from maternal plasma: introduction of a new noninvasive fetal RHD genotyping service. Transfusion 2002;42(8):1079–85.

[55] Minon JM, Gerard C, Senterre JM, Schaaps JP, Foidart JM. Routine fetal RHD genotyping with maternal plasma: a four-year experience in Belgium. Transfusion 2008;48(2):373–81.

[56] Müller S, Bartels I, Stein W, Emons G, Gutensohn K, Köhler M, et al. The determination of the fetal D status from maternal plasma for decision making on Rh prophylaxis is feasible. Transfusion 2008;48(11):2292–301.

[57] Macher HC, Noguerol P, Medrano-Campillo P, Garrido-Marquez MR, Rubio-Calvo A, Carmona-Gonzalez M, et al. Standardization non-invasive fetal RHD and SRY determination into clinical routine using a new multiplex RT-PCR assay for fetal cell-free DNA in pregnant women plasma: results in clinical benefits and cost saving. Clin Chim Acta 2012;413(3–4):490–4.

[58] Scheffer P, van der Schoot C, Page-Christiaens G, de Haas M. Noninvasive fetal blood group genotyping of rhesus D, c, E and of K in alloimmunised pregnant women: evaluation of a 7-year clinical experience. BJOG 2011;118(11):1340–8.

[59] Wikman AT, Tiblad E, Karlsson A, Olsson ML, Westgren M, Reilly M. Noninvasive single-exon fetal RHD determination in a routine screening program in early pregnancy. Obstet Gynecol 2012;120(2):227–34.

[60] Smid M, Galbiati S, Vassallo A, Gambini D, Ferrari A, Restagno G, et al. Fetal DNA in maternal plasma in twin pregnancies. Clin Chem 2003;49(9):1526–8.

[61] Soothill P, Lo YMD. Non-invasive prenatal testing for chromosomal abnormality using maternal plasma DNA. London: RCOG; 2014.

[62] ACOG/SMFM. Cell-free DNA screening for fetal aneuploidy. Committee opinion no. 640. Obstet Gynecol 2015;126:e31–7.

[63] Audibert F, De Bie I, Johnson J-A, Okun N, Wilson RD, Armour C, et al. No. 348-joint SOGC-CCMG guideline: update on prenatal screening for fetal aneuploidy, fetal anomalies, and adverse pregnancy outcomes. J Obstet Gynaecol Can 2017;39(9):805–17.

[64] ISPD. Position statement from the chromosome abnormality screening committee on behalf of the board of the International Society for Prenatal Diagnosis. Charlottesville, VA: International Society for Prenatal Diagnosis; 2015. Available from: http://www.ispdhome.org/public/news/2015/PositionStatementFinal04082015.pdf.

[65] Shaw S, Chen C-Y, Hsiao C-H, Ren Y, Tian F, Tsai C. Non-invasive prenatal testing for whole fetal chromosome aneuploidies: a multi-center prospective cohort trial in Taiwan. Prenat Diagn 2013;33(Suppl):81.

GENOME-WIDE CELL-FREE DNA-BASED PRENATAL TESTING FOR RARE AUTOSOMAL TRISOMIES AND SUBCHROMOSOMAL ABNORMALITIES

Mark D. Pertile

Victorian Clinical Genetics Services, Murdoch Children's Research Institute, Melbourne, VIC, Australia

Department of Paediatrics, University of Melbourne, Melbourne, VIC, Australia

INTRODUCTION

Cell-free DNA-based noninvasive prenatal testing (cfDNA-based NIPT) has revolutionized prenatal care since its validation as a highly sensitive and specific mode of prenatal screening [1–4]. Unparalleled detection rates of >98%–99% for trisomies 13, 18, and 21, coupled with exceedingly low false-positive rates (<0.1%) [5], mean cfDNA-based NIPT outperforms combined first trimester screening (CFTS) using maternal serum biochemical markers and nuchal translucency (NT) ultrasound measurement [6, 7].

Despite this superior performance, the current narrow focus of cfDNA-based prenatal testing, which targets only chromosomes 13, 18, 21, X, and Y, has led to concerns that this screening methodology may hinder, rather than enhance our ability to detect pathogenic chromosome disease [8–12]. This partly stems from the fact that CFTS is known to identify other "atypical" chromosome conditions that cannot be detected using standard methods of cfDNA-based NIPT (e.g., rare trisomy mosaicism, segmental copy number abnormalities, and other incidental findings); these conditions sometimes being associated with abnormal serum analytes and/or increased NT measurement [11, 13]. Also, for women who elect prenatal diagnosis, the widespread application of chromosome microarray (CMA) has made high-resolution genome analysis the norm, with marked improvements in diagnostic yields in very high-risk and average-risk pregnancies, when compared with conventional chromosome analysis [14, 15]. Lastly, prenatal testing in the era of modern genomic medicine can utilize whole genome sequencing (WGS) and whole exome sequencing (WES) to harvest vast amounts of genetic information at the single nucleotide level [16]. Thus our ability to diagnose genetic conditions during pregnancy has expanded far beyond the screening capability of standard cfDNA-based NIPT. So what can be done to help bridge this gap?

One approach has been to incorporate panels that target a small number of known microdeletions with clinically severe phenotypes [17]. With the exception of the 22q11.2 deletion syndrome however, these conditions are exceedingly rare. Their low prevalence in average screening risk populations results in poor positive predictive values (PPV), averaging 7.4% in one large clinical laboratory study reporting on cases received for prenatal diagnosis [18]. Clinical performance for the rarer microdeletions can be particularly poor [18–20]. A single-nucleotide polymorphism (SNP)-based NIPT used to screen >34,000 women for 5 recurrent microdeletions had a screen positive rate for the 15q11.2 microdeletion of 0.34% (1 in every 295 women tested). The deletion was confirmed in only one patient with known outcome; a PPV of 1.4% [19]. Enhancements to screening methodologies and bioinformatics algorithms can improve performance. The same SNP-based assay obtained higher PPVs and lower false-positive rates by increasing confidence thresholds and reflex sequencing putative deletions at higher read depths [19]. However, the problem of multiple hypothesis testing, where each individual targeted region contributes to a small, but cumulatively higher false-positive rate, is a weakness of all targeted microdeletion panels used for cfDNA-based prenatal testing [21]. Despite these reservations, the large numbers of women who opt for microdeletion screening, usually at increased cost, speaks loudly to the fact that these women are seeking more, rather than less genetic information about their pregnancies, and in a noninvasive manner.

An alternative strategy for increasing detection rates for a broad range of chromosome conditions is to implement a genome-wide screening approach which is the focus of this Chapter. Genome-wide cfDNA-based NIPT aims to analyze and report on all chromosomes. This screening modality is analogous to classical karyotyping, and more specifically mimics the copy number data obtained from chromosome microarray (CMA). At very high read depths and with sufficient fetal fraction, the technique can deliver screening at CMA-level resolution [22]. The gain or loss of genomic material is reflected in statistically significant changes in sequence read counts (tags) that are mapped to discrete bins distributed across the genome (see Chapter 3). Using this approach, whole chromosome and segmental aneuploidies can be identified for any chromosome [23–25], without the constraint of testing for a small number of known, recurrent conditions.

Although genome-wide cfDNA-based NIPT is not yet universally available, several groups have reported on their early clinical experience [21, 25–32]. One of the more common anomalies detected are the rare autosomal trisomies [26, 27, 31]. Their presence can be associated with miscarriage at earlier gestations [26], but they are more likely to represent confined placental mosaicism (CPM) in ongoing pregnancies, or occasionally true fetal mosaicism (TFM). Other pregnancy complications include uniparental disomy (UPD), intrauterine fetal growth restriction (IUGR), and fetal demise [25–27]. Pathogenic copy number variants (CNVs), larger segmental aneuploidies, and more complex structural chromosomal aberrations including unbalanced translocations can also be successfully identified using this approach [21, 25, 27–32].

This chapter reports on the benefits and limitations of genome-wide cfDNA-based NIPT and discusses the interpretation and management of these results. To achieve this, an understanding of the complexity of chromosomal mosaicism is first required. Not only is mosaicism an important consideration for the interpretation of rare autosomal trisomy results obtained during genome-wide cfDNA screening, but it also has relevance for segmental aneuploidies that arise from postfertilization mutation events.

HISTORICAL BACKGROUND
RARE AUTOSOMAL TRISOMY MOSAICISM DURING PREGNANCY AND AT BIRTH

Chromosomal mosaicism is the presence of two or more distinct cell lines in an individual [33]. In a prenatal setting, chromosomal mosaicism most commonly affects only the placenta (confined placental mosaicism; CPM), but may occasionally extend to the fetus (true fetal mosaicism; TFM). The clinical consequences of chromosomal mosaicism identified during prenatal diagnosis can be difficult to predict, ranging from no apparent phenotypic effect to early fetal lethality. In the absence of fetal anomalies, the outcome of TFM in a prenatal setting remains uncertain [34].

Autosomal trisomy is a common cause of early miscarriage. Of the 10%–15% of pregnancies that end in clinical miscarriage, about half will do so because of a chromosome abnormality, and of these, the majority will involve an autosomal trisomy [35]. With very rare exceptions, only trisomy for chromosomes 13, 18, and 21 (the so-called live birth trisomies) is compatible with survival to term, recognizing that even these conditions are associated with a high rate of miscarriage and stillbirth [36, 37]. All other autosomal trisomies (the so-called rare autosomal trisomies) are lethal in nonmosaic form, notwithstanding occasional reports of survival into the second, and very rarely the third trimester of pregnancy; stillbirth or neonatal death is expected. A large study reporting on the prevalence and types of rare chromosome abnormalities notified to 16 European congenital anomaly registers recorded 58 nonmosaic rare trisomies from 2.3 million births (0.25 per 10,000), none of whom survived [38]. All were notified following prenatal testing or late fetal death ($\geq$20 weeks of pregnancy). In contrast, 141 mosaic rare trisomies were reported (0.6 per 10,000 births), of which 78% were identified prenatally. Of these, 41% were liveborn, 7% stillborn, and 49% resulted in pregnancy terminations associated with fetal anomalies. Mosaicism involving trisomies 8 and 9 were most commonly notified. These findings show that true mosaicism for rare autosomal trisomies contributes to a small but significant part of pre- and perinatal adverse pregnancy outcomes.

RARE AUTOSOMAL TRISOMIES IN AMNIOTIC FLUID

Amniocentesis for cytogenetic prenatal diagnosis has been in widespread use since the early 1970s. The cells isolated from amniotic fluid closely reflect the chromosome constitution of the fetus, being derived from sources such as the fetal skin, nasopharyngeal tract, and urogenital tract [39], with extraembryonic cells being contributed from the amniotic membrane (amnion) [40]. Historically, amniocentesis samples used for conventional chromosome analysis have been divided and grown across several independent culture dishes. Specimens that exhibit the same mosaic chromosome abnormality in at least two culture dishes are considered to exhibit true (Level III) mosaicism, which is present in approximately 0.1%–0.3% of amniocentesis samples analyzed by conventional karyotyping [41–43]. True mosaicism most commonly involves the autosomal trisomies (48%), followed by sex chromosome aneuploidies (40%) and extra structurally abnormal chromosomes (12%) [42]. Confirmation of mosaicism in fetal or newborn samples occurs in about 60%–70% of cases [42, 43] and in one US collaborative study, approximately 38% of autosomal trisomy mosaics were reported to be associated with noticeable phenotypic abnormalities [42].

Hsu et al. have reported phenotypic outcome data for 151 rare autosomal trisomy mosaics (excluding trisomy 20) ascertained following amniocentesis [44]. This series was recently updated by

Wallerstein et al., who reported summary outcomes for all mosaic autosomal trisomies [34], including 506 cases of rare trisomy mosaicism (Table 1). Cases with prior abnormal ultrasound findings were excluded to help remove ascertainment bias. With regard to recorded abnormal outcomes, mosaic trisomy for chromosomes 2, 9, 16, 20* [*see below], and 22 were classified as very high risk (>60% with abnormal outcomes); chromosomes 5, 14, and 15 were classified as high risk (40%–59% abnormal); chromosomes 7, 12, and 17 as moderately high risk (20%–39%); chromosomes 6 and 8 as moderate risk (up to 19%); and no rare mosaic trisomies were classified as low risk (0%). Mosaic trisomy for chromosomes 1 and 10 was not observed. Mosaic trisomy for chromosomes 3, 4, 11, and 19 was not assigned a risk in the previous study by Hsu et al. due to insufficient cases ($n < 5$). In the Wallerstein et al. series, abnormal outcomes were recorded in 3/4 cases of trisomy 3 mosaicism, 3/5 trisomy 4, 0/4 trisomy 11, and 0/1 trisomy 19. Therefore true mosaicism for trisomies 3 and 4 suggests a high to very high risk, based on these small numbers. Trisomy 20 mosaicism* appears to have been misclassified as very high risk, rather than moderate risk (11% of cases with abnormal outcome), which is consistent with the lower frequency of abnormal outcomes in an earlier study [45].

The assessment of abnormal outcomes was made mostly by postmortem examination following termination or after birth. The authors note that subtle anomalies may not have been recognized and that neurodevelopmental follow-up after birth was rare. Nonetheless, these summary data are invaluable for helping evaluate possible outcomes following a diagnosis of rare trisomy mosaicism with normal ultrasound and to help guide patient counseling. Genetic counseling in the setting of normal fetal ultrasound remains problematic, but the presence of ultrasound anomalies indicates a very high risk for developmental and physical disabilities following the detection of rare trisomy mosaicism [34].

One last consideration is the introduction of chromosome microarray (CMA) into prenatal diagnosis. Few data currently exist on the interpretation of chromosomal mosaicism found in uncultured amniotic fluid samples ascertained using CMA. In the State of Victoria, Australia, >85% of all samples received for cytogenetic prenatal diagnosis are now analyzed using CMA [46]; the majority of which use DNA extracted directly from uncultured cells. In my own laboratory, using a single-nucleotide polymorphism (SNP) CMA (Illumina Inc.), which has a lower limit of detection of 7%–12% for

Table 1 Rare Trisomy Mosaicism Identified During Amniocentesis and Risk for Abnormal Outcome

Risk Classification According to Wallerstein et al. [34]	Proportion of Cases Recorded With Abnormal Outcomes	Chromosome
Very high risk	>60%	2, 9, 16, 22, 4
High risk	40%–59%	5, 14, 15
Moderately high risk	20%–39%	7, 12, 17
Moderate risk	Up to 19%	6, 8, 20[a]
Low risk		None
Unclassified[b]		1, 10, 3, 11, 19

Chromosome 4 classified as very high risk based on minimum of 5 reported cases.
Only cases with normal ultrasound findings at the time of amniocentesis qualify.
[a]*Misclassified as very high risk in original source (see main text for details).*
[b]*Not observed (1, 10) or too few (<5 cases; 3, 11, 19).*

trisomy mosaicism [47], we sometimes observe discrepancies between the results of CMA on uncultured cells, and the results of conventional karyotyping on cultured cells. In some, but not all cases of discrepancy, the SNP CMA will exhibit mosaicism, while the cultured cells are karyotypically normal, or perhaps show only a single colony of abnormal cells. Insufficient data currently exist to determine which method more accurately reflects the true fetal karyotype or provides a better prediction of phenotype.

CONFINED PLACENTAL MOSAICISM IN CHORIONIC VILLI AND ITS RELATIONSHIP TO CELL-FREE DNA-BASED NIPT

Chorionic villus sampling (CVS) for cytogenetic prenatal diagnosis emerged in the mid-1980s [48–50]. The procedure enables prenatal testing from the first trimester of pregnancy, using samples of chorionic villi biopsied from the placenta at around 10 to 12 weeks of gestational age. The basis of CVS is that the karyotype of the placental chorionic villi represents the karyotype of the fetus [51].

Samples of chorionic villi for conventional chromosome analysis can be prepared using two methods: (i) a direct or short-term (24–48 h) culture method (STC) that analyzes rapidly dividing cells from an outer villi layer of cytotrophoblast, and (ii) a long-term culture method (LTC) that analyzes cells grown from the mesodermal core of the chorionic villi. Over the past two decades, some laboratories have substituted STC for other rapid methods of analysis, using techniques such as fluorescence in situ hybridization (FISH) [52] or quantitative fluorescence polymerase chain reaction (QF-PCR) [53]. Whereas STC provides a low-resolution G-banded karyotype, FISH and QF-PCR typically only target aneuploidy for chromosomes 13, 18, 21, X, and Y. More recently, CMA using DNA extracted from whole chorionic villi has replaced LTC in some laboratories [54].

Reports of discrepancies between the karyotype of cells from STC and/or LTC, and the chromosome constitution of fetus emerged shortly after CVS was implemented into clinical practice [49, 50, 55]. These discrepancies usually involved chromosomal mosaicism that was present in the chorionic villi but not in the fetus—a phenomenon known as confined placental mosaicism (CPM) [56], which affects up to 2% of CVS samples [57]. This frequency of mosaicism is at least 10 times higher than the rate of TFM seen after amniocentesis.

CPM can be present in STC only (CPM I), in LTC only (CPM II), or in both (CPM III) [58]. Trisomy for CPM types I and II usually has a mitotic origin, where postzygotic gain of the trisomic chromosome is confined to the cytotrophoblast or the mesenchyme, respectively. CPM type III is more likely to have a meiotic origin and involve a trisomic conception that has lost one of the trisomic chromosomes in the first few cell divisions after fertilization—so-called trisomy rescue [59–63]. Rarely, a false-negative result may be reported. Here, the chorionic villi have a normal karyotype, but the fetus has a chromosome abnormality, either as a full trisomy or with TFM. False-negative results for the common autosomal trisomies [13, 18, 21] occur almost exclusively during analysis of STC cytotrophoblast cells [64]. From a developmental view point, the cytotrophoblast cells are more distantly related to the embryo proper than are cells from the LTC mesenchyme; the mesenchyme cells being known to more accurately reflect the fetal karyotype [65]. This is because the LTC mesenchyme cells derive from the hypoblast of the inner cell mass (ICM); the ICM also giving rise to the epiblast and embryo proper [66].

Discrepancies involving mosaicism in chorionic villi are critically important to our understanding and management of cfDNA test results, as the origin of "fetal" cfDNA is apoptotic cytotrophoblast

Table 2 Types of Mosaicism Found During CVS and Expected Result From cfDNA-Based NIPT

Type of Mosaicism	Cytotrophoblast (CV-STC)	Mesenchyme (CV-LTC)	Amniotic Fluid/Fetal Tissue	Expected cfDNA Result
CPM I	Abnormal[a]	Normal	Normal	False positive
CPM II	Normal	Abnormal	Normal	True negative
CPM III	Abnormal[a]	Abnormal	Normal	False positive
TFM IV	Abnormal[a]	Normal	Abnormal	True positive
TFM V	Normal	Abnormal	Abnormal	False negative
TFM VI	Abnormal[a]	Abnormal	Abnormal	True positive

CPM, *confined placental mosaicism;* CV-LTC, *chorionic villi long-term culture;* CV-STC, *chorionic villi short-term culture;* TFM, *true fetal mosaicism.*
[a]*Assumes sufficient abnormal cells in cytotrophoblast to enable detection by cfDNA analysis.*
Based on Grati FR, Malvestiti F, Ferreira JC, Bajaj K, Gaetani E, Agrati C, et al. Fetoplacental mosaicism: potential implications for false-positive and false-negative noninvasive prenatal screening results. Genet Med 2014;16(8):620–624.

cells. Thus cfDNA-based prenatal testing is analogous to CVS STC and is essentially a liquid biopsy of these placental cells. Several groups have used this knowledge to review large databases of CVS test results to predict the frequency of false-positive cases that will occur during cfDNA analysis [65, 67, 68] and to help guide the choice of follow-up prenatal procedure. In particular, these large reviews are helpful for the interpretation and management of rare autosomal trisomy cases identified during cfDNA-based NIPT [26, 27]. A caveat here is that the cfDNA result is a proxy for the CVS STC result only. No information is provided on the LTC mesenchyme cells, which are available to aid interpretation during the analysis of a diagnostic CVS sample (see Table 2). Professional societies governing standards in cytogenetic testing recommend against the analysis of CVS STC alone, because of the increased chance of both false-positive and false-negative results [69–71]. This recommendation is a salient reminder that cfDNA-based NIPT should always be regarded as a screening test.

RARE AUTOSOMAL TRISOMIES IN CHORIONIC VILLI AND FETAL COMPROMISE

The clinical implications of CPM involving the rare autosomal trisomies are well documented, and pregnancy outcomes may vary greatly, even for the same chromosome [60–62]. Reported pregnancy complications include spontaneous miscarriage, IUGR, intrauterine fetal demise (IUFD), preterm birth and stillbirth, but many pregnancies also proceed to term uneventfully [72]. Pregnancy-induced hypertension and preeclampsia are frequently reported complications of trisomy 16 mosaicism [73, 74], while uniparental disomy [the inheritance of two homologous chromosomes from one parent without a contribution of that chromosome from the other parent] may lead to imprinting disorders for those chromosomes known to harbor imprinted genes (chromosomes 6, 7, 11, 14, 15, and 20), following trisomy rescue [75]. Residual TFM has been reported more commonly in pregnancies associated with trisomies 9, 16, and 22, but in practice mosaicism can involve almost any rare trisomy [34]. Cryptic mosaicism, where fetal malformations are seen in association with CPM, but where the rare trisomy mosaicism cannot be demonstrated in the fetus or newborn, may be present in up to 10% of cases [76], and has been frequently suspected in trisomy 16 CPM [77].

The likelihood of an adverse pregnancy outcome after the detection of rare trisomy mosaicism following CVS appears to be influenced by 3 key variables. These are: (i) the distribution of trisomic cells in the chorionic villi [CPM types I, II, or III], (ii) the actual trisomy involved, and (iii) the frequency of abnormal cells.

Early studies of CVS mosaicism reported an association between very high frequencies of trisomic cells in the cytotrophoblast cell lineage (CPM I), or both the cytotrophoblast and mesenchyme cell lineages (CPM III), and serious pregnancy complications [60, 62]. UPD was also more commonly ascertained when the frequency of trisomic cells in both cell lineages was high. Both Robinson et al. [60] and Wolstenholme et al. [62] correlated CPM III with a meiotic origin of the trisomy and an increase in propensity for pregnancy complications. CPM types I and II were more commonly benign, these being associated with a somatic, postzygotic origin of the trisomy, particularly when the frequency of trisomic cells was low.

Toutain et al. reported a high frequency of pregnancy complications involving CPM III in a review of 13,809 CVS samples with either CPM II (37 cases) or CPM III (20 cases) [61]. The authors found no difference in the frequency of low birth weight, prematurity, and other adverse pregnancy outcomes for CPM II, when compared with a control population. Nor were there any confirmed cases of UPD reported (0/6 cases investigated). In contrast, the mean birth weight was lower, and the incidence of prematurity, IUGR, and pregnancy loss higher for CPM III, compared with controls. UPD was reported in 4/13 (30.8%) CPM III cases, which is consistent with the 1 in 3 theoretical prevalence expected following trisomy rescue.

From these and other studies it is clear that trisomy 16 mosaicism is the most frequent rare trisomy associated with pregnancy complications. Almost all cases have their origins in an error of maternal meiosis [60, 78]. A placenta containing a high proportion of trisomy 16 cells appears to be particularly vulnerable to placental insufficiency leading to fetal growth restriction and other complications. Many pregnancies with CPM for trisomy 16 are associated with very low pregnancy-associated plasma protein A (PAPP-A) levels, with a median MoM of 0.13 reported, equivalent to the 0.2th percentile [79]. Pregnancies with a PAPP-A level below the 5th percentile have a higher likelihood for IUGR, premature birth, preeclampsia, and stillbirth [80], consistent with outcomes reported for trisomy 16 CPM. Other rare trisomies might also have a predisposition toward abnormal first trimester screening analytes, with a large Danish study reporting 77% of confirmed rare trisomies as being screen positive by CFTS, usually in association with low PAPP-A measurements [81].

The fact that CPM types I and II are more likely to have a somatic origin of trisomy also explains their much lower association with UPD. This is particularly true for trisomy 3 (CPM 1) and trisomy 2 (CPM II) mosaicisms, which are commonly reported as being confined to the cytotrophoblast and mesenchymal core of the chorionic villi, respectively. Based on the distribution of rare trisomies confined to these cell lineages, Wolstenholme et al. estimated a postzygotic origin for trisomies 2, 3, 7, and 8 of 81%, 95%, 87%, and >95%, respectively [62]. Therefore the risk of UPD for these chromosomes after a finding of CPM should be low. A large Italian study on mosaicism in CVS provides evidence for this, with 0/65 and 0/74 cases with UPD2 and UPD7, respectively [57]. If a meiotic origin was common for these trisomies, up to one-third of all cases would be expected to have UPD following trisomy rescue.

Interest in pregnancy outcomes for the rare autosomal trisomies has existed since their increased prevalence was first noted after the introduction of CVS for prenatal diagnosis. A renewed interest has emerged with the arrival of genome-wide cfDNA-based NIPT. While diagnostic CVS examines only a small, localized region of the placenta, cfDNA screening provides information on the entire placental cytotrophoblast. Therefore bioinformatics algorithms can be employed to identify pregnancies with very high proportions of trisomic cells in the placenta [26]. It is often these pregnancies that are conceived with trisomy. Several decades of CVS outcome data indicate these pregnancies are those most at risk for complications that include fetal growth restriction, premature delivery, residual trisomy mosaicism, and UPD.

RARE AUTOSOMAL TRISOMIES AT CELL-FREE DNA-BASED PRENATAL TESTING

RARE AUTOSOMAL TRISOMIES AS A CAUSE OF UNUSUAL, FALSE, OR FAILED CELL-FREE DNA RESULTS

Bioinformatics algorithms used for cfDNA-based NIPT are designed to detect aneuploidies in the test autosomes [13, 18, 21] by comparing and normalizing their sequence counts to nontest reference chromosomes (see Chapter 3). A potential disadvantage of this approach is that a true aneuploidy involving a reference chromosome may cause a false aneuploidy call on a test chromosome. If the reference chromosome is trisomic, a false monosomy call can occur for the test chromosome when the higher sequence counts are used for normalization. The opposite is true if the reference chromosome involves a monosomy; a false trisomy call may occur on the test chromosome.

A recent cfDNA-based NIPT study reporting on two large patient cohorts used a common quality metric (NCD; normalized chromosome denominator) to monitor for unusual sequence counts on reference and other nontest chromosomes [26]. The larger of the two cohorts recorded 246 rare trisomies of nontarget chromosomes in approximately 73,000 pregnancies. Of these, 172 involved a reference chromosome that produced 21 cases with one or more putative false monosomy calls on the test chromosomes. Pregnancy outcomes were not available except in a small number of cases previously reported in association with maternal malignancy; these cases were associated with multiple false aneuploidy calls on the test chromosomes. The second, smaller patient cohort was analyzed using the same quality metric but in this instance the algorithm was designed to cancel the analysis rather than generate a false call on the test chromosome. Of 60 trisomies involving a reference or nontest chromosome in 16,885 pregnancies, 24 resulted in test cancellations. Cytogenetic or pregnancy outcome data in these cases indicated the fetus and/or placenta was affected by a rare trisomy that was often associated with miscarriage or other complications including IUGR, TFM, and UPD. However, normal pregnancy outcomes were also recorded. One test cancellation was found to be caused by a low-grade maternal mosaicism for trisomy 8.

Snyder et al. reported on 113,415 cfDNA cases, of which 138 cases (0.12%) were associated with autosomal monosomy ($n = 65$), aneuploidy for both a common trisomy and a sex chromosome ($n = 36$), or multiple autosomal aneuploidies ($n = 37$) [82]. Not surprisingly, fetal involvement was most frequent in the group involving a common trisomy and sex chromosome aneuploidy (these being true fetal abnormalities) and was least frequent when an autosomal monosomy was involved (false calls as described previously). Several maternal malignancies were confirmed in the multiple aneuploidy group. It is now well recognized that genome-wide copy number imbalances can lead to multiple aneuploidy calls for the test chromosomes and a subsequent suspicion of maternal malignancy [83, 84], see also Chapter 10.

Dharajiya et al. reported 55 cases with widespread copy number abnormalities from about 450,000 cfDNA-based NIPT [85]. Of cases with follow-up, 18 confirmed maternal malignancies, 20 benign uterine fibroids (uterine leiomyoma), and 3 cases without evidence of disease were recorded. Of the 18 confirmed malignancies, 7 cases were known prior to NIPT (but not to the testing laboratory) and 11 cases were unknown. Thus at least 11 unknown malignancies were documented among 450,000 cfDNA-based prenatal screening tests. These findings account for a small but important cause of rare and unusual cfDNA screening results. An important advantage of a genome-wide cfDNA analysis approach is the ability to help differentiate false or misleading results involving the common aneuploidies from other genuine genome-wide copy number aberrations that may involve whole chromosomes and/or segmental aneuploidies.

RARE TRISOMIES IDENTIFIED DURING CELL-FREE DNA-BASED NIPT AND PREGNANCY OUTCOMES

Several investigators have documented their experience of reporting rare autosomal trisomies using genome-wide cfDNA screening [26–29, 31, 32, 86]. From these studies it is clear that many pregnancies with rare trisomies represent CPM and that these pregnancies may proceed to term uneventfully (Box 1). However, other pregnancies may be complicated by miscarriage, poor fetal growth associated with placental insufficiency, TFM (Box 2), and pathogenic UPD if an imprinted chromosome is involved (Box 3). Samples where the trisomic (affected) fraction of cfDNA is similar to the fetal fraction appear most at risk for adverse outcomes [26]. This is because the placental cytotrophoblast in these pregnancies is expected to be almost uniformly (100%) trisomic, in keeping with a meiotic origin for the trisomy. In these circumstances the consequences for the pregnancy can be catastrophic if the trisomy extends to the embryo proper.

BOX 1 CASE 1 VIGNETTE: TRISOMY 10 CONCEPTION ASSOCIATED WITH NORMAL, TERM DELIVERY

- Indication: Age 38 years, NIPT as primary screening test at 10 wks 5 days gestation
- Genome-wide NIPT: Increased risk T10
- Trisomic fraction (TF) = 4.6%
- Fetal Fraction (FF) = 5.8%
- Ratio TF/FF: 0.8 (80% T10)
- Follow-up ultrasound at 12 wks: viable fetus, no abnormalities on scan
- Amniocentesis at 16 wks: SNP microarray—normal female result; arr(1-22,X)x2
- Pregnancy continued to healthy live birth at 39 wks, weight 2940 g, 10–25th centile
- Placental biopsy after delivery: SNP microarray on whole chorionic villi—65% T10 (meiosis II error, biparental disomy), 25% T20 (mitotic), 25% XXX (mitotic). [T20 and XXX were evident on retrospective cfDNA analysis].
- Interpretation: T10 conception with trisomy rescue and likely post zygotic gain of chromosomes 20 and chromosome X. CPM involving all 3 chromosomes.
- Summary: T10 conception with high-grade mosaicism (80%) on cfDNA analysis but with trisomy rescue can be associated with an uncomplicated term delivery and normal live birth outcome.

Abbreviations: *CPM*, confined placental mosaicism; *SNP*, single nucleotide polymorphism; *T10*, trisomy 10; *wks*, weeks.

BOX 2 CASE 2 VIGNETTE: TRISOMY 16 CONCEPTION ASSOCIATED WITH TFM AND CONGENITAL ANOMALIES

- Indication: CFTS T21 risk 1/65 (PAPP-A 0.07 MoM), NIPT at 13 wks 0 days
- Genome-wide NIPT: Increased risk T16
- Trisomic fraction (TF) = 13.0% Fetal Fraction (FF) = 8.9% (likely underestimate)
- Ratio TF/FF: 1.5 (100% T16)
- Follow-up ultrasound at 15 wks 5 days: viable fetus, no abnormality on scan, small echogenic spaces up to 12 mm in placenta
- Amniocentesis at 15 wks 5 days: conventional karyotype: TFM T16 47,XX,+16[17]/46,XX[9]; SNP microarray not performed.
- At 19/20 ultrasound: small cystic placenta; small kidneys, one underdeveloped, CHD (VSD). TOP elected.
- Interpretation: TFM for T16.
- Summary: Likely T16 conception with high grade T16 (100%) on cfDNA analysis. Partial trisomy rescue with residual T16 mosaicism. Initial ultrasounds at 13 and 16 wks were falsely reassuring. Follow-up ultrasound at 19 wks indicated fetal anomalies. T16 mosaicism with multiple anomalies is more likely to be associated with developmental delay.

Abbreviations: *CFTS*, combined first trimester screening; *CHD*, congenital heart defect; *PAPP-A*, pregnancy-associated plasma protein-A; *TFM*, true fetal mosaicism; *TOP*, termination of pregnancy; *T16*, trisomy 16; *VSD*, ventricular septal defect; *wks*, weeks.

BOX 3 CASE 3 VIGNETTE: TRISOMY 15 CONCEPTION ASSOCIATED WITH MATERNAL UPD 15

- Indication: Age 36 years, NIPT as primary screening test at 11 wks 0 days
- Genome-wide NIPT: Increased risk T15
- Trisomic fraction (TF) = 6.6% Fetal Fraction (FF) = 6.8%
- Ratio TF/FF: 0.97 (97% T15)
- Follow-up ultrasound at 12 wks: viable fetus, no abnormality on scan
- Ultrasound at 16 wks: viable fetus, no abnormality on scan.
- Amniocentesis at 16 wks: SNP microarray normal male arr(1-22)x2,(XY)x1; comparative analysis of parental and fetal SNPs consistent with maternal UPD15 causing Prader-Willi syndrome (PWS).
- TOP elected.
- Placental biopsy after TOP: SNP microarray on whole chorionic villi—12%–15% T15 (meiosis I error, no recombination). CPM for trisomy 15 mosaicism confirmed.
- Summary: T15 conception with high-grade T15 (100%) on cfDNA analysis. Trisomy rescue associated with maternal UPD15 causing PWS. PWS is characterized by hypotonia and feeding difficulties at birth, poor growth, developmental delay, and mild to moderate intellectual disability. Hyperphagia during childhood leads to obesity.

Abbreviations: *PWS*, Prader-Willi syndrome; *SNP*, single nucleotide polymorphism; *TOP*, termination of pregnancy; *T15*, trisomy 15; *wks*, weeks.

Miscarriage, fetal growth restriction, fetal demise, true fetal mosaicism, and uniparental disomy associated with rare trisomies identified by cell-free DNA-based NIPT (Table 3)

In 2014, Lau et al. described one of the first clinical studies employing a genome-wide cfDNA prenatal screening approach [32], reporting 6 out of 1982 (0.3%) pregnancies with a rare autosomal trisomy. All 6 pregnancies resulted in live births but 2 were associated with early onset growth restriction requiring preterm delivery (trisomy 6 at 34 weeks and trisomy 9 at 33 weeks of gestation). The authors

Table 3 Number and Frequency of Rare Trisomies Reported During cfDNA-Based Prenatal Testing

Study First Author and Year	Study Size No.	Rare Trisomy No. (%)	Most Common Rare Trisomies Reported	Population Screened
Lau et al. (2014) [32]	1982	7 (0.35)	22	Elevated and average risk
Pescia et al. (2017) [29]	6388	50 (0.78)	7, 8, 22, 16	Elevated and average risk
Pertile et al. [Cohort 1] (2017)[a] [26]	72,932	246 (0.34)	7, 15, 16, 22	Not determined
Pertile et al. [Cohort 2] (2017) [26]	16,885	60 (0.36)	15, 7, 16, 22	Elevated and average risk
Fiorentino et al. (2017) [86]	12,078	17 (0.14)	22, 7, 15	Elevated and average risk
Van Opstal et al. (2018) [27]	2527	24 (0.95)	16, 7, 9	Elevated risk
Ehrich et al. (2017)[a] [28]	10,000	78 (0.78)	16, 7, 3	Elevated risk
Brison et al. (2018) [87]	19,735	58 (0.29)	7, 16, 22	Elevated and average risk
Total	142,527	540 (0.38)		

[a]*Outcome data not available.*

recommended that the likelihood of fetal mosaicism, CPM and UPD should be discussed when counseling patients with positive NIPT results for rare trisomies, and that serial ultrasound examinations are warranted to monitor for fetal IUGR [32].

In a series of 6388 pregnancies, Pescia et al. reported 50 NIPT results (0.78%) with rare trisomies [29]. Trisomies 7, 8, 22, and 16 were most commonly encountered. Amniocentesis follow-up was available in 19 pregnancies and 4 mosaic aneuploidies were confirmed (3/3 cases with trisomy 22 mosaicism and one with trisomy 12 mosaicism). There were no cases of UPD recorded, including 6 cases investigated following trisomy 7 CPM. One trisomy 16 result had an unfavorable outcome. The authors concluded the detection of rare trisomies was clinically useful for several reasons including high analytic accuracy, positive predictive values (PPVs) that closely predicted the CVS data, rare trisomy mosaicism as a cause of true fetal aneuploidy, increased risk for UPD, and lastly the increased risk for IUGR, small-for-gestational-age infants, and unfavorable pregnancy outcomes, like those seen for trisomy 16.

Brady et al. documented 11 out of 4000 (0.28%) results with rare trisomies after NIPT [31]. Trisomy 7 was most commonly encountered (3 cases). There were two cases of true fetal mosaicism: one with trisomy 16 and another with trisomy 15. The trisomy 15 mosaic was confirmed on amniocentesis to have maternal UPD15 causing Prader-Willi syndrome, consistent with a trisomy rescue event. The authors recommended amniocentesis as the preferred method of prenatal diagnosis for rare trisomies identified following cfDNA-based NIPT and advised UPD testing be considered for those cases involving an imprinted chromosome. The series has been updated recently by Brison et al., who reported 58 rare trisomies in 19,735 pregnancies [87]. The rare trisomy frequency of 0.29% remained unchanged from the previous report. Trisomies 7, 16, and 22 were reported most commonly. An amniocentesis was performed in 44 out of 58 pregnancies and TFM was reported for single cases of trisomies 2, 15, and

16 (3/44, 6.8%). UPD testing in 12 pregnancies was positive only for the maternal UPD 15 case reported from the earlier series. UPD findings were 0/7, 0/2, 1/1, and 0/2 for chromosomes 7, 11, 15, and 16, respectively.

Fiorentino et al. used a genome-wide approach to report 17 rare trisomies from 12,078 cfDNA samples (0.14%). Trisomies 22, 7, and 15 were seen most commonly. All 7 miscarriages with rare trisomy were confirmed on follow-up testing (4 cases of trisomy 22 and 3 cases of trisomy 15). The only viable pregnancy with trisomy 15 was disomic on amniocentesis but had maternal UPD 15 causing Prader-Willi syndrome. TFM was reported for single cases of trisomies 7, 9, and 22.

Pertile et al. have described the largest series of rare trisomies identified using cfDNA screening—306 cases from 89,817 pregnancies (0.34%) in two clinical laboratory patient cohorts [26]. Trisomy for chromosomes 7, 15, 16, and 22 was recorded most frequently (Table 3). Outcome data were available for 52 of 60 cases in cohort 2. Of known outcomes, 22 cases (42%) were associated with an early or missed miscarriage. Another 4 cases (8%) had fetal demise recorded between 17 and 40 weeks of gestational age, involving trisomies 2, 4, and 16 (2 cases). All had fetal growth restriction. Seven pregnancies (13%) were reported with putative or confirmed UPD ascertained using single-nucleotide polymorphism (SNP) microarray and 6 cases (12%) exhibited TFM. One maternal UPD 15 case causing Prader-Willi syndrome was from a proven trisomy 15 rescue event (Box 3); the 6 other UPD events involved nonimprinted chromosomes [2, 4, 16]. The authors also calculated the proportion of trisomic cfDNA (referred to as the "trisomic fraction") in each rare trisomy sample and compared this with the estimated fetal fraction (Ratio = trisomic fraction divided by fetal fraction). Cases complicated by miscarriage, TFM, UPD, and IUGR often had a trisomic fraction to fetal fraction ratio approaching 1.0. This high ratio is consistent with the placental cytotrophoblast being predominantly or uniformly trisomic, in keeping with a trisomic conception. Pregnancies with lower ratios were more likely to be associated with a normal outcome. It was postulated that cases with lower ratios may represent normal conceptions with postzygotic gain of the trisomic chromosome, leading to benign CPM and normal fetal karyotype. More data are needed, but this ratio might be helpful in predicting those pregnancies at increased risk for complications.

The high number of miscarriages (42%) reported in this cohort might reflect the early gestation at which cfDNA testing was undertaken, which on average was 11.0 weeks. Many failed pregnancies were missed miscarriages, where ultrasound scanning was not done in the period immediately prior to NIPT. Trisomy 15 was noted to be particularly common, with 13 of 14 pregnancies miscarrying prior to 11–12 weeks of gestational age. The sole ongoing pregnancy with trisomy 15 was confirmed with UPD following trisomy rescue, as described previously. Nonmosaic trisomy 15 is reported with embryonic, but not fetal development [88], which is consistent with demise prior to or around the time of cfDNA-based screening at 10 weeks of gestational age.

Since this study was completed, the cohort 2 series has expanded to 141 single rare autosomal trisomy cases from approximately 40,000 NIPT referrals (Fig. 1; Pertile MD, unpublished data). The prevalence of rare trisomy in this expanded series (0.35%) remains unchanged from the previous study. Cytogenetic or pregnancy outcome information is available for 120 pregnancies and includes 49 cases (41%) of early and missed miscarriages. Three rare trisomies comprise the majority of these miscarriage cases—trisomy 14 (5/5 miscarriages), trisomy 15 (26/30), and trisomy 22 (11/15)—indicating these nonmosaic rare trisomies can sometimes remain viable up to and around the time of cfDNA screening at 10 weeks of gestational age. In contrast, other trisomies like those for chromosome 2 (0/7 cases) and chromosome 4 (0/6 cases) were not seen as miscarriages, but were sometimes

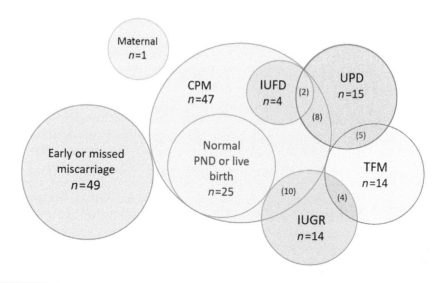

FIG. 1

Outcomes of 120 rare autosomal trisomies identified using cfDNA-based screening. Genome-wide screening was undertaken at the Victorian Clinical Genetics Services, Melbourne, Australia. A subset of the cohort has been reported previously [26].

seen in association with UPD (see below). This finding argues for early embryonic lethality of these nonmosaic aneuploidies unless they undergo trisomy rescue very early in pregnancy. Lastly, trisomy 7 ascertained by NIPT was not associated with early miscarriage (0/13 cases), nor with UPD (0/7 cases tested), consistent with previous reports suggesting trisomy 7 usually arises from a postfertilization gain of chromosome 7, with CPM and normal fetal karyotype [57, 62]. However, rare cases of TFM or UPD involving chromosome 7 have been documented, so a small proportion of trisomy 7 cases must be at risk for adverse outcomes [26, 60].

With regard to the TFM and UPD frequency in this expanded series, of 71 ongoing pregnancies with evidence for rare trisomy, there were 14 cases (14/71; 19.7%) of TFM and 15 cases (15/71; 21.1%) with confirmed or putative UPD. Cases of UPD ascertained using SNP microarray involved chromosomes 2 (2/7 cases), 4 (4/6 cases), 8 (1/4 cases), 12 (1/1 case), 15 (2/4 cases), 16 (4/13 cases), and 22 (1/4 cases). Both cases of UPD 15 were maternal in origin, causing Prader-Willi syndrome. The parental origin of putative UPD for other chromosomes was not routinely investigated because they are not known to harbor imprinted genes [75]. Lastly, 14/71 pregnancies (19.7%) were associated with IUGR, 4 of which were seen in conjunction with TFM (trisomies 2, 4, 7, and 16). Of note, trisomy 2 was documented with severe IUGR (<1st centile) in 4 of 5 ongoing pregnancies, one resulting in fetal demise at 17 weeks of gestation. The uncomplicated case had a low trisomic fraction to fetal fraction ratio of 0.2, while the 4 other pregnancies with severe growth restriction had ratios between 0.7 and 1.0 (median 0.9). These cases are in stark contrast to trisomy 2 confined to the mesenchyme only (CPM type II), which is reported to have a benign pregnancy outcome [62].

Finally, Van Opstal et al. reported on 2527 genome-wide cfDNA-based NIPT as part of the Dutch Trident study [27]. Trisomy for chromosomes 16, 7, and 9 was most commonly identified among

24 (0.9%) single rare trisomies. This series is comprised primarily of women referred for NIPT after an increased risk CFTS result of $\geq$1:200. Consequently, the cohort is enriched for cases with abnormal serum screening analytes, such as unusually low PAPP-A MoM. Indeed, 9 of 24 rare trisomies (40%) involved trisomy 16. TFM was described in 3 cases with multiple congenital anomalies (involving trisomies 9 and 22). Another 5 cases with apparent CPM were reported in association with multiple congenital anomalies, indicating possible cryptic mosaicism. Three of these cases involved trisomy 16.

In summary, several independent genome-wide cfDNA-based NIPT studies have reported rare autosomal trisomies in association with a range of pregnancy complications that include miscarriage, IUFD, TFM, and UPD, as well as IUGR associated with placental insufficiency. However, many normal pregnancy outcomes are also recorded. According to the study of Van Opstal et al. [27], which has near complete follow-up, normal pregnancy outcomes for single rare trisomies were documented in 9/24 (37.5%) pregnancies, increasing to 13/24 (54%) if small for gestational age but otherwise normal birth outcomes are included. In my own updated laboratory series, normal outcomes were recorded in 25/71 (35.2%) pregnancies once the large number of missed miscarriages or very early miscarriages are removed (Fig. 1), a frequency similar to the Van Opstal series. Thus at least one-third to one half of pregnancies that continue beyond the first trimester are likely to proceed uneventfully, while the remaining pregnancies will be complicated by fetal or placental involvement of the rare trisomy.

Finally, advances in the use of bioinformatics algorithms make it possible to estimate the degree of rare trisomy mosaicism that affects the cytotrophoblast cell lineage. This may prove helpful in quantifying the risk for adverse pregnancy outcomes for some or all rare trisomies identified using cfDNA screening. Although preliminary results appear promising, further cytogenetic and pregnancy outcome data are required from other centers before any firm recommendations can be made.

GENOME-WIDE CELL-FREE DNA-BASED NIPT AND COPY NUMBER VARIANTS

A natural extension of the use of genome-wide NIPT is to move beyond screening for whole aneuploidy of any chromosome and to also test for pathogenic segmental CNVs (Table 4).

The inherent advantage of this approach over targeted methods of NIPT is the capacity to report on nonrecurrent (unique) segmental chromosomal aneuploidies, which comprise a substantial proportion of pathogenic chromosome conditions seen during pregnancy and after birth [15]. Segmental aneuploidies visible at a karyotype level of resolution (on average 7–10 Mb) include nonrecurrent de novo

Table 4 Whole Chromosome and Structural Aberrations Detected by Genome-Wide cfDNA-Based NIPT

Abnormality
Common autosomal aneuploidy (trisomy of 13, 18, and 21)
Sex chromosome aneuploidy (monosomy X, XXX, XXY, and XYY)
Rare autosomal trisomy (trisomies other than 13, 18, and 21)
Copy number variations[a] (duplication and deletions, unbalanced translocations, etc.)

[a]*Sensitivity for a given size (Mb) is dependent on fetal fraction and sequence tag count (read depth).*

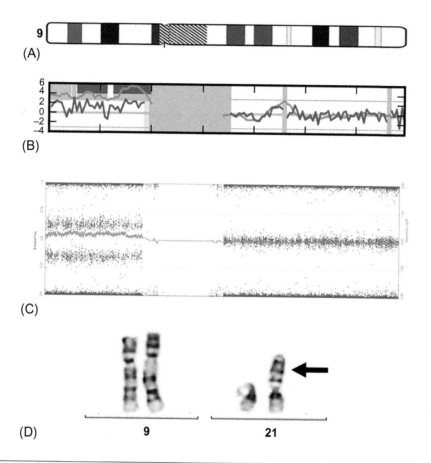

FIG. 2

Genome-wide analysis of cfDNA identifies chromosome 9p duplication using WISECONDOR algorithm.
(A) Chromosome 9 ideogram (p arm at left, q arm at right). (B) WISECONDOR plot showing gain of chromosome 9p. Vertical axis indicates z-score. *Red line* plots z-score using a sliding window. *Purple bars* indicate bins called by the algorithm across 9p. (C) SNP microarray analysis of whole chorionic villi DNA. *Blue dots* are genotyping calls plotted as B-allele frequency (BAF) on the vertical axis. Deviation of BAF across entire 9p, together with an increase in logR (sliding window of smoothed logR represented by *red line*), is consistent with trisomy of 9p. (D) Partial karyotype of long-term cultured chorionic villi (LT-CVS). Chromosome 9 homologs at left. At right, derivative chromosome 21 with an entire 9p arm (indicated by *arrow*) translocated onto chromosome 21p. The additional copy of 9p results in trisomy 9p; characterized by distinct craniofacial dysmorphism and intellectual disability. cfDNA was isolated from blood taken at 11 weeks of gestation for routine screening. Fetal fraction was estimated at 4.1% based on Y chromosome sequence count; 5% using Illumina platform specific method.

deletions and duplications, unbalanced translocations (Fig. 2) as well as other complex changes like ring or marker chromosomes, insertional rearrangements and isochromosomes (Fig. 3). Although these copy number changes occur at a lower prevalence than smaller, recurrent CNVs, collectively they have a birth prevalence of about 1 in 2500 [38].

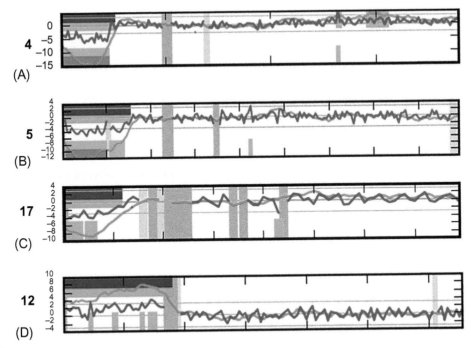

FIG. 3

Pathogenic segmental copy number abnormalities identified using genome-wide cfDNA-based NIPT. Each segmental aneuploidy was called using WISECONDOR algorithm and confirmed after SNP microarray analysis on DNA extracted from whole chorionic villi or uncultured amniotic fluid. (A) 24-Mb deletion of chromosome 4p16.3p15.2 causing Wolf-Hirschhorn syndrome (B) 34-Mb deletion of 5p15.33p13.3 causing Cri du chat syndrome (C) 6.3-Mb deletion of 17p13.3p13.2 causing Miller-Dieker Lissencephaly syndrome. (D) Whole chromosome 12p gain associated with mosaic tetrasomy 12p (isochromosome 12p) causing Pallister-Killian syndrome. WISECONDOR chromosome plots are not to scale.

The ability to accurately screen for segmental aneuploidies is determined by the segment size, the fetal fraction, and the number of sequence tags generated for statistical analyses [22, 30, 89, 90]. An increase in sequencing depth enables improved detection of CNVs at lower fetal fractions and allows for improved discovery of smaller CNVs [91]. Many early proof of principle studies obtained material for testing by using sonicated (sheared) genomic DNA with known copy number abnormalities to create synthetic samples to simulate placental cfDNA, and by using samples from pregnant women known to carry a fetus with a segmental copy number imbalance [17, 22, 30, 90].

Using these approaches, Srinivasan et al. were able to detect fetal CNVs as small as 300 kb by using an extremely high depth of sequencing of 600–1300 million reads [22]. Lo detected 15 out of 18 (83%) CNVs larger than 6 Mb using 4 to 12 million reads per sample [90]. Sensitivity fell for segment sizes below 6 Mb without the use of deeper sequencing. At higher read depths of up to 120 million reads, 29 out of 31 fetal segmental abnormalities were detected at sizes ranging from 3 Mb to 42 Mb, including 3 unbalanced translocations. The authors emphasized the impact of lower fetal fraction on test sensitivity. At 100 million reads, in silico analysis predicted 99% test sensitivity for 10-Mb CNVs with a fetal fraction of 5%, falling to 94% if approximately 5% of samples had a fetal fraction below 5%. Although the majority of CNVs >6 Mb in size could be identified at standard read depths of 4 to 12 million reads, for a false-positive rate of 0.4%, the authors concluded the test had limited clinical utility, in part because fetal fraction could not be measured for all samples.

Chen et al. used an in silico analysis to estimate close to 100% test sensitivity for segment sizes >10 Mb with 10% fetal fraction and 7 million sequence reads [30]. They also concluded the ability to detect rearrangements reduces rapidly with decreasing CNV size. Four samples with CNVs ≥10 Mb were identified from 1311 cfDNA samples with known fetal karyotype. Three of four cases were confirmed on fetal karyotyping while one result was a false-positive finding. No false-negative cases were observed. In another study, Li et al. [92] used an average of 3.6 million unique reads obtained from a semiconductor sequencing platform (SSP) [93] to detect 10 of 11 (90.9%) microarray-positive samples with a CNV size of >5 Mb. The one case not detected by NIPT had a fetal fraction of 4.7%. Only 1 of 7 (14.3%) CNVs <5 Mb were identified. The average gestational age in this retrospective series was 21 weeks, so the higher than average fetal fraction of 14.2% would have enhanced test performance. The bias toward later gestational age samples is evident in many proof-of-principle studies using retrospective clinical samples, as these pregnancies are usually selected on the basis of fetal sonographic abnormalities. Additionally, smaller (< 3 Mb) recurrent CNVs are sometimes identified because they are carried by the mother [94, 95], not necessarily through the resolving power of the cfDNA screening assay, particularly at lower sequence tag counts. This is because the mother is contributing on average 90% of the cfDNA, increasing to 100% of affected cfDNA if the CNV is inherited by the fetus.

Lefkowitz et al. employed a genome-wide cfDNA-based NIPT approach using 32 million reads per sample in a retrospective, blinded study of approximately 1200 samples to identify subchromosomal CNVs ≥7 Mb in size. Of 43 samples with rare autosomal trisomy (8 cases) or CNVs (35 cases), test sensitivity was reported as 97.7% and specificity as 99.9% [21]. This same group later reported on a larger series of 10,000 clinical samples [28]. A total of 154 results, representing 27.8% of all abnormal calls, were only detectable by genome-wide analysis. This included results for both rare autosomal trisomies and segmental aneuploidies. No outcome data were made available, so sensitivity and specificity estimates are unknown. Up to 25% of referrals were for abnormal fetal ultrasound findings, highlighting the fact that this genome-wide NIPT was being elected in preference to standard NIPT for high-risk referrals.

Fiorentino et al. used genome-wide NIPT to prospectively screen 12,114 clinical samples and argued for improved clinical utility over standard NIPT, with increased detection rates and only a minor decrease (0.1%) in test specificity [86]. Sequence tag counts were at least 25 million at a minimum 2% fetal fraction. A total of 13 segmental aneuploidies were reported, with 8 cases (PPV 61.5%) being confirmed after prenatal diagnosis. Two cases had maternally inherited unbalanced translocations without any prior history. The authors claimed a test sensitivity of 100% for all chromosome abnormalities screened, including rare autosomal trisomies and segmental aneuploidies, although no size limit for CNVs was specified. Genome-wide test specificity was 99.77%.

My own clinical laboratory employs WISECONDOR (WIthin SamplE COpy Number aberration DetectOR) algorithm for genome-wide cfDNA screening. This algorithm is designed to detect copy number imbalances of 10–20 Mb at fetal fractions above 5% using low sequence coverage data (23). Our standard sample workflow generates a sequence tag count of approximately 24 million unique reads per patient sample. Previously we have used WISECONDOR as part of our rare trisomy screening protocol [26] and now employ the algorithm for genome-wide segmental analysis (Figs. 2

and 3). Others have reported on the successful application of this algorithm for genome-wide cfDNA screening [27].

As part of our own validation study for genome-wide NIPT (unpublished) we prospectively screened 15,600 consecutive cfDNA samples from high- and average-risk women at 10–12 weeks of gestational age, and identified 18 pregnancies at increased risk for segmental aneuploidy (18/15,600; 0.12%), of which 11 cases (PPV 61.1%) were confirmed after prenatal diagnosis. True positive screening results were 4 unbalanced translocations without prior history (2 maternal, 1 de novo, and 1 inheritance unknown), 2 syndromic deletions (Wolf-Hirschhorn 4p- and Miller-Dieker 17p- syndromes), 1 novel de novo deletion (4.8 Mb within 16q23), 2 isochromosomes (mosaic isochromosome 12p—Pallister-Killian syndrome and nonmosaic isochromosome 18p syndrome), and 2 pathogenic de novo duplications (7.4 Mb from chromosome 10q22q23 and 12.5 Mb from chromosome 22q13).

The causes of the 7 false-positive results were varied and complex. A biological reason was identified in 3 cases—CPM was the cause in 2 pregnancies and maternal mosaicism in the third. One CPM case involved a mosaic isochromosome 18p (mosaic tetrasomy 18p) confined to cytotrophoblast. The other involved a large mosaic deletion of 18q confined to cytotrophoblast, together with a cell line showing mosaicism for segmental isodisomy of this same region. The 18q deletion was not present in LT-CVS mesenchyme, only the cell line with segmental isodisomy. The deletion had presumably been "rescued" by DNA repair from the nondeleted homolog. The maternal case was ascertained as a 22 Mb gain of terminal 8q using WISECONDOR. It was present in 23% of maternal blood lymphocytes as a mosaic supernumerary isochromosome, causing mosaic tetrasomy for region 8q24. The isochromosome had been stabilized by neocentromere formation [96]. A maternal origin was suspected based on a very high z-score relative to the fetal fraction. The patient declined prenatal diagnosis after normal ultrasound scans and a healthy baby was born. Large, mosaic maternal copy number abnormalities can have serious reproductive consequences if they are present in the patient's germ line. While the maternal carrier may have a normal phenotype, the fetus is at very high risk for abnormality if the unbalanced rearrangement is inherited in its nonmosaic form, as has been described previously in a cfDNA screening case report with abnormal outcome [97].

The origin of the segmental aneuploidy in the 4 remaining false-positive cases—3 deletions on 1p, 7q, and 11p, respectively, and a duplication from 11q—could not be determined. Segment sizes ranged from 20 to 78 Mb. Three patients elected amniocentesis and normal results were returned, so the false-positive calls might represent CPM or possibly maternal mosaicism. A SNP microarray analysis on blood from one patient was normal. The fourth patient elected CVS, which was normal, as was maternal blood analyzed using SNP microarray.

Finally, there were 3 false-negative cases recorded with large (>10 Mb) segmental aneuploidies. Placental mosaicism was a suspected or proven cause for each false-negative result. One case was a mosaic unbalanced reciprocal translocation between chromosomes 3q and 4q, present in 70% of uncultured amniotic fluid cells and the majority (93%) of cells after culture. Another involved mosaicism for a supernumerary isochromosome 8p (mosaic tetrasomy 8p), which was present at low level in whole chorionic villi and as a nonmosaic abnormality in LT-CVS mesenchyme. The final case was a 16-Mb deletion of 11q causing Jacobsen syndrome. SNP microarray analysis of DNA from whole chorionic villi showed a mosaic deletion of 11q together with mosaicism for segmental isodisomy of this same region. In contrast to the mosaic 18q deletion described previously, which corrected to isodisomy in the LT-CVS mesenchyme, in this instance the LT-CVS showed only the deletion of 11q. The fetus had a 7.3-mm nuchal fold at 13 weeks' gestation and the pregnancy was terminated. The two other false-negative cases were also ascertained after fetal abnormalities were observed on ultrasound scan, following low-risk genome-wide cfDNA screening results.

Performance statistics for segmental aneuploidy screening in this validation study were a test sensitivity of 78.6% (95% CI, 49.2–95.3), specificity 99.9% (99.8%–99.9%), PPV 61.1% (35.6–82.7), and negative predictive value (NPV) >99.9% (99.94–99.99). Specificity and PPV measures were similar to those reported for trisomy 13 screening using standard NIPT, while test sensitivity compared with trisomy 13 screening was reduced [98]. There was only a modest increase in cases recommended for prenatal diagnosis (18/15,600; 0.12%), and 61% of these calls were associated with a confirmed, pathogenic segmental aneuploidy in the fetus. In comparison, Van Opstal et al. used WISECONDOR for genome-wide cfDNA screening [27] and reported 12/2553 (0.47%) segmental aneuploidies among 2553 primarily high-risk CFTS referrals (risk >1:200). Six cases were confirmed after diagnostic testing (PPV 50%).

The broad range of cases described in our own study highlight the complex biology behind cfDNA-based NIPT. Biological factors such as CPM and maternal mosaicism can confound test sensitivity and specificity for segmental aneuploidy screening, as they do for standard cfDNA-based methods. Despite these challenges, the potential benefits of a genome-wide approach are apparent for women and their partners who seek a more comprehensive cfDNA screening test during pregnancy.

RECOMMENDATIONS FOR PREGNANCY MANAGEMENT USING GENOME-WIDE NIPT
PROFESSIONAL SOCIETY GUIDELINES

National and Professional Society guidelines do not currently recommend the use of genome-wide cfDNA-based NIPT [99–101]. Insufficient clinical validation studies were available at the time existing guidelines were formulated to allow for a rigorous assessment. Additional studies are needed to demonstrate analytical validity, clinical validity, and clinical utility before professional society recommendations in favor of genome-wide cfDNA-based prenatal screening can emerge. However, several commercial [26, 28, 30, 86], academic [26, 31], and national public health services [27] have already begun to incorporate or consider this approach in their cfDNA screening protocols. Based on this collective experience, and existing knowledge from the CVS literature, a number of recommendations can be made with respect to genome-wide cfDNA screening.

PRE- AND POSTTEST COUNSELING CONSIDERATIONS, LABORATORY AND CLINICAL RESEARCH RECOMMENDATIONS
Counseling considerations
Pretest counseling for genome-wide NIPT should include a discussion of the benefits and limitations of a genome-wide approach.

- Benefits include ascertainment of a broader range of cytogenetic conditions associated with birth defects and other adverse pregnancy outcomes.
- Limitations include an increased chance of false-positive findings, including clinically significant maternal findings. Although rare, maternal findings may have important reproductive or other

maternal health implications. The clinical significance of some rare trisomy findings may remain uncertain, even after an amniocentesis diagnosis of TFM.

- Genome-wide NIPT is not a substitute for CVS or amniocentesis. Patients who seek a high degree of certainty in the detection of pathogenic CNVs should consider prenatal diagnosis with chromosome microarray. Prenatal diagnosis should be advised when fetal abnormalities are identified on ultrasound, while respecting parental choice.
- Sensitivity estimates for segmental aneuploidy screening are not well defined but are high (>90%) based on statistical modeling. Detection rates for clinically significant rare trisomies are high based on the strong association between high trisomic fraction relative to fetal fraction and adverse pregnancy outcomes. Smaller CNV size, lower fetal fraction (<5%), lower sequence tag count, and placental mosaicism will limit the detection of segmental aneuploidies in some pregnancies.
- The chance of an NIPT result being called as increased risk for a rare trisomy or segmental aneuploidy in women with average to high risk is 0.3%–0.5%, and rarely exceeds 1%, depending on the reason for referral. Higher call rates may occur in pregnancies with high-risk CFTS results (>1:200).
- PPVs for larger segmental aneuploidies (>7–10 Mb) are estimated at 50%–60% (similar to the PPV for trisomy 13 in average- to high-risk women). PPVs for rare trisomies are frequently complicated by CPM. At least 30%–50% of pregnancies with rare trisomies ascertained at 10–13 weeks of gestational age will proceed uneventfully, whereas the remaining pregnancies may have complications such as miscarriage, TFM, UPD, and IUGR.

Posttest counseling

- For rare autosomal trisomy results ascertained at 10–12 weeks of pregnancy, a follow-up ultrasound scan can be recommended if this was not done immediately prior to blood draw. This may confirm a missed miscarriage or abnormal fetus.
- A rare autosomal trisomy result will usually require amniocentesis for a definitive prenatal diagnosis. Ultrasound investigation may help with decision making in the first trimester of pregnancy. Van Opstal and Srebniak have provided prudent advice on pregnancy management of rare trisomies identified using cfDNA-based NIPT, which is based on a review of a large dataset of abnormal CVS cytogenetic results with known outcome [65]. Although CVS was suggested as a diagnostic option for some rare trisomies that are usually found confined to the cytotrophoblast (e.g., 3, 7, 8, 9, 20), our own clinical service always recommends amniocentesis if a fetal anatomy scan appears normal. CVS may be helpful for some women who want more timely information about their pregnancy, although amniocentesis may also be required if the rare trisomy is identified in the LT-CVS mesenchyme.
- Some women may opt for fetal surveillance using expert ultrasound rather than amniocentesis after an NIPT finding of rare trisomy. Normal fetal ultrasound scans may sometimes be falsely reassuring.
- Closer fetal surveillance in ongoing pregnancies with rare trisomies is advised, particularly for trisomies 2 and 16, which have a substantially increased risk for IUGR and other pregnancy complications.

- UPD studies are recommended for trisomies involving imprinted chromosomes 6, 7, 11, 14, 15, and 20. The risk for maternal UPD15 appears to be higher than that predicted by the CVS literature. A high trisomic fraction relative to the fetal fraction is a strong predictor of a trisomic conception (and subsequently a 1 in 3 risk for UPD). The risk for UPD 7 appears lower and may be more in keeping with the low risk predicted by the CVS literature. Trisomy for imprinted chromosomes other than 7 and 15 is very rare in ongoing pregnancies identified at 10–12 weeks of gestational age, so the requirement for UPD studies will also be rare.
- CVS with chromosome microarray is recommended for segmental aneuploidies identified using cfDNA screening in the first trimester of pregnancy. Amniocentesis might also be required for rare or unusual mosaic abnormalities that are confirmed on CVS.

Laboratory considerations

cfDNA screening laboratories should provide clear and unambiguous recommendations on genome-wide findings.

- When reporting segmental aneuploidies, NIPT laboratories should provide approximate genomic coordinates of the affected region to help facilitate cytogenetic prenatal diagnosis.
- Chromosome microarray is advised for cytogenetic prenatal diagnosis of segmental aneuploidies. In particular, SNP microarray can be helpful for both analysis of segmental aneuploidy and for the investigation of rare trisomy mosaicism. Conventional chromosome analysis can supplement these findings where this is considered helpful.

Clinical research recommendations and conclusions

Genome-wide cfDNA screening provides an unparalleled opportunity to identify pregnancies affected by placental mosaicism, to quantify the degree of mosaicism affecting the placental cytotrophoblast, and to systematically monitor and record the impact of this mosaicism on placental health and fetal wellbeing in a way that has not previously been possible. Although rare trisomy mosaicism may not affect fetal tissues, its effect on placental health can be profound, as is evidenced by placental trisomy 16 mosaicism [74, 77], and by high-grade trisomy 2 mosaicism that also affects the cytotrophoblast (this report). For most other rare trisomies, the natural history of pregnancy and birth outcomes are not well known or are poorly understood.

Thus for amniocentesis investigations that are discordant with cfDNA-based NIPT results, or which show low-grade trisomy mosaicism in continuing pregnancies, we recommend systematically recording pregnancy history and birth details. Biopsies of chorionic villi can be taken from the smooth (fetal) side of the placenta after birth to allow for investigation of placental mosaicism. Our own clinical service finds patients are often very willing to provide consent and to donate placental samples for quality control and clinical research purposes. Others have reported women and their caregivers being similarly receptive [102]. A minimum of two physically separated placental biopsies is recommended but some researchers obtain up to three to five biopsies [27, 87, 102]. Confirmation rates in chorionic villi for the rare trisomies are typically very high [26, 27, 87] and the chromosomal origin of the trisomy (meiotic vs mitotic) can often be determined by using SNP microarray [26]. In other circumstances testing maternal or newborn samples may be appropriate to determine the origin of discordant prenatal cfDNA findings.

Further validation studies are also needed for cfDNA screening of segmental aneuploidies to better determine sensitivity and specificity estimates in the clinical setting, as well as for determining PPVs. Our recent experience is that placental mosaicism will also confound screening in this setting. The cytogenetic findings in the placenta may often show an intermediate step toward a related but otherwise different pathogenic cytogenetic finding in the fetus. In other instances the abnormality will be confined to the placenta and the fetus will have a normal karyotype. For these reasons prenatal diagnosis using chromosome microarray is recommended and will always be required to confirm the presence or absence of suspected segmental aneuploidy in the fetus.

Finally, it is inevitable that the use of genome-wide cfDNA screening will expand as evidence builds for the sensitive and specific noninvasive detection of a much broader range of clinically significant genomic conditions. This momentum is being driven by academic centers that combine cfDNA screening with molecular-based prenatal diagnosis enabling rapid accumulation of new knowledge to help inform clinical practice. We are witness to a profound and fundamental change—the dawn of a golden age of prenatal screening genomics.

ACKNOWLEDGMENTS

I am indebted to my clinical laboratory and analytical teams and thank them for their support—Nicola Flowers, Olivia Giouzeppos, Grace Shi, Clare Love, Rebecca Manser, Ian Burns, Shelley Baeffel, Sera Tsegay, Tom Harrington, and LaiEs Carver.

Research conducted at the Murdoch Children's Research Institute was supported by the State Government of Victoria's Operational Infrastructure Support Program.

REFERENCES

[1] Palomaki GE, Kloza EM, Lambert-Messerlian GM, Haddow JE, Neveux LM, Ehrich M, et al. DNA sequencing of maternal plasma to detect down syndrome: an international clinical validation study. Genet Med 2011;13(11):913–20.

[2] Bianchi DW, Platt LD, Goldberg JD, Abuhamad AZ, Sehnert AJ, Rava RP, et al. Genome-wide fetal aneuploidy detection by maternal plasma DNA sequencing. Obstet Gynecol 2012;119(5):890–901.

[3] Norton ME, Brar H, Weiss J, Karimi A, Laurent LC, Caughey AB, et al. Non-invasive chromosomal evaluation (NICE) study: results of a multicenter prospective cohort study for detection of fetal trisomy 21 and trisomy 18. Am J Obstet Gynecol 2012;207(2). 137 e1-8.

[4] Chiu RW, Akolekar R, Zheng YW, Leung TY, Sun H, Chan KC, et al. Non-invasive prenatal assessment of trisomy 21 by multiplexed maternal plasma DNA sequencing: large scale validity study. BMJ 2011;342: c7401.

[5] Gil MM, Accurti V, Santacruz B, Plana MN, Nicolaides KH. Analysis of cell-free DNA in maternal blood in screening for aneuploidies: updated meta-analysis. Ultrasound Obstet Gynecol 2017;50(3):302–14.

[6] Bianchi DW, Parker RL, Wentworth J, Madankumar R, Saffer C, Das AF, et al. DNA sequencing versus standard prenatal aneuploidy screening. N Engl J Med 2014;370(9):799–808.

[7] Norton ME, Jacobsson B, Swamy GK, Laurent LC, Ranzini AC, Brar H, et al. Cell-free DNA analysis for noninvasive examination of trisomy. N Engl J Med 2015;372(17):1589–97.

[8] Evans MI, Evans SM, Bennett TA, Wapner RJ. The price of abandoning diagnostic testing for cell-free fetal DNA screening. Prenat Diagn 2018;38:243–5.

[9] Evans MI, Wapner RJ, Berkowitz RL. Noninvasive prenatal screening or advanced diagnostic testing: caveat emptor. Am J Obstet Gynecol 2016;215(3):298–305.

[10] Vogel I, Petersen OB, Christensen R, Hyett J, Lou S, Vestergaard EM. Chromosomal microarray as primary diagnostic genomic tool for pregnancies at increased risk within a population-based combined first-trimester screening program. Ultrasound Obstet Gynecol 2017;50(Suppl.1):44.

[11] Petersen OB, Vogel I, Ekelund C, Hyett J, Tabor A. Danish fetal medicine study G, et al. potential diagnostic consequences of applying non-invasive prenatal testing: population-based study from a country with existing first-trimester screening. Ultrasound Obstet Gynecol 2014;43(3):265–71.

[12] Srebniak MI, Knapen M, Polak M, Joosten M, Diderich KEM, Govaerts LCP, et al. The influence of SNP-based chromosomal microarray and NIPT on the diagnostic yield in 10,000 fetuses with and without fetal ultrasound anomalies. Hum Mutat 2017;38(7):880–8.

[13] Lindquist A, Poulton A, Halliday J, Hui L. Prenatal diagnostic testing and atypical chromosome abnormalities following combined first-trimester screening: Implications for contingent models of non-invasive prenatal testing. Ultrasound Obstet Gynecol 2018;51:487–92.

[14] Shaffer LG, Rosenfeld JA, Dabell MP, Coppinger J, Bandholz AM, Ellison JW, et al. Detection rates of clinically significant genomic alterations by microarray analysis for specific anomalies detected by ultrasound. Prenat Diagn 2012;32(10):986–95.

[15] Wapner RJ, Martin CL, Levy B, Ballif BC, Eng CM, Zachary JM, et al. Chromosomal microarray versus karyotyping for prenatal diagnosis. N Engl J Med 2012;367(23):2175–84.

[16] Hayward J, Chitty LS. Beyond screening for chromosomal abnormalities: advances in non-invasive diagnosis of single gene disorders and fetal exome sequencing. Semin Fetal Neonatal Med 2018;23(2):94–101.

[17] Wapner RJ, Babiarz JE, Levy B, Stosic M, Zimmermann B, Sigurjonsson S, et al. Expanding the scope of noninvasive prenatal testing: detection of fetal microdeletion syndromes. Am J Obstet Gynecol 2015;212(3). 332 e1-9.

[18] Schwartz S, Kohan M, Pasion R, Papenhausen PR, Platt LD. Clinical experience of laboratory follow-up with noninvasive prenatal testing using cell-free DNA and positive microdeletion results in 349 cases. Prenat Diagn 2018;38(3):210–8.

[19] Martin K, Iyengar S, Kalyan A, Lan C, Simon AL, Stosic M, et al. Clinical experience with a single-nucleotide polymorphism-based non-invasive prenatal test for five clinically significant microdeletions. Clin Genet 2018;93(2):293–300.

[20] Petersen AK, Cheung SW, Smith JL, Bi W, Ward PA, Peacock S, et al. Positive predictive value estimates for cell-free noninvasive prenatal screening from data of a large referral genetic diagnostic laboratory. Am J Obstet Gynecol 2017;217(6). 691 e1-e6.

[21] Lefkowitz RB, Tynan JA, Liu T, Wu Y, Mazloom AR, Almasri E, et al. Clinical validation of a noninvasive prenatal test for genomewide detection of fetal copy number variants. Am J Obstet Gynecol 2016;215(2). 227 e1-e16.

[22] Srinivasan A, Bianchi DW, Huang H, Sehnert AJ, Rava RP. Noninvasive detection of fetal subchromosome abnormalities via deep sequencing of maternal plasma. Am J Hum Genet 2013;92(2):167–76.

[23] Straver R, Sistermans EA, Holstege H, Visser A, Oudejans CB, Reinders MJ. WISECONDOR: detection of fetal aberrations from shallow sequencing maternal plasma based on a within-sample comparison scheme. Nucleic Acids Res 2014;42(5):e31.

[24] Zhao C, Tynan J, Ehrich M, Hannum G, McCullough R, Saldivar JS, et al. Detection of fetal subchromosomal abnormalities by sequencing circulating cell-free DNA from maternal plasma. Clin Chem 2015;61(4):608–16.

[25] Bayindir B, Dehaspe L, Brison N, Brady P, Ardui S, Kammoun M, et al. Noninvasive prenatal testing using a novel analysis pipeline to screen for all autosomal fetal aneuploidies improves pregnancy management. Eur J Hum Genet 2015;23:1286–93.

[26] Pertile MD, Halks-Miller M, Flowers N, Barbacioru C, Kinnings SL, Vavrek D, et al. Rare autosomal trisomies, revealed by maternal plasma DNA sequencing, suggest increased risk of feto-placental disease. Sci Transl Med 2017;9(405).

[27] Van Opstal D, van Maarle MC, Lichtenbelt K, Weiss MM, Schuring-Blom H, Bhola SL, et al. Origin and clinical relevance of chromosomal aberrations other than the common trisomies detected by genome-wide NIPS: results of the TRIDENT study. Genet Med 2018;20:480–5.

[28] Ehrich M, Tynan J, Mazloom A, Almasri E, McCullough R, Boomer T, et al. Genome-wide cfDNA screening: Clinical laboratory experience with the first 10,000 cases. Genet Med 2017;19:1332–7.

[29] Pescia G, Guex N, Iseli C, Brennan L, Osteras M, Xenarios I, et al. Cell-free DNA testing of an extended range of chromosomal anomalies: clinical experience with 6,388 consecutive cases. Genet Med 2017;19:169–75.

[30] Chen S, Lau TK, Zhang C, Xu C, Xu Z, Hu P, et al. A method for noninvasive detection of fetal large deletions/duplications by low coverage massively parallel sequencing. Prenat Diagn 2013;33(6):584–90.

[31] Brady P, Brison N, Van Den Bogaert K, de Ravel T, Peeters H, Van Esch H, et al. Clinical implementation of NIPT—technical and biological challenges. Clin Genet 2016;89(5):523–30.

[32] Lau TK, Cheung SW, Lo PS, Pursley AN, Chan MK, Jiang F, et al. Non-invasive prenatal testing for fetal chromosomal abnormalities by low-coverage whole-genome sequencing of maternal plasma DNA: review of 1982 consecutive cases in a single center. Ultrasound Obstet Gynecol 2014;43(3):254–64.

[33] Biesecker LG, Spinner NB. A genomic view of mosaicism and human disease. Nat Rev Genet 2013;14(5):307–20.

[34] Wallerstein R, Misra S, Dugar RB, Alem M, Mazzoni R, Garabedian MJ. Current knowledge of prenatal diagnosis of mosaic autosomal trisomy in amniocytes: karyotype/phenotype correlations. Prenat Diagn 2015;35(9):841–7.

[35] RJM G, Sutherland GR, Shaffer LG. Chromosome abnormalities and genetic counseling. 4th ed. Oxford: Oxford University Press; 2012. xiv, 634 pp.

[36] Houlihan OA, O'Donoghue K. The natural history of pregnancies with a diagnosis of trisomy 18 or trisomy 13; a retrospective case series. BMC Pregnancy Childbirth 2013;13:209.

[37] Morris JK, Wald NJ, Watt HC. Fetal loss in down syndrome pregnancies. Prenat Diagn 1999;19(2):142–5.

[38] Wellesley D, Dolk H, Boyd PA, Greenlees R, Haeusler M, Nelen V, et al. Rare chromosome abnormalities, prevalence and prenatal diagnosis rates from population-based congenital anomaly registers in Europe. Eur J Hum Genet 2012;20(5):521–6.

[39] Gosden CM. Amniotic fluid cell types and culture. Br Med Bull 1983;39(4):348–54.

[40] Robinson WP, McFadden DE, Barrett IJ, Kuchinka B, Penaherrera MS, Bruyere H, et al. Origin of amnion and implications for evaluation of the fetal genotype in cases of mosaicism. Prenat Diagn 2002;22(12):1076–85.

[41] Bui TH, Iselius L, Lindsten J. European collaborative study on prenatal diagnosis: mosaicism, pseudomosaicism and single abnormal cells in amniotic fluid cell cultures. Prenat Diagn 1984;4. Spec No:145-62.

[42] Hsu LY, Perlis TE. United States survey on chromosome mosaicism and pseudomosaicism in prenatal diagnosis. Prenat Diagn 1984;4. Spec No:97-130.

[43] Worton RG, Stern R. A Canadian collaborative study of mosaicism in amniotic fluid cell cultures. Prenat Diagn 1984;4. Spec No:131-44.

[44] Hsu LY, Yu MT, Neu RL, Van Dyke DL, Benn PA, Bradshaw CL, et al. Rare trisomy mosaicism diagnosed in amniocytes, involving an autosome other than chromosomes 13, 18, 20, and 21: karyotype/phenotype correlations. Prenat Diagn 1997;17(3):201–42.

[45] Wallerstein R, Yu MT, Neu RL, Benn P, Lee Bowen C, Crandall B, et al. Common trisomy mosaicism diagnosed in amniocytes involving chromosomes 13, 18, 20 and 21: karyotype-phenotype correlations. Prenat Diagn 2000;20(2):103–22.

[46] Hui L, Norton M. What is the real "price" of more prenatal screening and fewer diagnostic procedures? Costs and trade-offs in the genomic era. Prenat Diagn 2018;38(4):246–9.

[47] Bruno DL, White SM, Ganesamoorthy D, Burgess T, Butler K, Corrie S, et al. Pathogenic aberrations revealed exclusively by single nucleotide polymorphism (SNP) genotyping data in 5000 samples tested by molecular karyotyping. J Med Genet 2011;48(12):831–9.

[48] Brambati B, Simoni G, Danesino C, Oldrini A, Ferrazzi E, Romitti L, et al. First trimester fetal diagnosis of genetic disorders: clinical evaluation of 250 cases. J Med Genet 1985;22(2):92–9.

[49] Hogge WA, Schonberg SA, Golbus MS. Chorionic villus sampling: experience of the first 1000 cases. Am J Obstet Gynecol 1986;154(6):1249–52.

[50] Simoni G, Gimelli G, Cuoco C, Romitti L, Terzoli G, Guerneri S, et al. First trimester fetal karyotyping: one thousand diagnoses. Hum Genet 1986;72(3):203–9.

[51] Simoni G, Brambati B, Danesino C, Terzoli GL, Romitti L, Rossella F, et al. Diagnostic application of first trimester trophoblast sampling in 100 pregnancies. Hum Genet 1984;66(2–3):252–9.

[52] Bryndorf T, Christensen B, Vad M, Parner J, Carelli MP, Ward BE, et al. Prenatal detection of chromosome aneuploidies in uncultured chorionic villus samples by FISH. Am J Hum Genet 1996;59(4):918–26.

[53] Holgado E, Liddle S, Ballard T, Levett L. Incidence of placental mosaicism leading to discrepant results between QF-PCR and karyotyping in 22,825 chorionic villus samples. Prenat Diagn 2011;31(11):1029–38.

[54] Dugoff L, Norton ME, Kuller JA. The use of chromosomal microarray for prenatal diagnosis. Am J Obstet Gynecol 2016;215(4):B2–9.

[55] Callen DF, Korban G, Dawson G, Gugasyan L, Krumins EJ, Eichenbaum S, et al. Extra embryonic/fetal karyotypic discordance during diagnostic chorionic villus sampling. Prenat Diagn 1988;8(6):453–60.

[56] Kalousek DK, Dill FJ, Pantzar T, McGillivray BC, Yong SL, Wilson RD. Confined chorionic mosaicism in prenatal diagnosis. Hum Genet 1987;77(2):163–7.

[57] Malvestiti F, Agrati C, Grimi B, Pompilii E, Izzi C, Martinoni L, et al. Interpreting mosaicism in chorionic villi: results of a monocentric series of 1001 mosaics in chorionic villi with follow-up amniocentesis. Prenat Diagn 2015;35(11):1117–27.

[58] Kalousek DK, Barrett IJ, Gartner AB. Spontaneous abortion and confined chromosomal mosaicism. Hum Genet 1992;88(6):642–6.

[59] Lestou VS, Kalousek DK. Confined placental mosaicism and intrauterine fetal growth. Arch Dis Child Fetal Neonatal Ed 1998;79(3):F223–6.

[60] Robinson WP, Barrett IJ, Bernard L, Telenius A, Bernasconi F, Wilson RD, et al. Meiotic origin of trisomy in confined placental mosaicism is correlated with presence of fetal uniparental disomy, high levels of trisomy in trophoblast, and increased risk of fetal intrauterine growth restriction. Am J Hum Genet 1997;60(4):917–27.

[61] Toutain J, Labeau-Gauzere C, Barnetche T, Horovitz J, Saura R. Confined placental mosaicism and pregnancy outcome: a distinction needs to be made between types 2 and 3. Prenat Diagn 2010;30(12–13):1155–64.

[62] Wolstenholme J. Confined placental mosaicism for trisomies 2, 3, 7, 8, 9, 16, and 22: their incidence, likely origins, and mechanisms for cell lineage compartmentalization. Prenat Diagn 1996;16(6):511–24.

[63] Toutain J, Goutte-Gattat D, Horovitz J, Saura R. Confined placental mosaicism revisited: Impact on pregnancy characteristics and outcome. PLoS One 2018;13(4)e0195905.

[64] Hahnemann JM, Vejerslev LO. European collaborative research on mosaicism in CVS (EUCROMIC)–fetal and extrafetal cell lineages in 192 gestations with CVS mosaicism involving single autosomal trisomy. Am J Med Genet 1997;70(2):179–87.

[65] Van Opstal D, Srebniak MI. Cytogenetic confirmation of a positive NIPT result: evidence-based choice between chorionic villus sampling and amniocentesis depending on chromosome aberration. Expert Rev Mol Diagn 2016;16(5):513–20.

[66] Bianchi DW, Wilkins-Haug LE, Enders AC, Hay ED. Origin of extraembryonic mesoderm in experimental animals: relevance to chorionic mosaicism in humans. Am J Med Genet 1993;46(5):542–50.

[67] Grati FR, Malvestiti F, Ferreira JC, Bajaj K, Gaetani E, Agrati C, et al. Fetoplacental mosaicism: potential implications for false-positive and false-negative noninvasive prenatal screening results. Genet Med 2014;16(8):620–4.

[68] Grati FR, Malvestiti F, Branca L, Agrati C, Maggi F, Simoni G. Chromosomal mosaicism in the fetoplacental unit. Best Pract Res Clin Obstet Gynaecol 2017;42:39–52.

[69] American College of Medical Genetics and Genomics (ACMG). Standards and Guidelines for Clinical Genetics Laboratories, 2018 ed., Revised January 2018, Available from. http://www.acmg.net/docs/Section_E_of_the_ACMG_Lab_Standards_and_Guidelines_(Jan%202018%20rev).pdf; 2018.

[70] Association for Clinical Cytogenetics. Prenatal diagnosis best practice guidelines V1.00. Available from, http://www.acgs.uk.com/media/765666/acc_prenatal_bp_dec2009_1.00.pdf; 2009.

[71] ECA. European cytogeneticists association newsletter No. 30 July 2012, specific constitutional cytogenetic guidelines, Available from, http://www.e-c-a.eu/files/downloads/Guidelines/Specific_Constitutional_Guidelines_NL30.pdf; 2012.

[72] Wolstenholme J, Rooney DE, Davison EV. Confined placental mosaicism, IUGR, and adverse pregnancy outcome: a controlled retrospective U.K. collaborative survey. Prenat Diagn 1994;14(5):345–61.

[73] Yong PJ, Langlois S, von Dadelszen P, Robinson W. The association between preeclampsia and placental trisomy 16 mosaicism. Prenat Diagn 2006;26(10):956–61.

[74] Sparks TN, Thao K, Norton ME. Mosaic trisomy 16: what are the obstetric and long-term childhood outcomes? Genet Med 2017;19(10):1164–70.

[75] Eggermann T, Soellner L, Buiting K, Kotzot D. Mosaicism and uniparental disomy in prenatal diagnosis. Trends Mol Med 2015;21(2):77–87.

[76] Daniel A, Wu Z, Darmanian A, Malafiej P, Tembe V, Peters G, et al. Issues arising from the prenatal diagnosis of some rare trisomy mosaics–the importance of cryptic fetal mosaicism. Prenat Diagn 2004;24(7):524–36.

[77] Benn P. Trisomy 16 and trisomy 16 mosaicism: a review. Am J Med Genet 1998;79(2):121–33.

[78] Hassold T, Merrill M, Adkins K, Freeman S, Sherman S. Recombination and maternal age-dependent nondisjunction: molecular studies of trisomy 16. Am J Hum Genet 1995;57(4):867–74.

[79] Spencer K, Pertile MD, Bonacquisto L, Mills I, Turner S, Donalson K, et al. First trimester detection of trisomy 16 using combined biochemical and ultrasound screening. Prenat Diagn 2014;34(3):291–5.

[80] Smith GC, Stenhouse EJ, Crossley JA, Aitken DA, Cameron AD, Connor JM. Early pregnancy levels of pregnancy-associated plasma protein a and the risk of intrauterine growth restriction, premature birth, preeclampsia, and stillbirth. J Clin Endocrinol Metab 2002;87(4):1762–7.

[81] Torring N, Petersen OB, Becher N, Vogel I, Uldbjerg N. First trimester screening for other trisomies than trisomy 21, 18, and 13. Prenat Diagn 2015;35(6):612–9.

[82] Snyder HL, Curnow KJ, Bhatt S, Bianchi DW. Follow-up of multiple aneuploidies and single monosomies detected by noninvasive prenatal testing: implications for management and counseling. Prenat Diagn 2016;36(3):203–9.

[83] Bianchi DW, Chudova D, Sehnert AJ, Bhatt S, Murray K, Prosen TL, et al. Noninvasive prenatal testing and incidental detection of occult maternal malignancies. JAMA 2015;314(2):162–9.

[84] Amant F, Verheecke M, Wlodarska I, Dehaspe L, Brady P, Brison N, et al. Presymptomatic identification of cancers in pregnant women during noninvasive prenatal testing. JAMA Oncol 2015;1(6):814–9.

[85] Dharajiya NG, Grosu DS, Farkas DH, McCullough RM, Almasri E, Sun Y, et al. Incidental detection of maternal neoplasia in noninvasive prenatal testing. Clin Chem 2018;64(2):329–35.

[86] Fiorentino F, Bono S, Pizzuti F, Duca S, Polverari A, Faieta M, et al. The clinical utility of genome-wide non invasive prenatal screening. Prenat Diagn 2017;37(6):593–601.

[87] Brison N, Neofytou M, Dehaspe L, Bayindir B, Van Den Bogaert K, Dardour L, et al. Predicting fetopla- cental chromosomal mosaicism during non-invasive prenatal testing. Prenat Diagn 2018;38(4):258–66.

[88] Philipp T, Terry J, Feichtinger M, Grillenberger S, Hartmann B, Jirecek S. Morphology of early intrauterine deaths with full trisomy 15. Prenat Diagn 2018;38(4):267–72.

[89] Benn P, Cuckle H. Theoretical performance of non-invasive prenatal testing for chromosome imbalances using counting of cell-free DNA fragments in maternal plasma. Prenat Diagn 2014;34(8):778–83.

[90] Lo KK, Karampetsou E, Boustred C, McKay F, Mason S, Hill M, et al. Limited clinical utility of non- invasive prenatal testing for subchromosomal abnormalities. Am J Hum Genet 2016;98(1):34–44.

[91] Fan HC, Quake SR. Sensitivity of noninvasive prenatal detection of fetal aneuploidy from maternal plasma using shotgun sequencing is limited only by counting statistics. PLoS One 2010;5(5)e10439.

[92] Li R, Wan J, Zhang Y, Fu F, Ou Y, Jing X, et al. Detection of fetal copy number variants by non-invasive prenatal testing for common aneuploidies. Ultrasound Obstet Gynecol 2016;47(1):53–7.

[93] Liao C, Yin AH, Peng CF, Fu F, Yang JX, Li R, et al. Noninvasive prenatal diagnosis of common aneu- ploidies by semiconductor sequencing. Proc Natl Acad Sci U S A 2014;111(20):7415–20.

[94] Helgeson J, Wardrop J, Boomer T, Almasri E, Paxton WB, Saldivar JS, et al. Clinical outcome of subchro- mosomal events detected by whole-genome noninvasive prenatal testing. Prenat Diagn 2015; 35(10):999–1004.

[95] Yin AH, Peng CF, Zhao X, Caughey BA, Yang JX, Liu J, et al. Noninvasive detection of fetal subchromo- somal abnormalities by semiconductor sequencing of maternal plasma DNA. Proc Natl Acad Sci U S A 2015;112(47):14670–5.

[96] Marshall OJ, Chueh AC, Wong LH, Choo KH. Neocentromeres: new insights into centromere structure, disease development, and karyotype evolution. Am J Hum Genet 2008;82(2):261–82.

[97] Flowers N, Kelley J, Sigurjonsson S, Bruno DL, Pertile MD. Maternal mosaicism for a large segmental duplication of 18q as a secondary finding following non-invasive prenatal testing and implications for test accuracy. Prenat Diagn 2015;35(10):986–9.

[98] Taneja PA, Snyder HL, de Feo E, Kruglyak KM, Halks-Miller M, Curnow KJ, et al. Noninvasive prenatal testing in the general obstetric population: clinical performance and counseling considerations in over 85 000 cases. Prenat Diagn 2016;36(3):237–43.

[99] American College of Obstetricians and Gynecologists. Committee opinion no. 640: Cell-free DNA screen- ing for fetal aneuploidy. Obstet Gynecol 2015;126. e31–7.

[100] Gregg AR, Skotko BG, Benkendorf JL, Monaghan KG, Bajaj K, Best RG, et al. Noninvasive prenatal screening for fetal aneuploidy, 2016 update: a position statement of the American College of Medical Genetics and Genomics. Genet Med 2016;18(10):1056–65.

[101] Salomon LJ, Alfirevic Z, Audibert F, Kagan KO, Paladini D, Yeo G, et al. ISUOG updated consensus state- ment on the impact of cfDNA aneuploidy testing on screening policies and prenatal ultrasound practice. Ultrasound Obstet Gynecol 2017;49(6):815–6.

[102] Wilkins-Haug L, Zhang C, Cerveira E, Ryan M, Mil-Homens A, Zhu Q, et al. Biological explanations for discordant noninvasive prenatal test results: preliminary data and lessons learned. Prenat Diagn 2018;38:445–58.

NONINVASIVE FETAL BLOOD GROUP TYPING

8

C. Ellen van der Schoot*,†, Dian Winkelhorst*,‡, Frederik B. Clausen§

Department of Experimental Immunohematology, Sanquin Research, Amsterdam, The Netherlands* Landsteiner Laboratory, Academic Medical Centre, University of Amsterdam, Amsterdam, The Netherlands† Department of Obstetrics and Gynaecology, Leiden University Medical Center, Amsterdam, The Netherlands‡ Department of Clinical Immunology, Rigshospitalet, Copenhagen University Hospital, Copenhagen, Denmark§

HEMOLYTIC DISEASE OF THE FETUS AND NEWBORN

Hemolytic disease of the fetus and newborn (HDFN) is a disease caused by maternal IgG alloantibodies against paternally inherited RBC alloantigens, for which the maternal cells are negative. The maternal alloantibodies are induced by fetomaternal hemorrhage (FMH) in a previous or current pregnancy or by prior incompatible blood transfusion. The risk for immunization by FMH is highest in the third trimester and during labor, but immunization can also occur in the first and second trimester. Alloantibodies of the IgG class are actively transported to the fetal circulation by the neonatal Fc receptor (FcRn) expressed on the placenta and bind to fetal erythroid cells carrying the alloantigen. Dependent on characteristics of the antibody, the opsonized RBCs can be destroyed in the reticuloendothelial cells in the spleen and/or Kupffer cells in the liver. Antibodies against antigens of the Kell and also the MNS (anti-Mur) blood group system induce anemia mainly by suppressing erythropoiesis [1,2]. HDFN has been a major cause of fetal and neonatal death throughout history. The clinical picture of untreated HDFN is very variable and starts to develop in fetal life. In some cases, there is hardly any sign of fetal anemia, whereas in more severe cases there is an anemia, which can cause cardiomegaly and hepatosplenomegaly in the fetus. In the most severe cases fetal hydrops develops, a condition consisting of edema in the fetal skin and serous cavities. Hemolysis of the fetal red cells results in raised bilirubin levels, but because bilirubin can pass the placenta, excess of bilirubin is cleared via the maternal circulation during pregnancy. After birth, the hemolytic process continues, but the neonate's liver is not yet capable of sufficiently conjugating the excess bilirubin. This may result in severe hyperbilirubinemia and, when untreated, in irreversible damage to the central nervous system by bilirubin deposition in the basal ganglia and brain stem nuclei, a condition known as "kernicterus," in most cases a fatal disease. Children who survive have permanent cerebral damage, characterized by choreoathetosis, spasticity, and hearing problems.

TREATMENT OF HDFN

If alloimmunization is detected in time, the disease can be prevented in most cases. Treatment of HDFN depends on the severity of fetal anemia. In the past, bilirubin levels in amniotic fluid were used to assess the anemia. Today, advances in Doppler ultrasonography have made noninvasive detection of fetal anemia possible. Mari et al. showed that ultrasonic Doppler determination of the middle cerebral artery peak systolic velocity (MCA-PSV) can be used as a surrogate measurement of fetal anemia [3]. In an international multicenter study it was demonstrated that MCA-PSV is superior to amniocentesis, with a sensitivity of 88% and a specificity of 82% [4]. In case of severe fetal anemia, the anemia can be corrected by intrauterine blood transfusions (IUT) and/or preterm labor might be induced followed by neonatal treatment. The first IUT is on average given around 26 weeks, range 16–35 weeks, the earlier transfusions are especially given in pregnancies complicated with anti-K antibodies. The perinatal survival rate of cases treated with IUT is around 95% [5] with a normal neurodevelopmental outcome in >95% [6]. The development of kernicterus after birth can be prevented by exchange transfusion after birth. In mild cases of HDFN phototherapy is sufficient. For timely treatment, all developed countries have screening programs in place to identify alloimmunized pregnancies at risk. Without a screening program, HDFN during pregnancy can only be suspected by decreased fetal movements or sudden fetal death, or by (early) neonatal jaundice after birth.

ANTI-D IMMUNOPROPHYLAXIS TO PREVENT IMMUNIZATION

Before the introduction of postnatal anti-D immunoprophylaxis HDFN was one of the major causes of perinatal death. In the Netherlands *postnatal* immunoprophylaxis was introduced in 1969 and resulted in a drastic decrease of the prevalence of anti-D immunization from 3.5% to 0.5% in the 1990s [7]. Postnatal prophylaxis has to be given within 72 h after delivery and traditionally on the basis of postnatal RhD typing of cord blood from the newborn [8]. Already in 1978, Bowman and colleagues demonstrated that the use of *antenatal* anti-D prophylaxis can further reduce immunization incidents [9]. The effects of combined antenatal and postnatal prophylaxis have been substantiated by more recent meta-analyses [10,11]. Combined prophylaxis reduces the immunization risk by half as compared to postnatal prophylaxis only, with a parallel reduction by half in severe HDFN cases [12,13]. Thus in several countries, antenatal prophylaxis is currently combined with postnatal prophylaxis to further minimize immunization risk. Antenatal prophylaxis is routinely given either as a single dose of 250–300 μg of immunoglobulin at gestational weeks 28–31, or as two 150-μg doses, respectively, at gestational weeks 29 and 34 [14]. In addition, it is given after invasive procedures during pregnancy, abdominal trauma, late miscarriage, or termination of pregnancy [8].

We recently obtained strong evidence that anti-D immunoprophylaxis is also preventing immunization toward other RBC antigens. Only 1 out of 99 pregnant women who became alloimmunized against non-Rh antigens during their first pregnancy had received anti-D immunoprophylaxis in this first pregnancy, which was significantly less often than the 10 women expected based on calculations for the general population [15]. Also ABO incompatibility can protect against immunization against other blood group antigens [15,16]. Theoretically, also immunoprophylaxis against Rhc or K, or by giving antibodies against a universal fetal antigen could have the same "wider" preventive effect, but so far, only anti-D immunoprophylaxis is available.

HDFN MEDIATED BY RhD AND NON-RhD ALLOANTIBODIES

Despite adequate antenatal and postnatal anti-D immunoprophylaxis still 1–3 in 1000 D-negative women develop anti-RhD antibodies, and in about 30% of these cases severe HDFN develops [12,17]. The immunization against RBC antigens other than RhD became relatively more important when the incidence of anti-D immunization decreased. The prevalence of non-D immunization as determined in large-scale (>70,000 women) studies of unselected pregnant women is calculated to be about 3 in 1000 women (Table 1). The specificities of non-D alloantibodies detected in a predominantly European population of pregnant women are, in order of frequency, anti-E, anti-K, anti-c, anti-C^w, anti-Fy^a, anti-S, anti-Jk^a, and anti-e [24]. Occasionally antibodies against Kp^a, anti-f, anti-Jk^b, and anti-s are detected. Antibodies against other blood group antigens are extremely rare. Anti-M antibodies are frequently found, but almost always of the IgM subclass and therefore harmless.

Any antibody that can pass the placenta and react with an antigen expressed by fetal RBCs can theoretically result in HDFN. However, only in a minority of immunized pregnancies, the presence of the antibody results in clinical disease. In 2008 we performed a Dutch nationwide prospective index-cohort study on 305,000 consecutive pregnancies screened in the 12th week of pregnancy, and determined the risk that the presence of non-RhD antibodies will lead to clinical signs of HDFN [22]. In Table 2 the results of this study are summarized. In total non-RhD antibodies were detected in 1279 pregnancies and in 403 cases the fetus was positive for the antigen. Severe HDFN, defined as need for intrauterine treatment and/or blood transfusion after birth, was almost exclusively seen in pregnancies with anti-c and anti-K antibodies, and rarely in pregnancies with other (non-D/c) Rh-antibodies. The risk of severe HDFN in case of anti-K antibodies has probably been underestimated in Table 2, because at present we observe that 50% of cases with anti-K antibodies lead to severe HDFN. Others have reported similar data [18–21,23,25]. All other antibodies (observed in this single center study) did not cause severe HDFN. In a cohort of 426 IUT-treated fetuses from the reference center for fetal treatment in the Netherlands after the introduction of antenatal prophylaxis, the vast majority of IUTs were still given because of alloimmunization against D (78%), compared to K in 15%, c in 3.5% and

Table 1 Prevalence of Non-RhD Antibodies and Non-RhD-Mediated HDFN in Screened Pregnant Women

Study	Screened Women	Immunization	Prevalence	HDFN	Prevalence
	n	*n*	*%*	*n*	*%*
Bowell, UK, 1968 [18]	70,000	315	0.45	NR	NR
Gottvall, Sweden, 1993 [19]	78,300	188	0.24	8	0.010
Filbey, Sweden, 1995 [20]	111,939	171	0.15	6	0.005
Gottvall, Sweden, 2008 [21]	78,145	196	0.25	8	0.010
Koelewijn, The Netherlands, 2008 [22]	305,700	1002	0.33	21	0.007
Dajak, Croatia, 2011 [23]	84,000	143	0.17	11	0.013
Total	728,084	2015	0.28	54	0.007

NR, *not reported.*

Table 2 Incidence of Severe HDFN in 403 Pregnancies Complicated With Non-RhD Antibodies and Antigen-Positive Fetuses

		Severe HDFN			
		Total	IUT	ET	BT
	Number	n (%)	n	n	n
Anti-K	19	5 (26)	4	1	0
Anti-c	118	12 (10)	1	6	5
Anti-E	95	2 (2)	0	1	1
Anti-Rh, other than -c, -D, -E	40	2 (5)	0	2	0
Anti-Duffy	42	0	–	–	–
Anti-A/B	30	0	–	–	–
Other	59	0	–	–	–

Severe HDFN was defined as perinatal death, intrauterine transfusion (IUT), exchange transfusion (ET), or blood transfusion (BT) within 168 h after birth because of RBC alloantibodies [22].

occasionally, antibodies against other antigens. [5] Similar observations were done in Denmark and Germany. [26,27]

In literature case reports of severe HDFN caused by a wide variety of other specificities have been described, but in population studies these specificities are seldom found as cause of severe HDFN. Anti-Duffy, anti-Kidd, and anti-MNS antibodies are relatively frequently found, but rarely result in severe HDFN [2,23,28], although they may lead to slightly decreased hemoglobin levels at birth and need for phototherapy. For antibodies against I, Le, P1, Lu, and Yt there is no risk for developing disease, because of very low expression of these antigens by fetal cells. Albeit the high rate of ABO incompatibility between the mother and her fetus, clinically significant hemolysis in neonates due to anti-A or anti-B is relatively low. And, if it occurs, it is usually mild [29]. For these reasons in Western countries pregnant women are not screened for the presence of anti-A or anti-B of IgG class. Because placental tissue expresses A and B antigens the maternal alloantibodies may be absorbed to some extent by the placenta. Furthermore, fetal RBCs show weak expression of A and B blood group antigens. There is a striking, and unexplained, difference in the incidence of ABO-mediated HDFN between populations. The incidence is around 0.3%–0.8% in the Caucasian population, vs 3%–5% in Black or Asian populations, also with a more severe clinical course [30,31]. Other alloantibodies detected in East Asian populations from China and Taiwan are antibodies specific for hybrid glycophorins of the MNS system, especially antibodies reactive with RBC expressing GP.Mur. This phenotype is virtually absent in European populations, whereas it occurs in 6%–8% of Chinese and Thai individuals, and 0.13% of pregnant Chinese women have anti-Mur antibodies [32]. The majority of anti-MUR antibodies are IgM, but IgG anti-MUR antibodies can result in very severe HDFN and might cause, in the absence of antibody screening, fatal hydrops fetalis [2].

In conclusion, in the Western world especially pregnancies complicated with anti-D, anti-c, anti-K, and possibly also anti-E should be carefully monitored. If fetal blood group typing demonstrates that the fetus is antigen positive, the titer of the antibody has to be investigated at regular intervals to check whether a critical titer is exceeded [33–35]. In that case the woman can be referred to a special care center for further clinical examinations like Doppler ultrasonography to determine the severity of anemia, and to start IUT in time or to induce labor, if needed.

MOLECULAR BASIS OF BLOOD GROUP SYSTEMS

Noninvasive fetal genotyping is especially indicated when alloantibodies are detected that might cause severe HDFN and therefore possibly indicate the need for active prenatal medical interventions during pregnancy, thus anti-D, anti-c, and anti-K. For pregnancies complicated with all other alloantibodies, fetal genotyping is mainly performed to either reassure the women and their care-givers, or to facilitate timeliness of postnatal medical interventions. In this chapter we will focus on the Rh- and Kell blood group systems.

RH SYSTEM

The Rh blood group system consists of two genes *RHD* and *RHCE* on chromosome 1 (1p36.11), positioned in opposite directions and separated by 31.8 kb, in which the TMEM50A gene (previously SMP1) is located [36]. The *RHD* gene arose as a duplication of the *RHCE* gene in the common ancestors of humans, chimpanzees, and gorillas [37]. Both genes have 10 exons and share an overall 93.8% gene sequence identity and 96.4% exon sequence identity [38,39]. The *RHD* gene is flanked by two 9 kb regions of 98.6% homology, the so-called Rh boxes [36]. The function of the RhD and RhCE polypeptides is unknown. They belong to the Rh family, but in contrast to three other members of this family (RhAG, RhBG, and RhCG) they most likely do not transport ammonia [40,41]. It has been suggested that they might play a role in CO_2 transport in RBC [42] but are functionally redundant. The *RHD* gene (NG_007494) encodes the RhD protein (CD240D), carrying the D antigen (RH1). The *RHCE* gene (NG_009208) encodes the RhCE protein (CD240CE), carrying the C (RH2) or c (RH4) and E (RH3) or e (RH5) antigens. Both proteins are red cell specific, consist of 417 amino acids, are not glycosylated, and have 12 membrane spanning domains. As shown in Fig. 1A, 37 nucleotides are specific for the coding sequence of *RHD* and not present in any of the *RHCE* alleles. Five nucleotides (in exon 2) are specific for *RHc*, of which the 307C encoding 103Pro is best correlated with c expression [43]. Exon 2 of *RHC* is identical to exon 2 of *RHD*, therefore only 48C (in exon 1) (16Cys) is specific for *RHC*. However, 16Cys is not correlated to C-expression and 74% of African blacks have 48C in a ce-allele with normal expression of Rhc [44]. Therefore *RHC* genotyping assays have to be based on a 109-bp insert in intron 2, that is, only present in RHC [45]. The E/e polymorphism is caused by a single nucleotide variation 676C > G (Pro226Ala) in exon 5 of *RHCE*. The 676G is also present in the *RHD* allele, but is in *RHD* surrounded by seven D specific nucleotides [46]. The frequency of the different alleles in the different populations is indicated in Fig. 1A.

In the RBC membrane the Rh proteins are complexed with RhAG, another member of the Rh family. The most likely in vivo subunit configuration of the Rh core complex is RhD-(RhAG)$_2$ and RhCcEe-(RhAG)$_2$ [47]. RhAG is essential for the assembly of the Rh complex in the erythrocyte membrane [48]. Mutations in the *RHAG* gene can result in the complete absence (regulator Rh$_{null}$) or severe reduction (Rh$_{mod}$) of both RhD and RhCE antigens [49]. The Rh complex is directly anchored to the cytoskeleton via Ankyrin R, and also mutations in *ANK1* can result in weakened expression of RhD and

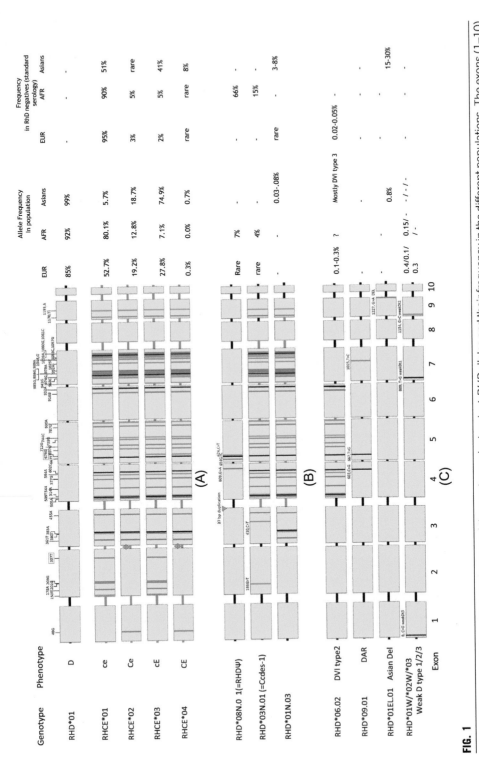

FIG. 1

Schematic representation of the *RH* genes and the most prevalent variant *RHD* alleles and their frequency in the different populations. The exons (1–10) are represented as yellow boxes, on scale; the introns are black for *RHD* and red for *RHCE* and are on scale with each other. The *RHD*01 gene, encoding the RhD polypeptide, is given as the reference sequence. The positions of the nucleotides differing between *RHD* and *RHCE* are listed above the top row, with the *RHD* nucleotide at that position, within red boxes the c and E/e determining nucleotides. Nucleotide substitutions are shown as red (T), black (G), green (A), or blue (C) lines, and if not given in the top row also the position is indicated. The ISBT allele name and the encoded phenotype are given at the left, and the frequencies in different populations at the right. (A) Five major alleles encoding the D and CcEe polypeptides. (B) Three most frequently occurring D-negative *RHD* alleles. (C) Most frequently occurring variant D-positive *RHD* alleles. In the bottom row three different weak D alleles, each characterized by a single nucleotide substitution. The *blue diamond* (◆) represents an intronic insert and the *blue triangle* (▼) represents a 37-bp

RhCE antigens [50]. The *RH* locus is highly polymorphic, and many hybrid *RHD-RHCE* gene variants have been described. The Rh antigens and *RH* alleles are numbered by the Working Party on Red Cell Immunogenetics and Blood Group Terminology of the International Society of Blood Transfusion and published on their website (http://www.isbtweb.org/working-parties/red-cell-immunogenetics-and-blood-group-terminology/). A complete database for *RHD* variants has been established by Wagner and Flegel and is continuously updated (http://www.rhesusbase.info/) [51]. At present (December 2017), 54 antigens, 378 *RHD* alleles, and 116 *RHCE* alleles have been recognized in the Rh blood group system. The main mechanism responsible for the generation of hybrid *RH* genes is thought to be gene conversion, explained by the opposite orientation of the highly homologous *RHD* and *RHCE* genes. Multiple exons can be converted, but often microconversion events lead to single amino acid changes. In addition, gene conversion events can be associated with untemplated mutations, in which the mutated nucleotides are not derived from either gene [52].

RHD NEGATIVITY

The frequency of the D negative phenotype is 15%–17% in populations of European descent, 5% in Africa, and very rare in Asian populations. Because D negativity will reduce reproductive fitness due to HDFN, it is surprising that D negativity has become so frequent in Europe, suggesting a positive selection for an as-of-yet unknown fitness benefit of the *RHD* deletion. Perry et al. found, however, no evidence that positive natural selection affected the frequency of the *RHD* deletion [53]. Thus the emergence of the *RHD* deletion in European populations may simply be explained by genetic drift/founder effect. Mourant indeed proposed already in 1954 a mixing of two populations, one essentially D− (Paleolithic peoples from the Basque region, which survived the last ice age) and the other D+ (Neolithic migrants) as cause for the high frequency of D negativity in Europeans [54]. Genetic analysis using mtDNA and Y chromosome markers has now provided strong evidence for this hypothesis [55,56]. In European and Asian people the D negative phenotype is usually caused by deletion of *RHD*, which has occurred between the Rh boxes, flanking the *RHD* gene [36]. Although many different nonsense mutations or splice site mutations have been described, each of them is rare. In contrast, in the African population deletion of *RHD* gene is not the only mechanism responsible for D negativity. Approximately 66% carry the *RHDΨ* gene and 15% carry a *RHD-CE-Ds* (r's) hybrid allele (Fig. 1B) [57]. The *RHDΨ* gene (RHD*08N.01) contains a 37 base pair duplication causing a frameshift, consisting of the last nucleotides of intron 3 and the first nucleotides of exon 4, and a nonsense mutation in exon 6 (807T>G, Tyr269stop) [57]. The African *RHD-CE-Ds* allele consists of exon 1, 2, and 3 of *RHD*DIIIa*; exons 4–7 of *RHCE*; and exons 8–10 of *RHD* (type 1 = RHD*03N.01) [58], but occasionally it is a hybrid *RHD* (exon1–2)-*RHCE*(exon3–7)-*RHD*(exon8–10) allele (type 2 = RHD*01N.06) [59]. These two genes produce no D, but do produce an abnormal C antigen. Because the *RHD-CE-Ds* alleles miss intron 2 on which *RHC* genotyping assays are based, reliable RhC prediction is also complicated in Africans. The observation that in Africa different mechanisms resulting in D negativity have emerged might indicate that there has been an ancient (unknown) selective pressure in Africa [55].

In Asian populations (China, Korea, Japan), 15%–30% of serologically typed D negative individuals carry the RHD*01EL.01 (*RHD*1227A*) gene, harboring a single mutation (rs549616139) causing a splice site defect resulting in the DEL phenotype [60,61]. In DEL individuals D expression on RBCs is extremely weak and can only be recognized by sensitive serological techniques such as the adsorption-elution test. These individuals are not at risk for RhD immunization [62]. Between 3% and 8% of truly D negative Asian individuals carry the D-negative hybrid *RHD*D-CE(2–9)-D* allele (RHD*01N.03) (Fig. 1B). In addition, the silent *RHD*711delC* (RHD*01N.16) allele is relatively frequent (>1%) in Chinese D negative individuals [61,63–65].

The presence of silent *RHD* genes in D negative individuals, shown in Fig. 1B, might hamper noninvasive fetal *RHD* genotyping assays. If the mother carries a silent *RHD* allele that is detected in the fetal genotyping assay, the genotype of the fetus cannot be reliably predicted. If the fetus inherited a silent *RHD* gene from the father, this will result in false-positive results. For that reason, it is important to take these variant *RHD* genes into account when designing fetal *RHD* genotyping assays, or at least in the interpretation of the assay results. This might also be true for the Asian silent alleles. In Asia, routine fetal *RHD* genotyping for D negative pregnant women seems to be not that relevant, because <5% of fathers are hemizygous for *RHD*. However, in a multiethnic society fetal genotyping assays should also be reliable for Asian D negative pregnant women, thus assays based on *RHD* exon 10 should be preferably avoided.

RHD VARIANT GENES WITH D EXPRESSION

As described previously the RH locus is highly polymorphic and numerous *RHD* variants have been described (see http://www.isbtweb.org/working-parties/red-cell-immunogenetics-and-blood-group-terminology/). *RHD* variants were originally subdivided into so-called DEL, weak D, and partial D. The most frequently occurring variant alleles are shown in Fig. 1C.

The Del phenotype is characterized by a very weak expression of the complete RhD protein, which expression is only detectable with the very sensitive absorption-elution technique [66]. The Del phenotype is most often caused by mutations that disturb splice sites or mutations in the C-terminal region of the RhD protein. DEL types are common in Asia (see earlier) and less common in Europe. The DEL frequency in the European population is 1:350 to 1:2000 [67].

Individuals with weak D expression express the complete RhD protein, however, in low quantities [52]. 0.2%–1% of populations of European descent express weak D antigens. Weak D expression is caused mostly by single mutations in *RHD* that cause amino acid change in the transmembrane or intracellular parts of the protein [68]. At present (December 2017), 139 different weak D antigens have been described. The most frequent weak D types are weak D type 1 (rs121912763; allele frequency 0.3% in ExaC database), weak D type 2 (rs71652374; allele frequency 0.1%), and weak D type 3 (rs144969459; allele frequency 0.3%).

RBCs of individuals carrying a partial D variant express an RhD protein that lacks 1 or several of the 30 D-epitopes and/or express new RhD epitopes, due to amino acid changes in (an) extracellular loop(s) of the RhD protein. Most partial variants arise from hybrid alleles in which parts of *RHD* are exchanged with the very homologous RHCE gene. In many of the partial D variants the expression is also weakened compared to normal RhD expression [52,69].

Fetal RBC expressing any of the gene products of these *RHD* variant genes can theoretically induce alloimmunization in the mother. DEL individuals cannot make anti-D, since they express the complete protein, so pregnant women do not need immunoprophylaxis. Furthermore, it was supposed that also weak D individuals cannot make anti-D, in contrast to individuals carrying partial D variants that miss some D epitopes. However, since the serological recognition of partial

D variants can be difficult and it became clear that some weak D individuals, for example, with weak D type 4.2 (DAR), type 11, 15, 21, or 57 (reviewed in [70]) and also DEL individuals carrying the RHD*DEL8 allele or the RHD*DEL5 [71] can become immunized against D, the DEL/weak D/ partial D nomenclature is misleading and nowadays the term "D variant" is applied for all *RHD* alleles that result in qualitatively and/or quantitatively aberrant expression [72]. The general consensus is that pregnant women or transfusion recipients with weak D type 1, 2, and 3 and Asian type DEL should be considered as RhD positive and not at risk for alloimmunization [70,72]. Pregnant women carrying any of the other *RHD* variants should be categorized as D negative and included in prophylaxis programs [73]. As outlined previously, the presence of these variant *RHD* genes will mask the fetal *RHD*, thereby hindering a straightforward prediction of the fetal D type. Furthermore, fetuses carrying a D-expressing-*RHD* variant might be missed by fetal genotyping and result in false-negative results.

KELL BLOOD GROUP SYSTEM

The Kell antigens are located on a single-pass type II transmembrane glycoprotein (CD238) of 93 kD (732 amino acids), which functions as a metalloendopeptidase that processes endothelin-3 [74,75]. On RBCs the Kell glycoprotein is covalently linked to the XK protein. The absence of XK protein leads to the McLeod syndrome that is characterized by mild hemolysis, late onset forms of muscular and neurological defects, red cell acanthocytosis, and a greatly reduced amount of the Kell protein [76]. The *KEL* gene, consisting of 19 exons distributed over 21.5 kb, is located at chromosome 7q33. Kell appears early in erythropoiesis but may also be expressed on myeloid progenitors and weak expression is found in other tissues including brain and muscles. Kell antigens are already expressed in the fetus from 6 to 7 weeks of gestation onward [77].

The Kell system consists of 34 antigens. The clinically most relevant antigen is K (KEL1). As described previously anti-K antibodies are second to anti-D antibodies the most frequently encountered antibodies in severe HDFN. A single nucleotide variation (rs8176058) 578T > C in exon 6 resulting in Met193Thr is responsible for the Kk (KEL1, KEL2) polymorphism [78]. Ninety-one percent of Europeans, 98% of Africans and South Asians, and almost all East Asians are K negative (kk or homozygous KEL2), and thus at risk for alloimmunization if carrying a K+ fetus. But also antibodies against other antigens of the KEL blood group system (anti-k (KEL2), anti-Kpa (KEL3, rs8176059), and anti-Jsb (KEL7, rs8176038)) may result in severe HDFN [79]. Recently, a single fatal case has been described against a novel, low frequency Kel alloantigen (KEAL, rs557358978) [80]. Js^a is a low frequency antigen in Caucasians (<0.01%), whereas 20% of Africans are positive. Consequently, anti-Js^b is more frequently found in Africans [81].

In the KEL system 29 silent alleles, so-called K$_o$ alleles, have been identified and numbered by the ISBT (http://www.isbtweb.org/working-parties/red-cell-immunogenetics-and-blood-group-terminology/). These alleles are caused by many different nonsense mutations, single nucleotide insertions or deletions, splice site mutations, and missense mutations. In addition, 12 so-called K$_{mod}$ alleles resulting in weak Kell expression have been annotated. The majority of the presently known mutations (29 K$_o$ and 11 K$_{mod}$) occur in a *k* allele. Although these mutations are rare, they are relatively frequent in individuals with RBCs that are reactive only with anti-K and not with anti-k. In Europeans only 2 out of 1000 individuals have this phenotype, but 3.5%–7% of them were shown to be genotypically *KEL*1/KEL*2*, in which the

KEL*2 allele was a K_o or K_{mod} allele [82,83]. As a fetus carrying K_o alleles and K_{mod} alleles does not express K and hence is not at risk for HDFN, fetal K genotyping is relevant in all anti-K alloimmunized kk women, even if the assured father is phenotypically homozygous K. Their fetus will not be at risk if it inherited a paternal K_o or K_{mod} allele, and this aberrant k allele will not be detected in the K PCR.

NONINVASIVE FETAL GENOTYPING, INDICATIONS, AND REPORTED ACCURACY

NONINVASIVE FETAL *RHD* TYPING TO GUIDE PROPHYLAXIS

In countries where antenatal prophylaxis is part of routine care, it is common practice that all D negative pregnant women are offered antenatal prophylaxis because the fetal RhD type remains unknown until birth. In individuals of European descent, approximately 40% of D negative pregnant women carry a D negative fetus. These women are not at risk for D immunization. With the advent of noninvasive analysis of cell-free DNA in maternal plasma (cfDNA), it has become possible to target antenatal RhD prophylaxis to those nonimmunized D negative women carrying an *RHD* positive fetus [84]. Such targeting prevents unnecessary exposure of RhD negative pregnant women to anti-D immunoglobulin, in Western countries this is >6% of all pregnant women. The prophylaxis product, that is, anti-D immunoglobulin, is a human-derived blood product that carries a theoretical risk of contaminants or unknown infective agents, such as prions, that may be detrimental to the woman or the fetus. Furthermore, a certain risk is also posed to the volunteer donors who allow to be hyperimmunized with RBCs to produce anti-D. Thus it has become an ethical necessity to offer pregnant women fetal *RHD* genotyping to avoid unnecessary immunoprophylaxis and the waste of a gift of donors [85]. In addition, access to the prophylaxis product is limited in some countries, and these countries would thus benefit from restricting the product only to pregnant women with a demonstrated indication for its use. Nevertheless, the argumentation is quite opposite from a US perspective. The consensus in the United States is that the current regimen with untargeted and thus unnecessary use of antenatal prophylaxis should be left unchanged. The arguments are that the supply of anti-D immunoglobulin is not limited in the United States, the antenatal prophylaxis is perfectly safe, and that introducing targeted antenatal prophylaxis will cause more immunizations unless the sensitivity of the cfDNA testing is 100% [86].

In Europe, the implementation of noninvasive fetal *RHD* typing to guide targeted antenatal and postnatal prophylaxis is in progress. In Table 3 an overview of large-scale trials [87–95] is given, the studies demonstrate high sensitivities, above 99% as early as GA 10–11 weeks [90–92,95] and 99.9% at a GA of 25–28 weeks. At present, all the studies are based on multiplex real-time PCR assays and do not include positive controls for the presence of fetal DNA. It is important to note that similar high sensitivities have been attained in the three countries (Denmark, the Netherlands, and Finland) where fetal *RHD* typing and targeted prophylaxis were implemented in a nationwide program [96–98]. The high sensitivity justified the abolition of cord blood serology for postnatal prophylaxis. In the studies evaluating the performance of implemented screening programs, only 21 false-negative cases in almost 50,000 (0.04%) screening tests were observed. The false-negative (FN) results are typically caused by low levels of cfDNA, failed DNA extraction, or human error [91,97,99]. None of the FN results were caused by *RHD* variants in the fetus. It seems feasible to perform antenatal *RHD* screening early in pregnancy [91–93], although the sensitivity decreases in very early pregnancy.

Table 3 Fetal *RHD* Typing Results

References	Samples (n)	GA, Weeks (Median)	RHD Exons	Inconclusive (n)	TP (n)	FP (n)	TN (n)	FN (n)	Sensitivity	Specificity	Accuracy
A. Hemolytic disease of the fetus and newborn											
Van der Schoot et al. [87]	1268	30	7	15	787	5	458	3	99.62	98.92	99.36
Finning et al. [88]	1869	8–28 (28)	5,7	64	1118	14	670	3	99.73	97.95	99.06
Muller et al. [89]	1022	6–32 (28)	5,7	0	659	3	358	2	99.7	99.17	99.51
Akelokar et al. [90]	586	11–13	5,7	84	332	0	164	6	98.22	100	98.8
Wikman et al. [91]	3291	8–40 (10)[a]	4	13	2045	14	1196	23	98.89[b]	98.84	98.87
Chitty et al. [92]	865	<11	5,7	111	400	1	337	16	96.15	99.7	97.75
	956	11–13		75	535	4	341	1	99.81	98.84	99.43
	542	14–17		43	272	1	225	1	99.63	99.56	99.6
	888	18–23		69	492	5	321	1	99.8	98.47	99.27
	1662	>24		95	864	7	696	0	100	99	99.55
Moise et al. [93]	467	11–13	4,5,7	26	310	2	128	1	99.68	98.46	99.32
	458	16–19		26	301	2	129	0	100	98.47	99.54
	425	28–29		26	277	1	121	0	100	99.18	99.75
Soothill et al. [94,95]	509	15–17	5,7	61	277	1	170	0	100	99.42	99.78
Vivanti et al. [95]	416	10–14	10	9	252	7	148	0	100	95.48	98.28
B. Nationwide fetal RHD screening programs with targeted immunoprophylaxis											
Clausen et al. [96]	12,688	25	5,7; 7,10; 5,10	274	7636	41	4706	11	99.86	99.14	99.42

Continued

Table 3 Fetal *RHD* Typing Results—cont'd

References	Samples (n)	GA, Weeks (Median)	RHD Exons	Inconclusive (n)	TP (n)	FP (n)	TN (n)	FN (n)	Sensitivity	Specificity	Accuracy
De Haas et al. [97]	25,789	27–29	5,7	0	15,816	225	9739	9	99.94	97.74	99.09
Haimila et al. [98]	10,814	24–26	5,7	86	7080	7	3640	1	99.99	99.81	99.93
C. Collated results of studies											
Total	49,291	8–40		360	30,532	273	18,085	21	99.93	98.51	99.36
Total 11–13 weeks	5716	11–13		207	3474	27	1977	31	99.12	98.65	98.95

Data from this table has been published in Current Opinion of Hematology.
Shaded lines are samples tested early in pregnancy (<14 weeks).
FN, false negative; FP, false positive; TN, true negative; TP, true positive.
Accuracy = (TP + TN)/(TP + TN + FP + FN).
[a]*80% in first trimester.*
[b]*99.3% for samples >10 weeks.*

Implementation of early testing enables the targeted use of prophylaxis for potential sensitizing events, such as amniocentesis and chorionic villus sampling.

In the Dutch study, nine screening results were false negative and in 225 cases fetal *RHD* results were positive and cord blood serology was negative. Because in a screening program especially FN have to be prevented, the scoring algorithm is less stringent toward false-positive (FP) results. Thus, firstly, low levels of nonspecific amplification, or from DNA derived from a vanishing twin, [100] will result in an *RHD* positive typing. Therefore in all studies listed in Table 3 the specificity is lower than the sensitivity. A second cause of FP or inconclusive results are variant genes in the mother or the child. This was the case in 100 out of 225 cases in the Dutch nationwide screening program evaluation. And thirdly, discrepancies between fetal *RHD* typing and cord blood serology can be caused by FN cord blood serology results [89,97]. In 10 of the 225 presumed FP cases analysis of backup plasma showed that the "cord blood serology" (generating RhD negative results) had mistakenly been performed on maternal blood samples or cord blood mixed with maternal blood. In an additional 22 FP cases with negative cord blood serology, this serology result appeared to be false as the newborn carried an *RHD* variant gene with weak D expression [97]. Thus the net effect of implementation of fetal *RHD* genotyping might even be fewer immunizations.

NONINVASIVE FETAL BLOOD GROUP TYPING IN ALLOIMMUNIZED WOMEN

As described previously in most countries pregnant women are serologically screened for the detection of alloantibodies against RBC antigens. If an anti-RBC antibody is detected, the specificity of the alloantibody(ies) is determined to assess the risk of HDFN. As listed in Table 2 especially anti-Rh antibodies and anti-K are frequently found and clinically relevant. Before fetal blood group genotyping was possible, the supposed father was serologically typed for the presence of the involved antigen and the antithetical allele. If the father was either homozygous positive or negative for the involved alloantigen, the fetal phenotype could be predicted. In particular in case of anti-K, the father is in the far majority of cases K-negative, and the mother has been immunized by a previous incompatible blood transfusion [101,102]. In the past, homozygosity or hemizygosity for *RHD* was predicted based on CcEe phenotype, because of the linkage disequilibrium between the different *RHD* and *RHCE* alleles. However, these predictions can be false and nowadays genotyping assays are used to determine *RHD* homo- or hemizygosity in the father. These assays can be based on the presence of a hybrid Rh-box, which is only present in an Rh locus from which the *RHD* is deleted [36]. Owing to the high frequency of mutated Rhesus boxes in African individuals this assay is not accurate in Africans [103,104], whereas it is in China [61]. The *RHD* gene copy number can be determined by quantitative PCR [105], double Amplification Refractory Mutation System [106], ddPCR, [107] MLPA, [108] or Malditof assays [109].

However, direct fetal blood group typing is to be preferred. Before the discovery of cell-free DNA, fetal blood group genotyping was performed with fetal DNA from amniotic fluid [110]. Since the introduction of cell-free DNA-based fetal blood group typing, invasive procedures for blood group typing have become obsolete for this indication. They have a small risk of fetal loss and a risk of FMH which might boost the maternal immune response. Noninvasive fetal blood group typing is offered by specialized diagnostic laboratories in many countries. For noninvasive fetal *RHD* genotyping in D immunized women, the accuracy was reported to be 98.5% in a meta-analysis of 41 publications. However, this meta-analysis included also publications on screening as well as studies from the early development of the method. Thirty of the 41 publications reported 100% accuracy [111].

Several of the laboratories have reported on the accuracy of their diagnostic tests for the other clinically relevant antibodies [112–121], and as shown in Table 4 in most studies the accuracy is 100% for RHC, RHc, and RHE, only for KEL1 detection some false results are obtained. Only in the Polish and Dutch studies the presence of fetal DNA was controlled [116,119], see for a discussion on the necessity of adding a positive control for blood group genotyping the review of Scheffer et al. [122].

TECHNICAL APPROACHES FOR NONINVASIVE FETAL BLOOD GROUP TYPING ASSAYS

All laboratories involved in fetal *RHD* screening for guiding prophylaxis apply real-time PCR with Taqman probes. But also for laboratories performing fetal blood group typing in a diagnostic setting for alloimmunized pregnant women this is still often the method of choice. The major drawback of real-time PCRs is the low level of multiplexing, which makes it difficult to include a control for the presence of fetal DNA other than Y chromosome markers. Therefore a false-negative result might be caused by the failure to isolate fetal DNA. In an alloimmunized pregnant woman this can have very serious clinical consequences, and therefore in a diagnostic setting the inclusion of fetal controls is warranted. Real-time PCRs for polymorphic markers like short tandem repeats, insertion/deletion polymorphisms, or single nucleotide polymorphisms are being used for this purpose, but the workload is considerable [116,123]. Hypermethylated RASSF1a has been suggested as a universal fetal control [124], but sensitivity and specificity of this marker in real-time PCR assays is not optimal. In other technical approaches such as single-base extension, followed by demonstration of the specific products by GeneScan [125] or mass spectrometry [126] the level of multiplexing is much higher, which allows the inclusion of a fetal DNA control. More recently the use of digital droplet PCR has been applied for fetal *RHD* typing [127]. With that approach also the fetal fraction can be exactly and reproducibly determined using hypermethylated RASSF1a as fetal marker.

Several next-generation sequencing approaches for blood group genotyping have been reported. The same assays can be used for fetal blood group genotyping, as already described by Rieneck et al. [121,128].

FETAL *RHD* GENOTYPING

So far, all laboratories involved in fetal *RHD* screening apply real-time PCR with Taqman probes, mostly a duplex PCR. *RHD* exons 4, 5, 7, and 10 are used as targets (Table 3). Detecting at least two *RHD* exons is highly recommended to increase the chance of a positive detection among most variants. The Danish developed an elegant method to further increase the sensitivity of the assay by combining the two signals for exon 7 and exon 10 into one signal by applying the same reporting dye for each assay [129].

The most widely applied assay is the duplex exon 5/exon 7 PCR that has been validated in many European laboratories in the EU granted SAFE project [130,131]. This assay examines the *RHD* sequence that encodes the portion that harbors most exofacial D epitopes. Nevertheless, also this

Table 4 Accuracy of Fetal RHC, RHc, RHE, and K Blood Group Typing

References	RHC		RHc		RHE		KEL1		Methods
	Samples (n)	Accuracy (%)	Samples (n)	Accuracy (%)	Samples (n)	Accuracy (%)	Samples (n)	Accuracy (%)	
Legler [112]	23	100	1	100	35	100			RQ-PCR[b]
Hromadnikova [113]			41	100	45	100			RQ-PCR[b]
Finning [114][a]	13	100	44	100	46	100	70	98.6	RQ-PCR[b]
Li [115]							32	93.8	Maldi-TOF
Orzinska [117]			11	100					RQ-PCR[b]
Gutensohn [118]	46	100	87	100	100	100			RQ-PCR[b]
Scheffer [119][a]			19	100	21	100	33	100	RQ-PCR[b]
Rieneck [121]							2	100	NGS
Böhmova [120]							128	100	Minisequencing
Orzinska [116][a]	64	100	24	100	26	100	43	95.5	RQ-PCR[b]
Total	146	100	227	100	273	100	308	98.4	

$Accuracy = (TP + TN)/(TP + TN + FP + FN)$.
[a]Results obtained in a diagnostic setting.
[b]Real-Time Quantitative PCR.

duplex assay will give false-negative results with some rare hybrid *RHD* variants: DBT-1 (RHD-Ce(5–7)-D), DBT-2(RHD-Ce(5–9)-D) [132], and DBU (RHD-cE(5–7;226P)-D) [133]. The relatively frequent African alleles DAR, DAR-E, and Weak D type 29 have single mutations both in exon 5 (667T>G) and exon 7(957G>A), and this might theoretically hamper their detection. In the SAFE exon 5/exon 7 PCR these variants appeared to be amplified however [97].

Exon 7 and the untranslated 3′-end of exon 10 harbor multiple D specific nucleotides, which makes it relatively easy to develop specific assays. The theoretical disadvantage of exon 10 as compared to exon 7 is that it does not encode for D epitopes, which is exemplified by its presence in D negative *RHD* variants with gross deletions such as the African (C)ce^s and the Asian RHD*01N.03 genes (Fig. 1B).

PCR assays in exon 4 or exon 5 can be designed in such a way that they will not amplify the most common *RHD* variant genes in Africans and Europeans, *RHD*Ψ* and *RHD*DVI*, respectively (Fig. 1B and C). However, in duplex assays with exon 7 or exon 10, the recognition of an RhD positive fetus in women carrying the *RHD*Ψ* and *RHD*DVI* will depend on the amplification of only a single exon 4 or 5. This makes the assay less sensitive and all fetal *RHD* variants with mutations in, respectively, exon 4 or 5 might be missed. Furthermore, the concomitant amplification of another exon derived from the maternal variant gene might interfere with the amplification of the fetal exon 4 or 5. At present, the reliability of fetal *RHD* genotyping for screening in mixed ethnic populations has not been established yet.

FETAL *RHc* GENOTYPING

RHc genotyping is rather straightforward and does not suffer from nonspecific amplification as is the case for Kell genotyping (see later) because of the five *RHc* specific nucleotides in exon 2 (Fig. 1A). The 307A nucleotide encoding 203Pro is strictly correlated with Rhc expression [43]. Although *RHC* and *RHc* alleles also differ in exon 1 (48C in *RHC* and 48G in *RHc*, Fig. 1A), this nucleotide should not be used, because in most of the African population the *RHCE*ce* allele has a 48C at this position but Rhc and not *RHC* expression [44]. All laboratories reporting results of diagnostic assays for fetal *RHc* genotyping in alloimmunized pregnant women listed in Table 4 applied real-time PCR. Three [114,116,119] obtained results with a similar assay, originally developed in Bristol, UK [114]. In this assay both the forward and reverse primers are specific for the *RHc* allele at their 3′-end (201A, 203A, and 307C, respectively). The German group applied an approach in which only the reverse primer is specific (307C) and in this case the probe harbor the c-specific nucleotides 201A and 201A [118]. Theoretically the assay developed in the UK might be superior, also because of the lower amplicon size (106 vs 175 bp).

FETAL *K* GENOTYPING

Expression of the K antigen is determined by only a single nucleotide substitution (c.578C>T, rs8176058), and the development of a specific real-time PCR assay is, among others for that reason, more challenging. The amplification of the maternal k-allele by mispriming of the K-specific primer has been tackled in several ways. Finning et al. designed a K-specific primer with an extra mismatch and two locked nuclei (LNA) acid bases, to increase specificity, however, at the expense of lower sensitivity [114]. Scheffer et al. included a k allele-specific peptide nucleic acid (PNA) probe in the PCR reaction. Clamping of this PNA probe to the maternal k-allele diminishes the nonspecific amplification of the maternal allele. In this study 24 Kell positive and 32 Kell negative newborns were identified by antenatal fetal *K* genotyping. In 4 out of 60 cases results were inconclusive because of weak amplification signals or because presence of fetal DNA could not be confirmed. In one case with low amplification signals, an intrauterine demise was diagnosed (and potentially already present at the time of the first blood draw) before a second sample could be taken, and a Kell positive child was born [119]. Several labs have developed alternative approaches. Rieneck et al. showed proof of principle using a targeted NGS approach. Li et al. applied Maldi-TOF for K-genotyping but missed 2 out of 13 K-positive fetuses [128]. Bohmova successfully applied a minisequencing approach, but reported remarkably low relative levels of fetal DNA ($<1\%$) and only 4 out of 128 tested women carried a K positive fetus [120]. In conclusion, from all noninvasive blood group genotyping PCRs the K-assay is the most cumbersome and awaits technical improvement, for example, by digital droplet PCR.

In addition to these challenges with the assay, it is worth mentioning that extremely rare variants can lead to false-negative results with deleterious clinical consequences. Poole et al. described a variant *K* gene in which another SNP at codon 193 (c.577A $>$ T, rs61729031) results in K expression, and which will be missed by most K genotyping assays [134]. However, the frequency of this SNP is below 0.01% (ExaC database).

FETAL AND NEONATAL ALLOIMMUNE THROMBOCYTOPENIA

Fetal and neonatal alloimmune thrombocytopenia is a form of alloimmunization during pregnancy that can arise after exposure of the mother to an incompatible, paternally derived human platelet antigen (HPA) on fetal platelets. The maternal alloantibodies that are formed are of the IgG subclass and can enter the fetal circulation by active transport across the placental barrier by the neonatal Fc-receptor (FcRn), which is expressed in placental syncytiotrophoblasts [135]. The predominantly involved alloantibodies in fetal neonatal alloimmune thrombocytopenia (FNAIT) are targeted against HPA-1a, an epitope present on glycoprotein IIIa (or integrin β3). The clinical consequences of this alloimmunization for fetuses and neonates can vary from a chance finding of asymptomatic thrombocytopenia to bleeding complications. These bleeding complications can be relatively harmless skin bleedings such as hematomas or petechiae or more severe hemorrhages [136]. Due to its high mortality and morbidity rates, with many surviving children suffering from neurodevelopmental impairment, an intracranial hemorrhage (ICH) is the most feared bleeding complication in FNAIT. The vast majority of these ICHs occur in utero, with two-third starting before 34 weeks' gestation and 54% even before 28 weeks' gestation [137].

Estimates of the incidence of FNAIT are preferably extracted from nonintervention, observational, prospective cohort studies. These are scarce and therefore rough estimates of incidence and prevalence of FNAIT are usually based on intervention prospective studies and retrospective data as well. HPA-1a is the predominant cause of FNAIT and a limited number of prospective studies attempted to estimate its incidence. The largest prospective screening study included 100,488 pregnant women in Norway, reported an incidence of HPA-1a negative women of 2.1% and HPA-1a alloimmunization in 10.7% of these [138]. The authors reported 58% of alloimmunizations to result in FNAIT (platelet count $<150 \times 10^9/L$) and 33% in severe FNAIT (platelet count $<50 \times 10^9/L$). Two % of all alloimmunizations resulted in an ICH. Unfortunately, this cohort study was not observational, but all alloimmunized women underwent elective cesarean section 2–4 weeks before term, whereas the highest anti-HPA-1a antibody levels and lowest platelet numbers might be reached at term. Therefore the incidence of the risk of bleeding problems in FNAIT may have been underestimated in this study. Another, much smaller, nonintervention prospective screening study reported a comparable overall incidence of HPA-1a related FNAIT of 1 in 1163 (0.09%) live births and an incidence of severe FNAIT of 1 in 1695 (0.06%) [139].

The most accurate estimation of the population incidence of FNAIT has been made by Kamphuis and colleagues, who performed a systematic review including the earlier prospective studies on HPA-1a screening [140]. They concluded that HPA-1a alloimmunization occurred in 9.7% (1 in 500 live births) of pregnancies at risk, leading to severe FNAIT in 31% (1 in 1000–2000 live births) of the cases and to perinatal ICH in 10% of the severe FNAIT cases (1 in 10,000–20,000 live births).

Another approach used, for estimating incidence, is assessing platelet counts in unselected, consecutive newborns. A systematic review pooled the results of six postnatal screening studies, including 59,425 newborns [141]. In all studies, cord blood samples were used to assess platelet counts and confirmed with neonatal capillary or venous blood. Two studies [142] performed diagnostic workup for FNAIT (HPA-1a typing and HPA-1a antibody screening) in all cases. Three studies only did so in case of confirmed neonatal thrombocytopenia [143–146]. In one study on 933 mother/child pairs no cases of FNAIT were found [147]. Definitions of thrombocytopenia, and therefore threshold for testing for FNAIT, ranged from a platelet count below 50×10^9/L to a platelet count below 150×10^9/L. Overall, the reported incidence of severe FNAIT, with a platelet count $<50 \times 10^9$/L, was 0.04% (1 in 2500 newborns), leading to an intracranial hemorrhage in 25% of them.

In current practice, a discrepancy exists between the estimated prevalence of FNAIT reported in literature and the actual number of diagnoses. An Irish cohort study estimated that only 7% of expected cases are detected clinically [148]. In the absence of antenatal screening programs FNAIT is most likely underdiagnosed.

TREATMENT OF FNAIT

In absence of population-based screening FNAIT is usually diagnosed after the onset of symptoms, and treatment is aimed at minimizing complications rather than preventing the symptoms from occurring. The current clinical challenge lies in the prevention of symptoms in subsequent pregnancies with known alloimmunization. Weekly infusion of intravenous immunoglobulins has a 98.7% effectiveness to prevent severe bleeding complications with minimal side effects [149]. This noninvasive strategy, with or without the addition of corticosteroids, has almost completely replaced invasive management with serial fetal blood sampling (FBS), followed by an intrauterine platelet transfusion (IUPT), if necessary [150]. Fewer and fewer centers perform this invasive strategy that is best avoided because of the high risk of complications associated with puncturing the umbilical cord of a potentially thrombocytopenic fetus, and the need for repetitive procedures considering the short half-life of platelets as compared to red blood cells.

ANTI-HPA IMMUNOPROPHYLAXIS TO PREVENT IMMUNIZATION

The implementation of anti-D prophylaxis has led to a great decrease of mortality and morbidity caused by red blood cell immunization. Naturally, with all its pathophysiologic similarities to HDFN, an anti-D prophylactic equivalent for FNAIT is a highly debated topic and an important focus for research groups. Whereas RhD was and is the predominant cause of (severe) HDFN, prophylactic studies for FNAIT focus on HPA-1a, the most frequent cause of (severe) FNAIT [24,151].

In vivo animal studies have shown that the antibody mediated immune suppression that is induced with anti-D prophylaxis can also be generated in FNAIT mouse models. In murine studies, β3 integrin-deficient (β3 −/−) mice are used as the equivalent for HPA-1a negative women. The equivalent for HPA-1a alloantibodies in humans is therefore β3 antibodies in mice. In β3 −/− mice, first injected with HPA-1a positive human platelets or their equivalent β3 +/+ platelets, and immediately thereafter with a human anti-HPA-1a or a murine anti-β3 serum, a strong reduction of the β3 antibody response was seen in mice who had received platelets and immunoprophylaxis as compared to controls that had only received the platelets. In the experiments where the mice became pregnant after the immunization with β3 +/+ platelets, fewer miscarriages, fewer stillborn pups, and fewer pups with ICH were seen in the mice that received the murine anti-β3 serum than in the control group, and the pups had significantly higher platelet counts [152]. These promising results from animal models call for follow-up human studies. So far, a recombinant anti-HPA-1a antibody (B2G1Δnab) has been tested in healthy human volunteers [153]. Previous in vitro studies had shown that B2G1 did not affect platelet function and could block binding of maternal polyclonal HPA-1a antibodies. B2G1Δnab was produced to add a nondestructive constant region to the human immunoglobulin IgG1 [154]. It was shown that B2G1Δnab effectively clears HPA-1a-positive platelets in vivo and it is proposed to be an ideal candidate for prophylaxis after delivery in HPA-1a negative women. Studies with anti-HPA-1a IgG obtained from alloimmunized women, as done for anti-D immunoprophylaxis, are planned. However, without population-based screening programs in place it is currently impossible to identify women that will benefit from a potential anti-HPA-1a prophylaxis. In addition, no follow-up in vivo studies on administration of B2G1Δnab during pregnancy have been performed. Although it has been suggested that the administration of B2G1Δnab after delivery would be effective, no consensus has been reached on the timing of the administration of a potential immunoprophylaxis. The largest prospective screening study reported that most alloimmunizations in HPA-1a negative primigravid women occurred at or soon after delivery [138]. This might suggest that prophylaxis is effective when administered soon after birth. Retrospective data, however, which are probably skewed toward the more severe cases, report a high number of affected first newborns. An analysis of 43 cases of an ICH caused by FNAIT showed that almost half of these occurred in first pregnancies [137]. A prophylaxis postpartum could not have prevented these cases, and prophylaxis during pregnancy seems more adequate.

MOLECULAR BASIS OF HUMAN PLATELET ANTIGEN SYSTEMS

Human platelet antigens are expressed on glycoproteins (GP), present in complexes on the platelet membrane [155]. Of these complexes, GPIIb/IIIa carries the majority of all HPAs and has the largest amount of copies per platelet (Table 5). To date, 35 different HPAs have been identified (current list: https://www.ebi.ac.uk/ipd/hpa/table1.html). The most frequently involved are clustered into six biallelic groups (Table 6). Other HPAs are called low frequency and are mainly private antigens that are found in single families.

Table 5 Glycoprotein Complexes.

Glycoprotein	CD, Integrin	HPAs	Copy Number per Platelet	Encoding Gene, Chromosome
A. Complex GPIa/IIa				
GPIa	CD49b, α2	5, 13, 18, 25	3000–5000	*ITGA2*, 5
GPIIa	CD29, β1	–		*ITGB1*, 9
B. Complex GPIIb/IIIa				
GPIIb	CD41, α2b	3, 9, 20, 22, 24, 27	80,000	*ITGA2B*, 17
GPIIIa	CD61, β3	1, 4, 6, 7, 8, 10, 11, 14, 16, 17, 19, 21, 23, 26		*ITGB3*, 17
C. Complex GPIb/V/IX				
GP1bα	CD42b	2	25,000	*GP1BA*, 17
GP1bβ	CD42c	12		*GP1BB*, 22
GPV	CD42d	–		*GP5*, 3
GPIX	CD42a	–		*GP9*, 3
D. Other				
GPIV	CD36	–	6000	*CD36*, 7
CD109	CD109	15	1000	*CD109*, 6

In the Caucasian population the alloantibody that is most frequently involved in FNAIT is targeted against HPA-1a. Anti-HPA-1a is responsible for approximately 80% of all cases of FNAIT, anti-HPA-5b for approximately 9%, anti-HPA-1b for 4%, and anti-HPA-3a for 2% [163]. As can be expected from the differences in allele and phenotype frequencies, this distribution differs between the various ethnic populations. In non-Caucasian populations the HPA-1a negative phenotype is less prevalent and therefore in these populations HPA-1a alloantibodies are rarely involved in FNAIT. In the Asian population, frequent causers of FNAIT are anti-HPA-4b, anti-HPA-5b, and anti-Nak(a) [164]. The latter are also called anti-CD36 or anti-GPIV and are elicited in CD36 null individuals in response to CD36 positive platelets [165]. CD36-deficient platelets are more frequent in Asian (approximately 3%–11%) and African populations (approximately 7%–8%) compared to Caucasian populations (<0.3%) [166,167]. However, in 90% of individuals with CD36-deficient platelets, CD36 is expressed on monocytes, thus these individuals are not at risk for alloimmunization [168]. The genetic causes of CD36 negativity are variable although some mutations are more frequent, such as the deletion 329-330delAC [169].

NONINVASIVE FETAL HPA-1A GENOTYPING

The fetal status can be determined using invasive diagnostic procedures, such as amniocentesis or chorionic villus sampling. As with every invasive procedure this carries the risk of boosting the immune response and causing pregnancy loss, especially in case of pregnancies with potential thrombocytopenic fetuses. After the introduction of cell-free DNA-based prenatal tests for fetal *RHD* typing, Scheffer

Table 6 Human Platelet Antigens and Their Prevalence

Antigen	Gene, Chromosome, Nucleotide Change	Amino Acid Change	Allele or Phenotype	Frequencies				
				Caucasian[a]	North-African[b]	Sub-Sahara African[d]	African-American[e]	Asian[f]
HPA-1	ITGB3, 17, 196T>C, rs5918	L33P	HPA-1a	0.848	0.748	0.904	0.92	0.994
			HPA-1b	0.152	0.252	0.096	0.08	0.006
			HPA-1b/b	0.02	0.084	0.008	0	0
HPA-2	GP1BA, 17, 482C>T, rs6065	T145M	HPA-2a	0.92	0.818	0.776	0.82	0.952
			HPA-2b	0.08	0.182	0.224	0.18	0.049
			HPA2b/b	0.006	0.028	0.040	0.03	0.001
HPA-3	ITGA2B, 17, 2621T>G, rs5911	I843S	HPA-3a	0.62	0.616	0.596	0.63	0.595
			HPA-3b	0.38	0.384	0.434	0.37	0.406
			HPA-3b/b	0.15	0.125	0.168	0.15	0.169
HPA-4	ITGB3, 17, 506G>A, rs5917	R143Q	HPA-4a	1	1	1	1	0.996
			HPA-4b	0	0	0	0	0.005
			HPA-4b/b	<0.001	<0.001	0	0	0
HPA-5	ITGA2, 5, 16000G>A, rs10471371	E505K	HPA-5a	0.874	0.902	0.732	0.79	0.986
			HPA-5b	0.126	0.098	0.268	0.21	0.014
			HPA-5a/a	0.813	0.732	0.944	0.62	0.973
HPA-15	CD109, 6, 2108C>A, rs10455097	S682Y	HPA-15a	0.455	0.861	0.701	—	0.532
			HPA-15b	0.545	0.139	0.299	—	0.468
			HPA-15b/b	0.23	0.221[c]	0.094	—	0.217
CD36, GPIV[g]	CD36, 7, variable	variable	GPIV negative	—	<0.004	0.08	0.02	0.11

[a]Caucasian (French) population, n=525–6135 [156].
[b]North-African (Moroccan Berber) population, n=104–112 [157].
[c]African (Egyptian) population, n=367 [158].
[d]Sub-Sahara African (Congo), n=125 [159].
[e]African-American population (USA), n=100 [160].
[f]Asian (Chinese Han) population, n=1000 [161].
[g]All phenotype numbers extracted from Curtis et al. [162].
Source: Immuno Polymorphism Database, https://www.ebi.ac.uk/ipd/hpa/freqs_2.html

Table 7 Accuracy of Fetal *HPA-1a* Typing Assays

References	Antigen	Assay	*n*	Accuracy
Scheffer [170]	HPA-1a	Real-time PCR	34	100%
Le Toriellec [171]	HPA-1a	Real-time PCR	44	100%
		HRM analysis	44	100%
Wienzek-Lischka [172]	HPA-1a	Massive parallel sequencing	4	100%
Ferro [173]	HPA-1a	HRM analysis	12	100%

Accuracy = (TP + TN)/(TP + TN + FP + FN).

et al. were the first to describe the possibility of fetal HPA-1a genotyping. In a series of 34 women, no false-positive or false-negative results were obtained [170]. Since this first report, a couple of series have been published reporting high performance of fetal HPA-1a typing assays (Table 7).

In contrast to fetal RHD typing, noninvasive fetal HPA-1a typing is currently exclusively performed in the diagnostic setting, in alloimmunized women. All incompatible pregnancies with known alloimmunization against anti-HPA1a will receive antenatal therapy with weekly infusions of IVIG. In many cases this will be overtreatment since intracranial hemorrhage occurs only in about 1 out of 30 immunized pregnancies. There currently is no reliable diagnostic tool or assay to determine which alloimmunized pregnancies are at risk of developing antenatal bleeding complications and hence will benefit from (the expensive) IVIG treatment.

Once anti-HPA1a immunoprophylaxis will have been shown effective, routine maternal, eventually followed by paternal and fetal HPA-1a typing for screening purposes will become relevant. In case of paternal homozygosity to the involved alloantibody, as a rule, every pregnancy of this couple is incompatible and no additional fetal genotyping has to be performed. In about 30% the father will be heterozygous, then there is a 50% chance that the fetus inherited the "safe" allele and is therefore is not at risk for FNAIT. In case of a heterozygous father fetal genotyping is indicated to determine the fetal platelet blood group, and the need for antenatal preventive measures.

Paternal genotyping is essential in the workup of HPA-1a negative pregnant women as well in the diagnostic as in the screening pathway.

TECHNICAL APPROACHES FOR NONINVASIVE FETAL *HPA-1a* TYPING

Because a single nucleotide substitution is responsible for the HPA-1 antigens, a real-time PCR assay has to be based on an allele-specific primer. Similar to fetal *K* genotyping assays, the HPA-1a assay is hampered by nonspecific amplification of the maternal HPA-1b allele. Scheffer et al. [170] increased the specificity by predigestion of the isolated cfDNA with the restriction enzyme *MspI*, of which the restriction site is only present in the maternal HPA-1b allele and not in the HPA-1a allele. Using real-time PCR as well, Le Toriellec also obtained reliable results without this predigestion step [171]. High-resolution melting analysis was also successfully applied, however on only a limited number of samples [171,173]. Wienzek-Lischka et al. showed a proof of principle applying NGS for fetal HPA-1a detection, in this targeted NGS approach an Ampliseq panel was used that also amplified regions encoding common other blood group and platelet antigen systems [172].

CONCLUSION

Fetal blood group genotyping has been one of the first applications of cfDNA-based noninvasive diagnosis (NIPD) [174,175]. Twenty years after the first publication on noninvasive fetal *RHD* genotyping, blood group genotyping assays are routinely offered in a diagnostic setting for alloimmunized women and in several countries immunoprophylaxis is guided by fetal *RHD* genotyping. The accuracy is invariably high. Evaluation of already implemented nationwide screening programs shows a sensitivity of 99.9%, even in the absence of a control for the presence of fetal DNA. For the design as well as for the correct interpretation of genotyping assays it is important to take the variant genes as described in this chapter into account. Especially in non-European populations variant *RH* genes are frequently encountered. After anti-D antibodies, anti-K antibodies are the second most important cause of HDFN, indicating the importance of reliable fetal *K* genotyping. Although fetal *K* genotyping is difficult because only a single nucleotide variation is responsible for the K epitope and thus mispriming of the K-specific primer to the maternal k-allele has to be circumvented, reliable assays have been developed. The 100% accuracy in a diagnostic setting, where the inclusion of fetal control markers is highly recommended, has made invasive blood group typing obsolete.

HPA-1a alloimmunization is the most common cause of fetal and neonatal alloimmune thrombocytopenia. In current practice, no population-based screening programs are in place to timely detect, prevent, and/or reduce its disease burden. Therefore for now, noninvasive fetal blood group typing for platelets solely exists in a diagnostic setting. In this setting fetal *HPA-1a* typing can be performed using a real-time PCR assay. However, similar to fetal *K* genotyping assays, the HPA-1a assay is hampered by nonspecific amplification of the maternal *HPA-1b* allele, which can be overcome by using a predigestion step of the isolated cfDNA with the restriction enzyme *MspI*. Most recently, NGS has been described to accurately determine the fetal *HPA-1a* type. This practice might pave the way for more extensive use of NGS not just for typing different platelet antigens, but for future application in a screening setting as well.

REFERENCES

[1] Vaughan JI, Manning M, Warwick RM, Letsky EA, Murray NA, Roberts IA. Inhibition of erythroid progenitor cells by anti-Kell antibodies in fetal alloimmune anemia. N Engl J Med 1998;338(12):798–803.

[2] Heathcote DJ, Carroll TE, Flower RL. Sixty years of antibodies to MNS system hybrid glycophorins: what have we learned? Transfus Med Rev 2011;25(2):111–24.

[3] Mari G, Deter RL, Carpenter RL, Rahman F, Zimmerman R, Moise Jr KJ, et al. Noninvasive diagnosis by Doppler ultrasonography of fetal anemia due to maternal red-cell alloimmunization. Collaborative Group for Doppler Assessment of the blood velocity in anemic fetuses. N Engl J Med 2000;342(1):9–14.

[4] Oepkes D, Seaward PG, Vandenbussche FP, Windrim R, Kingdom J, Beyene J, et al. Doppler ultrasonography versus amniocentesis to predict fetal anemia. N Engl J Med 2006;355(2):156–64.

[5] Zwiers C, van Kamp I, Oepkes D, Lopriore E. Intrauterine transfusion and non-invasive treatment options for hemolytic disease of the fetus and newborn—review on current management and outcome. Expert Rev Hematol 2017;10(4):337–44.

[6] Lindenburg IT, Smits-Wintjens VE, van Klink JM, Verduin E, van Kamp IL, Walther FJ, et al. Long-term neurodevelopmental outcome after intrauterine transfusion for hemolytic disease of the fetus/newborn: the LOTUS study. Am J Obstet Gynecol 2012;206(2):141.e1–8.

[7] Dutch Health Council Committee Prevention Pregnancy Immunisation. Prevention of Pregnancy Immunization. Publication no 1992/08, 1992, 47.

[8] Urbaniak SJ, Greiss MA. RhD haemolytic disease of the fetus and the newborn. Blood Rev 2000;14(1):44–61.

[9] Bowman JM, Pollock JM. Antenatal prophylaxis of Rh isoimmunization: 28-weeks'-gestation service program. Can Med Assoc J 1978;118(6):627–30.

[10] McBain RD, Crowther CA, Middleton P. Anti-D administration in pregnancy for preventing Rhesus alloimmunisation. Cochrane Database Syst Rev 2015;(9):Cd000020.

[11] Turner RM, Lloyd-Jones M, Anumba DO, Smith GC, Spiegelhalter DJ, Squires H, et al. Routine antenatal anti-D prophylaxis in women who are Rh(D) negative: Meta-analyses adjusted for differences in study design and quality. PLoS ONE 2012;7(2)e30711.

[12] Koelewijn JM, de Haas M, Vrijkotte TG, Bonsel GJ, van der Schoot CE. One single dose of 200 microg of antenatal RhIG halves the risk of anti-D immunization and hemolytic disease of the fetus and newborn in the next pregnancy. Transfusion 2008;48(8):1721–9.

[13] Tiblad E, Taune Wikman A, Ajne G, Blanck A, Jansson Y, Karlsson A, et al. Targeted routine antenatal anti-D prophylaxis in the prevention of RhD immunisation—outcome of a new antenatal screening and prevention program. PLoS ONE 2013;8(8)e70984.

[14] Davies J, Chant R, Simpson S, Powell R. Routine antenatal anti-D prophylaxis—is the protection adequate? Transfus Med 2011;21(6):421–6.

[15] Zwiers C, Koelewijn JM, Vermij L, van Sambeeck J, Oepkes D, de Haas M, van der Schoot CE. ABO incompatibility and RhIg immunoprophylaxis protect against non-D alloimmunization by pregnancy. Transfusion 2018; https://doi.org/10.1111/trf.14606.

[16] Levine P. The influence of the ABO system on Rh hemolytic disease. Hum Biol 1958;30(1):14–28.

[17] Koelewijn JM, de Haas M, Vrijkotte TG, van der Schoot CE, Bonsel GJ. Risk factors for RhD immunisation despite antenatal and postnatal anti-D prophylaxis. BJOG 2009;116(10):1307–14.

[18] Bowell PJ, Allen DL, Entwistle CC. Blood group antibody screening tests during pregnancy. Br J Obstet Gynaecol 1986;93(10):1038–43.

[19] Gottvall T, Selbing A, Hilden JO. Evaluation of a new Swedish protocol for alloimmunization screening during pregnancy. Acta Obstet Gynecol Scand 1993;72(6):434–8.

[20] Filbey D, Hanson U, Wesstrom G. The prevalence of red cell antibodies in pregnancy correlated to the outcome of the newborn: a 12 year study in Central Sweden. Acta Obstet Gynecol Scand 1995;74(9):687–92.

[21] Gottvall T, Filbey D. Alloimmunization in pregnancy during the years 1992-2005 in the central west region of Sweden. Acta Obstet Gynecol Scand 2008;87(8):843–8.

[22] Koelewijn JM, Vrijkotte TG, van der Schoot CE, Bonsel GJ, de Haas M. Effect of screening for red cell antibodies, other than anti-D, to detect hemolytic disease of the fetus and newborn: a population study in the Netherlands. Transfusion 2008;48(5):941–52.

[23] Dajak S, Stefanovic V, Capkun V. Severe hemolytic disease of fetus and newborn caused by red blood cell antibodies undetected at first-trimester screening (CME). Transfusion 2011;51(7):1380–8.

[24] de Haas M, Thurik FF, Koelewijn JM, van der Schoot CE. Haemolytic disease of the fetus and newborn. Vox Sang 2015;109(2):99–113.

[25] Caine ME, Mueller-Heubach E. Kell sensitization in pregnancy. Am J Obstet Gynecol 1986;154(1):85–90.

[26] Nordvall M, Dziegiel M, Hegaard HK, Bidstrup M, Jonsbo F, Christensen B, et al. Red blood cell antibodies in pregnancy and their clinical consequences: synergistic effects of multiple specificities. Transfusion 2009;49(10):2070–5.

[27] Legler T. Prenatal rhesus testing. ISBT Science Serie 2010;5:7–11.

[28] Hughes LH, Rossi KQ, Krugh DW, O'Shaughnessy RW. Management of pregnancies complicated by anti-Fy(a) alloimmunization. Transfusion 2007;47(10):1858–61.

[29] Brouwers HA, Overbeeke MA, van Ertbruggen I, Schaasberg W, Alsbach GP, van der Heiden C, et al. What is the best predictor of the severity of ABO-haemolytic disease of the newborn? Lancet 1988;2(8612):641–4.

[30] Bhat YR, Pavan Kumar CG. Morbidity of ABO haemolytic disease in the newborn. Paediatr Int Child Health 2012;32(2):93–6.

[31] Vos GH, Adhikari M, Coovadia HM. A study of ABO incompatibility and neonatal jaundice in Black South African newborn infants. Transfusion 1981;21(6):744–9.

[32] Lee CK, Ma ES, Tang M, Lam CC, Lin CK, Chan LC. Prevalence and specificity of clinically significant red cell alloantibodies in Chinese women during pregnancy—a review of cases from 1997 to 2001. Transfus Med 2003;13(4):227–31.

[33] Moise Jr KJ. Management of rhesus alloimmunization in pregnancy. Obstet Gynecol 2008;112(1):164–76.

[34] Oepkes D, van Kamp IL, Simon MJ, Mesman J, Overbeeke MA, Kanhai HH. Clinical value of an antibody-dependent cell-mediated cytotoxicity assay in the management of Rh D alloimmunization. Am J Obstet Gynecol 2001;184(5):1015–20.

[35] Moise KJ. Red blood cell alloimmunization in pregnancy. Semin Hematol 2005;42(3):169–78.

[36] Wagner FF, Flegel WA. RHD gene deletion occurred in the rhesus box. Blood 2000;95(12):3662–8.

[37] Blancher A, Apoil PA. Evolution of RH genes in hominoids: characterization of a gorilla RHCE-like gene. J Hered 2000;91(3):205–10.

[38] Le van Kim C, Mouro I, Cherif-Zahar B, Raynal V, Cherrier C, Cartron JP, et al. Molecular cloning and primary structure of the human blood group RhD polypeptide. Proc Natl Acad Sci USA 1992;89(22):10925–9.

[39] Avent ND, Ridgwell K, Tanner MJ, Anstee DJ. cDNA cloning of a 30 kDa erythrocyte membrane protein associated with Rh (rhesus)-blood-group-antigen expression. Biochem J 1990;271(3):821–5.

[40] Hemker MB, Ligthart PC, Berger L, van Rhenen DJ, van der Schoot CE, Wijk PA. DAR, a new RhD variant involving exons 4, 5, and 7, often in linkage with ceAR, a new Rhce variant frequently found in African blacks. Blood 1999;94(12):4337–42.

[41] Westhoff CM, Ferreri-Jacobia M, Mak DO, Foskett JK. Identification of the erythrocyte Rh blood group glycoprotein as a mammalian ammonium transporter. J Biol Chem 2002;277(15):12499–502.

[42] Soupene E, Inwood W, Kustu S. Lack of the rhesus protein Rh1 impairs growth of the green alga *Chlamydomonas reinhardtii* at high CO_2. Proc Natl Acad Sci U S A 2004;101(20):7787–92.

[43] Faas BH, Beuling EA, Ligthart PC, van Rhenen DJ, van der Schoot CE. Partial expression of RHc on the RHD polypeptide. Transfusion 2001;41(9):1136–42.

[44] Westhoff CM, Silberstein LE, Wylie DE, Skavdahl M, Reid ME. 16Cys encoded by the RHce gene is associated with altered expression of the e antigen and is frequent in the R_0 haplotype. Br J Haematol 2001;113(3):666–71.

[45] Tax MG, van der Schoot CE, van Doorn R, Douglas-Berger L, van Rhenen DJ, Maaskant-vanWijk PA. RHC and RHc genotyping in different ethnic groups. Transfusion 2002;42(5):634–44.

[46] Faas BH, Simsek S, Bleeker PM, Overbeeke MA, Cuijpers HT, von dem Borne AE, et al. Rh E/e genotyping by allele-specific primer amplification. Blood 1995;85(3):829–32.

[47] Gruswitz F, Chaudhary S, Ho JD, Schlessinger A, Pezeshki B, Ho CM, et al. Function of human Rh based on structure of RhCG at 2.1 A. Proc Natl Acad Sci U S A 2010;107(21):9638–43.

[48] Mouro-Chanteloup I, D'Ambrosio AM, Gane P, Le Van Kim C, Raynal V, Dhermy D, et al. Cell-surface expression of RhD blood group polypeptide is posttranscriptionally regulated by the RhAG glycoprotein. Blood 2002;100(3):1038–47.

[49] Cartron JP. RH blood group system and molecular basis of Rh-deficiency. Baillieres Best Pract Res Clin Haematol 1999;12(4):655–89.

[50] Satchwell TJ, Bell AJ, Hawley BR, Pellegrin S, Mordue KE, van Deursen CT, et al. Severe Ankyrin-R deficiency results in impaired surface retention and lysosomal degradation of RhAG in human erythroblasts. Haematologica 2016;101(9):1018–27.

[51] Wagner FF, Flegel WA. The rhesus site. Transfus Med Hemother 2014;41(5):357–63.

[52] Daniels G. Rh and RHAG blood group systems. In: Daniels G, editor. Human blood groups. 3rd ed. Oxford, UK: Wiley-Blackwell; 2013. p. 182–258.

[53] Perry GH, Xue Y, Smith RS, Meyer WK, Caliskan M, Yanez-Cuna O, et al. Evolutionary genetics of the human Rh blood group system. Hum Genet 2012;131(7):1205–16.

[54] Mourant A. The distribution of the human blood groups. Oxford: Blackwell Scientific Publications; 1954.

[55] Anstee DJ. The relationship between blood groups and disease. Blood 2010;115(23):4635–43.

[56] Achilli A, Rengo C, Battaglia V, Pala M, Olivieri A, Fornarino S, et al. Saami and Berbers—an unexpected mitochondrial DNA link. Am J Hum Genet 2005;76(5):883–6.

[57] Singleton BK, Green CA, Avent ND, Martin PG, Smart E, Daka A, et al. The presence of an RHD pseudogene containing a 37 base pair duplication and a nonsense mutation in Africans with the Rh D-negative blood group phenotype. Blood 2000;95(1):12–8.

[58] Faas BH, Beckers EA, Wildoer P, Ligthart PC, Overbeeke MA, Zondervan HA, et al. Molecular background of VS and weak C expression in blacks. Transfusion 1997;37(1):38–44.

[59] Pham BN, Peyrard T, Juszczak G, Dubeaux I, Gien D, Blancher A, et al. Heterogeneous molecular background of the weak C, VS+, hr B-, Hr B- phenotype in black persons. Transfusion 2009;49(3):495–504.

[60] Shao CP, Maas JH, Su YQ, Kohler M, Legler TJ. Molecular background of Rh D-positive, D-negative, D(el) and weak D phenotypes in Chinese. Vox Sang 2002;83(2):156–61.

[61] Xu Q, Grootkerk-Tax MG, Maaskant-van Wijk PA, van der Schoot CE. Systemic analysis and zygosity determination of the RHD gene in a D-negative Chinese Han population reveals a novel D-negative RHD gene. Vox Sang 2005;88(1):35–40.

[62] Shao CP. Transfusion of RhD-positive blood in "Asia type" DEL recipients. N Engl J Med 2010;362 (5):472–3.

[63] Ji YL, Luo H, Wen JZ, Haer-Wigman L, Veldhuisen B, Wei L, et al. RHD genotype and zygosity analysis in the Chinese southern Han D+, D- and D variant donors using the multiplex ligation-dependent probe amplification assay. Vox Sang 2017;112(7):660–70.

[64] Ye L, Yue D, Wo D, Ding X, Guo S, Li Q, et al. Molecular bases of unexpressed RHD alleles in Chinese D-persons. Transfusion 2009;49(8):1655–60.

[65] Kim JY, Kim SY, Kim CA, Yon GS, Park SS. Molecular characterization of D-Korean persons: development of a diagnostic strategy. Transfusion 2005;45(3):345–52.

[66] Kwon DH, Sandler SG, Flegel WA. DEL phenotype. Immunohematology 2017;33(3):125–32.

[67] Wagner FF. RHD PCR of D-negative blood donors. Transfus Med Hemother 2013;40(3):172–81.

[68] Wagner FF, Gassner C, Müller TH, Schönitzer D, Schunter F, Flegel WA. Molecular basis of weak D phenotypes. Presented at the 25th congress of the International Society of Blood Transfusion held in Oslo on June 29, 1998 and published in abstract form in Vox Sang 74:1998 (suppl) 55. Blood 1999;93(1):385–93.

[69] Avent ND, Reid ME. The Rh blood group system: a review. Blood 2000;95(2):375–87.

[70] Sandler SG, Chen LN, Flegel WA. Serological weak D phenotypes: a review and guidance for interpreting the RhD blood type using the RHD genotype. Br J Haematol 2017;179(1):10–9.

[71] Kormoczi GF, Gassner C, Shao CP, Uchikawa M, Legler TJ. A comprehensive analysis of DEL types: partial DEL individuals are prone to anti-D alloimmunization. Transfusion 2005;45(10):1561–7.

[72] Daniels G. Variants of RhD—current testing and clinical consequences. Br J Haematol 2013;161 (4):461–70.

[73] Sandler SG, Queenan JT. A guide to terminology for Rh Immunoprophylaxis. Obstet Gynecol 2017;130 (3):633–5.

[74] Lee S, Zambas ED, Marsh WL, Redman CM. Molecular cloning and primary structure of Kell blood group protein. Proc Natl Acad Sci 1991;88(14):6353–7.

[75] Lee S. Molecular basis of Kell blood group phenotypes. Vox Sang 1997;73(1):1–11.

[76] Redman CM, Russo D, Lee S. Kell, Kx and the McLeod syndrome. Baillieres Best Pract Res Clin Haematol 1999;12(4):621–35.

[77] Toivanen P, Hirvonen T. Antigens Duffy, Kell, Kidd, Lutheran and Xg a on fetal red cells. Vox Sang 1973;24(4):372–6.

[78] Lee S, Wu X, Reid M, Zelinski T, Redman C. Molecular basis of the Kell (K1) phenotype. Blood 1995;85 (4):912–6.

[79] Bowman JM, Pollock JM, Manning FA, Harman CR, Menticoglou S. Maternal Kell blood group alloimmunization. Obstet Gynecol 1992;79(2):239–44.

[80] Scharberg EA, Wieckhusen C, Luz B, Rothenberger S, Sturzel A, Rink G, et al. Fatal hemolytic disease of the newborn caused by an antibody to KEAL, a new low-prevalence Kell blood group antigen. Transfusion 2017;57(1):217–8.

[81] Al Riyami AZ, Al Salmani M, Al Hashami S, Al Mahrooqi S, Al Hinai S, Al Balushi H, et al. Successful management of severe hemolytic disease of the fetus due to anti-Jsb using intrauterine transfusions with serial maternal blood donations: a case report and a review of the literature. Transfusion 2014;54(1):238–43.

[82] Kormoczi GF, Wagner T, Jungbauer C, Vadon M, Ahrens N, Moll W, et al. Genetic diversity of KELnull and KELel: a nationwide Austrian survey. Transfusion 2007;47(4):703–14.

[83] Ji Y, Veldhuisen B, Ligthart P, Haer-Wigman L, Jongerius J, Boujnan M, et al. Novel alleles at the Kell blood group locus that lead to Kell variant phenotype in the Dutch population. Transfusion 2015;55(2):413–21.

[84] van der Schoot CE, de Haas M, Clausen FB. Genotyping to prevent Rh disease: has the time come? Curr Opin Hematol 2017;24(6):544–50.

[85] Kent J, Farrell AM, Soothill P. Routine administration of anti-D: the ethical case for offering pregnant women fetal RHD genotyping and a review of policy and practice. BMC Pregnancy Childbirth 2014;14:87.

[86] Ma KK, Rodriguez MI, Cheng YW, Norton ME, Caughey AB. Should cell-free DNA testing be used to target antenatal rhesus immune globulin administration? J Matern Fetal Neonatal Med 2016;29 (11):1866–70.

[87] Van der Schoot CE, Soussan AA, Koelewijn J, Bonsel G, Paget-Christiaens LG, de Haas M. Non-invasive antenatal RHD typing. Transfus Clin Biol 2006;13(1–2):53–7.

[88] Finning K, Martin P, Summers J, Massey E, Poole G, Daniels G. Effect of high throughput RHD typing of fetal DNA in maternal plasma on use of anti-RhD immunoglobulin in RhD negative pregnant women: prospective feasibility study. BMJ 2008;336(7648):816–8.

[89] Muller SP, Bartels I, Stein W, Emons G, Gutensohn K, Kohler M, et al. The determination of the fetal D status from maternal plasma for decision making on Rh prophylaxis is feasible. Transfusion 2008;48 (11):2292–301.

[90] Akolekar R, Finning K, Kuppusamy R, Daniels G, Nicolaides KH. Fetal RHD genotyping in maternal plasma at 11-13 weeks of gestation. Fetal Diagn Ther 2011;29(4):301–6.

[91] Wikman AT, Tiblad E, Karlsson A, Olsson ML, Westgren M, Reilly M. Noninvasive single-exon fetal RHD determination in a routine screening program in early pregnancy. Obstet Gynecol 2012;120(2 Pt 1):227–34.

[92] Chitty LS, Finning K, Wade A, Soothill P, Martin B, Oxenford K, et al. Diagnostic accuracy of routine antenatal determination of fetal RHD status across gestation: population based cohort study. BMJ 2014;g5243:349.

[93] Moise Jr KJ, Gandhi M, Boring NH, O'Shaughnessy R, Simpson LL, Wolfe HM, et al. Circulating cell-free DNA to determine the fetal RHD status in all three trimesters of pregnancy. Obstet Gynecol 2016;128 (6):1340–6.

[94] Soothill PW, Finning K, Latham T, Wreford-Bush T, Ford J, Daniels G. Use of cffDNA to avoid administration of anti-D to pregnant women when the fetus is RhD-negative: implementation in the NHS. BJOG 2015;122(12):1682–6.

[95] Vivanti A, Benachi A, Huchet FX, Ville Y, Cohen H, Costa JM. Diagnostic accuracy of fetal rhesus D genotyping using cell-free fetal DNA during the first trimester of pregnancy. Am J Obstet Gynecol 2016;215(5):606.e1–5.

[96] Clausen FB, Steffensen R, Christiansen M, Rudby M, Jakobsen MA, Jakobsen TR, et al. Routine noninvasive prenatal screening for fetal RHD in plasma of RhD-negative pregnant women-2 years of screening experience from Denmark. Prenat Diagn 2014;34(10):1000–5.

[97] de Haas M, Thurik FF, van der Ploeg CP, Veldhuisen B, Hirschberg H, Soussan AA, et al. Sensitivity of fetal RHD screening for safe guidance of targeted anti-D immunoglobulin prophylaxis: prospective cohort study of a nationwide programme in the Netherlands. BMJ 2016;i5789:355.

[98] Haimila K, Sulin K, Kuosmanen M, Sareneva I, Korhonen A, Natunen S, et al. Targeted antenatal anti-D prophylaxis program for RhD-negative pregnant women - outcome of the first two years of a national program in Finland. Acta Obstet Gynecol Scand 2017;96(10):1228–33.

[99] Clausen FB, Christiansen M, Steffensen R, Jorgensen S, Nielsen C, Jakobsen MA, et al. Report of the first nationally implemented clinical routine screening for fetal RHD in D- pregnant women to ascertain the requirement for antenatal RhD prophylaxis. Transfusion 2012;52(4):752–8.

[100] Thurik FF, Ait Soussan A, Bossers B, Woortmeijer H, Veldhuisen B, Page-Christiaens GC, et al. Analysis of false-positive results of fetal RHD typing in a national screening program reveals vanishing twins as potential cause for discrepancy. Prenat Diagn 2015;35(8):754–60.

[101] Koelewijn JM, Vrijkotte TG, de Haas M, van der Schoot CE, Bonsel GJ. Risk factors for the presence of non-rhesus D red blood cell antibodies in pregnancy. BJOG 2009;116(5):655–64.

[102] Dajak S, Culic S, Stefanovic V, Lukacevic J. Relationship between previous maternal transfusions and haemolytic disease of the foetus and newborn mediated by non-RhD antibodies. Blood Transfus 2013;11 (4):528–32.

[103] Grootkerk-Tax MG, Maaskant-van Wijk PA, van Drunen J, van der Schoot CE. The highly variable RH locus in nonwhite persons hampers RHD zygosity determination but yields more insight into RH-related evolutionary events. Transfusion 2005;45(3):327–37.

[104] Wagner FF, Moulds JM, Flegel WA. Genetic mechanisms of rhesus box variation. Transfusion 2005;45 (3):338–44.

[105] Krog GR, Clausen FB, Dziegiel MH. Quantitation of RHD by real-time polymerase chain reaction for determination of RHD zygosity and RHD mosaicism/chimerism: an evaluation of four quantitative methods. Transfusion 2007;47(4):715–22.

[106] Chiu RW, Murphy MF, Fidler C, Zee BC, Wainscoat JS, Lo YM. Determination of RhD zygosity: comparison of a double amplification refractory mutation system approach and a multiplex real-time quantitative PCR approach. Clin Chem 2001;47(4):667–72.

[107] Sillence KA, Halawani AJ, Tounsi WA, Clarke KA, Kiernan M, Madgett TE, et al. Rapid RHD Zygosity determination using digital PCR. Clin Chem 2017;63(8):1388–97.

[108] Haer-Wigman L, Veldhuisen B, Jonkers R, Loden M, Madgett TE, Avent ND, et al. RHD and RHCE variant and zygosity genotyping via multiplex ligation-dependent probe amplification. Transfusion 2013;53 (7):1559–74.

[109] Gassner C, Meyer S, Frey BM, Vollmert C. Matrix-assisted laser desorption/ionisation, time-of-flight mass spectrometry-based blood group genotyping—the alternative approach. Transfus Med Rev 2013;27(1):2–9.

[110] van der Schoot CE, Tax GH, Rijnders RJ, de Haas M, Christiaens GC. Prenatal typing of Rh and Kell blood group system antigens: the edge of a watershed. Transfus Med Rev 2003;17(1):31–44.

[111] Zhu YJ, Zheng YR, Li L, Zhou H, Liao X, Guo JX, et al. Diagnostic accuracy of non-invasive fetal RhD genotyping using cell-free fetal DNA: a meta analysis. J Matern Fetal Neonatal Med 2014;27(18):1839–44.

[112] Legler TJ, Lynen R, Maas JH, Pindur G, Kulenkampff D, Suren A, et al. Prediction of fetal Rh D and Rh CcEe phenotype from maternal plasma with real-time polymerase chain reaction. Transfus Apher Sci 2002;27(3):217–23.

[113] Hromadnikova I, Vesela K, Benesova B, Nekovarova K, Duskova D, Vlk R, et al. Non-invasive fetal RHD and RHCE genotyping from maternal plasma in alloimmunized pregnancies. Prenat Diagn 2005;25 (12):1079–83.

[114] Finning K, Martin P, Summers J, Daniels G. Fetal genotyping for the K (Kell) and Rh C, c, and E blood groups on cell-free fetal DNA in maternal plasma. Transfusion 2007;47(11):2126–33.

[115] Li Y, Finning K, Daniels G, Hahn S, Zhong X, Holzgreve W. Noninvasive genotyping fetal Kell blood group (KEL1) using cell-free fetal DNA in maternal plasma by MALDI-TOF mass spectrometry. Prenat Diagn 2008;28(3):203–8.

[116] Orzinska A, Guz K, Debska M, Uhrynowska M, Celewicz Z, Wielgo M, et al. 14 years of polish experience in non-invasive prenatal blood group diagnosis. Transfus Med Hemother 2015;42(6):361–4.

[117] Orzinska A, Guz K, Brojer E, Zupanska B. Preliminary results of fetal Rhc examination in plasma of pregnant women with anti-c. Prenat Diagn 2008;28(4):335–7.

[118] Gutensohn K, Muller SP, Thomann K, Stein W, Suren A, Kortge-Jung S, et al. Diagnostic accuracy of non-invasive polymerase chain reaction testing for the determination of fetal rhesus C, c and E status in early pregnancy. BJOG 2010;117(6):722–9.

[119] Scheffer PG, van der Schoot CE, Page-Christiaens GC, de Haas M. Noninvasive fetal blood group genotyping of rhesus D, c, E and of K in alloimmunised pregnant women: evaluation of a 7-year clinical experience. BJOG 2011;118(11):1340–8.

[120] Bohmova J, Vodicka R, Lubusky M, Holuskova I, Studnickova M, Kratochvilova R, et al. Clinical potential of effective noninvasive exclusion of KEL1-positive fetuses in KEL1-negative pregnant women. Fetal Diagn Ther 2016;40(1):48–53.

[121] Rieneck K, Bak M, Jonson L, Clausen FB, Krog GR, Tommerup N, et al. Next-generation sequencing: proof of concept for antenatal prediction of the fetal Kell blood group phenotype from cell-free fetal DNA in maternal plasma. Transfusion 2013;53(11 suppl 2):2892–8.

[122] Scheffer PG, de Haas M, van der Schoot CE. The controversy about controls for fetal blood group genotyping by cell-free fetal DNA in maternal plasma. Curr Opin Hematol 2011;18(6):467–73.

[123] Scheffer PG, van der Schoot CE, Page-Christiaens GC, Bossers B, van Erp F, de Haas M. Reliability of fetal sex determination using maternal plasma. Obstet Gynecol 2010;115(1):117–26.

[124] Chan KC, Ding C, Gerovassili A, Yeung SW, Chiu RW, Leung TN, et al. Hypermethylated RASSF1A in maternal plasma: a universal fetal DNA marker that improves the reliability of noninvasive prenatal diagnosis. Clin Chem 2006;52(12):2211–8.

[125] Doescher A, Petershofen EK, Wagner FF, Schunter M, Muller TH. Evaluation of single-nucleotide polymorphisms as internal controls in prenatal diagnosis of fetal blood groups. Transfusion 2013;53(2):353–62.

[126] Bombard AT, Akolekar R, Farkas DH, VanAgtmael AL, Aquino F, Oeth P, et al. Fetal RHD genotype detection from circulating cell-free fetal DNA in maternal plasma in non-sensitized RhD negative women. Prenat Diagn 2011;31(8):802–8.

[127] Sillence KA, Roberts LA, Hollands HJ, Thompson HP, Kiernan M, Madgett TE, et al. Fetal sex and RHD genotyping with digital PCR demonstrates greater sensitivity than real-time PCR. Clin Chem 2015;61 (11):1399–407.

[128] Rieneck K, Clausen FB, Dziegiel MH. Next-generation sequencing for antenatal prediction of KEL1 blood group status. Methods Mol Biol 2015;1310:115–21.

[129] Clausen FB, Krog GR, Rieneck K, Rasmark EE, Dziegiel MH. Evaluation of two real-time multiplex PCR screening assays detecting fetal RHD in plasma from RhD negative women to ascertain the requirement for antenatal RhD prophylaxis. Fetal Diagn Ther 2011;29(2):155–63.

[130] Chitty LS, van der Schoot CE, Hahn S, Avent ND. SAFE—the special non-invasive advances in fetal and neonatal evaluation network: aims and achievements. Prenat Diagn 2008;28(2):83–8.

[131] Legler TJ, Liu Z, Mavrou A, Finning K, Hromadnikova I, Galbiati S, et al. Workshop report on the extraction of foetal DNA from maternal plasma. Prenat Diagn 2007;27(9):824–9.

[132] Beckers EA, Faas BH, Simsek S, Overbeeke MA, van Rhenen DJ, Wallace M, et al. The genetic basis of a new partial D antigen: DDBT. Br J Haematol 1996;93(3):720–7.

[133] Flegel WA, von Zabern I, Wagner FF. Six years' experience performing RHD genotyping to confirm D- red blood cell units in Germany for preventing anti-D immunizations. Transfusion 2009;49(3):465–71.

[134] Poole J, Warke N, Hustinx H, Taleghani BM, Martin P, Finning K, et al. A KEL gene encoding serine at position 193 of the Kell glycoprotein results in expression of KEL1 antigen. Transfusion 2006;46(11):1879–85.

[135] Simister NE, Story CM, Chen HL, Hunt JS. An IgG-transporting fc receptor expressed in the syncytiotrophoblast of human placenta. Eur J Immunol 1996;26(7):1527–31.

[136] Ghevaert C, Campbell K, Walton J, Smith GA, Allen D, Williamson LM, et al. Management and outcome of 200 cases of fetomaternal alloimmune thrombocytopenia. Transfusion 2007;47(5):901–10.

[137] Tiller H, Kamphuis MM, Flodmark O, Papadogiannakis N, David AL, Sainio S, et al. Fetal intracranial haemorrhages caused by fetal and neonatal alloimmune thrombocytopenia: an observational cohort study of 43 cases from an international multicentre registry. BMJ Open 2013;3(3):e002490.

[138] Kjeldsen-Kragh J, Killie MK, Tomter G, Golebiowska E, Randen I, Hauge R, et al. A screening and intervention program aimed to reduce mortality and serious morbidity associated with severe neonatal alloimmune thrombocytopenia. Blood 2007;110(3):833–9.

[139] Turner ML, Bessos H, Fagge T, Harkness M, Rentoul F, Seymour J, et al. Prospective epidemiologic study of the outcome and cost-effectiveness of antenatal screening to detect neonatal alloimmune thrombocytopenia due to anti-HPA-1a. Transfusion 2005;45(12):1945–56.

[140] Kamphuis MM, Paridaans N, Porcelijn L, De Haas M, Van Der Schoot CE, Brand A, et al. Screening in pregnancy for fetal or neonatal alloimmune thrombocytopenia: systematic review. BJOG 2010;117(11):1335–43.

[141] Kamphuis MM, Paridaans NP, Porcelijn L, Lopriore E, Oepkes D. Incidence and consequences of neonatal alloimmune thrombocytopenia: a systematic review. Pediatrics 2014;133(4):715–21.

[142] Uhrynowska M, Niznikowska-Marks M, Zupanska B. Neonatal and maternal thrombocytopenia: incidence and immune background. Eur J Haematol 2000;64(1):42–6.

[143] Burrows RF, Kelton JG. Fetal thrombocytopenia and its relation to maternal thrombocytopenia. N Engl J Med 1993;329(20):1463–6.

[144] Dreyfus M, Kaplan C, Verdy E, Schlegel N, Durand-Zaleski I, Tchernia G. Frequency of immune thrombocytopenia in newborns: a prospective study. Immune Thrombocytopenia Working Group. Blood 1997;89(12):4402–6.

[145] de Moerloose P, Boehlen F, Extermann P, Hohfeld P. Neonatal thrombocytopenia: incidence and characterization of maternal antiplatelet antibodies by MAIPA assay. Br J Haematol 1998;100(4):735–40.

[146] Sainio S, Jarvenpaa AL, Renlund M, Riikonen S, Teramo K, Kekomaki R. Thrombocytopenia in term infants: a population-based study. Obstet Gynecol 2000;95(3):441–6.

[147] Panzer S, Auerbach L, Cechova E, Fischer G, Holensteiner A, Kitl EM, et al. Maternal alloimmunization against fetal platelet antigens: a prospective study. Br J Haematol 1995;90(3):655–60.

[148] Davoren A, McParland P, Barnes CA, Murphy WG. Neonatal alloimmune thrombocytopenia in the Irish population: a discrepancy between observed and expected cases. J Clin Pathol 2002;55(4):289–92.

[149] Winkelhorst D, Murphy MF, Greinacher A, Shehata N, Bakchoul T, Massey E, et al. Antenatal management in fetal and neonatal alloimmune thrombocytopenia: a systematic review. Blood 2017;129:1538–47.

[150] Daffos F, Forestier F, Muller JY, Reznikoff-Etievant M, Habibi B, Capella-Pavlovsky M, et al. Prenatal treatment of alloimmune thrombocytopenia. Lancet 1984;2(8403):632.

[151] Zwiers C, Lindenburg ITM, Klumper FJ, de Haas M, Oepkes D, Van Kamp IL. Complications of intrauterine intravascular blood transfusion: lessons learned after 1678 procedures. Ultrasound Obstet Gynecol 2017;50(2):180–6.

[152] Tiller H, Killie MK, Chen P, Eksteen M, Husebekk A, Skogen B, et al. Toward a prophylaxis against fetal and neonatal alloimmune thrombocytopenia: induction of antibody-mediated immune suppression and prevention of severe clinical complications in a murine model. Transfusion 2012;52(7):1446–57.

[153] Ghevaert C, Herbert N, Hawkins L, Grehan N, Cookson P, Garner SF, et al. Recombinant HPA-1a antibody therapy for treatment of fetomaternal alloimmune thrombocytopenia: proof of principle in human volunteers. Blood 2013;122(3):313–20.

[154] Ghevaert C, Wilcox DA, Fang J, Armour KL, Clark MR, Ouwehand WH, et al. Developing recombinant HPA-1a-specific antibodies with abrogated Fcgamma receptor binding for the treatment of fetomaternal alloimmune thrombocytopenia. J Clin Invest 2008;118(8):2929–38.

[155] Curtis BR, McFarland JG. Human platelet antigens - 2013. Vox Sang 2014;106(2):93–102.

[156] Merieux Y, Debost M, Bernaud J, Raffin A, Meyer F, Rigal D. Human platelet antigen frequencies of platelet donors in the French population determined by polymerase chain reaction with sequence-specific primers. Pathol Biol (Paris) 1997;45(9):697–700.

[157] Ferrer G, Muniz-Diaz E, Aluja MP, Arilla M, Martinez C, Nogues R, et al. Analysis of human platelet antigen systems in a Moroccan Berber population. Transfus Med 2002;12(1):49–54.

[158] Husebekk A, El Ekiaby M, Gorgy G, Killie MK, Uhlin-Hansen C, Salma W, et al. Foetal/neonatal alloimmune thrombocytopenia in Egypt; human platelet antigen genotype frequencies and antibody detection and follow-up in pregnancies. Transfus Apher Sci 2012;47(3):277–82.

[159] Halle L, Bigot A, Mulen-Imandy G, M'Bayo K, Jaeger G, Anani L, et al. HPA polymorphism in sub-Saharan African populations: Beninese, Cameroonians, Congolese, and pygmies. Tissue Antigens 2005;65(3):295–8.

[160] Kim HO, Jin Y, Kickler TS, Blakemore K, Kwon OH, Bray PF. Gene frequencies of the five major human platelet antigens in African American, white, and Korean populations. Transfusion 1995;35(10):863–7.

[161] Feng ML, Liu DZ, Shen W, Wang JL, Guo ZH, Zhang X, et al. Establishment of an HPA-1- to −16-typed platelet donor registry in China. Transfus Med 2006;16(5):369–74.

[162] Curtis BR, Ali S, Glazier AM, Ebert DD, Aitman TJ, Aster RH. Isoimmunization against CD36 (glycoprotein IV): description of four cases of neonatal isoimmune thrombocytopenia and brief review of the literature. Transfusion 2002;42(9):1173–9.

[163] Davoren A, Curtis BR, Aster RH, McFarland JG. Human platelet antigen-specific alloantibodies implicated in 1162 cases of neonatal alloimmune thrombocytopenia. Transfusion 2004;44(8):1220–5.

[164] Ohto H, Miura S, Ariga H, Ishii T, Fujimori K, Morita S. The natural history of maternal immunization against foetal platelet alloantigens. Transfus Med 2004;14(6):399–408.

[165] Tomiyama Y, Take H, Ikeda H, Mitani T, Furubayashi T, Mizutani H, et al. Identification of the platelet-specific alloantigen, Naka, on platelet membrane glycoprotein IV. Blood 1990;75(3):684–7.

[166] Curtis BR, Aster RH. Incidence of the Nak(a)-negative platelet phenotype in African Americans is similar to that of Asians. Transfusion 1996;36(4):331–4.

[167] Wu G, Zhou Y, Li L, Zhong Z, Li H, Li H, et al. Platelet immunology in China: research and clinical applications. Transfus Med Rev 2017;31(2):118–25.

[168] Yamamoto N, Akamatsu N, Sakuraba H, Yamazaki H, Tanoue K. Platelet glycoprotein IV (CD36) deficiency is associated with the absence (type I) or the presence (type II) of glycoprotein IV on monocytes. Blood 1994;83(2):392–7.

[169] Li R, Qiao Z, Ling B, Lu P, Zhu Z. Incidence and molecular basis of CD36 deficiency in shanghai population. Transfusion 2015;55(3):666–73.

[170] Scheffer PG, Ait Soussan A, Verhagen OJ, Page-Christiaens GC, Oepkes D, de Haas M, et al. Noninvasive fetal genotyping of human platelet antigen-1a. BJOG 2011;118(11):1392–5.

[171] Le Toriellec E, Chenet C, Kaplan C. Safe fetal platelet genotyping: new developments. Transfusion 2013;53(8):1755–62.

[172] Wienzek-Lischka S, Krautwurst A, Frohner V, Hackstein H, Gattenlohner S, Brauninger A, et al. Noninvasive fetal genotyping of human platelet antigen-1a using targeted massively parallel sequencing. Transfusion 2015;55(6 Pt 2):1538–44.

[173] Ferro M, Macher HC, Noguerol P, Jimenez-Arriscado P, Molinero P, Guerrero JM, et al. Non-invasive prenatal diagnosis of Feto-maternal platelet incompatibility by cold high resolution melting analysis. Adv Exp Med Biol 2016;924:67–70.

[174] Faas BH, Beuling EA, Christiaens GC, von dem Borne AE, van der Schoot CE. Detection of fetal RHD-specific sequences in maternal plasma. Lancet 1998;352(9135):1196.

[175] Lo YM, Hjelm NM, Fidler C, Sargent IL, Murphy MF, Chamberlain PF, et al. Prenatal diagnosis of fetal RhD status by molecular analysis of maternal plasma. N Engl J Med 1998;339(24):1734–8.

GLOSSARY

Epitope Specific portion of an antigen, also called the antigenic determinant, to which the antibody binds and that stimulates an immune response.

Integrin Heterodimeric, transmembrane receptors with an alpha and beta subtype, that mediate cell adhesion.

Isoantibody An antibody against an antigen that is not present on the individual's own cells.

Alloantibody An antibody against a polymorphic variant of an antigen that is present as another variant on the individual's own cells.

Accuracy (TP+TN)/(TP+TN+FP+FN).

NONINVASIVE PRENATAL DIAGNOSIS OF MONOGENIC DISORDERS

Stephanie Allen*,†, Elizabeth Young*, Amy Gerrish*

Birmingham Women's and Children's NHS Foundation Trust, Edgbaston, Birmingham, United Kingdom*
West Midlands Regional Genetics Laboratory, Birmingham Women's and Children's NHS Foundation Trust,
Birmingham, United Kingdom†

INTRODUCTION

The groundbreaking discovery of the presence of cell-free DNA in maternal plasma was made in 1997 by Dennis Lo [1]. In the years since that, gradual technological development has meant that we can now use a maternal plasma sample as a source of material for prenatal testing for many different applications. One obvious potential application for this is noninvasive prenatal diagnosis (NIPD) of monogenic disorders—so-called Mendelian single gene disorders; however, the clinical application of this testing has lagged behind in comparison to other areas such as cfDNA-based noninvasive prenatal testing, commonly called NIPT, for aneuploidy.

Initial testing for this group of patients was limited to sex-linked and sex-limited disorders, and the use of fetal sexing by detection of Y chromosome markers. This testing was developed and implemented first as it was technologically much simpler to detect a marker that was not present in the mother's blood. Testing for many monogenic disorders is now becoming possible and available to patients, and this is described in more detail in this chapter.

It is also worth noting at this point that circulating fetal cells have also been explored as a possible source of testing material. They offer certain advantages and disadvantages over analysis of cell-free DNA and are dealt with in more detail in Chapter 19. At the time of writing this chapter, this is not considered a clinically practical approach to NIPD of monogenic disorders and so is not dealt with further here.

CLINICAL APPLICATIONS OF NIPD FOR MONOGENIC DISORDERS
CONDITIONS WITH A FAMILY HISTORY

The majority of NIPD assays reported to date have been for conditions where there is a known family history of a disorder, typically with a high recurrence risk in each pregnancy. For example in the case of an autosomal recessive disorder where both parents are carriers there is a one in four risk of an affected child with each subsequent pregnancy. The methods used for variant detection can vary depending upon the mode of inheritance. In the UK, next-generation sequencing (NGS)-based approaches to NIPD are already in routine use, allowing around 30% of all molecular prenatal diagnosis to be performed noninvasively [2].

Autosomal dominant disorders

NIPD for autosomal dominant conditions that are paternally inherited or have arisen de novo is the least technically challenging scenario; the causative mutation is not present in the maternal DNA, but if present in fetal DNA, would be detectable at low levels in maternal plasma. Several approaches have been taken to develop assays for a variety of autosomal dominant disorders and it should be noted that the same technologies described in the examples later could be modified for a range of conditions. For further technical details of these methods, refer to section "Technical Approaches to NIPD."

Droplet digital polymerase chain reaction assays (ddPCR) have been described to detect the presence of paternal *KCNJ11* [3] and *FGFR3* [4] mutations, which cause neonatal diabetes and achondroplasia, respectively. Using this technique, the authors were able to detect paternal alleles at a concentration of <1%. Orhant et al. demonstrated that it is also possible to detect the same *FGFR3* variant through a minisequencing assay, but this technique was only able to detect the disease-causing allele if it is present at >3% in the maternal plasma.

These genotyping assays can only detect the variant for which they are specifically designed. Although this approach may be appropriate for common mutations or to provide bespoke assays for families with a history of a rare disease, it may not be the most cost-effective way to provide NIPD for common disorders that have a spectrum of associated disease-causing mutations. Other publications have described an alternative approach by designing a single assay that could be used for many families, such as an NGS panel that covers 29 known *FGFR3* disease-causing mutations associated with both achondroplasia and thanatophoric dysplasia [5] and a targeted capture-sequencing method that is capable of screening the coding regions of 16 genes known to be associated with skeletal dysplasias [6].

Monogenic disorders which are caused by expansion of large polymorphic tracts, such as the trinucleotide repeat disorders, pose a particular challenge for NIPD. Van den Oever et al. describe a fragment length analysis technique to determine paternally inherited CAG repeat length in pregnancies at risk of Huntington disease [7]. The authors were able to detect disease causing and intermediate CAG repeats in all patients tested, but could only detect transmission of a normal paternal repeat in 50% of cases as it could not always be distinguished from the maternal allele. In addition, the authors advised caution when testing for very large repeats, as these may not be detectable due to the fragmented nature of cfDNA.

Autosomal recessive disorders

For autosomal recessive conditions, familial mutations may be the same in both parents (homozygous proband), or both parents may carry different mutations in the same gene (compound heterozygous proband). These situations require different approaches to NIPD. An exception to this is when haplotype-based assays are used these methods determine the inheritance of parental mutations in a linkage-type manner and as they do not directly detect the mutations, can be used for any scenario. To date, haplotype-based assays for use in pregnancies at risk of *GJB2*-associated hearing loss [8], congenital adrenal hyperplasia (CAH) [2,9,10], spinal muscular atrophy (SMA) [11,12], alpha- and beta-thalassemia [13,14], and Gaucher disease [15] have been published. However, these assays are limited to use for couples who have already had an affected child, as DNA from a proband is required to "phase" the analysis. This obstacle has been overcome by Vermeulen et al. [16], who combined targeted haplotyping of both parents using TLA (targeted locus amplification) with targeted deep sequencing of cfDNA extracted during pregnancy. Using this method, the authors were able to predict the inherited allele with >98% accuracy in 18 pregnancies at risk of cystic fibrosis (CF), CAH, or beta-thalassemia [16].

Detection of paternally inherited mutations

In a family where each parent is a carrier of a different mutation in the same gene, the most straightforward approach to NIPD is exclusion of a paternal mutation. If the paternal mutation is excluded, no further testing is indicated as the fetus may be a carrier of the condition but will not be affected. If the paternal mutation is detected by NIPD, further testing by traditional invasive methods is required to look for the presence of the maternal mutation.

As with paternally inherited dominant mutations, this is the least challenging technical scenario for autosomal recessive conditions, as if the mutation is detected in the maternal plasma, it must be fetal in origin. As such, the same techniques that have been used in autosomal dominant conditions can also been used for exclusion of paternal mutations in autosomal recessive conditions.

Debrand et al. describe a ddPCR assay that can detect the common CF causing mutation, DF508, at a level as low as 1.3% of total DNA in pregnancies where the father is a carrier of this mutation [17]. Further demonstrating the range of techniques that can be utilized for NIPD for the same conditions, Galbiati et al. developed a COLD-PCR assay that could detect both the DF508 mutation, as well as an additional three common CF-causing variants [18]. The authors reported that the mutations could not be detected by traditional PCR methods, but that using the COLD-PCR technique, they were able to detect mutations down to a level of <1%. As mentioned previously, these techniques can be applied to many different conditions, demonstrated by the fact that the same COLD-PCR methodology used for NIPD of CF was also validated for seven common beta-thalassemia-causing mutations [18]. In order to show that different methodologies can be used to confirm NIPD results, the authors also validated a microarray-based methodology for NIPD of CF and beta-thalassemia, which was in complete concordance with the COLD-PCR results [18]. This range of techniques that have been reported in the literature for the same conditions demonstrates the versatility of NIPD.

Detection of maternally inherited mutations

At present, assays to directly detect maternally inherited mutations, either in autosomal recessive or autosomal dominant conditions are not in routine clinical use because of the technical difficulties involved in differentiating between the maternal and fetal DNA in the maternal plasma sample. However, improvements in sensitivity of available techniques mean that advances are being made in this area.

A proof-of-principle study used ddPCR to perform precise allele quantification in maternal plasma and calculate relative dosage (relative mutation dosage, RMD) [19]. This demonstrates the potential of ddPCR for NIPD studies of fetal mutations independently of their parental origin, with 100% accuracy for the detection of paternal alleles and 96% accuracy for the detection of maternal alleles. By using single-nucleotide polymorphisms (SNPs) rather than disease-causing variants, the authors were able to demonstrate the utility of this technique for NIPD analysis of fetal mutations for any inheritance pattern.

An alternative approach was taken by Lv et al., who developed and validated an assay termed circulating single-molecule amplification and resequencing technology (cSMART). In their initial publication, this technique was successfully applied to pregnancies at risk of Wilson disease [20] where inheritance of different maternal and paternal mutant alleles was determined. Further validation work was performed to determine the inheritance of SNP genotypes to demonstrate that this assay can also be used for pregnancies when the parents are both carriers of the same mutation. The authors have since further developed this methodology to determine fetal genotypes in pregnancies where one or both partners were known carriers of an autosomal recessive nonsyndromic hearing loss (ARNSHL) disease causing mutation in either the *GJB2*, *GJB3*, and *SLC26A4* gene [21,22].

X-linked and sex-limited disorders

NIPD to determine fetal sex can be offered as a frontline test in pregnancies at risk of serious X-linked conditions, such as Duchenne muscular dystrophy. If this analysis shows the presence of a male fetus, further testing for a definitive diagnosis of the condition in question can be offered. NIPD to directly detect a maternally inherited mutation is technically challenging due to the high background of maternal cfDNA in maternal plasma (see previously), however, haplotype-based approaches have been described for NIPD of Duchenne and Becker muscular dystrophies [23–25] and hemophilia [26].

Congenital adrenal hyperplasia (CAH) is a sex-limited disorder where NIPD is used to guide treatment in pregnancy. In affected pregnancies with a female fetus there is a risk of in utero virilization. Antenatal dexamethasone administered very early in pregnancy may prevent or decrease virilization; however, it is important to target its use to at-risk pregnancies. NIPD to determine fetal sex can be used as a first step to target female pregnancies only and this is available from 7 weeks' gestation. Reports in the literature also describe methods that could be used to determine whether the fetus is affected with CAH, that is, carries the familial mutations or high-risk parental haplotype [2,9,10].

PGD confirmation

It is recommended that prenatal diagnosis is performed to confirm preimplantation genetic diagnosis (PGD), yet many patients are reluctant to undergo testing, due to the risks associated with traditional invasive procedures. In 2010 the ESHRE (European Society for Human Reproduction and Embryology) best practice guidelines [27] included NIPD as an appropriate method for confirming PGD pregnancies. One group has described a customized care trajectory for a family with a history of GJB2-associated hearing impairment that combined PGD with cfDNA-based noninvasive prenatal testing (NIPT) for fetal aneuploidy and a custom-designed NIPD assay, with the results confirmed by amniocentesis [8]. Others have described a combination of haplotype analysis and direct analysis of a paternal mutation to confirm the results of PGD in a pregnancy at risk of Marfan syndrome [28]. Although these management strategies would be prohibitively expensive for standard clinical practice, as different NIPD methodologies become part of routine prenatal diagnosis, these can become incorporated into the PGD care pathway.

For pregnancies at risk of an X-linked disorder where a specific NIPD assay is not available for the condition in question, noninvasive methods can be used to confirm fetal sex following transfer of a female embryo [28].

PREGNANCIES WITH ULTRASOUND SCAN FINDINGS

Another clinical application of NIPD for monogenic disorders is where there is no known family history of a disorder, but when a pregnancy presents with ultrasound scan findings that point toward the possibility of a monogenic disorder. A good example of this is skeletal dysplasia where NIPD provides a useful aid to clinical management, as it allows a definitive diagnosis, differentiation between lethal and nonlethal forms of the disease, and the option of a surgical termination as a postmortem is not required.

Multiple noninvasive methods for the detection of individual common mutations within the FGFR3 gene that cause either achondroplasia [29–33] or thanatophoric dysplasia [34] have been described; however, skeletal dysplasias are a group of heterogenous diseases and the detection rate is dependent on the accuracy of ultrasound diagnosis. With this in mind, Chitty et al. describe an NGS panel that covers 29 known FGFR3 disease-causing mutations that are associated with both achondroplasia and thanatophoric dysplasia [5] and Dan et al. have developed a targeted capture-sequencing method for the detection of de novo mutations in 16 lethal dysplasia genes [6].

These same assays allow testing to be performed very early in pregnancy in cases where there is a history of a skeletal dysplasia, either where the mutation is de novo (i.e., not detected in either parent) or paternally inherited.

NIPD where there are abnormalities on ultrasound scan is likely to increase in the future as technology becomes cheaper and it becomes feasible to screen for mutations in panels of genes. One could envisage testing panels of genes associated with particular ultrasound scan features such as "cardiac abnormality," "brain abnormality," or even ultimately the whole exome or genome.

PREGNANCIES AT RISK FOLLOWING PARENTAL CARRIER SCREENING

An increasing number of couples are finding that even in the absence of a family history, they are at risk of having a pregnancy affected with a monogenic disorder. This is due to the development of carrier screening programs and the introduction of commercially available direct-to-consumer carrier testing. With the exception of the haplotype-based approaches, where a proband is required to "phase" analysis, the techniques described previously could also be used to provide NIPD for these pregnancies.

TECHNICAL APPROACHES TO NIPD

As already described, many different technical approaches have been employed for the NIPD of monogenic disorders. These are summarized in Table 1.

Table 1 A Summary of Technical Approaches Employed in the NIPD of Monogenic Disorders Along With Some Recent Publications That Have Used These Techniques

Technique	Disorders	References
Amplicon-based NGS	ACH	[5]
cSMART	Wilson disease	[20]
	Hearing loss	[21,22]
Capture-based NGS	Skeletal dysplasias	[6]
ddPCR	CF	[17]
	Neonatal diabetes	[3]
	ACH	[4]
	Multiple SNPs	[19]
	Hemophilia	[26]
COLD-PCR	CF and beta-thalassemia	[18]
Microarray	CF and beta-thalassemia	[18]
Minisequencing	ACH	[4]
Fragment length analysis	HD	[7]
	CF	[35]
Relative mutation dosage (RMD)	Beta-thalassemia	[36]
Haplotype based NGS (RHDO and MG-NIPD)	CF	
	DMD	[23–25]
	Hearing Loss	[8]
	CAH	[2,9,10,16]
	Gaucher disease	[15]
	SMA	[11,12]
	Alpha- and beta-thalassemia	[13,14,16]
	Hemophilia	[26]

ACH, *achondroplasia;* CAH, *congenital adrenal hyperplasia;* CF, *cystic fibrosis;* DMD, *Duchenne muscular dystrophy;* HD, *huntington disease;* SMA, *spinal muscular atrophy;* SNP, *single-nucleotide polymorphism.*

DETECTION OF PATERNAL OR DE NOVO MUTATIONS

As described in section "Clinical Applications of NIPD for Monogenic Disorders," the least challenging NIPD test is determining the presence or absence of a pathogenic variant in maternal plasma that is not present in the maternal genome such as paternally inherited or de novo mutations in cfDNA. As a result, several methods have been developed for this type of assay. The most common of which are described as follows.

Bespoke/targeted NGS

Along with many other areas of genetics, NGS has revolutionized NIPD of monogenic disorders, mainly due to the huge increase in sensitivity NGS provides. Bespoke or targeted NGS involves the selection of a specific chromosomal region or regions of interest for sequencing. This selection of

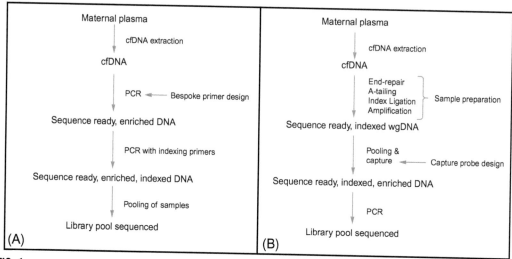

FIG. 1

An overview of the technical stages involved in amplicon- (A) and capture-based (B) NGS sample preparation and enrichment. Capture-based NGS requires a multistep sample preparation.

region(s) for sequencing, also known as enrichment, allows for high read depth of a single or series of mutations and can be done using either amplicon- or capture-based methodology.

The term "amplicon-based" NGS is used to describe an NGS protocol where sample preparation and enrichment is solely performed using a polymerase chain reaction (PCR) using PCR primers that are designed to amplify the specific region(s) of interest. Amplicon-based enrichment is relatively straightforward compared to other NGS sample preparation methods (see Fig. 1). However, this style of enrichment is limited by the same issues that face conventional PCR and PCR bias can potentially be an issue. These can especially be problematic when designing an assay to analyze multiple variants in a single test.

"Capture-based" NGS is where enrichment is performed via the hybridization of capture probes, designed against the region(s) of interest, to template DNA. This often provides a more uniform enrichment than amplicon-based methods and therefore the capacity for analyzing multiple variants/regions at once is increased. However unlike amplicon-based assays, a sample preparation stage is necessary prior to capture, which often requires multiple steps (see Fig. 1). Furthermore, if gDNA sequencing is required alongside cfDNA, then the gDNA must be fragmented prior to the sample preparation stage. This is not necessary when performing amplicon-based NGS as the PCR can be designed so that the PCR product is the correct size for sequencing. The additional sample preparation requirements in capture-based NGS and the use of probes mean it is more time consuming and expensive than amplicon-based NGS.

cSMART

As mentioned previously, an issue with amplicon-based NGS using conventional forward and reverse primers is the possibility of PCR bias. One cause of PCR bias is the preferential PCR amplification of smaller fragments which can skew allelic ratios.

Using a combination of amplicon circularization and the barcoding of individual library molecules, cSMART removes any size bias from individual fragments [20,22]. The workflow starts with a sample preparation protocol similar to that used in capture-based NGS. Individual library molecules are circularized and uniquely barcoded using a bridging oligonucleotide, and a second PCR is then performed using back to back inverse primers located adjacent to the mutation site(s) of interest. This avoids any size bias from the individual fragments. Duplicate molecules generated by PCR can also be identified and removed due to their identical barcodes and start/stop position.

Droplet digital PCR

As an alternative approach to NGS, several studies have utilized ddPCR for the NIPD of monogenic disorders [3,4,17,19,26]. ddPCR is a modified form of real-time PCR, which utilizes a water–oil emulsion droplet system. Each water droplet separates template DNA molecules into individual PCR reactions. As a result, thousands of independent amplification events are able to take place within a single sample. These amplification events are then analyzed individually on a droplet reader which counts whether droplets are positive or negative for the mutation of interest (Fig. 2). ddPCR allows the absolute quantification of target DNA copies without the need for standard curves, allowing for significant enhanced sensitivity of standard real-time PCR. One drawback of real time PCR and ddPCR in comparison to NGS is that individual assays are required for each mutation assayed, and multiplexing can be difficult. Therefore for individual mutations ddPCR is a cost-effective technique, but NGS should be considered where panels of mutations are being tested.

Cold-PCR

While many technical approaches aim to avoid PCR bias, coamplification at lower denaturation temperature-PCR (COLD-PCR) has used PCR bias to its advantage for the NIPD of monogenic disorders [18,37]. Unless modified, PCR will amplify all alleles with approximately equal efficiency comparable to their initial concentration. In contrast, COLD-PCR preferentially enriches minority alleles from samples containing a mixture of wild-type and mutation-containing sequences by exploiting the small, but critical and reproducible difference in amplification melting temperature (Tm). COLD-PCR uses an initial denaturation and hybridization step to create heteroduplexes between the mutant and wild-type allele. Heteroduplexes melt at lower temperatures than homoduplexes, the majority of which will only contain the wild-type allele. As a result heteroduplexes can be selectively denatured and subsequently preferentially amplified throughout the course of PCR, thus enriching for the mutant allele. For NIPD, COLD-PCR can potentially be used as a basis for subsequent Sanger Sequencing, NGS, or genotyping assays.

Minisequencing

Minisequencing is a genotyping method, which uses primer extension and standard capillary electrophoresis. Following amplification of the region of interest, a single-base extension reaction is performed using primer(s) that bind up to the mutation site of interest (Fig. 3). This extension reaction also incorporates fluorescently labeled dideoxynucleotides triphosphates (ddNTPs). The ddNTP complementary to either the wild-type or mutation allele will bind to the next base after the annealing primer. ddNTPs lack the 3′-OH group of dNTPs and as a result the extension reaction is terminated after the addition of just that one base. The product of this reaction is then sequenced on a genetic analyzer. In the case of NIPD this determines the presence or absence of the mutant allele. Orhant et al.

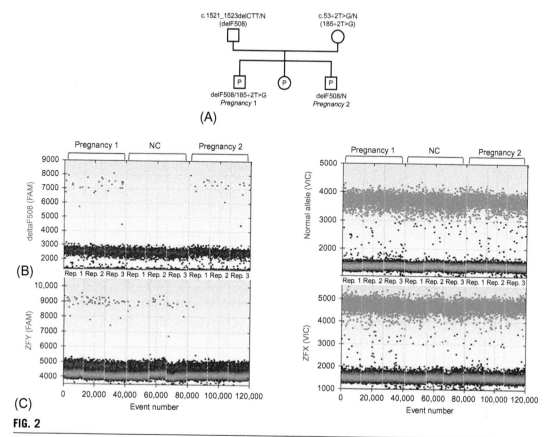

FIG. 2

Droplet digital PCR (ddPCR): (A) Pedigree of a family with a paternally inherited deltaF508 CFTR mutation; (B) and (C) Examples of ddPCR results with ZFX/ZFY and deltaF508MUT/NOR assays for pregnancies 1 and 2 indicated in the pedigree and for a normal control pregnancy; left-hand side: FAM probes, deltaF508 (Top) and ZFY (bottom); positive droplets are depicted in blue; right-hand side: VIC channel for loading control amplicons, ZFX and Normal allele at CFTR position c.1521_1523, respectively; positive droplets are shown in green, negative droplets in black. The x-axis shows the event number, that is, each droplet as counted sequentially by the QX100 reader; dotted vertical *yellow lines* indicate each replicate well.

used this technique to detect the FGFR3 1138 mutation. While their results were concordant with a ddPCR assay run in parallel, it should be noted that the sensitivity of minisequencing may be limited and restricted to NIPD at a later gestation, as was the case in this paper where plasma samples were taken at 18 weeks' gestation and above [4].

Fragment length analysis

The majority of bespoke/targeted NIPD has focused on the detection of single-nucleotide variants (SNVs). However, one study has used fragment length analysis to detect the paternally inherited Huntington's disease (HD) expanded CAG repeat [7]. In fragment length analysis a PCR is performed using

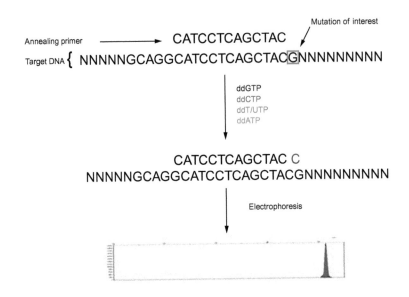

FIG. 3

Overview of minisequencing. The sequences of the annealing primer, target sequence, and mutation of interest are depicted in the top diagram, and then an example of the resulting electropherogram in the bottom diagram, with a *blue peak* showing incorporation of a ddCTP at the mutation site.

fluorescently labeled primers, which capture a chromosomal region where sequence variation has been shown to affect fragment length. The PCR product is run on a genetic analyzer along with a size standard to determine fragment length (bp). For NIPD it is necessary to run maternal gDNA and plasma cfDNA together so they can be compared in order to detect the paternal allele. Van de Oever and colleagues then used an empirically determined and validated panel to convert fragment length into the number of CAG repeats [7]. They were able to detect paternally derived normal and intermediate repeats where there was a size difference between maternal and paternal repeats.

An adapted version of fragment length analysis, known as Mutant Enrichment with 3′-Modified Oligonucleotide PCR coupled to Fragment Length Analysis (MEMO-PCR-FLA), has been developed for the NIPD of CF [35]. MEMO-PCR uses chemically modified blocking oligonucleotides to impede the amplification of specific sequence(s), such as maternal cfDNA in NIPD. Using this technique, coupled with standard fragment length analysis, Guissart et al. were able to enrich for paternally inherited Phe508del CFTR alleles and reported a sensitivity of detection of 2%.

Microarray

Microarrays are genotyping platforms which are formed of single-stranded DNA (ssDNA) probes on a solid surface. These probes consist of DNA sequence up to the mutation site of interest. Template DNA is hybridized to these probes and incubated with fluorescently labeled dNTPs. The fluorescence subsequently detected indicates the genotype of the template DNA at the mutation site. Due to the low levels of "fetal" DNA in maternal plasma, standard microarray technology is generally not sensitive enough for NIPD. However, Galbiati et al. developed a highly sensitive microarray, which provides a strong fluorescent signal coupled with low background, optimal for the detection of low abundant sequences in complex admixtures, such as fetal DNA in maternal plasma [18]. Using these specialized microarrays, the authors were able to detect the presence of several paternally inherited beta-thalassemia and CF mutations to a level of 0.8% without the need for further enrichment.

DETECTION OF MATERNAL MUTATIONS

Relative mutation dosage

The earlier methods have been used for the detection of paternally inherited or de novo mutations, absent in maternal plasma. If the mother is a carrier of an X-linked or autosomal dominant disorder, or alternatively, parents carry the same mutation for an autosomal recessive condition, then detection of these variants in fetal cfDNA is more difficult. Newer genomic technologies with greater sensitivity such as NGS and ddPCR have started to provide the possibility of analyzing maternally inherited fetal alleles. Both methods facilitate more precise allele quantification in maternal plasma and its relative dosage. This RMD can then be used to determine if a fetus has inherited the maternal mutant allele (see Fig. 4).

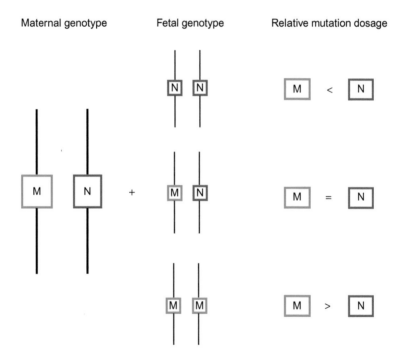

FIG. 4

Principles of relative mutation dosage (RMD) which can be used to determine fetal genotypes, despite the coexistence of the background maternal DNA in the maternal plasma. If a pregnant woman and her fetus are both heterozygous for a mutation, the amount of the mutant allele (M) and wild-type allele (N) will be in allelic balance in maternal plasma. However, when the fetus is homozygous for the wild-type or mutant allele, there would be an underrepresentation or overrepresentation of the mutant allele, respectively. A sequential probability ratio test (SPRT) is used to statistically evaluate the dosage imbalance between the M and N alleles.

Haplotype-based NGS

The detection and quantification of specific mutations for NIPD, whether de novo, paternally or maternally inherited can be a challenge whichever technical approach is used. Firstly, some types of genetic variants, for example large deletions, can be difficult to genotype/sequence and other variants can be located in regions not suitable for genetic analysis (see section "Challenges"). Secondly, for maternally inherited mutations, the high read/allele count required to gain adequate sensitivity can make NIPD assays challenging and often prohibitively expensive. Finally, the nature of bespoke/targeted NIPD assays means that the work involved in developing an individual test can be considerable and not cost effective for very rare disorders or for diseases, which are caused by a variety of individually rare mutations.

An alternative approach to bespoke/targeted NIPD is haplotype-based NGS where the fetal inheritance of a whole chromosomal region is determined rather than a specific variant/variants. In the majority of reports using this approach, genomic DNA from both the parents and a previous affected child (or another specific individual in the pedigree) is sequenced in addition to cfDNA derived from maternal plasma. This allows for the phasing of parental haplotypes and the identification of which haplotype(s) the mutation(s) reside upon.

Haplotype-based NGS requires the analysis of hundreds if not thousands of SNPs along a chromosomal region. Capture-based enrichment is preferable to minimize amplification bias, and probes are designed to capture highly heterozygous SNPs across the region of interest, in order to maximize informativity.

Once the mutant haplotype(s) have been identified, relative haplotype dosage analysis (RHDO) is used to determine whether the fetus has inherited the mutant or normal haplotype [38]. Fig. 5 shows the principle of RHDO in both X-linked disorders and autosomal recessive disorders.

While haplotype-based NGS offers considerable advantages for NIPD, there are some limitations. Due to coverage of sequencing required, it is relatively expensive compared to other assays. However, Parks and colleagues [12,14] have shown it can be cost effective in a clinical setting. By sequencing the proband and parental genomic DNA alongside the cfDNA as well as multiplexing up to three patients on a run, the group have determined that NIPD by RHDO is feasible for implementation into clinical service.

At present, RHDO analysis is limited to couples who have a previous affected child, and a DNA sample available from that child. This is because it is necessary to determine which haplotype(s) segregates with the pathogenic mutation. In some cases it may be possible to use DNA from other key family members in the pedigree dependent on the inheritance pattern; however, there are still a significant number of couples who cannot access the testing for this reason. However, recent publications have shown that haplotype-based NGS may also be available to patients without affected children in the future. Hui et al. [39] were able to directly resolve the haplotype of parental genomes using long-read sequencing. This technology requires the input of long DNA molecules which are separated, uniquely barcoded, and then fragmented. Postsequencing, reads carrying the mutant or wild-type allele are then linked and phased to the same haplotype. Using this approach Hui et al. resolved the haplotypes of the 12 of out 13 parental genomes and subsequently correctly classified the mutational status of these 12 fetuses by RHDO. Alternatively, Vermeulen et al. [16] combined targeted haplotyping of both parents with targeted deep sequencing of cfDNA extracted during pregnancy. This method involves targeted locus amplification-based phasing of heterozygous variants at a gene of interest in the parental samples,

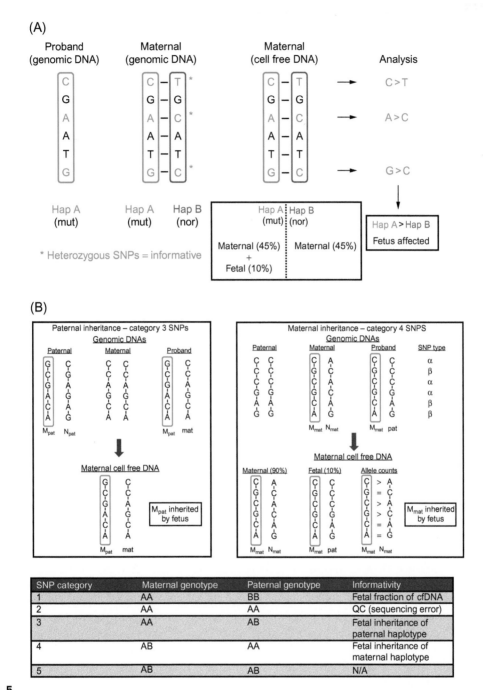

FIG. 5

Principles of relative haplotype dosage. (A) RHDO for X-linked disorders showing how overrepresentation of informative single-nucleotide polymorphisms on the X chromosome is used to determine whether the fetus has inherited the mutant or normal copy of the gene. (B) RHDO for autosomal recessive disorders. Once the mutant haplotype has been identified, RHDO analysis of category 3 and 4 SNPs is used to determine which haplotypes the fetus has inherited. RHDO measures the allelic imbalance between two haplotypes in plasma cfDNA. *Hap*, haplotype; *mat*, maternal; *mut*, mutant; *nor*, normal; *pat*, paternal.

followed by capture probes-based targeted sequencing of cfDNA from the pregnant mother and statistical analysis to predict fetal gene inheritance. Using this technique, the authors were able to predict the alleles inherited by the fetus in 18 pregnancies, with >98% confidence, from as early as 8 weeks' gestation [16].

FUTURE TECHNICAL APPROACHES—EXOME SEQUENCING

The majority of monogenic disorders are caused by exonic variants. Therefore another approach that could be used for NIPD in the future is exome sequencing. Several studies [38,40–42] have been able to sequence the entire fetal genome using cell-free DNA in maternal plasma, therefore exome sequencing is also theoretically possible. This approach is currently prohibitively expensive for clinical application due to the read depth required. As technology improves and costs decrease it is expected that NIPD using exome sequencing could become an attractive option as it would allow the combination of multiple NIPD assays into one streamlined service. Once exome sequencing of cfDNA from maternal plasma is carried out, samples could undergo a tailored bioinformatic analysis of a specific variant or panel of variants related to the disorder of interest or clinical presentation. As with other bespoke/targeted approaches, it is likely that exome sequencing would first be used for the NIPD of de novo or paternally inherited mutations. This is currently being offered by a company in the United States who carry out sequencing of 30 genes to detect de novo mutations associated with skeletal, cardiac, and neurological disorders with a combined incidence of 1:600 pregnancies such as Noonan syndrome, osteogenesis imperfecta, achondroplasia, and Rett syndrome [43].

CHALLENGES
INHERITANCE PATTERN, MUTATION TYPE, INCIDENCE OF DISORDER

As described previously, NIPD may be requested for autosomal dominant, autosomal recessive, and X-linked disorders, and also where there is an apparently de novo mutation in an index case with a low risk of recurrence. For autosomal recessive disorders, mutations may be the same in both parents, or there may be compound heterozygosity in the index case with two different mutations in the same gene. As the cell-free DNA in maternal plasma is largely maternally derived, it is technically much simpler to detect paternally derived or de novo mutations, that is, mutations that are not present in the maternal genome. This can be done by the methods described previously as long as there is a fetal fraction assay (see section "Fetal Fraction"). In situations such as X-linked disorders, maternally derived dominant disorders and autosomal recessive disorders with both parents carrying the same mutation more complex and costly techniques such as RHDO are required as the differences in mutation levels that the assay is required to detect are much more subtle.

Pathogenic mutations causing monogenic disorders include single-base changes, small insertions and deletions of a few bases, much larger duplications and deletions, and more complex rearrangements such as inversions. Assays designed for NIPD need to be able to detect the particular repertoire of mutations causing the disorder. For some disorders this is simpler as there is only one mutation causing the disorder, for example, Sickle cell anemia, for other disorders there are many thousands of mutations some of which are unique to particular families, and it may be better to adopt a more generic linkage-based approach, for example, for cystic fibrosis (CFTR gene mutations).

When developing NIPD, it is important to take into account the incidence of the disorder and requirement for prenatal diagnosis. The current lag in development of NIPD for monogenic disorders is largely due to the huge number of very rare disorders and the cost of setting up testing for individual disorders with relatively small number of patients and prenatal diagnoses. This accounts for the low commercial drive in companies offering this testing in comparison to cfDNA NIPT for aneuploidy. For disorders that are very rare there will be a need for some centralization of testing to enable efficient method development, cost effectiveness, and to meet the rapid turnaround times required.

FETAL FRACTION

Cell-free DNA in maternal plasma consists predominantly of circulating DNA derived from maternal cells, with a variable proportion of DNA fragments derived from placental trophoblast cells. The proportion of placentally derived DNA is due to many factors such as sample handling, gestation of pregnancy, placental size and health, maternal weight, chorionicity, ethnic origin, smoking, and so on [44]. As the proportion of so-called free fetal DNA is so variable, it is important when performing NIPD to have an assay for quality control to confirm that there is sufficient cfDNA in the sample tested. This is particularly important in cases where there is a negative result in the fetus, for example, testing for a paternal mutation that appears to be absent in the fetus.

There are numerous methods that have been described in the literature to assay and estimate fetal fraction (Table 2). This is very difficult to do accurately, particularly at low fetal fractions. Methods have addressed differences in fragment size, methylation, presence of Y chromosome for male fetuses, and SNPs (reviewed in [45]). It is generally better, if possible, to have a method that measures fetal fraction in the same assay as the NIPD. Many of the methods in the literature involve bioinformatic analysis of NGS data from cfDNA NIPT assays. Dependent upon the NIPD assay being used, it may not be appropriate or cost effective to use an NGS-based assay, for example, with a real time PCR or ddPCR-based assay. Of note, for RHDO methods, the fetal fraction is measured as an integral part of the assay by analysis of the SNP data generated as part of the assay. Whatever method is used for estimation of the fetal fraction, it is important that the limits of detection are validated alongside the assay that is being used, as estimations do vary between methods.

Table 2 A Summary of Some of the Approaches Used for Determination of the Fetal Fraction

Basis of Assay	Method	Advantages/Limitations
Presence of Y chromosome sequences	Real-time PCR/NGS	Simple; only suitable for male pregnancies
Methylation differences	Bisulfite sequencing/methylation-sensitive restriction enzyme digestion	Complex methodology; methylation may not be complete
SNPs	NGS	Accurate; usually requires parental samples; need lots of SNPs to always be informative
Fragment size	Paired-end NGS	Shallow-depth sequencing; moderate accuracy
Nucleosome based	NGS	Shallow-depth sequencing; lower accuracy

TRINUCLEOTIDE REPEAT DISORDERS

NIPD for dominant disorders caused by trinucleotide repeat expansions are particularly challenging—these include Huntington disease (HD), myotonic dystrophy, and fragile X syndrome. As detailed in section "Fragment length analysis," NIPD has successfully been performed for HD where CAG repeat sizes expand to a range of approximately 30–40 repeats (100–200 bp) [7]. For other disorders such as fragile X and myotonic dystrophy, trinucleotide repeats expand to 1000 s of repeats and therefore NIPD by direct PCR will not be possible as cfDNA is highly fragmented. Technical development for these disorders, if assaying cfDNA, will need to be by a linkage-based approach such as RHDO.

PSEUDOGENES

Some genes causing monogenic disorders have very closely related redundant gene copies (pseudogenes). These have evolved over time, and although nonfunctional can have a huge impact on testing due to the sequence homology with the disease-causing gene. Examples of this include congenital adrenal hyperplasia (CAH) due to CYP21A2 deficiency and spinal muscular atrophy (SMA). The gene causing SMA is called SMN1 (survival of motor neurone 1), and it has a pseudogene called SMN2 which has >99.9% sequence homology to SMN1 [46]. Any NIPD assays for disorders where there is a pseudogene would have to use sequences where there is a difference between the gene and pseudogene to attempt direct detection of the mutation. This can be difficult as cell-free DNA fragments are very short, and therefore there would have to be sequence differences in the same fragment as the disease-causing mutation. For this reason, assays for these disorders have tended to use linkage-based approaches, for example, using SNPs or linked markers [2,10–12].

CONSANGUINITY

Development and availability of NIPD for consanguineous couples has been particularly challenging. Typically pregnancies may be at an increased risk of rare autosomal recessive disorders where both parents carry the same mutation on a similar haplotype background. For these couples it is not possible to develop an individual assay to detect the paternal mutation as the mutation is the same in both partners. Equally RHDO is difficult using the SNPs described in section "Haplotype-based NGS" because there will be very small numbers of informative SNPs. Analysis of a particular subset of SNPs called class 5 SNPs, where both parents are heterozygous, is likely to be helpful in this situation [38]. A similar complementary approach has also been described looking at NIPD by universal founder haplotype analysis in inbred subpopulations where there is genetic relatedness in the disease locus [15].

IMPLEMENTING TESTING

Although there have been many papers reporting methods for NIPD of monogenic disorders, translation of the testing into routine clinical practice has been limited due to inadequate sensitivity, specificity, and reproducibility for clinical application. Cost and turnaround times are also challenging—any assays that are introduced clinically must be able to generate reportable results within an appropriate timeframe for the pregnancy, and at a cost that is affordable for the healthcare system.

As individual disorders are so rare, it is also difficult to accumulate sufficient numbers of samples to validate a new assay with plasmas from pregnancies at risk of that particular disorder. cfDNA-based

NIPT for aneuploidy has been validated on huge cohorts of women in various populations, however this is not possible for individual monogenic disorders. Biobanks have been developed to attempt to address this issue with some success, for example, the RAPID project [47]. It is also possible to validate testing using linkage-based techniques such as RHDO on more generic samples as they test for SNPs and not pathogenic mutations.

Once testing is implemented it is necessary to continuously monitor quality of testing, and one component of this is participation of diagnostic laboratories in external quality assurance schemes (EQA), and development of best practice. As there are so many different rare monogenic disorders with few laboratories offering testing, this will present a particular challenge in initiating this quality assurance and sourcing sufficient plasma from at-risk pregnancies to enable such schemes.

NIPD (unlike NIPT) is diagnostic as there is not the same risk of confined placental mosaicism. There is therefore not the same requirement for confirmation of results as there is with cfDNA-based NIPT. Definitive results can be actioned without the need for an invasive procedure. In the case of paternal exclusion for an autosomal recessive disorder, a follow-up chorionic villus sampling (CVS) or amniocentesis is required where the result shows that the paternal mutation has been inherited, to determine whether or not the fetus is affected. With RHDO, as it is carried out by linkage analysis rather than direct mutation detection, there is an extremely small theoretical risk of misdiagnosis due to recombination or de novo events. This is however minimal as SNPs are tested across the whole gene region, and it is usually clear if a recombination has occurred. Patients should be counseled about such minor risks when deciding to undertake NIPD as an alternative to invasive testing, as would be the case if linkage analysis were performed on chorionic villi or amniotic fluid cells.

COUNSELING
ETHICAL, LEGAL, SOCIAL CONSIDERATIONS

It is important that consideration is given to the ethical, legal, and social issues that arise with the implementation of NIPD into routine clinical practice. Several studies have investigated the views of stakeholders including adult patients, support group representatives, and carriers of autosomal recessive diseases [48–50], including those who have an affected child (living or deceased) and those who have used NIPD [51–53]. There was a general consensus that NIPD allows early, low-risk testing, but several common concerns were identified, including accuracy of results, routinization of testing, informed consent, increased pressure to have prenatal testing, and a resulting increase in terminations [48–53].

In addition to these concerns, the financial impact of the introduction of NIPD to routine antenatal care has to be considered. Not only can the total costs of NIPD for some conditions exceed those of invasive testing [54], it is anticipated that the reduced risks associated with NIPD will result in an increased uptake in prenatal testing overall. One study has shown that 90% of potential service users would have NIPD if it were available whereas only 43.5% would consider invasive testing [48].

It is also important to consider the ethical implications for the future of NIPD. At present, much of the testing available is for disorders that affect the infant at birth or early in life, but it is likely that in the near future, NIPD will be available for a much wider range of conditions, including those with varying penetrance and/or adult onset. This raises questions around which conditions is it ethically justifiable to test for prenatally and would parents who use these results "for information only" be invading the privacy of that future child's right to make their own decisions about genetic testing?

These questions highlight the need for guidelines surrounding what testing should be available and when it is appropriate to be offered to families.

EDUCATION AND INFORMATION

It is likely that NIPD will transform prenatal care for many couples with pregnancies at risk of a genetic disorder; however, the technical requirements of current technologies mean that testing is not yet possible for all families. In addition, interpretation of results generated by NIPD requires considerations that are not relevant in the provision of invasive prenatal testing. It is therefore imperative that couples considering NIPD are fully counseled prior to testing and that results are given by an appropriately trained member of their antenatal care team.

CONCLUSIONS

NIPD is finally becoming a reality for many couples with a pregnancy at risk of a monogenic disorder, either due to a family history or where there are abnormalities noted on ultrasound scan. There are still many conditions, however, where this is still not possible. It is also very challenging and costly to offer this testing, and therefore there is not yet equity of access. Future technological development and a reduction in price of testing will lead to an increase in the repertoire of disorders where testing can be offered and also an increase in access. It is important that NIPD is introduced in a well-validated quality-controlled manner. In addition, it is expected that population-based screening for monogenic disorders may increase, and it is important to continue to evaluate the ethical, social, and cost implications of this testing.

REFERENCES

[1] Lo YM, Corbetta N, Chamberlain PF, Rai V, Sargent IL, Redman CW, Wainscoat JS. Presence of fetal DNA in maternal plasma and serum. Lancet 1997;350(9076):485–7.

[2] Drury S, Mason S, McKay F, Lo K, Boustred C, Jenkins L, et al. Implementing non-invasive prenatal diagnosis (NIPD) in a National Health Service Laboratory; from dominant to recessive disorders. Adv Exp Med Biol 2016;924:71–5.

[3] De Franco E, Caswell R, Houghton JA, Iotova V, Hattersley AT, Ellard S. Analysis of cell-free fetal DNA for non-invasive prenatal diagnosis in a family with neonatal diabetes. Diabet Med 2017;34(4):582–5.

[4] Orhant L, Anselem O, Fradin M, Becker PH, Beugnet C, Deburgrave N, et al. Droplet digital PCR combined with minisequencing, a new approach to analyse fetal DNA from maternal blood: application to the non-invasive prenatal diagnosis of achondroplasia. Prenat Diagn 2016;36(5):397–406.

[5] Chitty LS, Mason S, Barrett AN, McKay F, Lench N, Daley R, et al. Non-invasive prenatal diagnosis of achondroplasia and thanatophoric dysplasia: next-generation sequencing allows for a safer, more accurate, and comprehensive approach. Prenat Diagn 2015;35(7):656–62.

[6] Dan S, Yuan Y, Wang Y, Chen C, Gao C, Yu S, et al. Non-invasive prenatal diagnosis of lethal skeletal dysplasia by targeted capture sequencing of maternal plasma. PLoS ONE 2016;11(7):e0159355.

[7] van den Oever JM, Bijlsma EK, Feenstra I, Muntjewerff N, Mathijssen IB, Bakker E, et al. Noninvasive prenatal diagnosis of Huntington disease: detection of the paternally inherited expanded CAG repeat in maternal plasma. Prenat Diagn 2015;35(10):945–9.

[8] Xiong W, Wang D, Gao Y, Gao Y, Wang H, Guan J, et al. Reproductive management through integration of PGD and MPS-based noninvasive prenatal screening/diagnosis for a family with GJB2-associated hearing impairment. Sci China Life Sci 2015;58(9):829–38.

[9] Kazmi D, Bailey J, Yau M, Abu-Amer W, Kumar A, Low M, et al. New developments in prenatal diagnosis of congenital adrenal hyperplasia. J Steroid Biochem Mol Biol 2017;165(Pt. A):121–3.

[10] Ma D, Yuan Y, Luo C, Wang Y, Jiang T, Guo F, et al. Noninvasive prenatal diagnosis of 21-hydroxylase deficiency using target capture sequencing of maternal plasma DNA. Sci Rep 2017;7(1):7427.

[11] Chen M, Lu S, Lai ZF, Chen C, Luo K, Yuan Y, et al. Targeted sequencing of maternal plasma for haplotype-based non-invasive prenatal testing of spinal muscular atrophy. Ultrasound Obstet Gynecol 2017;49(6):799–802.

[12] Parks M, Court S, Bowns B, Cleary S, Clokie S, Hewitt J, et al. Non-invasive prenatal diagnosis of spinal muscular atrophy by relative haplotype dosage. Eur J Hum Genet 2017;25(4):416–22.

[13] Saba L, Masala M, Capponi V, Marceddu G, Massidda M, Rosatelli MC. Non-invasive prenatal diagnosis of beta-thalassemia by semiconductor sequencing: a feasibility study in the Sardinian population. Eur J Hum Genet 2017;25(5):600–7.

[14] Wang W, Yuan Y, Zheng H, Wang Y, Zeng D, Yang Y, et al. A pilot study of noninvasive prenatal diagnosis of alpha- and beta-thalassemia with target capture sequencing of cell-free fetal DNA in maternal blood. Genet Test Mol Biomarkers 2017;21(7):433–9.

[15] Zeevi DA, Altarescu G, Weinberg-Shukron A, Zahdeh F, Dinur T, Chicco G, et al. Proof-of-principle rapid noninvasive prenatal diagnosis of autosomal recessive founder mutations. J Clin Invest 2015;125(10):3757–65.

[16] Vermeulen C, Geeven G, de Wit E, Verstegen MJAM, et al. Sensitive monogenic noninvasive prenatal diagnosis by targeted haplotyping. Am J Hum Genet 2017;101(3):326–39.

[17] Debrand E, Lykoudi A, Bradshaw E, Allen SK. A non-invasive droplet digital PCR (ddPCR) assay to detect paternal CFTR mutations in the cell-free fetal DNA (cffDNA) of three pregnancies at risk of cystic fibrosis via compound heterozygosity. PLoS ONE 2015;10(11):e0142729.

[18] Galbiati S, Monguzzi A, Damin F, Soriani N, Passiu M, Castellani C, et al. COLD-PCR and microarray: two independent highly sensitive approaches allowing the identification of fetal paternally inherited mutations in maternal plasma. J Med Genet 2016;53(7):481–7.

[19] Perlado S, Bustamante-Aragonés A, Donas M, Lorda-Sánchez I, Plaza J, Rodríguez de Alba M. Fetal genotyping in maternal blood by digital PCR: towards NIPD of monogenic disorders independently of parental origin. PLoS ONE 2016;11(4):e0153258.

[20] Lv W, Wei X, Guo R, Liu Q, Zheng Y, Chang J, et al. Noninvasive prenatal testing for Wilson disease by use of circulating single-molecule amplification and resequencing technology (cSMART). Clin Chem 2015;61(1):172–81.

[21] Chen Y, Liu Y, Wang B, Mao J, Wang T, Ye K, et al. Development and validation of a fetal genotyping assay with potential for noninvasive prenatal diagnosis of hereditary hearing loss. Prenat Diagn 2016;36(13):1233–41.

[22] Han M, Li Z, Wang W, Huang S, Lu Y, Gao Z, et al. A quantitative cSMART assay for noninvasive prenatal screening of autosomal recessive nonsyndromic hearing loss caused by GJB2 and SLC26A4 mutations. Genet Med 2017;19(12):1309–16.

[23] Xu Y, Li X, Ge HJ, Xiao B, Zhang YY, Ying XM, et al. Haplotype-based approach for noninvasive prenatal tests of Duchenne muscular dystrophy using cell-free fetal DNA in maternal plasma. Genet Med 2015;17(11):889–96.

[24] Yoo SK, Lim BC, Byeun J, Hwang H, Kim KJ, Hwang YS, et al. Noninvasive prenatal diagnosis of Duchenne muscular dystrophy: comprehensive genetic diagnosis in carrier, proband, and fetus. Clin Chem 2015;61 (6):829–37.

[25] Parks M, Court S, Cleary S, Clokie S, Hewitt J, Williams D, et al. Non-invasive prenatal diagnosis of Duchenne and Becker muscular dystrophies by relative haplotype dosage. Prenat Diagn 2016;36(4):312–20.

[26] Hudecova I, Jiang P, Davies J, Lo YMD, Kadir RA, Chiu RWK. Noninvasive detection of F8 int22h-related inversions and sequence variants in maternal plasma of hemophilia carriers. Blood 2017;130(3):340–7.

[27] Harton GL, de Rycke M, Fiorentino F, Moutou C, SenGupta S, Traeger-Synodinos J, et al. European Society for Human Reproduction and Embryology (ESHRE) PGD consortium ESHRE PGD consortium best practice guidelines for amplification-based PGD. Hum Reprod 2011;26:33–40.

[28] Bustamante-Aragones A, Perlado-Marina S, Trujillo-Tiebas MJ, Gallego-Merlo J, Lorda-Sanchez I, Rodríguez-Ramirez L, et al. Non-invasive prenatal diagnosis in the management of preimplantation genetic diagnosis pregnancies. J Clin Med 2014;3(3):913–22.

[29] Saito H, Sekizawa A, Morimoto T, Suzuki M, Yanaihara T. Prenatal DNA diagnosis of a single-gene disorder from maternal plasma. Lancet 2000;356(9236):1170.

[30] Li Y, Page-Christiaens GC, Gille JJ, Holzgreve W, Hahn S. Non-invasive prenatal detection of achondroplasia in size-fractionated cell-free DNA by MALDI-TOF MS assay. Prenat Diagn 2007;27(1):11–7.

[31] Lim JH, Kim MJ, Kim SY, Kim HO, Song MJ, Kim MH, et al. Non-invasive prenatal detection of achondroplasia using circulating fetal DNA in maternal plasma. J Assist Reprod Genet 2011;28(2):167–72.

[32] Chitty LS, Griffin DR, Meaney C, Barrett A, Khalil A, Pajkrt E, et al. New aids for the non-invasive prenatal diagnosis of achondroplasia: dysmorphic features, charts of fetal size and molecular confirmation using cell-free fetal DNA in maternal plasma. Ultrasound Obstet Gynecol 2011;37(3):283–9.

[33] Chen J, Yu C, Zhao Y, Niu Y, Zhang L, Yu Y, et al. A novel non-invasive detection method for the FGFR3 gene mutation in maternal plasma for a fetal achondroplasia diagnosis based on signal amplification by hemin-MOFs/PtNPs. Biosens Bioelectron 2017;91:892–9.

[34] Chitty LS, Khalil A, Barrett AN, Pajkrt E, Griffin DR, et al. Safe, accurate, prenatal diagnosis of thanatophoric dysplasia using ultrasound and free fetal DNA. Prenat Diagn 2013;33(5):416–23.

[35] Guissart C, Dubucs C, Raynal C, Girardet A, Tran Mau Them F, Debant V, et al. Non-invasive prenatal diagnosis (NIPD) of cystic fibrosis: an optimized protocol using MEMO fluorescent PCR to detect the p.Phe508del mutation. J Cyst Fibros 2017;16(2):198–206.

[36] Lun FMF, Tsui NBY, Allen Chan KC, Leung TY, Lau TK, Charoenkwan P, Chow KCK, Lo WYW, Wanapirak C, Sanguansermsri T, Cantor CR, Chiu RWK, Lo YMD. Noninvasive prenatal diagnosis of monogenic diseases by digital size selection and relative mutation dosage on DNA in maternal plasma. PNAS 2008;105(50):19920–5.

[37] Ferro M, Macher HC, Noguerol P, Jimenez-Arriscado P, Molinero P, Guerrero JM, et al. Non-invasive prenatal diagnosis of feto-maternal platelet incompatibility by cold high resolution melting analysis. Adv Exp Med Biol 2016;924:67–70.

[38] Lo YM, Chan KC, Sun H, Chen EZ, Jiang P, Lun FM, et al. Maternal plasma DNA sequencing reveals the genome-wide genetic and mutational profile of the fetus. Sci Transl Med 2010;2(61):61ra91.

[39] Hui WW, Jiang P, Tong YK, Lee WS, Cheng YK, New MI, et al. Universal haplotype-based noninvasive prenatal testing for single gene diseases. Clin Chem 2017;63(2):513–24.

[40] Fan HC, Gu W, Wang J, Blumenfeld YJ, El-Sayed YY, Quake SR. Non-invasive prenatal measurement of the fetal genome. Nature 2012;487(7407):320–4.

[41] Kitzman JO, Snyder MW, Ventura M, Lewis AP, Qiu R, Simmons LE, et al. Noninvasive whole-genome sequencing of a human fetus. Sci Transl Med 2012;4(137):137ra76.

[42] Best S, Wou K, Vora N, Van der Veyver IB, Wapner R, Chitty LS. Promises, pitfalls and practicalities of prenatal whole exome sequencing. Prenat Diagn 2018;38(1):10–19.

[43] https://prenatal.natera.com/vistara.

[44] Zhou Y, Zhu Z, Gao Y, Yuan Y, Guo Y, Zhou L, et al. Effects of maternal and fetal characteristics on cell-free fetal DNA fraction in maternal plasma. Reprod Sci 2015;22(11):1429–35.

[45] Peng XL, Jiang P. Bioinformatics approaches for fetal DNA fraction estimation in noninvasive prenatal testing. Int J Mol Sci 2017;18(2):453.

[46] Monani UR, Lorson CL, Parsons DW, Prior TW, Androphy EJ, Burghes AH, et al. A single nucleotide difference that alters splicing patterns distinguishes the SMA gene SMN1 from the copy gene SMN2. Hum Mol Genet 1999;8(7):1177–83.

[47] http://www.rapid.nhs.uk.

[48] Hill M, Twiss P, Verhoef TI, Drury S, McKay F, Mason S, et al. Non-invasive prenatal diagnosis for cystic fibrosis: detection of paternal mutations, exploration of patient preferences and cost analysis. Prenat Diagn 2015;35(10):950–8.

[49] Hill M, Suri R, Nash EF, Morris S, Chitty LS. Preferences for prenatal tests for cystic fibrosis: a discrete choice experiment to compare the views of adult patients, carriers of cystic fibrosis and health professionals. J Clin Med 2014;3(1):176–90.

[50] Hill M, Compton C, Karunaratna M, Lewis C, Chitty L. Client views and attitudes to non-invasive prenatal diagnosis for sickle cell disease, thalassaemia and cystic fibrosis. J Genet Couns 2014;23(6):1012–21.

[51] Lewis C, Hill M, Chitty LS. Non-invasive prenatal diagnosis for single gene disorders: experience of patients. Clin Genet 2014;85(4):336–42.

[52] Skirton H, Goldsmith L, Chitty LS. An easy test but a hard decision: ethical issues concerning non-invasive prenatal testing for autosomal recessive disorders. Eur J Hum Genet 2015;23(8):1004–9.

[53] Pisnoli L, O'Connor A, Goldsmith L, Jackson L, Skirton H. Impact of fetal or child loss on parents' perceptions of non-invasive prenatal diagnosis for autosomal recessive conditions. Midwifery 2016;34:105–10.

[54] Verhoef TI, Hill M, Drury S, Mason S, Jenkins L, Morris S, et al. Non-invasive prenatal diagnosis (NIPD) for single gene disorders: cost analysis of NIPD and invasive testing pathways. Prenat Diagn 2016;36(7):636–42.

MATERNAL CONSTITUTIONAL AND ACQUIRED COPY NUMBER VARIATIONS

10

Maria Neofytou, Joris Robert Vermeesch

Center for Human Genetics, KU Leuven, Leuven, Belgium

INTRODUCTION

Extracellular nucleic acids have already been known to exist in the blood circulation since 1948 [1]. The observation of abnormally high concentrations of DNA in serum of cancer patients lead Stroun et al. to suggest that DNA found in the serum of cancer patients might be released by tumor cells [2,3]. This hypothesis was confirmed when tumor-derived oncogene mutations were identified in the plasma and serum of patients with hematological malignancies [4]. Since then, analysis of cell-free DNA in plasma identified copy number alterations, single nucleotide variants, and viruses associated with several types of cancers [5–7]. Similarly, the placenta is considered by many a pseudomalignant tissue, invading the uterine epithelium to enable embryo implantation while releasing fragments of DNA in the maternal plasma as part of a physiological placental cell turnover during pregnancy [8]. Inspired by this theory, Lo et al. discovered Y-chromosomal sequences in the plasma of pregnant women with male fetuses in 1997 [9]. In healthy pregnant women, the majority of plasma DNA is a result of apoptosis or necrosis of maternal cells from multiple tissues [10]. In contrast, "fetal" DNA constitutes only a minor fraction of approximately 5%–20% of cfDNA in plasma during the first trimester, and originates mainly from the cytotrophoblast cells of the placenta [11]. Therefore any noninvasive prenatal test contains a major signal from the maternally derived DNA and a minor signal from the placenta-derived DNA. When applying genome-wide sequencing of plasma cfDNA not only the placental genome and potential aneuploidies in the conceptus are identified, but also the maternal genome is scanned for differences compared to the normal.

Early reports on the size of cfDNA using qPCR have shown that placenta-derived DNA fragments were generally shorter than the maternal ones [12]. More recently, paired-end sequencing has allowed to determine the exact size of millions of cell-free DNA fragments in plasma and distinguished unique placental reads from the high background of maternal DNA. The predominant size of plasma DNA of pregnant women is 166 bp followed by a series of smaller peaks that present at a periodicity of 10 bp at sizes of 143 bp and shorter [12–14]. It has been shown that the shorter placental DNA fragments are overrepresented in the fraction with fragments smaller than 150 bp [13]. It has also been shown that DNA degradation is not random but occurs by enzymatic processes at specific sites between nucleosomes, the internucleosomal linker sites. These patterns of cfDNA release have been studied extensively by two groups. Snyder et al. demonstrated that nucleosome footprints in cfDNA are tissue specific. They inferred nucleosome spacing from cfDNA in healthy individuals and demonstrated that it correlates with epigenetic features of lymphoid and myeloid cells. This finding reinforces the idea that hematopoietic cell death is the main source of cfDNA [15]. The group of Lo performed plasma DNA tissue mapping, a technique based on genome-wide bisulfite sequencing of plasma DNA. This analysis revealed the major tissue contributors of circulating DNA in both healthy nonpregnant and pregnant women as well as in patients with cancer. Similarly to Snyder et al. they found that >70% of the circulating DNA is derived from white blood cells, neutrophils, and lymphocytes. The remaining fraction is mainly derived from the liver and, in pregnant women, the placenta. The authors speculate that with further refinements in the deconvolution algorithm, it might become possible to identify the contribution of other tissues to the DNA pool [10]. Hence, localization of the cfDNA fragments can reveal the tissue of origin that contributes to cfDNA in both healthy and pathological conditions.

MATERNAL CONSTITUTIONAL COPY NUMBER VARIATIONS

Because maternal DNA constitutes the major DNA fraction of cfDNA, it may not be surprising that discrepancies of the maternal genome with respect to the reference genome can interfere with the interpretation of cfDNA-based prenatal testing results. Maternal constitutional copy number variations (CNVs) have been a major contribution to false positive cfDNA-based prenatal test results [16,17]. Our group was the first to show that CNVs as small as 500 kb can cause trisomy z-scores to pass the accepted cutoff of three leading to false positive or false negative results [18,19]. The group of Snyder demonstrated that a trisomy 18 discordancy between noninvasive and invasive prenatal test results was caused by a constitutional large duplication on the maternal chromosome 18 and, in another case, by a smaller maternal CNV of approximately 500 kb [20]. In a recent systematic review of the literature it was compiled that among the 60 false positive trisomy cases, 29 (48%) had an underlying maternal CNV [21]. Zhou et al. showed that there is a strong correlation of higher z-scores with increasing size of maternal CNVs [22]. Similarly, maternal X chromosomal CNVs have been reported to be the cause of false positive sex chromosome aneuploidy cfDNA-based prenatal test results [23].

Because placental DNA constitutes only a small fraction of plasma cfDNA, not only constitutional (i.e., present in all cells of the body), but also somatic CNVs (i.e., present in only a subset of all cells) can cause false positive and, theoretically, false negative results. Maternal (segmental) chromosomal mosaicism is another major cause of discordant results of cfDNA screening and even low-grade mosaicism in the maternal genome can skew the z-statistics significantly. Bayindir et al. identified a maternal mosaic segmental deletion on chromosome 13 by applying genome-wide analysis. This deletion was present in only 17% of the maternal blood cells analyzed by FISH [17,18]. Although rare, similar large mosaic maternal duplications of the chromosome 18 long arm have been observed and they caused false positive cfDNA test results [24]. In another case with a false positive cfDNA test for fetal trisomy 18, the mother was identified with a supernumerary ring chromosome 18 in 35% of her cells [25].

Cell-free DNA-based prenatal testing in pregnancies with sex chromosome aneuploidies (SCAs) has lower sensitivities and specificities as compared to testing for the common fetal trisomies

[26,27]. In a prospective study, Wang et al. have shown that maternal chromosome X mosaicism was the main cause of discordant sex chromosomal aneuploidy [28]. In another study by the same group, cfDNA Noninvasive Prenatal Testing (NIPT) findings suggested a case with lower X chromosome concentration. Maternal karyotyping showed mosaicism for monosomy X (mosaic Turner syndrome), indicating that this maternal mosaicism masked the true contribution of fetal chromosome X to the cell-free DNA pool [29].

The relatively high frequency of maternal CNVs and maternal sex chromosome mosaicism warrants maternal white blood cell testing with array-CGH or FISH to avoid unnecessary invasive testing [22]. Most maternal CNVs are benign or of unknown clinical significance [30].

MATERNAL MALIGNANCY

Almost 20 years after the discovery of tumor-derived cell-free DNA in plasma of cancer patients these notable findings were also observed in the pregnant population. In 2013, false positive results with unexplained trisomy 13 and monosomy 18 lead scientists to look beyond the fetal genome and identified a metastatic small cell carcinoma postpartum [31]. A unique pattern of segmental chromosomal imbalances identified in cfDNA-based NIPT and persistent in consecutive samples prompted Vandenberghe et al. to suspect a malignancy. Whole body MRI and subsequent molecular characterization identified a non-Hodgkin lymphoma as the underlying cause of the aberrant pattern [32]. Amant et al. showed that plasma DNA profiling during cfDNA NIPT identifies approximately 1 cancer per 2000 women who were not yet aware of the malignancy. Molecular characterization of the CNVs confirmed the cfDNA variation to be tumor derived. [33]. Sun et al. used plasma DNA tissue mapping and identified a case of follicular lymphoma based on the observation that the contribution of B-cells was unexpectedly higher in circulating plasma DNA [10]. More case reports describing the detection of cancers during pregnancies are emerging. A woman with a positive cfDNA NIPT result for full or partial monosomies of chromosomes, 13, 18, 21, and X was subsequently found to have hepatic lesions and was postpartum diagnosed with late stage colon cancer [34]. Similarly, a multiple myeloma was identified during routine cfDNA-NIPT showing subchromosomal imbalances involving seven autosomal chromosomes [35]. In a larger retrospective study by Bianchi et al. seven out of the 10 cases with occult maternal malignancies had a cfDNA NIPT result with unique patterns of nonspecific copy-number gains and losses across multiple chromosomes [36]. In an effort to determine the underlying biological cause in 37 abnormal cfDNA NIPT reports with multiple aneuploidies, follow-up analysis revealed a maternal malignancy in 19% of the cases [37].

Amant et al. estimate the incidence of cancers in pregnant woman, unsuspected to have a cancer, to be at least 1/2000. Hartwig et al. reviewed the literature and estimate that 15% of unexplained false positive cases were caused by maternal malignancies [21]. Diagnosis of a malignant disease in pregnancy is often delayed because symptoms are attributed to pregnancy [38]. Detection of malignancy and its management during pregnancy presents a conflict between optimal maternal therapy and fetal wellbeing [39]. Recent data from a multicenter control study suggest that prenatal exposure to maternal cancer with or without treatment did not impair the health and general development of children [40]. Therefore early diagnosis and appropriate management of pregnancy-associated cancer can accelerate treatment and improve both maternal and neonatal outcome [41].

OTHER MATERNAL DISORDERS

Because the cell-free DNA is derived from multiple tissues, diseases in other organs or organ systems that result in apoptosis and DNA shedding in the circulation may well skew prenatal testing results.

It has been demonstrated that upon solid organ transplantation, DNA from the transplant can be detected in the cfDNA [14]. More importantly, it has been shown that this DNA fraction increases upon graft rejection [42]. Not surprisingly, pregnant women with an organ transplant are at risk of shedding large amounts of graft DNA in the plasma. This, in turn, will reduce the overall "fetal" DNA fraction which could subsequently cause a misdiagnosis. Discordances in fetal sex determination between cfDNA prenatal testing and ultrasound imaging have been described in two separate reports so far [43,44]. In both cases, the pregnant women had undergone a transplant from a male donor prior to pregnancy. cfDNA NIPT results reported a male pregnancy in contrast to fetal anatomy ultrasound showing female genitalia. Follow-up testing confirmed the fetuses to be female. Hence, apoptosis of cells derived from the transplanted organ lead to the detection of Y-chromosome sequences and, as a consequence, to incorrect calling of the fetal sex. This type of erroneous fetal sex assessment may be avoided using bioinformatics cfDNA analysis pipelines that can identify unusual fetal fractions or which compare different approaches of fetal fraction estimation [44].

Systemic lupus erythematosus (SLE) was one of the first diseases reported to be associated with elevated circulating plasma DNA concentrations [45]. Paired-end massively parallel sequencing of plasma DNA of nonpregnant patients with active SLE showed aberrations in measured genomic representations (MGRs), this is significantly more or less DNA fragments aligned to the reference genome as compared to controls. In addition, the size distribution profiles of SLE patients show elevated proportions of shorter fragments (<115 bp) and hypomethylation compared to healthy controls [46]. The aberrations in MGR and shortening of plasma DNA fragments correlated with disease activity. The authors hypothesize that preferential binding of antidouble strand DNA antibodies present in SLE patients with active disease hinders clearance of these short fragments from the circulation. SLE is one of the most common autoimmune disorders in women during their childbearing years and therefore another possible cause of abnormal cell-free DNA genomic profiles.

Finally, abnormal genome-wide representation profiles have been reported in a pregnant woman with severe vitamin B12 deficiency. After treatment with parenteral vitamin B12, the abnormal genome-wide profiles normalized [47]. Vitamin B12 deficiency is known to cause ineffective erythropoiesis and intramedullary hemolysis. Although this is a single case report, the authors speculated that the abnormal profile was a consequence of the vitamin B12 deficiency. In Table 1 we summarize the main causes of abnormal or unexpected findings identified by cell-free DNA prenatal screening caused by maternal genomic alterations.

OPPORTUNITIES FOR MATERNAL PERSONALIZED GENOMIC MEDICINE

The initial observations of maternal constitutional and acquired CNVs focused attention on the risk of making a fetal aneuploidy misdiagnosis. Awareness of the causes leading to false positive or negative results have led to improvements in the analysis methods [19,49]. However, those incidental, or rather unsolicited findings in the context of fetal aneuploidy screening, offer novel opportunities to identify maternal health risks that should be communicated to the pregnant women as they may benefit maternal and fetal health and alter pregnancy management. We consider this true maternal genomic medicine opportunities.

Table 1 **Overview of Maternal Acquired or Constitutional Genomic Alterations That Can Cause Discordant cfDNA-Based Prenatal Screening Results**

Condition	References
Maternal CNV	Zhang et al. [16], Bayindir et al. [18], Snyder et al. [20], Brison et al. [19], Meschino et al. [48], Clark-Ganheart et al. [25]
Maternal mosaicism	Ganheart [25], Bayindir et al. [18], Flowers et al. [24], Wang et al. [28]
Maternal malignancy	Osborne et al. [31], Vandenberghe et al. [32], Bianchi et al. [36], Snyder et al. [15], Smith et al. [34], Imbert-Bouteille et al. [35]
Nonmalignant maternal medical disorder	Schuring-Blom et al. [47] Chan et al. [46]
Transplantation	Neufeld-Kaiser et al. [43], Neofytou et al. [44]

By applying genome-wide sequencing and analysis of cell-free DNA in pregnant women, Brison et al. were able to identify five cases with clinically actionable maternal CNVs. These included a maternal deletion that encompassed RUNX1, a low-grade (20%) mosaic segmental deletion of chromosome 13, an unbalanced translocation, and two cases with interstitial deletion of chromosome X in pregnancies of female fetuses [19]. Haploinsufficiency for *RUNX1* is known to cause a platelet disorder with associated myeloid malignancy. The thrombocytopenia and platelet dysfunction put mother and infant at risk for extensive bleeding at delivery. Knowledge of the disorder allowed appropriate pregnancy and delivery management. Given the autosomal dominant inheritance of the disease, postnatal follow-up testing was performed in the newborn. This confirmed the presence of the deletion in the newborn. Clinical follow-up was advised for both the mother and the son for early diagnosis of a potential myeloid malignancy [19]. The two cases of a large maternal interstitial deletion in chromosome X pose a risk for X-linked disorders in the patient's male offspring and eventual female family members, and warrant an offer of invasive prenatal diagnosis for these conditions [19]. Such cases would remain undiagnosed with targeted cfDNA NIPT strategies that only detect common aneuploidies. These cases illustrate that the detection of maternal CNVs can enhance pregnancy management of both current and future pregnancies.

Another group reported the incidental finding of a familial amyloid beta (A4) precursor protein (APP) duplication causing a false positive cfDNA NIPT result for trisomy 21. Duplication of the APP gene causes early onset Alzheimer disease (AD) cerebral amyloid angiopathy. Such incidental findings highlight the potential for cfDNA prenatal testing to predict late-onset genetic conditions in both fetus and mother and therefore the need for and complexity of fully informed patient consent prior to genome-wide testing [48].

FROM LOCALIZED INITIATIVES TO PROFESSIONAL GUIDELINES

A challenge in noninvasive prenatal testing is how to deal with maternal incidental findings during cfDNA screening and how these results should be reported to the pregnant woman. Incidental findings are defined as findings which are not directly related to the indication for which the cfDNA prenatal testing was performed (trisomy 21, 18, and 13). The ACMG has published guidelines on how to manage such unanticipated findings [50] and recommends to: "(1) prior to testing inform patients of the possibility of identifying maternal genomic imbalances and that this possibility depends on the methodology used (2) when cfDNA prenatal testing identifies a maternal genomic imbalance refer the patient to a trained genetics professional. (3) Offer aneuploidy screening other than cfDNA-based testing for patients with a history of bone marrow or organ transplantation from a male donor or donor of uncertain biologic sex. (4) Discuss the possibility of discordant fetal biologic sex if maternal blood transfusion was given <4 weeks prior to the blood draw."

A more directive approach is provided by the Belgian Society of Human Genetics.

The society has published guidelines on managing incidental findings detected by cfDNA prenatal testing which were approved by the Belgian College for Medical Genetics (http://www.beshg.be/download/guidelines/20170126_belgian_guidlines_for_managing_incidental_findings_detected_by_NIPT.pdf) [51].

According to these guidelines five main categories of maternal incidental findings can be distinguished: (1) CNVs causing highly penetrant disorders with validated evidence on the phenotype associated with the deletion or duplication. These CNVs are considered clinically relevant and will be reported. (2) CNVs proven to be risk factors for developmental disorders with reduced penetrance and/or variable expression. The predictability of the future phenotype resulting from such CNVs remains very poor. Therefore these susceptibility CNVs will not be reported. (3) CNVs causing late-onset genetic disorders that are still asymptomatic in the mother. Disorders with clinical utility, typically cancer caused by the deletion of a tumor suppressor gene, will be reported according to the latest guidelines from the American College of Medical Genetics (ACMG) [52] as undeniable health benefit can be expected for the mother and/or relatives when communicated. (4) CNVs that have no consequence for the mother but, if inherited, are potentially harmful for the fetus in the current or in a future pregnancy are categorized based on inheritance: (a) Carriership for autosomal recessive disorders will not be communicated, unless the disorder is frequent, that is, carrier frequency >1/50 (CFTR, SMA and Connexin 26). (b) Carriership for X-linked recessive disorders will be communicated, irrespective of the sex of the fetus. (c) Carriership of a mosaic CNV will be communicated if it poses a risk for highly penetrant developmental disorders in the fetus. (5) cfDNA-based NIPT may lead to the incidental finding of a malignancy in the mother and this will be communicated.

The incidental detection of maternal malignancies should be part of pre- and posttest counseling to avoid women receiving unexpected prenatal test results. Health care providers should decide on what findings will be reported [53]. In the Leuven hospital, we have established a specific care path: pregnant women with cfDNA prenatal testing profiles suggesting the presence of malignancy are referred to the oncology unit. There, often, a whole body MRI is offered to search for the malignant lesions. A joint consultation including gynecologists, oncologists, and genetic care providers assures proper counseling and follow up. Further decisions about treatment are approached interdisciplinary and are made together with the patient.

FUTURE DIRECTIONS

Cell-free DNA-based prenatal testing with genome-wide sequencing and analysis offers several advantages: it does not only allow the detection of the common trisomies 13, 18, and 21, but also the detection of other fetal whole chromosomal or segmental aneuploidies. This can be important for pregnancy management and may provide insights into the biological causes of various obstetric pathologies such as intrauterine growth retardation, intrauterine death, fetal mosaicism, and uniparental disomy [54–56]. Because of these advantages, we envision that genome-wide profiling will become the main approach for cell-free DNA-based prenatal testing. As a consequence, unsolicited maternal genomic anomalies will be detected. Despite the challenges imposed upon counseling, we advocate this genome-wide approach, as the current and the anticipated technological advancements provide opportunities to improve overall pregnancy management and maternal care [30].

In order to assure proper use of the information generated it is important that professional caregivers are well aware of these clinical different aspects. There is an important role for the providers of the tests and the professional societies to teach the caregivers and to provide adequate pre- and post-test counseling tools. It seems timely for the different professional bodies involved in prenatal testing to take the initiative and draft global guidelines, which can be further fine-tuned by national societies to meet local regulatory, societal and ethical standards. In parallel, undertaking pilot multidisciplinary studies will prove invaluable for the assessment of the social, health, and economic benefits of genome-wide cell-free DNA-based prenatal testing and ultimately lead to modification of existing health system policies.

REFERENCES

[1] Mandel P, Metais P. Les acides nucléiques du plasma sanguin chez l'homme. C R Seances Soc Biol Ses Fil 1948;142:241–3.

[2] Leon SA, Shapiro B, Sklaroff DM, Yaros MJ. Free DNA in the serum of cancer patients and the effect of therapy. Cancer Res 1977;37(3):646–50.

[3] Stroun M, Anker P, Maurice P, Lyautey J, Lederrey C, Beljanski M. Neoplastic characteristics of the DNA found in the plasma of Cancer patients. Oncology 2009;46(5):318–22.

[4] Vasioukhin V, Anker P, Maurice P, Lyautey J, Lederrey C, Stroun M. Point mutations of the N-ras gene in the blood plasma DNA of patients with myelodysplastic syndrome or acute myelogenous leukaemia. Br J Haematol 1994;86(4):774–9.

[5] Lo YMD, Chan LY, Lo KW, Leung SF, Zhang J, Chan AT, et al. Quantitative analysis of cell-free Epstein-Barr virus DNA in plasma of patients with nasopharyngeal carcinoma. Cancer Res 1999;59(6):1188–91.

[6] Leary RJ, Sausen M, Kinde I, Papadopoulos N, Carpten JD, Craig D, et al. Detection of chromosomal alterations in the circulation of cancer patients with whole-genome sequencing. Sci Transl Med 2012;4 (162):162ra154.

[7] Chan KCA, Jiang P, Zheng YWL, Liao GJW, Sun H, Wong J, et al. Cancer genome scanning in plasma: detection of tumor-associated copy number aberrations, single-nucleotide variants, and tumoral heterogeneity by massively parallel sequencing. Clin Chem 2013;59(1):211–24.

[8] Strickland S, Richards WG. Invasion of the trophoblasts. Cell 1992;71(3):355–7.

[9] Lo YMD, Corbetta N, Chamberlain PF, Rai V, Sargent IL, Redman CW, et al. Presence of fetal DNA in maternal plasma and serum. Lancet 1997;350(9076):485–7.

[10] Sun K, Jiang P, Chan KCA, Wong J, Cheng YKY, Liang RHS, et al. Plasma DNA tissue mapping by genome-wide methylation sequencing for noninvasive prenatal, cancer, and transplantation assessments. Proc Natl Acad Sci 2015;112(40):E5503–12.

[11] Faas BHW, de Ligt J, Janssen I, Eggink AJ, Wijnberger LDE, van Vugt JMG, et al. Non-invasive prenatal diagnosis of fetal aneuploidies using massively parallel sequencing-by-ligation and evidence that cell-free fetal DNA in the maternal plasma originates from cytotrophoblastic cells. Expert Opin Biol Ther 2012;12 (Suppl. 1):S19–26.

[12] Chan KCA, Zhang J, Hui ABY, Wong N, Lau TK, Leung TN, et al. Size distributions of maternal and fetal DNA in maternal plasma. Clin Chem 2004;50(1):88–92.

[13] Lo YMD, Chan KCA, Sun H, Chen EZ, Jiang P, Lun FMF, et al. Maternal plasma DNA sequencing reveals the genome-wide genetic and mutational profile of the fetus. Sci Transl Med 2010;2(61):61ra91.

[14] Zheng YWL, Chan KCA, Sun H, Jiang P, Su X, Chen EZ, et al. Nonhematopoietically derived DNA is shorter than hematopoietically derived DNA in plasma: A transplantation model. Clin Chem 2012;58(3):549–58.

[15] Snyder MW, Kircher M, Hill AJ, Daza RM, Shendure J. Cell-free DNA comprises an in vivo nucleosome footprint that informs its tissues-of-origin. Cell 2016;164(1–2):57–68.

[16] Zhang H, Gao Y, Jiang F, Fu M, Yuan Y, Guo Y, et al. Non-invasive prenatal testing for trisomies 21, 18 and 13: Clinical experience from 146 958 pregnancies. Ultrasound Obstet Gynecol 2015;45(5):530–8.

[17] Brady P, Brison N, Van Den Bogaert K, de Ravel T, Peeters H, Van Esch H, et al. Clinical implementation of NIPT - technical and biological challenges. Clin Genet 2016;89:523–30.

[18] Bayindir B, Dehaspe L, Brison N, Brady P, Ardui S, Kammoun M, et al. Noninvasive prenatal testing using a novel analysis pipeline to screen for all autosomal fetal aneuploidies improves pregnancy management. Eur J Hum Genet 2015;23(10):1286–93.

[19] Brison N, Van Den Bogaert K, Dehaspe L, Van Den Oever JME, Janssens K, Blaumeiser B, et al. Accuracy and clinical value of maternal incidental findings during noninvasive prenatal testing for fetal aneuploidies. Genet Med 2017;19(3):306–13.

[20] Snyder MW, Simmons LE, Kitzman JO, Coe BP, Henson JM, Daza RM, et al. Copy-number variation and false positive prenatal aneuploidy screening results. N Engl J Med 2015;1–7.

[21] Hartwig TS, Ambye L, Sørensen S, Jørgensen FS. Discordant non-invasive prenatal testing (NIPT) – a systematic review. Prenat Diagn 2017;37:527–39.

[22] Zhou X, Sui L, Xu Y, Song Y, Qi Q, Zhang J, et al. Contribution of maternal copy number variations to false-positive fetal trisomies detected by noninvasive prenatal testing. Prenat Diagn 2017;37(4):318–22.

[23] Wang S, Huang S, Ma L, Liang L, Zhang J, Zhang J, et al. Maternal X chromosome copy number variations are associated with discordant fetal sex chromosome aneuploidies detected by noninvasive prenatal testing. Clin Chim Acta 2015;444:113–6.

[24] Flowers N, Kelley J, Sigurjonsson S, Bruno DL, Pertile MD. Maternal mosaicism for a large segmental duplication of 18q as a secondary finding following non-invasive prenatal testing and implications for test accuracy. Prenat Diagn 2015;35(10):986–9.

[25] Clark-Ganheart CA, Iqbal SN, Brown DL, Black S, Fries MH. Understanding the limitations of circulating cell free fetal DNA: an example of two unique cases. J Clin Gynecol Obstet 2014;3(2):38–70.

[26] Zhang B, Lu B-Y, Yu B, Zheng F-X, Zhou Q, Chen Y-P, et al. Noninvasive prenatal screening for fetal common sex chromosome aneuploidies from maternal blood. J Int Med Res 2017;45(2):621–30.

[27] Porreco RP, Garite TJ, Maurel K, Marusiak B, Ehrich M, Van Den Boom D, et al. Noninvasive prenatal screening for fetal trisomies 21, 18, 13 and the common sex chromosome aneuploidies from maternal blood using massively parallel genomic sequencing of DNA. Am J Obstet Gynecol 2014;211(4):365. e1–365.e12.

[28] Wang Y, Chen Y, Tian F, Zhang J, Song Z, Wu Y, et al. Maternal mosaicism is a significant contributor to discordant sex chromosomal aneuploidies associated with noninvasive prenatal testing. Clin Chem 2014; 60(1):251–9.

[29] Wang L, Meng Q, Tang X, Yin T, Zhang J, Yang S, et al. Maternal mosaicism of sex chromosome causes discordant sex chromosomal aneuploidies associated with noninvasive prenatal testing. Taiwan J Obstet Gynecol 2015;54(5):527–31.

[30] Van Opstal D, van Maarle MC, Lichtenbelt K, Weiss MM, Schuring-Blom H, Bhola SL, et al. Origin and clinical relevance of chromosomal aberrations other than the common trisomies detected by genome-wide NIPS: results of the TRIDENT study. Genet Med 2018;20:480–5.

[31] Osborne CM, Hardisty E, Devers P, Kaiser-Rogers K, Hayden MA, Goodnight W, et al. Discordant noninvasive prenatal testing results in a patient subsequently diagnosed with metastatic disease. Prenat Diagn 2013;33:609–11.

[32] Vandenberghe P, Wlodarska I, Tousseyn T, Dehaspe L, Dierickx D, Verheecke M, et al. Non-invasive detection of genomic imbalances in Hodgkin/reed-Sternberg cells in early and advanced stage Hodgkin's lymphoma by sequencing of circulating cell-free DNA: a technical proof-of-principle study. Lancet Haematol 2015;2(2):e55–65.

[33] Amant F, Verheecke M, Wlodarska I, Dehaspe L, Brady P, Brison N, et al. Presymptomatic identification of cancers in pregnant women during noninvasive prenatal testing. JAMA Oncol 2015;1(6):814–9.

[34] Smith J, Kean V, Bianchi DW, Feldman G, Petrucelli N, Simon M, et al. Cell-free DNA results lead to unexpected diagnosis. Clin Case Reports 2017;5(8):1323–6.

[35] Imbert-Bouteille M, Chiesa J, Gaillard J-B, Dorvaux V, Altounian L, Gatinois V, et al. An incidental finding of maternal multiple myeloma by non invasive prenatal testing. Prenat Diagn 2017;37(12):1257–60.

[36] Bianchi DW, Chudova D, Sehnert AJ, Bhatt S, Murray K, Prosen TL, et al. Noninvasive prenatal testing and incidental detection of occult maternal malignancies. JAMA 2015;314(2):162.

[37] Snyder HL, Curnow KJ, Bhatt S, Bianchi DW. Follow-up of multiple aneuploidies and single monosomies detected by noninvasive prenatal testing: Implications for management and counseling. Prenat Diagn 2016;36(3):203–9.

[38] Mahmoud HK, Samra MA, Fathy GM. Hematologic malignancies during pregnancy: A review. J Adv Res 2016;7:589–96.

[39] Aziz Karim S, Shafi MI. Malignancy in pregnancy. Curr Obstet Gynaecol 2005;15(6):414–6.

[40] Amant F, Vandenbroucke T, Verheecke M, Fumagalli M, Halaska MJ, Boere I, et al. Pediatric outcome after maternal Cancer diagnosed during pregnancy. N Engl J Med 2015;373(19):1824–34.

[41] Shim MH, Mok C-W, Chang KH-J, Sung J-H, Choi S-J, Oh S-Y, et al. Clinical characteristics and outcome of cancer diagnosed during pregnancy. Obstet Gynecol Sci 2016;59(1):1–8.

[42] Snyder TM, Khush KK, Valantine HA, Quake SR. Universal noninvasive detection of solid organ transplant rejection. Proc Natl Acad Sci U S A 2011;108(15):6229–34.

[43] Neufeld-Kaiser WA, Cheng EY, Liu YJ. Positive predictive value of non-invasive prenatal screening for fetal chromosome disorders using cell-free DNA in maternal serum: independent clinical experience of a tertiary referral center. BMC Med 2015;13(1):129.

[44] Neofytou M, Brison N, Van den Bogaert K, Dehaspe L, Devriendt K, Geerts A, et al. Maternal liver transplant: another cause of discordant fetal sex determination using cell free DNA. Prenat Diagn 2018;38(2):148–50.

[45] Chen JA, Meister S, Urbonaviciute V, Rödel F, Wilhelm S, Kalden JR, et al. Sensitive detection of plasma/serum DNA in patients with systemic lupus erythematosus. Autoimmunity 2007;40(4):307–10.

[46] Chan RWY, Jiang P, Peng X, Tam L-S, Liao GJW, Li EKM, et al. Plasma DNA aberrations in systemic lupus erythematosus revealed by genomic and methylomic sequencing. Proc Natl Acad Sci 2014;111(49):E5302–11.

[47] Schuring-Blom H, Lichtenbelt K, van Galen K, Elferink M, Weiss M, Vermeesch JR, et al. Maternal vitamin B12 deficiency and abnormal cell-free DNA results in pregnancy. Prenat Diagn 2016;36(8):790–3.

[48] Meschino WS, Miller K, Bedford HM. Incidental detection of familial APP duplication: An unusual reason for a false positive NIPT result of trisomy 21. Prenat Diagn 2016;36(4):382–4.

[49] Brison N, Neofytou M, Dehaspe L, Van De Bogaert K, Darbour L, Van Esch H, et al. Predicting fetoplacental chromosomal mosaicism during non-invasive prenatal testing. Prenat Diagn 2018; [accepted for publication].

[50] Gregg AR, Skotko BG, Benkendorf JL, Monaghan KG, Bajaj K, Best RG, et al. Noninvasive prenatal screening for fetal aneuploidy, 2016 update: a position statement of the American College of Medical Genetics and Genomics. Genet Med 2016;18(10):1056–65.

[51] Vanakker O, Vilain C, Janssens K, Van der Aa N, Smits G, Bandelier C, et al. Implementation of genomic arrays in prenatal diagnosis: The Belgian approach to meet the challenges. Eur J Med Genet 2014;57:151–6.

[52] Green RC, Berg JS, Grody WW, Kalia SS, Korf BR, Martin CL, et al. CORRIGENDUM: ACMG recommendations for reporting of incidental findings in clinical exome and genome sequencing. Genet Med 2017;19(5):606.

[53] Grace MR, Hardisty E, Dotters-Katz SK, Vora NL, Kuller JA. Cell-free DNA screening: Complexities and challenges of clinical implementation. Obstet Gynecol Surv 2016;71(8):477–87.

[54] Wilkins-Haug L, Quade B, Morton CC. Confined placental mosaicism as a risk factor among newborns with fetal growth restriction. Prenat Diagn 2006;26(5):428–32.

[55] Wolstenholme J, Rooney DE, Davison EV. Confined placental mosaicism, IUGR, and adverse pregnancy outcome: A controlled retrospective U.K. collaborative survey. Prenat Diagn 1994;14(5):345–61.

[56] Pertile MD, Halks-Miller M, Flowers N, Barbacioru C, Kinnings SL, Vavrek D, et al. Rare autosomal trisomies, revealed by maternal plasma DNA sequencing, suggest increased risk of feto-placental disease. Sci Transl Med 2017;9:1–12.

CLINICAL INTEGRATION

BEST PRACTICES FOR INTEGRATING CELL-FREE DNA-BASED NIPT INTO CLINICAL PRACTICE

Zandra C. Deans*, **Nicola Wolstenholme*,†**, **Ros J. Hastings†**

UK NEQAS for Molecular Genetics, GenQA Edinburgh Office, The Royal Infirmary of Edinburgh, Edinburgh, United Kingdom* Cytogenomics External Quality Assessment (CEQAS), GenQA Oxford Office, Women's Centre, Oxford University Hospitals NHS Trust, Oxford, United Kingdom†

INTRODUCTION

External Quality Assessment (EQA), which is also referred to as proficiency testing (PT) in some countries, is a means of assessing clinical analytical and interpretation performance with interlaboratory comparison and against international standards (ISO). EQA allows an independent appraisal of the laboratory's results and interpretation compared to the validated results and to the performance criteria. The aim of EQA is to be educational and to improve quality of care where a laboratory is failing to meet the required standard. EQA is also an external verification of quality of service and gives confidence to the laboratory director and host institution that the laboratory's performance is satisfactory. However, if the performance is poor, it will alert the laboratory to investigate and perform a root cause analysis to identify and subsequently rectify the problem; participation in an EQA may also enable a laboratory to identify areas of concern before an error leading to patient harm is made. EQA can be qualitative (presence of absence of a variant) or quantitative (uses arithmetic parameters to set limits of acceptability) but should reflect the nature of the testing. In genetics and specifically cell-free DNA-based prenatal testing and prenatal diagnosis, the test results are qualitative. The EQAs are designed to check and sometimes challenge laboratory screens and tests for a particular genetic disease/disorder or gene/target combination. Ideally the EQA covers the whole preanalytical, analytical, and interpretation of results process through the distribution of the relevant clinical samples, although for very limited or rare genetic samples, the EQA provider may only offer an EQA for part of the process, for example, analytical and interpretative.

EQA is recognized by ISO and consequently, accreditation bodies use EQA performance as a tangible measure of the quality of a laboratory's performance. The ISO15189 [1] standard requires laboratory participation in EQA(s) for all of the diagnostic services offered and this participation should be continuous for all aspects of the service.

There are many benefits to EQA. Firstly, it is educational and may highlight deficiencies and/or improvements needed to the diagnostic service. EQA is also useful for benchmarking a laboratories' own performance. Over the years Cytogenomic External Quality Assessment Scheme (CEQAS), European Molecular Quality Network (EMQN), UK National External Quality Assessment Service (UK NEQAS) for Molecular Genetics, and other EQA providers have seen improvement in the quality of analysis and reporting in genetics laboratories over time through regular EQA participation [2–4]. EQA also identifies gaps in the Quality Management system or in installation of updated software packages thus enabling laboratories to improve their service provision. In addition, the delivery of EQA can highlight where there is a lack of Professional Guidelines and EQA providers are proactive in collaborating with the professional bodies to publish new guidelines thereby assisting laboratories further in reporting genetic results [5–7]. Guidelines are established based upon the opinion and experience of experts working within the field of the subject, for example, cell-free DNA (cfDNA)-based noninvasive prenatal testing (NIPT) and noninvasive prenatal diagnosis (NIPD) [7]. Consensus opinion is essential as it allows different points of views to be expressed and evidence to be heard in a mutually beneficial way to establish best practice. Finally, EQA providers occasionally identify flaws in commercial kits and software, which may either result in the withdrawal of the kit or an improvement of the kit/software, and as a consequence improves the accuracy of the patient test result (unpublished data).

DEVELOPING NEW EQA FOR CELL-FREE DNA-BASED NIPT

Having demonstrated the benefits of establishing EQA in this field and ascertaining the need from the diagnostic community, EQA providers have been developing assessments to address the unique challenges of cfDNA NIPT EQA.

PROBLEMS ASSOCIATED WITH DEVELOPING MATERIALS FOR PLASMA-BASED EQAs

The major challenge for any cfDNA NIPT EQA is to provide sufficient material that is of a suitable quality and has the relevant genetic aberration; an ideal EQA would provide every EQA participant with a batch of material which is identical to the type of sample they normally receive for routine testing (i.e., the same material type provided in the same quantity, collected, stored, and transported in the optimal manner). To allow interlaboratory comparisons the material provided to all participants should ideally come from a single or small number of batches validated by two independent laboratories. The EQA provider also needs to ensure that the samples have not deteriorated in the interval between validation and use by the participant. Finally, the EQA provider must consider the cost of the materials and transportation requirements which might affect those costs. For plasma-based EQA there are difficulties in all of these aspects and the provider therefore needs to carefully consider the advantages and disadvantages of both real clinical samples and artificial reference materials.

Using patient-based clinical material has the main advantage that a result can be obtained by any relevant methodologies and is therefore applicable to all possible participants; patient-based material represents the gold standard being truly representative of what a laboratory is likely to encounter day to day. However, for cfDNA NIPT where laboratories usually require approximately 4 mL of plasma to complete testing it is not possible to obtain the amount of donated plasma required for all participants from a single pregnant patient. Hence there will be a need to use multiple samples in the EQA, limiting the reliability of interlaboratory comparisons. Additionally, any validation requirements prior to distribution further limit the volume of sample available for participants to test. As cell-free DNA in plasma is very susceptible to degradation during handling, transit time/conditions and freeze/thawing are all issues which will require careful consideration to ensure samples arrive in the testing laboratory in an optimal condition. In order to ensure stability of real samples it may be necessary to consider low temperature distribution increasing the costs and logistical difficulties of exporting the materials. Finally, limited availability of materials from pregnancies with the rarer trisomies, for example, Edwards/ Patau syndrome is an additional limiting factor for including these into the EQA.

Artificial reference materials have the main advantage that large quantities can be manufactured and validated prior to distribution. When designing an artificial material, it might also be possible to include an artificial freeze-dried plasma component, providing a more stable sample. In this case handling and shipping problems/costs may be reduced. As stated previously having a single batch of material removes variables, leading to more reliable interlaboratory comparisons. Another benefit of artificial materials is the ability to engineer specific characteristics of interest into the sample, for example, in a Chinese cfDNA NIPT EQA the effect of low fetal fraction was investigated by limiting the amount of sheared "fetal" DNA in the plasma [8]. However, artificial materials are not identical to real samples and any laboratory obtaining a poor performance will always question if their performance was due to differing characteristics in the artificial sample. For example, firstly mechanical shearing of DNA produces fragments which differ than those found in maternal plasma with smaller DNA fragments (<100 bp) [8]. Secondly, the fragments produced by mechanical shearing may be problematic for paired-end NGS approaches. Thirdly, at the current time, commercially available artificial reference materials are not matched for the parent/child relationship, that is, the "maternal" and "fetal" components are not from related individuals, and therefore SNP analyses are not possible. For some of these problems there is potential to overcome these deficiencies, for example, by providing patient materials or producing artificial materials that include matched mother/"fetal" DNA to allow SNP analysis. One future approach might be to include both real and artificial materials in an EQA to access the benefits of each type of material.

BACKGROUND TO THE JOINT EQA

"NIPT" is the term used for a cell-free DNA (cfDNA) based prenatal genetic screening test while NIPD is the term used for a rapid cfDNA-based prenatal genetic diagnostic test. NIPT and NIPD techniques are both based on the analysis of circulating cfDNA within maternal plasma. The cfDNA is derived from the placenta in early gestation [9] and at/after 10 weeks' gestation between 3% and 20% of the cell-free DNA in maternal plasma is of "fetal" origin [10]. The fetal fraction generally increases with advancing gestational age but is also affected by maternal weight and multiple pregnancies [11].

Cell-free DNA-based NIPT for aneuploidy has been shown to be an effective prenatal screening test for the common trisomies (chromosomes 13, 18, and 21) and initially was introduced through the commercial sector and private healthcare providers [12]. Some of these additionally screen for sex chromosome aneuploidies and for other chromosomal rearrangements, for example, microdeletion syndromes. Although sensitivity and specificity is high for trisomy 21, the test performance for other aneuploidies (involving chromosomes 13 and 18, X & Y) is more variable [13]. cfDNA-based NIPT is a screening test and so false positives and negatives can occur. Consequently, a high-risk cfDNA NIPT result should always be followed up with a diagnostic invasive test, for example, chorionic villus sampling or amniocentesis. In addition, current cfDNA NIPT does not identify the range of chromosome abnormalities detected by microarray analysis of placental or fetal cells obtained following invasive testing [14]. In contrast, cfDNA-based NIPD is a diagnostic test used for sex determination, and a range of monogenic disorders and does not require any additional invasive testing.

cfDNA NIPT results may be discordant with diagnostic prenatal tests due to a variety of parameters including: a statistical/technical false positive/negative, low fetal fraction, confined placental mosaicism (vanishing), twin pregnancies, maternal mosaicism [15,16] and in rare circumstances even maternal malignancy [17]. This is discussed in more detail in Chapter 5. Most performance data reported to date only relates to singleton pregnancies; however, there is increasing data becoming available for twin pregnancies [13]. NIPD is used for identifying male fetuses, Rhesus D status, and other disorders where the presence of chromosome-specific sequences (e.g., SRY sequences or a paternal autosomal recessive or dominant mutation) in the mothers' blood can only be of fetal origin. False positives are rare but may occur, for example, if a previous pregnancy was male and there is still cfDNA circulating in the maternal blood. The challenges of delivering EQA for NIPD are discussed in Jenkins et al. [18].

CURRENT EQA
FETAL SEXING FROM CELL-FREE DNA IN MATERNAL PLASMA (EMQN AND UK NEQAS FOR MOLECULAR GENETICS)

A pilot EQA for noninvasive prenatal diagnosis (NIPD) for fetal sex determination was introduced by EMQN and UK NEQAS for Molecular Genetics in 2013 and ran for 3 years with up to 46 laboratories participating in the EQA, each receiving 3 samples to test. Although genotyping error rates were always low (e.g., 0.7%, 1/138 genotypes in 2015) it was noted that there was a considerable improvement in the quality of submitted reports over the duration of this EQA. It is noteworthy that when this EQA was designed, a majority of laboratories were performing testing based on the detection of Y chromosome markers (e.g., using real-time PCR) and confirming the presence of "fetal" DNA in the sample (e.g., *RASSF1A* assay). Since that time there has been an expansion in the number of laboratories analyzing free "fetal" DNA using Next-Generation Sequencing (NGS) or using a single-nucleotide polymorphism (SNP) based approach. In designing this EQA it was found, despite extensive testing, that no synthetic material performed adequately for inclusion in the EQA. Consequently, to ensure all participants received material of similar quality, pooled plasma samples were used. Although this approach worked well at the beginning, an increasing number of participants were unable to use the pooled plasma samples as their methods relied on comparing maternal/fetal SNP data. Additionally, there was an increasing demand for a cfDNA NIPT EQA for trisomy testing and the decision was taken to temporarily suspend the EQA on fetal sexing to allow the development of the 2016 cfDNA NIPT pilot EQA for trisomy testing. However, participants subsequently recognized the need for a continuing specific fetal sex determination EQA for laboratories performing sexing for X-linked genetic disorders therefore the EQA was reinstated during 2017.

PILOT EQA FOR CELL-FREE DNA-BASED NIPT FOR TRISOMY 13/18/21 AND SEX CHROMOSOME ANEUPLOIDIES (CHINA)

An EQA for cfDNA-based NIPT for trisomy 13/18/21 and sex chromosome aneuploidy was introduced in The People's Republic of China in 2014 [8]. In this pilot EQA artificial materials (13 in total) were created using either cell lines or placental DNA from pregnancies as the source of aneuploidy. Fragmented DNA from the cell lines/placental DNA was sonicated and mixed with DNA from healthy female donors. In this EQA only 55% laboratories correctly identified all of the samples, although there were no false positive results reported. The problems arose because this EQA included challenging samples which had simulated fetal fractions below the 4% quality threshold value recommended for many assays [19]; there were a number of laboratories who reported false negative results after failing to identify the low fetal fraction. The authors did note that there were a number of limitations of the materials used: mechanical shearing of DNA necessitated end repair for use in some assays; the material could not be used by laboratories using SNP-based assays as the female donor (simulated maternal plasma) and the simulated "fetal" DNA were not from related individuals, the mechanically sheared DNA samples (unlike real materials) were deficient in fragments <100 bp and the plasma from nonpregnant females yielded shorter fragments than those generally seen in pregnant females.

JOINT PILOT NIPT EQA

In 2016, a pilot EQA for cfDNA-based NIPT for common aneuploidy was introduced as collaboration between three EQA providers: CEQAS, EMQN, and UK NEQAS for Molecular Genetics. The plan was to learn from the experience of existing EQAs, to survey laboratories to understand current practice and to facilitate the writing of Best Practice Guidelines to improve clinical practice and to allow the development of EQA marking criteria.

NIPT/NIPD SERVICE PROVIDER SURVEY

In order to provide an overview of the needs and approaches employed by laboratories across the globe and to develop a plan for the pilot EQA(s) to meet the needs of the participants, a survey of current practice, including reporting for cfDNA-based NIPT and NIPD, was undertaken.

In February 2016, approximately 3000 laboratories registered with CEQAS, EMQN, and UK NEQAS for Molecular Genetics were invited to complete an online survey focused on services provided for cfDNA NIPT and NIPD. Many of those do not perform cfDNA-based NIPT and therefore were less likely to respond.

One hundred and twenty-one laboratories responded (5% return rate). Four of these laboratories offered a service for NIPD but were not included as one only offered NIPD for achondroplasia, and two only for fetal Rhesus D genotyping and one was currently validating the NIPD service. Seventy-five (62%) of the responding laboratories performed cfDNA NIPT for aneuploidy at the time of the survey, and 43% of the responding laboratories submitted sample reports. A further 25 laboratories were in the process of validating a cfDNA NIPT assay and 17 were offering cfDNA NIPT on a research basis.

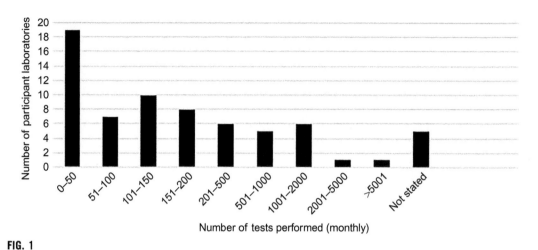

FIG. 1

Number of cfDNA NIPT tests performed per month by participating laboratories.

The volume of blood laboratories stated to require for undertaking cfDNA NIPT/NIPD testing varied considerably: from 4 mL to a total of 30 mL of blood. The majority of laboratories (43%) required 10 mL of blood and a further 40% of centers required <10 mL. The volume of plasma required was more consistent with most centers (86%) requiring 5 mL or less. Centers used either Streck tubes (76%) or EDTA tubes (24%) for collecting the blood sample and the vast majority of centers received these blood samples via a courier.

Of the laboratories delivering a cfDNA NIPT service, 19 centers performed up to 50 tests per month (Fig. 1) and a small number (eight) of centers >1000 per month. The methods used varied: the majority (74%) were based on NGS-based counting methods, 11% used the SNP-based sequencing approach, and 15% used other methods including Sanger sequencing, array based, MLPA, and TaqMan. Of the centers that used an NGS platform, 70% followed a whole genome approach rather than a targeted one.

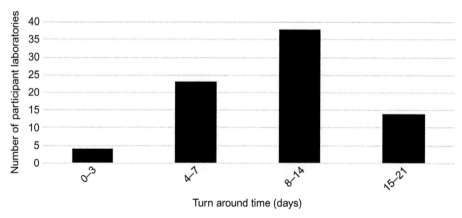

FIG. 2

cfDNA NIPT turnaround time for participating laboratories.

Six different NGS platforms were used and included Illumina NextSeq, HiSeq, and MiSeq and ThermoFisher Ion Proton and Ion Proton Personal Genome Machine (PGM). The aim of this type of testing is to deliver quick accurate results therefore it is important to monitor turnaround times. This TAT varied from 3 to 21 days with the majority reporting within 14 days (Fig. 2).

The fetal fraction (FF) is the amount of the cell-free DNA in the maternal plasma that is of placental (fetal) origin and was used as a quality parameter by some providers. Some algorithms use a minimum FF below which results are not reported while others combine FF with other quality metrics to determine if a result is reportable. The fetal fraction was measured by 61% of laboratories. Given the number of factors that affect the FF, the current NIPT guidelines [3] recommend that this quality parameter is used in conjunction with other quality parameters and test limitations.

It is expected that the scope of cfDNA NIPT and NIPD will expand with time but at the time of the survey, 73 centers tested for aneuploidy for chromosomes 13, 18, and 21; 46% tested for sex chromosome abnormalities; and 38% centers also performed testing for trisomies other than 13, 18, and 21 and subchromosomal abnormalities. In the absence of sex chromosome abnormalities fetal sex was always reported by 23% of laboratories, only when requested by 48.9% of laboratories and not at all by 28.4%.

Twenty two percent of laboratories outsourced cfDNA NIPT testing to 11 different companies/institutes. The majority of these outsourcing laboratories would check that the provider participates in an EQA to ensure the standard of service being commissioned and would even participate in EQA themselves and forward the EQA samples to their outsourced provider. This indicates that laboratories are aware of the importance of participating in EQA to monitor either their own in-house service or the test provided by external partners.

BEST PRACTICE GUIDELINES FOR REPORTING NIPT ANEUPLOIDY RESULTS

EQA and best practice guidelines have a synergistic relationship: In order to assess the results of an EQA it is important to develop marking criteria based on current best practice. EQA allows the initial, often limited, consensus to become the established norm facilitating the development of further and more detailed guidance over time. The three EQA providers therefore facilitated drafting best practice guidelines [7].

These best practice guidelines support laboratories to assure that the service being delivered is in line with consensus opinion. It is important that guidance documents are produced by multiple experts in the field and not biased by individual points of view. There are many different approaches to providing a clinical service for cfDNA NIPT, therefore any guidance in this area must cover all testing strategies and highlight issues to be addressed in order to provide a high quality and accurate test.

In preparation for the production of the first guidance document on reporting results of cfDNA-based prenatal testing for common aneuploidies, the three EQA providers (CEQAS, EMQN, and UK NEQAS for Molecular Genetics) performed an audit of the content of reports. In addition to completing the current practice survey described previously, laboratories were invited to submit two example reports (a) high-risk aneuploidy result and (b) a low-risk aneuploidy result. The content of the reports was reviewed anonymously by a "Rapid Prenatal Specialist Advisory Group" and the information present and absent in reports was collated. The data was presented and discussed as part of a "NIPT quality workshop" held during the International Society for Prenatal Diagnosis (ISPD) conference in 2016 and consensus opinion was obtained on essential items to be included within a cfDNA

NIPT test result report. The minimum guidelines agreed on for reporting cfDNA NIPT test results for trisomy 13, 18, and 21 were subsequently published [7] and are summarized later.

INDIVIDUAL INFORMATION

ISO 15189 standards [1] recommend that at least two unique patient identifiers are used on a report and this should include, for example, the patient's full name and date of birth or hospital number. All the submitted reports included the patient/mother's name and 81% also provided the mother's date of birth. However, 19% only provided the mother's age (Fig. 3). The NIPT guidelines therefore incorporated the recommendation that two unique identifiers are the minimum standard that reports should adhere to [7].

SAMPLE INFORMATION

ISO 15189 clause 5.8.3 [1] requires the type of primary sample and the date of primary sample collection. However, sample information on the submitted reports was variable. Approximately half of the reports reviewed provided the sample type being tested, which is essential information for the interpretation of the results and may be helpful for traceability. Sixty-two percent of centers included the date of sample collection and slightly more provided the date of sample receipt in their laboratory (Fig. 3). This information should be present according to ISO 15189 [1]. As maternal cell lysis occurs over time thereby reducing the FF, it is important for the laboratory to monitor the transport time. Any suboptimal results can potentially be explained by a delay between sample collection and processing. Therefore we included in the guidelines that the sample type, date of sample collection, and date the sample is received within the laboratory should be stated on the report [7].

CLINICAL INFORMATION

There was substantial variation in the clinical information provided in the reports (Fig. 3). Most but not all laboratories (77%) reported gestational age. The current Guidelines [7] state that gestational age must be defined by fetal ultrasound scan at the time when the sample is taken and mentioned on the report. It was agreed at the workshop that the following information was important in interpreting the results but there was no majority view on whether or not the following items should be included on the laboratory report: maternal body mass index (BMI), singleton or multiple pregnancy, any relevant sonographic data, whether in vitro fertilization (IVF) had been performed, the combined test or serum screening result, and the referral reason. ISO 15189 [1] does require the referral indication to be reported.

REPORTING INFORMATION

Several laboratories failed to provide the contact details of the referrer or the addressee of the report, and the report date (Fig. 3). As ISO 15189 clause 5.8.3d [1] requires these details we recommended in the NIPT guidelines [7] that the report states the date of reporting and contact details of the individual requesting the test and/or the location to where the authorized report should be issued.

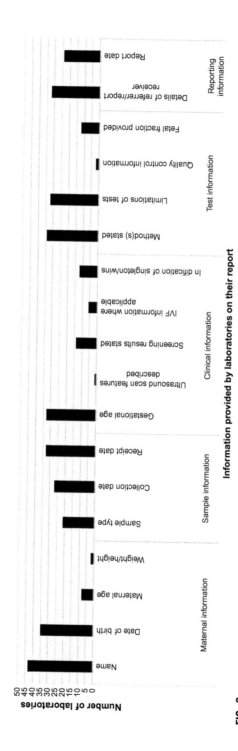

FIG. 3

Information included in cfDNA test result reports. The information is divided into five categories: maternal information, sample information, clinical information, test information, and reporting information.

TESTING INFORMATION

For enabling the clinician or the receiving laboratory to interpret the test results, it is important they have information on which test was performed as well as the methodology of this test. Nine laboratory reports did not include any information on the test(s) being reported (Fig. 3). We recommend in the guidelines that the report states the methodology and platform used, for example, NGS.

Both ISO15189 [1] and the NIPT guidelines [7] require laboratories to state the limitations of the test undertaken. This is particularly important when no abnormality has been detected, so the recipient of the report can interpret and act upon the result as appropriate. Eleven laboratories did not include any information with regards to the limitations of the test (Fig. 3). In addition, fetal fraction and quality parameters were not included in the report content by most laboratories (Fig. 3).

REPORTING THE RESULT

The terminology used to report the result is very important and should be clear and unambiguous. If the result is "hidden" with the text of the report it can be missed and a summary result at the top, or predominantly placed within the report is always helpful. The EQA survey of submitted reports demonstrated significant variation between centers in reporting the same result. As NIPT is a screening test not a diagnostic test it is important that the term "abnormal or normal" is not used to describe the result. Instead it is recommended in the guidelines that the terms "high risk/low risk of aneuploidy" are used to describe the result. In the survey all laboratories recommended that a high-risk trisomy result must be confirmed by invasive testing as cfDNA-based NIPT is a screening test.

Some laboratories reported that a trisomy result had been detected using the International System for Cytogenomic Nomenclature (ISCN) [20]. As cfDNA NIPT is a screening test, the guidelines do not recommend the use of ISCN in reports as it implies that a diagnostic test has been performed.

It is discretionary whether a cfDNA NIPT report should include negative and positive predictive values and no consensus was reached by the reviewers nor by the workshop participants.

DELIVERING THE JOINT EQA PILOT FOR NIPT FOR TRISOMY 13, 18, AND 21

As previously stated the delivery of EQA in the field of cfDNA NIPT is challenging, primarily due to the limited availability and volumes of patient samples, the instability of such material and the low incidence of some of the diseases tested for. The first EQA pilot delivered by three EQA providers for the assessment of the quality of cfDNA NIPT for the detection of trisomy 13, 18, and 21 in maternal plasma was provided during 2016. Based on the results of the preceding laboratory survey, 32 participants were registered to take part (eight laboratories per EQA provider).

The EQA providers commissioned a Reference Material commercial company to manufacture bespoke artificial material for use in the EQA pilot. Sheared DNA (to approximately 170 base pairs) was supplied in 1 mL aliquots of real plasma for the purposes of this EQA. The samples were validated prior to EQA distribution by four laboratories using different testing protocols. The limitations of the EQA material were that (a) the maternal and "fetal" cfDNA was not matched and (b) fragmentation did not reflect the difference in size of the maternal and fetal DNA. Therefore it was not appropriate for testing by any method relying on comparison between maternal and fetal SNPs, or pair-ended NGS. Due to these limitations SNP assays and pair-ended NGS were not included in the validation process prior to the EQA distribution. Validation by laboratories performing other techniques did report the correct genotypes but often the samples did not perform as well as routine clinical samples. In the absence of any current EQAs and in order to meet the overwhelming demand for external assessment of cfDNA NIPT, an EQA restricted to 32 laboratories able to test the artificial material was delivered in a pilot phase despite the known limitations.

Three EQA samples were sent to each center. Each sample was accompanied by a clinical case scenario with mock patient names, identifiers, and reason for referral. The laboratories were requested to perform NIPT for trisomy 13, 18, and 21, and report each case according to local protocols and policies. The expected results were low risk for trisomy 13, 18, and 21; high risk of trisomy 18; and high risk of trisomy 21. Respective fetal fractions of the three samples of the artificial material were 12%, 10%, and 4%. A summary of the EQA process is summarized in Fig. 4.

Overall, the standard of genotyping accuracy was high. An incorrect genotype was classed as critical genotyping error and four laboratories reported a genotyping error in one of the samples (4/32 laboratories reporting results; 12%): two laboratories reported the low risk for trisomy 13, 18, and 21 sample to be trisomy 21, and two other laboratories did not report the correct results for the high-risk trisomy 21 sample (one reported the case as normal and the second reported a borderline trisomy 13 results as well as a trisomy 21).

Two laboratories reported partial results for one case each, that is, an inconclusive result for one of the chromosomes tested. The EQA providers did not assign scores for those cases for those laboratories as the cause of the failure to obtain reportable results could not be determined.

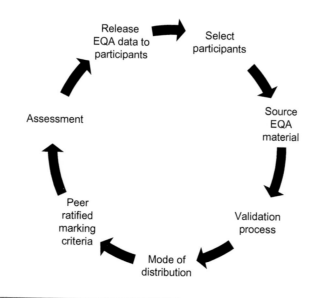

FIG. 4

Pilot EQA scheme for cfDNA-based NIPT for common trisomies.

As discussed previously, a fetal fraction (FF) is important to correctly interpret results and FF was reported by 51% of participants. The lack of FF on the report does not indicate that it was not measured and used to interpret the genotyping result, merely that it was not stated on the report. A small number of participants either reported the FF only for the low-risk case or only when FF was too low.

The range of FFs for the EQA sample with a FF of 12% was wide with a mean of 10.6%. The case with a FF of 10% resulted in a mean of 9.7% and the case with a FF manufactured at 4% had a mean of 7.6%. Overall there were 61% of laboratories reporting FFs ±2% of the expected value without consisting trends by laboratory or by testing methodology.

The EQA run demonstrated a variable approach in reporting cfDNA NIPT results. Many reports only reported on the NGS platform used and not on the method used. It was not stated whether the approach was whole genome sequencing and analysis or a targeted approach. Limitations such as the effect of (vanishing) twins were often not outlined nor whether or not partial trisomy was detectable by the method used. For "high-risk" results, often only the chromosome involved in the high-risk result was reported whereas the low-risk results for the other chromosomes tested were not reported. Additionally, and overall, the essential information (e.g., confirmation of result by invasive testing) was either buried in report rider or not placed in a prominent position on the report.

Although the test may be accurate, the mechanism of reporting the results is critical in ensuring the results are not misinterpreted and the key interpretative elements are not missed by the reader of the report. Interestingly, one EQA case was referred for cfDNA NIPT because the individual had a previous child with trisomy 21. This was often not included in the reason for referral on the report, and some laboratories overinterpreted the result as a diagnostic test and referred to a "recurrent trisomy 21". It is important to clearly state that a diagnostic test has not been performed and if the trisomy 21 was confirmed subsequently by invasive testing then further a referral to Clinical Genetic was recommended.

The variable results and way of reporting indicate the need for more EQA in this field and with the publication of consensus guidance subsequent assessments can apply them to address the omissions and improve the standard of cfDNA NIPT.

FUTURE EQA
UK NEQAS/CEQAS/EMQN JOINT EQAs

Due to the limitations noted in the 2016 EQA, the three EQA providers (CEQAS, EMQN, and UK NEQAS for Molecular Genetics) have decided to continue EQA provision with the focus on the use of real patient samples. Further pilot EQAs are being provided, the first covering cfDNA NIPT for trisomy 13, 18, and 21 and the second covering NIPD for fetal sexing; both EQAs will use patient plasma samples. In these EQAs the samples are not pooled to ensure material suitability for laboratories using a SNP-based approach. Overall EQA is being delivered to 96 laboratories for the aneuploidy testing EQA and 43 for the sex determination. These EQAs are underway as we go to press.

Use of artificial reference materials awaits development of materials which more closely resemble real patient samples. If suitable materials are produced in the future, then piloting alongside real samples using methodology tested through the current EQA will allow proper assessment of their merits.

COLLEGE OF AMERICAN PATHOLOGISTS PILOT EQA FOR TRISOMY 21,18, AND 13

The College of American Pathologists (CAP) have indicated that they will also pilot an EQA for cfDNA NIPT for trisomy during 2018, based on three real maternal samples sent biannually. Due to the difficulties of obtaining samples this will be limited to 10–30 laboratories and the costs of providing the EQA may be high. As 4 mL volunteer plasma samples will be used most laboratories will receive different samples limiting interlaboratory comparison. The scheme organizers state that this disadvantage will be outweighed by the benefits of using real materials.

FUTURE TOPICS FOR GUIDELINE DEVELOPMENT

Future rounds of EQA should lead to increased consensus over some issues not covered in the recently published guidelines. For example, in the joint pilot from CEQAS/UKNEQAS for Molecular Genetics/ EMQN reporting of the FF was found to be variable with some laboratories performing tests but not reporting the results or only reporting them when no trisomy was detected. Additionally, there was variation in the FF values returned by different laboratories both in this joint pilot EQA and in the EQA in China [8]. As FF is a key quality indicator [21–23] and low FF may itself be associated with trisomy [24,25] improving consistency of measurements or validating its measurement/cut offs for particular tests are goals for both future EQA and for Best Practice guidance.

Regarding clinical information there is considerable variation in current practice and the current minimum guidelines focus on ensuring the inclusion of referral reasons and gestational age into the clinical information provided. However, if consensus can be obtained then future guidance might look to include further relevant clinical information affecting the interpretation/reliability of the results, for example, maternal BMI, multiple pregnancy, and maternal hypertension, have been postulated to affect the fetal fraction [25,26]. Inclusion of other aspects of maternal health in the referral information might also be investigated as cancer and systemic lupus erythematosus have been associated with changes to cfDNA in plasma [27,28].

In addition, the use of anticoagulants by the mother can adversely affect the fetal fraction leading to failed or false results [29,30].

Finally, there is no current consensus on what risk information should be included in the reports; it is important to remember that cfDNA NIPT is a screening test with the possibility of both false positive and false negative results, for example, via vanishing twin, placental mosaicism, or other mechanism. The discussions which lead to the current guidance identified a lack of consensus around the reporting of positive and negative predictive values with revised individualized risks. While including risks was thought to be ideal, the attendees of the workshop felt that both aspects required accurate prior risk assessment and data from large series. It was therefore not practical to include these risks in current guidance but as the field develops and more data is available it might be possible to revisit this issue.

EQA providers work in a global setting and this in itself can cause some issues with development of both EQA and Best Practice Guidelines applicable across international borders. For example, what might be considered "clinical advice" or "clinical recommendations" to refer for counseling, may be obligatory in some jurisdictions but be outside the remit of testing laboratories in other countries. What the EQA provider seeks is to ensure that patients receive the correct results, correctly interpreted and in an appropriate timeframe; for most providers this is most reliably delivered by a comprehensive, fully interpreted clinical report which can be assessed for both the accuracy and interpretation of the results.

REFERENCES

[1] ISO 15189:2012. Medical laboratories – requirements for quality and competence and analysis. http://www.iso.org/iso/catalogue_detail?csnumber=56115 [last accessed 19.11.16]

[2] Berwouts S, Girodon E, Schwarz M, Stuhrmann M, Dequeker E. Improvement of interpretation in cystic fibrosis clinical laboratory reports: longitudinal analysis of external quality assessment data. Eur J Hum Genet 2012;20:1209–15.

[3] Deans Z, Fiorentino F, Biricik A, Traeger-Synodinos J, Moutou C, De Rycke M, Renwick P, Sengupta S, Goossens V, Harton G. The experience of 3 years of external quality assessment of preimplantation genetic diagnosis for cystic fibrosis. Eur J Hum Genet 2013;21(8):800–6.

[4] Richman SD, Fairley J, Butler R, Deans ZC. RAS screening in colorectal cancer: a comprehensive analysis of the results from the UK NEQAS colorectal cancer external quality assurance schemes (2009-2016), Virchows Arch 2017;471(6):721–9. 28653203.

[5] Hastings R, Howell R, Bricarelli FD, Kristoffersson U, Cavani S. A common European framework for quality assessment for constitutional, acquired and molecular cytogenetic investigations. ECA Newsletter 2012;29:7–25.

[6] Claustres M, Kožich V, Dequeker E, Fowler B, Hehir-Kwa JY, Miller K, Oosterwijk C, Peterlin B, van Ravenswaaij-Arts C, Zimmermann U, Zuffardi O, Hastings RJ, Barton DE, European Society of Human Genetics. Recommendations for reporting results of diagnostic genetic testing (biochemical, cytogenetic and molecular genetic). Eur J Hum Genet 2014;22(2):160–70.

[7] Deans ZC, Allen S, Jenkins F, Khawaja F, Hastings R, Mann K, Patton S, Sistermans EA, Chitty LS. Recommended practice for laboratory reporting of non-invasive prenatal testing (NIPT) of trisomies 13, 18 and 21: A consensus opinion. Prenat Diagn 2017;37:699–704.

[8] Zhang R, Zhang H, Li Y, et al. External quality assessment for detection of fetal trisomy 21, 18, and 13 by massively parallel sequencing in clinical laboratories. J Mol Diagn 2016;18:244–52.

[9] Alberry M, Maddocks D, Jones M, et al. Free fetal DNA in maternal plasma in anembryonic pregnancies: confirmation that the origin is the trophoblast. Prenat Diagn 2007;27:415–8.

[10] Lo YM, Tein MS, Lau TK, et al. Quantitative analysis of fetal DNA in maternal plasma and serum: Implications for noninvasive prenatal diagnosis. Am J Hum Genet 1998;62:768–75.

[11] Wang E, Batey A, Struble C, et al. Gestational age and maternal weight effects on fetal cell-free DNA in maternal plasma. Prenat Diagn 2013;33:662–6.

[12] Minear MA, Lewis C, Pradhan S, Chandrasekharan S. Global perspectives on clinical adoption of NIPT. Prenat Diagn 2015;35:959–67.

[13] Gil MM, Quezada MS, Revello R, et al. Analysis of cell-free DNA in maternal blood in screening for fetal aneuploidies: updated meta-analysis. Ultrasound Obstet Gynecol 2015;45:249–66.

[14] Rao RR, Valderramos SG, Silverman NS, et al. The value of the first trimester ultrasound in the era of cell free DNA screening. Prenat Diagn 2016;36(13):1192–8.

[15] Wang JC, Sahoo T, Schonberg S, et al. Discordant noninvasive prenatal testing and cytogenetic results: a study of 109 consecutive cases. Genet Med 2015;17(3):234–6.

[16] Wang Y, Chen Y, Tian F, et al. Maternal mosaicism is a significant contributor to discordant sex chromosomal aneuploidies associated with noninvasive prenatal testing. Clin Chem 2014;60:251–9.

[17] Osborne CM, Hardisty E, Devers P, et al. Discordant noninvasive prenatal testing results in a patient subsequently diagnosed with metastatic disease. Prenat Diagn 2013;33:609–11.

[18] Jenkins LA, Deans ZC, Lewis C, et al. Delivering an accredited non-invasive prenatal diagnosis service for monogenic disorders, and recommendations for best practice. Prenat Diagn 2018;38(1):44–51.

[19] Ashoor G, Syngelaki A, Poon LC, et al. Fetal fraction in maternal plasma cell-free DNA at 11-13 weeks' gestation: relation to maternal and fetal characteristics. Ultrasound Obstet Gynecol 2013;41:26–32.

[20] ISCN. 2016: An International System for Human Cytogenomic Nomenclature (2016) Reprint of: In: McGowan-Jordan J, Simons A, Schmid M, (Eds.), Cytogenetic and Genome Research 2016, Vol. 149, No. 1-2.

[21] Canick JA, Palomaki GE, Kloza GM, et al. The impact of maternal plasma DNA fetal fraction on next generation sequencing tests for common fetal aneuploidies. Prenat Diagn 2013;33:667–74.

[22] Pe'er Dar KJ, Curnow SJ, Gross MP, et al. Clinical experience and follow-up with large scale single-nucleotide polymorphism based noninvasive prenatal aneuploidy testing. Am J Obstet Gynecol 2014;211:527.e1–527.e17.

[23] Gregg AR, Skotko BG, Benkendorf MS, et al. Noninvasive prenatal screening for fetal aneuploidy, 2016 update: a position statement of the American College of Medical Genetics and Genomics. Genet Med 2016;18(10):1056–65.

[24] Pergament E, Cuckle H, Zimmermann B, et al. Single-nucleotide polymorphism-based noninvasive prenatal screening in a high-risk and low-risk cohort. Obstet Gynecol 2014;124(201):210–8.

[25] Zhou Y, Zhu Z, Gao Y, et al. Effects of maternal and fetal characteristics on cell-free fetal DNA fraction in maternal plasma. Reprod Sci 2015;22(11):1429–35.

[26] Palomaki GE, Kloza EM, Lambert-Messerlian GM, et al. DNA sequencing of maternal plasma to detect down syndrome: an international clinical validation study. Genet Med 2011;13(11):913–20.

[27] Dharajiya NG, Grosu DS, Farkas DH, McCullough RM, Almasri E, Sun Y, Kim SK, Jensen TJ, Saldivar J-S, Topol EJ, van den Boom D, Ehrich M. Incidental detection of maternal neoplasia in noninvasive prenatal testing. Clin Chem 2018;64(2):329–35.

[28] Chan RWY, Jiang P, Peng X, Tam L-S, Liao GJW, Li EKM, Wong PCH, Sun H, Allen Chan KC, Chiu RWK, Dennis Lo YM. Plasma DNA aberrations in systemic lupus erythematosus revealed by genomic and methylomic sequencing. Proc Natl Acad Sci USA 2014;111(49):E5302–11.

[29] Burns W, et al. The association between anticoagulation therapy, maternal characteristics, and a failed cfDNA test due to a low fetal fraction. Prenat Diagn 2017;37(11):1125–9.

[30] Ma G-C, Wu W-J, Lee M-H, Lin Y-S, Chen M. Low-molecular-weight heparin associated with reduced fetal fraction and subsequent false-negative cell-free DNA test result for trisomy 21. Ultrasound Obstet Gynecol 2018;51:276–7.

QUALITY ASSURANCE AND STANDARDIZATION OF CELL-FREE DNA-BASED PRENATAL TESTING LABORATORY PROCEDURES

Peter W. Schenk*, Verena Haselmann[†]

Center for Human Genetics and Laboratory Diagnostics (AHC), Martinsried, Germany* Institute for Clinical Chemistry, Medical Faculty of the University of Heidelberg, University Hospital Mannheim, Mannheim, Germany[†]

CELL-FREE DNA-BASED PRENATAL TESTING AS A GENETIC TEST IN LABORATORIES

Cell-free DNA-based prenatal testing is a screening test that estimates the risk of fetal aneuploidy (trisomy 21, 18, and 13) by analyzing cell-free DNA (cfDNA) circulating in the maternal plasma [1]. The requirements for cfDNA tests are similar as for other screening and diagnostic tests in the medical domain [1–3]. This includes all steps from blood collection, transport and storage [4] to DNA extraction [5] and analysis, and interpretation of results [6,7].

STANDARDIZATION

Harmonization of procedures is needed to obtain reproducible and comparable results and minimize the risk of individual deviations and unexpected events. The term "standardization" refers to "standards." Standards are a published set of minimum requirements for the execution of processes, designed for harmonization and to ensure safety [8,9]. They are developed, written, and subsumed by experts representing all relevant parties. As standards are very often international, referral to national regulations is not made. A list of harmonized standards used to demonstrate compliance with relevant EU legislations has been published at http://ec.europa.eu/growth/single-market/european-standards/harmonised-standards_en.

Noninvasive Prenatal Testing (NIPT). https://doi.org/10.1016/B978-0-12-814189-2.00012-8

ISO 15189

ISO 15189 is an international standard that specifies the quality management system requirements particular to medical laboratories. The standard is provided by the International Organization for Standardization [10] and covers all steps relevant for laboratory tests done to obtain information about the health of a patient in relation to diagnosis, treatment, or prevention of disease.

Laboratories that implement a quality management system according to ISO 15189 must fulfill management as well as technical requirements. Management requirements include, for instance, specifications for personnel, facilities, and equipment. Technical requirements include preanalytical steps, analytical procedures, and postanalytical steps. Furthermore, ISO 15189 identifies the needs and expectations of customers. Through corrective and preventive actions, it focuses on continuous quality improvement and addresses prevention of errors and nonconformities. Some of the most important parts of ISO 15189 [11] are discussed later.

DOCUMENTS, DOCUMENT CONTROL, AND DOCUMENTATION

Written "standard operating procedures" (SOP) have to be in place and accessible for each staff member for their respective field. They have to be updated or reconfirmed at agreed points in time ("lifecycle" check; "document control"). Laboratories have to guarantee that the content of the SOP is correct during their development, implementation, and revision.

All working steps, from sample to report, must be documented.

INTERNAL AUDITS

A laboratory is required to perform internal audits in each section at least once a year. An audit aims at controlling compliance to the ISO guidelines and to the specified procedures, and at providing information about eventual actions needed to improve processes and quality.

VALIDATION

The validation process ensures that the results a laboratory issues are compared to a "standard" and are thus trustworthy. A framework for validation of clinical molecular genetic tests has been described by Mattocks et al. [6]. There are different possibilities for validating a method/process. The most common ways are to compare the results with an already validated method, to perform a cross-validation of results with another laboratory or reference institute, and to participate in an external quality assessment scheme (Chapter 11).

Validation studies should comprise all stages of a method or process and should especially focus on critical steps. In cfDNA-based testing preanalytical steps, which are operator dependent, and evaluation of results with bioinformatics programs, are part of the validation process.

QUALIFICATION

Almost all methods in laboratories require special instruments. These instruments must be qualified, which means that their functioning has been proven.

PERSONNEL AND TRAINING

Skilled staff has to be available. The laboratory is in charge of identifying the need for staff and to train the staff both in performing tests and in complying to and contributing to the quality environment of the laboratory.

OCCURRENCE MANAGEMENT

It is impossible to fully prevent nonconformities, deviations from instructions, and other adverse events in a laboratory. Therefore part of the ISO focuses on appropriate plans and options to correct errors and prevent them from reoccurring. To achieve this goal, detailed root-cause analyses have to be performed and appropriate corrective and preventive actions must be defined afterward. Individuals should not be blamed for errors and adverse events should be addressed as an opportunity to improve processes and procedures.

INTERNAL QUALITY CONTROL

Laboratories should implement internal quality control procedures that detect errors occurring during handling and check proper functioning of instruments. For example, including negative controls enables to detect contamination and including positive controls ascertains successful handling of samples.

In general, internal quality controls should provide information about potential nonconformities before a result is reported to a patient.

EXTERNAL QUALITY ASSESSMENT

The last aspect of ISO is EQA, also known as proficiency testing (PT). EQA is discussed in more detail in Chapter 11. In ISO 17043 EQA has been defined as "evaluation of participant performance against preestablished criteria by means of inter-laboratory comparison" [12]. In general, participation in EQA schemes represents the last major step in the entire validation process. It is educational and allows long-term, retrospective analysis of laboratory performance. It is a key strategy for comparing analytical test performance between laboratories [13,14], permits laboratories to compare results with their peers, and can reveal intermethod variability [15]. Regular participation in EQA schemes enables laboratories to improve their test performance and correlates with higher quality assurance scores [16]. A review by Hoeltge et al. on three rounds of PT results with data from 6300 accredited laboratories revealed that an increase in number of participations reduced the error rate [17]. Haselmann et al. reported that the error rate within a molecular genetic EQA scheme for genotyping significantly decreased over the years, demonstrating that the performance of laboratories increased. Additionally, the ability of EQA schemes to improve the quality of diagnostics was proven by the decreased error rate of samples that were provided twice [18].

In principle, every laboratory is allowed to participate in EQA schemes and the choice of a suitable EQA provider lies with the laboratory, unless otherwise dictated by legislation. Lists of EQA providers and their available schemes are available to help laboratories to identify an appropriate EQA provider. To the best of the author's knowledge, the listing on the Eurogentest and IFCC websites cover all European and globally active providers for molecular genetic diagnostics and their representative Schemes [19,20].

With respect to PT for cell-free DNA-based prenatal testing, ISO 15189 requests that an EQA covers the entire process from preanalytical sample handling to sample analysis and interpretation and reporting [10].

ACCREDITATION

The International Society for Prenatal Diagnosis has recommended that "providers of cfDNA testing utilize laboratory services that meet national guidelines for quality control and proficiency testing" [21]. Implementation of a quality management system according to ISO 15189, although voluntary, is the gold standard for medical laboratories and has been recommended for cfDNA testing [3]. It represents a "formal recognition by an authoritative body that a laboratory has the competence to carry out specific tasks" [22].

Each country has a relevant authoritative body that provides a list of all accredited laboratories. This authoritative body is responsible for the entire accreditation process including surveys by specialized auditors and correction of potential nonconformities. The specialized auditors have the expertise to authenticate the executed test validation and the reliability of results. Laboratories performing successfully receive a certificate that can be presented to the users.

Accreditation offers many benefits, not restricted to benefits to the laboratories. The most important is that the person requesting a test can trust the validity of the results because accreditation focuses on the use of validated methods and on the appropriate use of procedures and test reagents. The test-requesting person can be assured that proper procedures are in place to continuously guarantee and improve quality. On the side of the laboratory, accredited laboratories gain competitive advantages over nonaccredited laboratories by giving objective proof of quality and competence. In addition, accreditation can prevent costs associated with multiple assessments, because the authoritative bodies of other countries accept the national accreditation, hence leading to international recognition of the accredited laboratory [23].

Once a laboratory has been successfully accredited, it is subject to regular reassessments in order to ensure the retention of compliance to the ISO.

Quality management systems of medical laboratories are subject to regulations, standards, and accreditation. National and international regulations set the obligatory requirements. Standards, although usually voluntary, provide specific instructions. The implementation of standards leads to accreditation, which testifies of conformity assessed by an impartial party (Fig. 1).

QUALITY ASSURANCE OF CELL-FREE DNA-BASED PRENATAL TESTING LABORATORY PROCEDURES

In the following section, we take a closer look at the single steps of the cfDNA test laboratory process and the criteria to be fulfilled to assure quality according to ISO 15189 for cell-free DNA-based prenatal testing. The general requirements outlined in the previous sections also apply to cell-free DNA-based prenatal testing (Fig. 2).

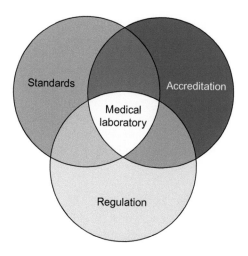

FIG. 1

Components relevant to quality systems of medical laboratories [9].

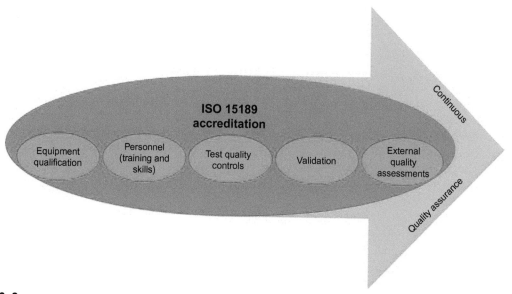

FIG. 2

Basic principles to ensure quality of cell-free DNA-based prenatal testing.

BLOOD DRAW AND TRANSPORTATION

Cell-free DNA-based prenatal testing begins with a blood draw. To prevent maternal blood cells from hemolysis, which would enlarge the pool of maternal cell-free DNA and decrease the "fetal" fraction, validated tubes containing a stabilizing reagent, such as the Cell-Free DNA Streck BCT tubes or the CellSafe Preservative Tubes, must be used [4].

The second step is correct labeling of the tube and hence, the sample. If the sample is not appropriately labeled, the laboratory is obliged to reject the sample. Other reasons for sample rejection are an inappropriate blood volume and visible hemolysis. Ideally, the laboratory offering the test should request a sufficient amount of blood to allow for repeating the test in case this is needed.

After the blood draw, the sample must be stored and shipped as to prevent the cfDNA from degrading and to avoid lysis of the maternal leucocytes. Optimal transport and storage conditions have been described [4]. In case validation studies have not been performed by the tube manufacturer, the laboratory has to perform its own validation, minimally covering the effect of temperature, storage, and transport conditions on, for example, total amount of cfDNA and "fetal" fraction.

EXTRACTION OF CELL-FREE DNA

Validation should also be performed for preanalytical variables that could potentially influence the test result, especially in cases where operator-dependent handling is required.

The extraction of cfDNA from maternal blood differs from genomic DNA extraction from blood cells for conventional genetic tests. Methods differ as to the quantity of cfDNA extracted and the fraction of "fetal" cfDNA [5,24]. Internal controls provide a positive control with known sequencing results for demonstrating successful cfDNA extraction, and a negative control allows to identify contamination. So far, one approach to standardized quality controls has been published [24]. This step provides a preanalytical proof of the samples suitability for cell-free DNA-based prenatal testing.

DETECTION/QUANTIFICATION OF CELL-FREE DNA

It is not sufficient to restrict the evaluation to the presence of cfDNA. The next step in the cfDNA-based prenatal testing workflow quantifies the amount of isolated cfDNA and "fetal" cfDNA. Too little "fetal" DNA can lead to false negative results. This point has been illustrated in a small case series where test requesters sent laboratories blood of nonpregnant women [25]. Validation can be performed by using a second, already validated method. We refer to Chapter 3 for a more detailed description on fetal fraction calculation.

ANALYSIS OF CELL-FREE DNA WITH SEQUENCING

The most common method used to analyze cfDNA is Next-Generation Sequencing (NGS) [2]. Chapter 2 describes the technology in detail. Quality standards and guidelines for implementing NGS in a laboratory have been published [26,27] and descriptions of NGS workflows and EQA schemes are available [17,19]. The quality requirements for NGS in the context of cfDNA-based prenatal testing from a quality manager's point of view are discussed here.

Equipment plays a crucial role in analyzing cfDNA by NGS. NGS-based techniques are more complex than conventional sequencing techniques such as Sanger sequencing and Pyrosequencing. To ensure reliability, the sequencer has to be qualified and all functions checked prior to the start of the analysis. Maintenance procedures must be in place, and function tests must be performed for other equipment, such as pipettes that have to be calibrated as undesirable volume differences will generate critical deviations, that could lead to false results.

The hands-on steps include the preparation of a DNA library and eventually the enrichment of target sequences, as well as the final preparation steps prior to loading the samples into the sequencer. These procedures are performed by technical assistants and demand precise and accurate sample handling including a four-eye principle to check the correctness of crucial functions such as adapter ligation, for which errors cannot be subsequently detected. The implementation of internal controls, signaling deviations during sample processing, should be considered. The hands-on time for cfDNA-based prenatal testing ends when samples have been loaded into the sequencer along with positive controls and the sequencing run starts.

BIOINFORMATICS

The last step in producing test results is the bioinformatics step, the analysis of data generated during sequencing [26,28]. As these data are large and complex, computational aid is essential. In addition, bioinformatics provides software solutions to deal with the backup of a huge amount of raw data.

MEDICAL VALIDATION, INTERPRETATION OF RESULTS, AND REPORTING

All results in a medical laboratory go through "medical validation." This ensures that results are interpreted and reported in a medically relevant manner and can be seen as a supplementary four-eye principle control for reliable results. ISO 15189 describes the required content for test reports [10] and Chapter 11 addresses "Best Practices" that resulted from EQA rounds. As to reporting, the American College of Medical Genetics and Genomics (ACMG) has advised to report patient-specific PPVs (positive predictive values) when reporting abnormal results, or population-derived PPVs in cases in which patient-specific PPV cannot be determined due to unavailable information [29]. Posttest genetic counseling in case of normal results is organized in different ways in different geographies.

ACCREDITATION vs CE-MARK

CE-IVD marking indicates that a device complies with the European-in-Vitro Diagnostic Devices Directive and that the device may be legally commercialized in the EU. In vitro diagnostic (IVD) devices include reagents, calibration materials, control materials, instruments, equipment, and systems that are intended for use in the examination of specimens taken from the human body (tissue, blood, urine, etc.) to diagnose diseases, to monitor a person's state of health, or to monitor therapeutic procedures. In April 2017 the European Parliament and Council *Directives* on Medical Devices have been replaced by *Regulations* (https://ec.europa.eu/growth/sectors/medical-devices/regulatory-framework_en) (https://www.bsigroup.com/en-GB/medical-devices/our-services/IVDR-Revision/).

The new regulations have come into force on May 26, 2017. They will ensure a consistently high level of health and safety protection for EU citizens using the products, their free and fair trade throughout the EU, and the adaptation of EU legislation to the significant technological and scientific progress of the last 20 years. The new rules will apply after a transitional period of 3 years for medical devices and 5 years for IVD. They provide a "more robust EU legislative framework to ensure better protection of public health and patient safety." The process of labeling is a self-certification process. [30]. To gain the permission to label IVD with a CE mark, the producer has to implement a quality management system and prove the suitability of products with regard to safety and capability [31]. Key elements of the regulatory approach are the supervision of notified bodies, risk classification, conformity assessment procedures, performance evaluation and performance studies, vigilance and market surveillance, transparency and traceability. Devices have been divided in risk categories: Class A covers devices with low personal and low public health risk, class B devices with low to moderate personal risk and low public health risk, class C high personal risk and low to moderate public health risk, and class D high personal and high public health risk. Human genetic tests and screening tests for congenital disorders in the embryo or fetus such as cfDNA testing are class C, devices related to blood safety and high-risk infectious diseases such as HIV are class D. Requirements for CE-IVD marking and ISO 15189 cover similar aspects, but ISO 15189 goes further in that it also focuses on the performance in the laboratory, whereas CE-IVD marking provides information on the medical device only and the information is provided by the manufacturer. The use of a CE-IVD marked test does not make complying to ISO accreditation requirements redundant. Under the new regulations health institutes that directly support the healthcare system will have the possibility of using "in-house" developed devices if "the specific needs of target patient groups cannot be met at the appropriate level of performance by an equivalent device available on the market" (IVDR Regulatory Text on Health Institute Exemption—Recital 30 and Article 2.36).

CONCLUSION

Standards, product marking, laboratory accreditation, guidelines, and external quality assessments have been put in place for IVD such as cfDNA-based prenatal testing. These all aim at harmonizing test offers and improving patient safety and will be translated into national rules and legislation in the coming years.

REFERENCES

[1] Gekas J, Langlois S, Ravitsky V, Audibert F, van den Berg DG, Haidar H, et al. Non-invasive prenatal testing for fetal chromosome abnormalities: review of clinical and ethical issues. Appl Clin Genet 2016;9:15–26.

[2] Kotsopoulou I, Tsoplou P, Mavrommatis K, Kroupis C. Non-invasive prenatal testing (NIPT): limitations on the way to become diagnosis. Diagnosi 2015;2(3):141–58.

[3] Schenk PW. Quality assurance and standardization in view of non-invasive prenatal testing (NIPT). LaboratoriumsMedizin 2016;40(5):307–12.

[4] Wong D, Moturi S, Angkachatchai V, Mueller R, DeSantis G, van den Boom D, et al. Optimizing blood collection, transport and storage conditions for cell free DNA increases access to prenatal testing. Clin Biochem 2013;46(12):1099–104.

[5] Jorgez CJ, Dang DD, Simpson JL, Lewis DE, Bischoff FZ. Quantity versus quality: optimal methods for cell-free DNA isolation from plasma of pregnant women. Genet Med 2006;8(10):615–9.

[6] Mattocks CJ, Morris MA, Matthijs G, Swinnen E, Corveleyn A, Dequeker E, et al. A standardized framework for the validation and verification of clinical molecular genetic tests. Eur J Hum Genet 2010;18(12):1276–88.

[7] McGuire AL, Fisher R, Cusenza P, Hudson K, Rothstein MA, McGraw D, et al. Confidentiality, privacy, and security of genetic and genomic test information in electronic health records: points to consider. Genet Med 2008;10(7):495–9.

[8] Blind K, Jungmittag A, Mangelsdorf A. The economic benefits of standardisation. An update of the study carried out by DIN in 2000. https://www.dinde/blob/89552/68849fab0eeeaafb56c5a3ffee9959c5/economic-benefits-of-standardization-en-data; 2000 [accessed 01.12.17].

[9] Woodcock S, Fine G, McClure K, Unger B, Rizzo-Price P. The role of standards and training in preparing for accreditation. Am J Clin Pathol 2010;134(3):388–92.

[10] International Organization for Standardization. ISO 15189:2014: medical laboratories - requirements for quality and competence [ISO 15189:2012, Corrected version 2014-08-15], 2014.

[11] Schneider F, Maurer C, Friedberg RC. International Organization for Standardization (ISO) 15189. Ann Lab Med 2017;37(5):365–70.

[12] European Committee for Standardization. Molecular in vitro diagnostic examinations - specifications for pre-examination processes for venous whole blood - Part 3: Isolated circulating cell free DNA from plasma. PD CEN/TS 16835-3:2015. 2015.

[13] Miller WG, Jones GR, Horowitz GL, Weykamp C. Proficiency testing/external quality assessment: current challenges and future directions. Clin Chem 2011;57(12):1670–80.

[14] Neumaier M, Braun A, Gessner R, Funke H. Experiences with external quality assessment (EQA) in molecular diagnostics in clinical laboratories in Germany. Working Group of the German Societies for Clinical Chemistry (DGKC) and Laboratory Medicine (DGLM). Clin Chem Lab Med 2000;38(2):161–3.

[15] Porto G, Brissot P, Swinkels DW, Zoller H, Kamarainen O, Patton S, et al. EMQN best practice guidelines for the molecular genetic diagnosis of hereditary hemochromatosis (HH). Eur J Hum Genet 2016;24 (4):479–95.

[16] Kalman LV, Lubin IM, Barker S, du Sart D, Elles R, Grody WW, et al. Current landscape and new paradigms of proficiency testing and external quality assessment for molecular genetics. Arch Pathol Lab Med 2013;137 (7):983–8.

[17] Hoeltge GA, Phillips MG, Styer PE, Mockridge P. Detection and correction of systematic laboratory problems by analysis of clustered proficiency testing failures. Arch Pathol Lab Med 2005;129(2):186–9.

[18] Haselmann V, Geilenkeuser WJ, Helfert S, Eichner R, Hetjens S, Neumaier M, et al. Thirteen years of an international external quality assessment scheme for genotyping: results and recommendations. Clin Chem 2016;62(8):1084–95.

[19] Eurogentest. Molecular genetic testing - external quality assessment scheme provision. http://www.eurogentest.org/index.php?id=706; 2007.

[20] IFCC - molecular genetic testing - external quality assessment schemes. http://www.ifcc.org/ifcc-scientific-division/sd-committees/c-md/externalqualityassessment-proficiencytestinginmoleculardiagnostics/; 2015.

[21] Benn P, Borrell A, Chiu RW, Cuckle H, Dugoff L, Faas B, et al. Position statement from the Chromosome Abnormality Screening Committee on behalf of the Board of the International Society for Prenatal Diagnosis. Prenat Diagn 2015;35(8):725–34.

[22] Dequeker E, Ramsden S, Grody WW, Stenzel TT, Barton DE. Quality control in molecular genetic testing. Nat Rev Genet 2001;2:717–23.

[23] Deutsche Akkreditierungsstelle GmbH. "What benefits does accreditation offer?" https://www.dakksde/en/content/what-benefits-does-accreditation-offer [accessed 24.11.17].

[24] Devonshire AS, Whale AS, Gutteridge A, Jones G, Cowen S, Foy CA, et al. Towards standardisation of cell-free DNA measurement in plasma: controls for extraction efficiency, fragment size bias and quantification. Anal Bioanal Chem 2014;406(26):6499–512.

[25] Takoudes T, Hamar B. Performance of non-invasive prenatal testing when fetal cell-free DNA is absent. Ultrasound Obstet Gynecol 2015;45(1):112.

[26] Vogl I, Eck SH, Benet-Pagès A, Greif PA, Hirv K, Kotschote S, et al. Diagnostic applications of next generation sequencing: working towards quality standards/Diagnostische Anwendung von next generation sequencing: Auf dem Weg zu Qualitätsstandards. LaboratoriumsMedizin 2012;36(4):227–39.

[27] Matthijs G, Souche E, Alders M, Corveleyn A, Eck S, Feenstra I, et al. Guidelines for diagnostic next-generation sequencing. Eur J Hum Genet 2016;24(1):2–5.

[28] Vogl I, Benet-Pagès A, Eck SH, Kuhn M, Vosberg S, Greif PA, et al. Applications and data analysis of next-generation sequencing. LaboratoriumsMedizin 2013;37(6):305–15.

[29] Gregg AR, Skotko BG, Benkendorf JL, Monaghan KG, Bajaj K, Best RG, et al. Noninvasive prenatal screening for fetal aneuploidy, 2016 update: a position statement of the American College of Medical Genetics and Genomics. Genet Med 2016;18(10):1056–65.

[30] Khan MU, Bowsher RR, Cameron M, Devanarayan V, Keller S, King L, et al. Recommendations for adaptation and validation of commercial kits for biomarker quantification in drug development. Bioanalysis 2015;7(2):229–42.

[31] European Commission. The 'blue guide' on the implementation of EU products rules 2016. Off J Eur Union 2016;59:58–64.

GLOSSARY

Accreditation Formal recognition by an authoritative body, proving that a laboratory has the competence to carry out specific tasks

CE-mark Marking required for medical devices and in vitro diagnostics in Europe and Iceland, Norway, and Liechtenstein

Documentation Written proof of execution

Guideline Written requirements that should be followed for a specific task

Internal audit Self-inspection of workflows and procedures

Internal quality control Controls implemented into procedures to check correct functioning/operation

ISO 15189 International Standard 15189: Describes the requirements for medical laboratories to assure quality and competence

Nonconformities Deviations from prior defined specifications for procedures and processes

Occurrence management Handling of nonconformities

Qualification Proof of proper functioning of a device and proof of capability of personnel

Quality assurance Set of implemented processes that ensure quality

Specification Definition of frames for processes and procedures

Standard operation procedure Detailed written instructions for procedures, functioning, and handling

Standardization Process of harmonizing procedures

Training Process of defining the demand for education of personnel, the briefing and further education and the survey if the education is successful or if additional education is needed

Validation Confirming that a product or service meets the needs of its users

DECISIONAL SUPPORT FOR EXPECTANT PARENTS

13

Jane Fisher

Antenatal Results and Choices (ARC), London, United Kingdom

To place in context the decisional support needs women and their partners may have that are specific to cfDNA NIPT, it is helpful to first describe the pre-cfDNA NIPT screening landscape. Prior to the introduction of commercial cfDNA NIPT in 2011 [1], most developed countries offered pregnant women screening tests for Down's syndrome [2]. For over a decade the recommended test for women who want screening has been first trimester combined testing using a combination of ultrasound and biochemical markers together with relevant maternal factors in the testing algorithm [3]. Risk calculation software generates a statistical chance of the fetus having trisomy 21 (and often trisomy 18 and 13). In the UK, a standardized cutoff of 1 in 150 is used to inform the offer of invasive diagnostic testing [4].

There has been much discussion in the literature regarding pretest counseling and how well women are supported in making decisions to take up or decline prenatal screening [5–7]. As embarking on screening may lead to the offer of diagnostic testing, which carries a risk of miscarriage, and may result in decisions about pregnancy termination, the emphasis of publicly funded prenatal screening programs has not been on uptake targets (as in most public health screening programs in other health contexts) but on providing reproductive choice [4]. This means such programs should make it clear to women that participation is optional and that the aim is to enable autonomous reproductive decision-making. In turn, this requires providers to present aneuploidy screening in a nondirective way and facilitate women in making an informed choice to accept or decline the offer.

INFORMED CHOICE IN PRENATAL SCREENING

If the aim of pretest counseling for prenatal aneuploidy screening is that a woman makes an "informed choice" to accept or decline the testing, this well-used phrase warrants scrutiny. While there is not unanimous agreement on a definition of informed choice, the one with most currency in the context of prenatal screening is the description used by Marteau et al. which defines an informed choice as "one that is based on relevant knowledge, consistent with the decision-maker's values and behaviourally implemented" [8]. There is some discussion on what might constitute "relevant knowledge" but there is broad consensus that information provided to women should include up-to-date, accurate information on the conditions being screened for, what the testing process involves, the potential benefits and harms of the test, and the possible outcomes. There is some evidence that there is greater satisfaction with their screening decision among those women who consider their choice to be informed, but it is more difficult to fully establish whether decisions throughout the screening and diagnostic journey are consistent with values expressed at the outset [9]. Women cannot always know what they will do in real life situations, for example, when surveyed fewer women say they would terminate a pregnancy on the receipt of a diagnosis than do so in reality [10]. This means professionals need to be mindful not to make assumptions about the decisions women may make after screening and to be ready to take a more stepwise approach.

The literature has highlighted how challenging the facilitation of informed choice in prenatal screening is in practice [11]. Some researchers suggest that professionals' views on women's information needs can be at odds with what women perceive they need [12]. Others present evidence that women are not being made sufficiently aware of the possible outcomes of screening and the range of decisions that may arise, particularly that they may be confronted with the offer of termination of pregnancy [13,14]. There is also evidence that women are not always offered accurate balanced information on life with screened for conditions [15].

The first trimester timing of prescreening decision-making creates another impediment to achieving informed choice. If women are to take up screening at the optimal time of 11–14 gestational weeks, they need to consider their options before this. The first trimester can be a "tumultuous" time for women [16]. They are often adjusting to the reality of impending motherhood; many will be excited and full of hope, others more ambivalent or anxious. Some women may be tentative in these early weeks or worried about miscarriage. All will be subject to distinct physical changes and the emotional fluctuations caused by pregnancy hormones. These factors can impair a woman's capacity to assimilate information and deliberate on her testing options. At this sensitive time, according to her psychological relationship to the pregnancy and the values she has internalized up to this point, she may find it difficult or impossible to tolerate thoughts of possible fetal anomaly, or as she might see it, something being "wrong" with her baby. It can be equally difficult to contemplate the possibility of confronting a decision about "risky" invasive testing or about the very future of the pregnancy. This is not to suggest that women should not be gently encouraged to reflect on potential outcomes, but staff discussing screening choices need to understand that some women will be unable to do so. Moreover, no amount of high-quality pretest information and counseling can fully prepare a woman for the psychological impact of a "positive" screening result [17,18].

PARENT DECISION-MAKING—CHALLENGES FOR CLINICIANS

Health care professionals face challenges, both systemic and personal, which can affect their ability to effectively support women in their prenatal screening decisions. In many public health systems (and this is certainly the case for the National Health Service in the UK) staffing in antenatal and maternity units is at suboptimal levels [19]. In the UK, midwives have long been tasked with discussing screening options with women and at the time of writing the profession is understaffed by approximately 3500 [20]. Screening is most often discussed at a first visit "booking" appointment with a midwife. This is a consultation that is at best an hour long and covers detailed history taking and many aspects of pregnancy care often including birthing options. As a result, the screening conversation is unlikely to extend much beyond 10 min. The limited availability of ultrasound appointments (another resourcing issue) can mean that women are encouraged to "book in" for a nuchal translucency scan and decide later if they want to proceed [21]. A useful starting point for examining the organizational constraints on women's screening decision-making in a UK setting is provided by the qualitative study by Ukuhor et al. [22]. They make some recommendations for changes, such as a separate appointment to discuss screening options, which may help address some of the issues but admit that more research is needed. Some maternity units in the UK have scheduled evening group information sessions for parents on screening and anecdotally the response has been positive.

The personal challenges include the complexity of the information professionals are asked to communicate to women. In order to make an informed choice, a woman will need to have at least a basic grasp of the test properties to know what the results might mean to her. Not all professionals feel sufficiently equipped to talk about the technicalities of the screening process such as test sensitivities, specificities, and positive predictive values [23]. Public understanding of relative and absolute risk values is notoriously poor, and professionals are asked to help a woman understand these complicated concepts, when they too may lack confidence [5,18]. As noted earlier, the woman may still be adjusting to her pregnancy and find it difficult to absorb complex information.

A further personal versus professional challenge for those working in antenatal screening relates to the potential outcome of pregnancy termination. Abortion laws vary across jurisdictions, but most countries allow for termination for fetal anomaly. However, even when working in a setting where abortion is available, some professionals may have personal values that conflict with the provision of abortion in any circumstances, or perhaps in the context of prenatal diagnosis of conditions they deem "less severe." Others may hold the personal view that it is preferable to avoid the birth of a child who may have a life-limiting or disabling condition [24]. While their professionalism should enable them to put aside their personal values, health care professionals with strong beliefs will need to be vigilant in order that they do not inadvertently exhibit bias.

There are further tensions inherent in maintaining the "nondirective" approach which underpins the aim of fostering women's autonomous decisions. A nondirective stance has long been seen as the best strategy to deal with the ethical sensitivities and counter the asymmetric power relationship between professional and "patient"; though the approach might also be seen as a tactic to keep at bay the historical specter of eugenics [25]. Furthermore, it can be a way for a health care professional to avoid being implicated in a decision that the woman may come to regret.

Midwives are often on the frontline in discussing a woman's prenatal screening choices. The ethos of midwifery is to be "with woman" to advocate for her and be by her side through pregnancy and birth. As Farsides et al. note: "Antenatal practitioners often nurture close relationships with the women they care for during the months of pregnancy, and see this as an important and even integral part of their work. These practitioners can feel that the non-directive approach compromises or is incompatible with their concept of a caring practitioner-client relationship" [25]. Moreover, midwives will often care for a diverse population and not all women may value the western model of making autonomous choices in the same way. For example, in Chinese culture more emphasis is attached to "relational autonomy" so while individual autonomy is deemed important, the family can also have a pivotal role in decision-making [26]. A Q-methodology questionnaire study carried out in an area of England with a diverse ethnic population explored attitudes to autonomous informed choice in antenatal screening in a group of 98 women of African, British white, Caribbean, Chinese, and Pakistani origin. Their findings suggested that while most wanted to make their "own" decisions, they also valued the advice of health professionals [6]. In view of this, health professionals may be safest avoiding an over rigid interpretation of nondirective counseling and adapting their decisional support techniques to the needs of the individual woman and ensuring that if she wishes to involve others in her decision-making she is able to do so [27].

HOW CELL-FREE DNA NIPT MIGHT IMPACT ON PARENT DECISION-MAKING

Having established that effectively supporting parent decisions in the prenatal screening context is complex and challenging, the next question to address is how the introduction of cfDNA NIPT might impact on this process for pregnant women and their partners. cfDNA NIPT has been available in the private sector since 2011 and at the time of writing many public health systems have or are introducing it into their standardized trisomy screening programs (either publicly funded or as an out-of-pocket option) [28]. In many public health systems cfDNA NIPT is not offered as a frontline screen but contingent on a designated risk assessment from conventional screening techniques. For example, from late 2018 in England, all pregnant women whose result from combined or biochemical screening of having a baby with trisomy 21, 18, or 13 assigns a chance higher than 1 in 150 will be given the option of cfDNA NIPT (but can also opt to have invasive diagnostic testing) [29].

Offering cfDNA NIPT in a contingent way adds another element to pretest counseling. With the introduction of cfDNA NIPT into the pathway, women must be made aware of another possible step in the screening process before they decide to opt in. This potential extra step involved when cfDNA NIPT is offered after initial screening will be significant to some women because of the delay it imposes on the timing of a diagnosis. If she opts for cfDNA NIPT as a second-tier screen she may not receive the result in time to schedule chorionic villus sampling (CVS) if she wishes to have earlier diagnosis. In some instances, clinicians advocate waiting for amniocentesis to avoid the possibility of the CVS replicating a false positive result due to a confined placental mosaicism [30]. Instead of having a confirmed result through CVS before 13 gestational weeks, a woman could find herself at least 16 weeks at the time of diagnosis. Women who know they want to terminate an affected pregnancy and wish to do so surgically may face difficulty accessing the required dilatation and evacuation (D&E) procedure as this is not always as widely available as earlier surgical methods [31]. Moreover, women from certain faith or cultural groups risk being close to or going beyond the gestational age at which termination is officially sanctioned (e.g., many Muslims will only consider termination before they have passed 140 days when it is believed ensoulment occurs) [32].

The procedure women have to undergo in order to have cfDNA NIPT is very straightforward. It requires a maternal blood draw, often described as a "simple" blood test. The issue of the "routinisation" of blood tests arose when biochemical screening for Down's syndrome was introduced. Some women reported that they had been unaware of the specific purpose of a blood draw, confused between those carried out for maternal indications (e.g. rhesus factor, infectious diseases) and those for fetal indications [33]. Importantly, this "simple" test will present expectant parents with much more accurate information than conventional screening. Whereas most women who have a positive screening result from combined testing will be found to be *false* positives, most women who have a positive result from cfDNA NIPT will be *true* positives. Those imparting the result will need to find the right balance in explaining its implications; this will be different from their previous experience of communicating either screening results that have a much lower sensitivity or conveying the more conclusive nature of results from diagnostic testing.

The impetus behind the widespread introduction of cfDNA NIPT has been that this technology enables a more accurate assessment of the chance of aneuploidy than other screening methods without putting the pregnancy at risk. There would be few that would deny the benefit of reducing the number of invasive diagnostic tests and so the number of associated procedure-related miscarriages. Indeed, the improved sensitivity of cfDNA NIPT potentially expands choice to those women who would value an accurate assessment of the chance of their baby having a trisomy, but have previously been deterred from screening tests because of relatively high false positive rates. Perhaps counterintuitively, the improved accuracy and safety of cfDNA NIPT has the potential to add a layer of difficulty for some women making pretest decisions. The "safety" aspect may make some women reticent to go directly for invasive procedures even when their personal preference is for definitive diagnosis [34]. It is important to acknowledge that the labeling of cell-free DNA-based screening as "NIPT" is not value neutral. The emphasis on "noninvasive" promotes the safety aspect and may make it more difficult for some women to decline the testing or proceed directly to diagnostic testing. In view of this, pretest counseling will need to give time to enabling women to articulate their individual priorities and preferences. Some women may need to be helped to give themselves permission to pursue diagnostic testing if this is right for her in her circumstances. It is equally important that no woman feels pressured to have "NIPT" because it poses no risk when she would prefer *not* to have a more accurate assessment of aneuploidy in the prenatal period because she does not want to confront further decisions.

There is evidence in the literature that the simplicity and safety of cfDNA NIPT may affect the importance health care providers place on women making fully informed choices to use the technology in contrast to how they might counsel around invasive diagnostic procedures [35]. While it is true that the majority of women who opt for cfDNA NIPT will receive a result telling them their baby is very unlikely to be affected, this should not obscure the fact that a significant minority will face potentially difficult news. It is also important that women are aware of the limitations of cfDNA NIPT and that they could receive a failed or inconclusive result which may provoke uncertainty and anxiety.

ETHICAL DEBATES SURROUNDING CELL-FREE DNA NIPT

The societal and ethical implications of cfDNA NIPT are explored elsewhere in this volume, but it is worth examining how some of the current ethical debates encroach on the decision-making space for women making decisions around cfDNA NIPT. Concerns have been raised that provision of cfDNA NIPT will lead to a rise in the numbers of women taking up screening, in turn resulting in more diagnoses and, as the majority decide to terminate an affected pregnancy [36], ultimately more terminations and fewer live births of babies with Down's syndrome. While such concerns have been contested [37], the spotlight on termination rates has triggered questions around how screening is presented to potential participants and the quality of the information provided on life with Down's syndrome. In an attempt to address these concerns, the Fetal Anomaly Screening Programme in England established a comprehensive implementation plan for cfDNA NIPT in the NHS in conjunction with stakeholders [38]. There is ongoing work on updating prescreening information literature. A comprehensive training program for health care professionals has been developed which encompasses the technical aspects of cfDNA NIPT and counseling issues. Both projects were achieved with input from a variety of stakeholders; these included national associations supporting families affected by trisomy 21, 18, and 13 and my own organization, ARC which supports women and couples throughout screening and provides specialized help to those who make the decision to terminate.

There is agreement that information provided about conditions must be up to date, balanced, and evidence based. Alongside medical information, there is a need for evidence of the lived experience of those living with the condition and their families. This is hampered somewhat by the limited data available on long-term outcomes, particularly for people with Down's syndrome. Brian Skotko's team at the Massachusetts General hospital for Children have established a patient database to track life span health of individuals with Down's syndrome which should improve information in this area [39].

There is contention around how balanced the condition-related information provided to parents might be and how this may affect decision-making. In 2006, ARC was involved in a project in conjunction with disability support organizations and academics to create a website which sought to provide balanced information on screened for conditions to inform parent decision-making through screening [40]. Testimonies from those living with conditions were given prominence. Unfortunately, the project was abandoned due to lack of sustainable funding. However, the early evaluations were mixed in relation to how "unbiased" those visiting the site considered it to be [40]. The experience demonstrated the slippery nature of the concept of "balance." This is further illustrated by the information provided on the Lettercase website in the United States for those who are seeking an overview of Down's syndrome after receiving a prenatal diagnosis [41]. The information is endorsed by the American College of Medical Genetics and Genomics (ACMG). The text covers many aspects of what living with the condition can mean, including associated challenges and is accompanied by compelling images of children and young people with Down's syndrome. Research has suggested that photographic images that accompany text may have an impact on the neutrality of the information [42].

It is necessary to present a realistic picture of the lived experience of people living with the conditions cfDNA NIPT screens for, but at the same time it is important to remember that a prenatal diagnosis means expectant parents are making decisions about the fetus in utero. This is a child not yet born. They try to imagine a future containing this child which can often be fraught with uncertainty. No prenatal test can predict with absolute certainty how the chromosomal change is going to affect an individual. In the context of a wanted pregnancy most women and couples conceptualize a baby and invest in that baby hopes and dreams for him or her and their family. They will often need the space to work out whether they want to make the adjustment to what could be a very different future for their child, themselves, and their family from the one they envisaged before the diagnosis.

Before and after prenatal screening, access to evidence-based information about what continuing the pregnancy and having a child with a chromosomal change might involve is essential to help women make the decisions they can best live with. However, women must also be reassured that screening is predicated on offering and supporting choice and that terminating the pregnancy is a valid and acceptable option. This is especially important when abortion in general is stigmatized and at a time when what has been called "abortion for disability" is held up as ethically questionable in some quarters [43].

Termination for fetal anomaly and particularly Down's syndrome has been problematized by some disability advocates and antiabortion campaigners as being inherently discriminatory toward disabled people [44]. They cite the "expressivist" argument which sees individual decisions to terminate a pregnancy because of Down's syndrome as denigrating the value of those living with the condition. This view is combined with the assertion that widespread implementation of cfDNA NIPT will lead to greater uptake of screening and an increase in termination, which could ultimately threaten the very existence of the community of people with Down's syndrome. Organizations such as "Saving Down Syndrome" (which was established in New Zealand but has groups in the United States and UK) are explicit in their aims [45]. The Saving Down Syndrome mission statement reads "We wish to ensure that prenatal screening exists only to provide unborn children with Down's syndrome and their parents with life-affirming unbiased care through education, support and understanding – worldwide."

Many disability rights advocates have expressed disquiet with prenatal screening and are uncomfortable with subsequent choices to end a pregnancy when a diagnosis is made [43], but not all favor restricting parental reproductive choice. The British academic and disability rights campaigner (who himself has achondroplasia) Professor Tom Shakespeare acknowledges the difficulties but supports parents being enabled to make autonomous choices: "I conclude that prenatal diagnosis is not straightforwardly eugenic or discriminatory. We should be on hand to offer counselling, good quality information and support, but we should not venture to dictate where the duties of prospective parents may lie. Nor should we interpret a decision or termination of pregnancy as expressing disrespect or discrimination towards disabled people. Choices in pregnancy are painful and may be experienced as burdensome but they are not incompatible with disability rights" [46].

These ethical debates around cfDNA NIPT are not restricted to academia but have also been taken up extensively by both mainstream and social media. They are important discussions and the open exchange of views is to be welcomed and can be helpful in the quest to negotiate how we can maximize societal benefit from cfDNA NIPT. At the same time, as with any debate played out in the media, there is the danger of oversimplification and polarization. Expectant parents are not making their decisions in a vacuum. Those caring for parents need to be aware that this coverage will often form part of the background to parents contemplating cfDNA NIPT or facing difficult decisions following an cfDNA NIPT result. Some will be aware of the ethical debates in the media, others may encounter strong opinions when seeking information online to aid their decision-making. Either way, when supporting decision-making, it is incumbent on their caregivers to ensure that nothing encountered in the media is allowed to inhibit expectant parents' ability to express their preferences and act according to their own values and unique circumstances. They are the ones who will live with the decisions they make.

DECISION-MAKING AROUND CELL-FREE DNA NIPT IN THE PRIVATE SECTOR

Private sector providers of cfDNA NIPT now extend across the world. It is offered in a variety of settings, from hospitals and clinician's private offices, to ultrasound scan clinics. Although there does not yet seem to be a "direct to consumer" option there are outlets on line which allow parents to purchase a testing kit and then simply seek clinical assistance to take the blood sample [47]. We can estimate that thousands of private tests are being performed annually but private providers will not necessarily be aware of pregnancy outcomes.

As results from cfDNA NIPT can lead to profound decisions for expectant parents, all providers have a responsibility to do their best to ensure that parents are making informed choices when using their services. This is especially true when a woman may seek private testing independently, before any discussions with her antenatal care provider. cfDNA NIPT has been widely reported as having a "greater than 99% accuracy" [48] which may mislead some women into thinking it is practically diagnostic. Any prospective user of cfDNA NIPT must be properly appraised of the limitations and potential outcomes of the testing as well as potential benefits.

A 2015 systematic review of UK cfDNA NIPT websites concluded: "we would recommend that companies offering prenatal testing services via the Internet should be required to review and maintain their information for prospective parents to ensure it is comprehensive, accurate and easily accessible and includes information recommended by national and international bodies" [49]. In 2017 an expert working group, convened by the UK-based nongovernmental organization the Nuffield Council on Bioethics, reviewed evidence and consulted widely to produce a comprehensive report on the ethical issues associated with cfDNA NIPT. They proposed that the information provided by private providers should be quality assured by a recognized national body. They have produced a downloadable leaflet as guidance. The report also recommends that providers "should only offer cfDNA NIPT as part of an inclusive package of care that should include, at a minimum, pre- and post-test counselling and follow-up invasive testing if required" [50].

It is likely that, for the foreseeable future and within public health settings, cfDNA NIPT will only be offered to women with elevated risk. At the same time in most countries access is possible via private providers for all patients willing and able to pay themselves, regardless of their a priori risk. Regulations for private tests vary across jurisdictions. Ideally, easily accessible online guidance for both expectant parents considering testing in the private sector, and providers, informed and endorsed by relevant international and national professional bodies and stakeholder organizations would be helpful. The Nuffield Council provides a good model [50].

CELL-FREE DNA NIPT BEYOND TRISOMY 21, 18, AND 13

As screening for Down's syndrome is long established, cfDNA NIPT in this context has garnered most attention. However, cfDNA NIPT is also available for other chromosomal conditions, fetal sexing, and increasingly for sex-linked disorders and other genetic variations including microdeletions and duplications [51]. Cell-free DNA techniques have also been validated in the detection of some single gene disorders (though it is commonly referred to as NIPD in this context). Research is currently underway to examine the efficacy of cell-free DNA techniques for whole genome and exome sequencing [52].

At present, international professional bodies advise against using cfDNA NIPT for other than the common trisomies without specialized genetic counseling [53]. However, it is not unusual for women to be offered cfDNA NIPT for sex chromosomal disorders or even be presented with the option of: "the only prenatal blood test available to date that can analyse every chromosome of your baby to identify extra or missing parts of chromosomes, or other whole chromosome changes" [54]. There is currently reason to urge expectant parents to exercise caution in the face of testing offers which have limited independent validation or may yield highly uncertain results. As the applications for cfDNA NIPT increase so must guidance and counseling for potential users, which should attempt to manage expectations and be explicit about the type of information testing might reveal.

DECISIONAL SUPPORT AROUND CELL-FREE DNA NIPT—DO DECISION AIDS HAVE A PLACE?

The complexities and ethical sensitivities around prenatal genetic screening already outlined, coupled with the time pressures on decision-making (for women confronted with it and their caregivers facilitating it) in pregnancy means there has been increasing interest in the role that patient decision aids (PDAs) may play in facilitating parent decision-making and promoting informed choices.

The Ottawa Hospital Research Institute has led the field in developing and reviewing PDAs and defines them as "tools that help people become involved in decision making by making explicit the decision that needs to be made, providing information about the options and outcomes, and by clarifying personal values. They are designed to complement, rather than replace, counselling from a health practitioner" [55]. Patient decision aids have been used by professionals to help support patient decision-making in a range of health contexts. The most recent Cochrane Review found them likely to improve knowledge and understanding of benefits and harms of options and found no evidence of harm [56]. However, most evidence on PDAs to date is in the context of a patient making treatment, screening, or care management decisions for themselves as an individual rather than in relation to pregnancy. A woman making decisions about prenatal screening is not just making choices that will impact on her life. Her decisions will have real consequences for her child if born, and dependent on her circumstances, will affect her partner's future and that of the wider family.

The potential advantage of using a decision aid for prenatal screening is that it is designed to help women consider her options and explore factors that are of relevance to her *personally* in relation to screening and the information it can provide. The use of a decision aid can enable the health care professional to "step back" and encourage women to examine their screening options on their own terms, according to their personal beliefs and values. A further advantage is that an individual woman can use the aid in her own time, at her own pace, and involve her partner should she wish.

A Dutch study examining use of a web-based multimedia decision aid for fetal chromosomal anomaly screening found the aid did increase women's decision relevant knowledge and informed decision-making, though the homogeneity of their sample limits the generalizability of their findings [57]. In a systematic review on decision aids in pregnancy, Vlemmix et al. suggest that their use can reduce anxiety and decisional conflict but noted that more research was needed on outcome measures to establish whether women's actual decisions across the whole pathway were consistent with their stated values [58]. They also noted that despite health care professionals being favorably disposed to PDAs, they were not consistently used in practice. Lepine et al. shed some light on the latter point in a qualitative study examining the factors influencing health professionals use of decision aids for Down syndrome prenatal screening in Quebec [59]. Interviews with 36 professionals (15 family physicians, 12 midwives, and 9 obstetricians) found that while the majority saw the value in validated PDAs for Down's syndrome screening, barriers to their application included availability and the perception that use of PDAs would be more time consuming.

Regarding women's attitudes to decision aids in screening there is limited evidence. A randomized control trial on a computer information aid for prenatal genetic testing reported that 100% of women who used the tool found it helpful (though the researchers concede that only 57% of women approached agreed to take part in the study and were all from one center) [60]. A qualitative study into a PDA for Down's syndrome screening found that women identified disadvantages as well as advantages, "suggesting a certain ambivalence overall." Some women expressed "confusion" at too much information and while many acknowledged that it was useful to have information they disclosed that: "knowing this information on the risks and benefits of the options would most likely increase their stress" [61]. This reminds us of the psychological burden for women inherent in being made to contemplate what she may perceive of as negative pregnancy outcomes. The same study highlighted that perhaps because of the "value-laden" nature of screening decisions most women were keen to involve their partners and health care professionals in the process.

Overall, research thus far shows that decision aids show promise as a tool to help women make informed decisions about prenatal testing which could include cfDNA NIPT. Clearly there remains work to be done. Firstly, there is a need for the development of more high quality and validated decision aid tools that are applicable to the complex domain of prenatal screening in general and cfDNA NIPT in particular. Careful consideration is required so as to provide the requisite amount of accurate balanced information without overloading or confusing women and taking account of varying levels of health literacy [62]. Online tools may have an advantage here as they can be designed to present information at various levels and make use of techniques to aid comprehension and accessibility such as graphics, illustrations, video, and animations. It is important that such products are validated and easily accessible to expectant parents and professionals, so open access and search engine optimized. There is scope for international collaboration between professional bodies and relevant stakeholders to create suitable resources and disseminate them effectively. A good example of collaboration on a national level between health services and a not-for-profit patient focused organization is a series of videos to help people understand and make decisions about prenatal genetic testing (including cfDNA NIPT) produced by the Washington State Department of Health, in partnership with the US-based Genetic Support Foundation [63]. These resources were made available in 2016 and up to now, no formal evaluation has been published.

Research also raises awareness of a need for professional education in optimal application of decision aids, to counter the perception that they will necessarily lengthen consultations. If women can be encouraged to use the aid before discussing options with their health care professional, it is likely that time will be used more effectively and be better focused on her personal preferences rather than screening more broadly. It is also important that professionals and women see decision aids for screening as complementary to their discussions together rather than a replacement for them and that partners or other family members are involved as much as the woman wishes. This is particularly pertinent to the context of prenatal screening which requires women to consider information and potential outcomes for her family's future that can provoke stress and anxiety. Furthermore, health care professionals will need to be alert to the possibility of women displaying "expectation bias" [64]. This is a phrase used in social science to describe those participants in qualitative research who are conscious of how their answers might be judged and tailor them accordingly, particularly if asked to consider morally controversial subjects. This can be mitigated by explicit assurances within any decision aid, reaffirmed by health care professionals, that choices to accept or decline screening and/or diagnosis and to continue or terminate a pregnancy are of equal validity and will be equally supported.

DECISIONAL SUPPORT AND CELL-FREE DNA NIPT—THE ROLE OF THE HEALTH CARE TEAM

Much of this chapter has been taken up with the complexities and challenges involved in supporting expectant parents through decisions about cfDNA NIPT. It is important that all health care professionals involved with the pathway through cfDNA NIPT are cognizant of these difficulties and accordingly pragmatic in finding effective ways to support parents in their choices. Most parents who opt to use cfDNA NIPT will not be faced with further dilemmas but for others cfDNA NIPT could be the catalyst for life-changing decisions. It is doubtful that we will ever have research evidence on how exactly parents make their decisions after a positive result, as the ethical and practical challenges in gathering this information at the time decisions are made are great. In the absence of hard evidence, ensuring the process is as "parent-centered" as possible is likely to be of most benefit to parents.

Doctors and midwives providing prenatal care are inevitably involved in parent decision-making regarding prenatal screening, including cfDNA NIPT. Most women value their involvement, recognizing that while their individual circumstances, preferences, and values are paramount, their clinicians hold medical and scientific expertise as well as experience with other parents that they can draw on. Therefore it is incumbent on professionals involved in the screening pathway to keep up to date with the fast-changing field of cfDNA NIPT [65] and feel competent to explain test properties in an unbiased and comprehensible way. It has been found that clinicians tend to be positive about the benefits of cfDNA NIPT which means they may have to be particularly careful to be clear about the limitations and any potential disbenefits to individuals [66]. For example, as noted previously, cfDNA NIPT may not be appropriate for women who would favor conclusive diagnostic information or for those who do not wish to have the "level" of information a positive cfDNA NIPT result represents.

Along with clear balanced information about the scientific elements of cfDNA NIPT, expectant parents will need to have some knowledge of the conditions screened for by cfDNA NIPT to decide whether it is of benefit to them to know prenatally if their baby is affected. Many clinicians may be confident in discussing the implications of genetic conditions from a phenotypical point of view, but less equipped to cover possible psychosocial implications and may vary in their ability to deal with the parental anxiety and distress associated with positive cfDNA NIPT results. In many settings in developed countries, genetic counselors often provide pre- and posttest counseling around cfDNA NIPT in a more holistic way [67]. However, it is not a given that all genetic counselors explore what the implications of having a child with a potential disability might be for individuals from a social as well as medical perspective [68]. Moreover, numbers of genetic counselors working in prenatal setting vary and will need to increase substantially to meet demand as cfDNA NIPT becomes more embedded in routine prenatal care.

We have highlighted the challenges in making sure parents deliberate on the consequences of opting into cfDNA NIPT and we know the majority will undertake the screening hoping to be (and ultimately being) reassured. Therefore although due attention should be given to pretest decision-making, there is onus on the health care team to offer carefully coordinated care if cfDNA NIPT brings a result that suggests a baby is likely to have a genetic condition, is "inconclusive" on this point, or there is no result because of test failure. As in most circumstances this news will provoke stress and anxiety, it is essential that parents' practical and psychological needs are met. They will need support in, for example, making a decision about diagnostic testing which may include the choice between chorionic villus sampling and amniocentesis. Some parents will choose to forego invasive testing and continue the pregnancy under the assumption their baby has the condition predicted by cfDNA NIPT. After a "positive" cfDNA NIPT, all parents are likely to value the offer of a multidisciplinary clinical team (MDT) approach with input, when appropriate, from obstetric, fetal medicine, pediatric, neonatal, and genetic specialists. Optimizing the coherence of the MDT will require open communication between all members to provide parents with good continuity and consistent information. When there are differences in clinical opinion, the reasons should be sensitively discussed with women. In the UK, women have found it helpful to have access to a specialist midwife to aid continuity and provide a source of support through post-cfDNA NIPT decision-making and the subsequent outcomes [69,70]. This can be important when women find themselves interacting with a number of different clinicians and sometimes asked to assimilate complex medical information.

If health services are to provide high-quality decisional support to parents, cfDNA NIPT must not be viewed as a test in isolation but as an element of a robust prenatal screening and diagnostic care pathway. All staff should be aware of their roles and responsibilities along the pathway. Clinical rigor and adherence to agreed quality standards are essential, but attention to the "human" side of screening and its psychosocial sequelae is also important. Staff who deliver results and then support parents through ensuing decisions will need to have the necessary communication skills to deliver sensitive care and the ability to manage distress.

THE ROLE OF INDEPENDENT SOURCES OF INFORMATION AND SUPPORT

Health care professionals will continue to have a pivotal role in supporting parent decision-making before and after cfDNA NIPT in the public and private sector. However, given the complex personal and ethical issues involved in this fast-moving area of testing, there is a place for the provision of independent high-quality information and support that is outside the health care realm. If we take the UK as an example, NHS screening services in England, Scotland, and Wales all provide links to national third sector organizations supporting people with genetic conditions. National genetic condition-specific organizations are often best placed to provide an accurate picture of what living with a condition can involve, for affected individuals and their families. Some organizations provide testimonies from people with the condition and their carers about their day-to-day lived experience. Some may be able to put expectant parents faced with decisions in contact with parents who have children with the condition in question. Not all parents will want to take up this opportunity, but it can be helpful that they know the option is available.

NHS screening services also signpost to Antenatal Results and Choices (ARC) [71]. ARC was formally established as an independent national charity in 1988 with the aim of addressing the practical and psychological needs of women and couples undergoing termination for fetal anomaly and to work in partnership with health care professionals to improve care in this area. Since then the charity has widened its remit to "provide non-directive information and support to parents throughout antenatal testing." ARC runs a national helpline service which is widely publicized within the NHS and private sector. ARC's trained helpline team handles around 6000 contacts every year by telephone and email. About 20% of those contacts currently concern cfDNA NIPT.

The ARC helpline team does not give medical or condition-specific advice, but can signpost to clinical experts and other sources of information as necessary. What they offer is a safe confidential space for parents to consider their options; all the time they might need to explore potential outcomes and support around the uncertainties, anxieties, and distress prenatal test results can evoke. The aim is to help parents gather all the information they need, consider what implications this information has for them, and come to decisions that are congruent with their individual situations and values. The emphasis is on process not outcome, as the crucial factor is that they are able to manage the long-term consequences of the eventual decisions they make. ARC can provide a "containing" space, independent of their health care professionals, family, and close community which enables them to be open, to express difficult emotions freely and ask questions which they may not always feel comfortable doing with those emotionally or professionally connected to them.

ARC maintains a close working relationship with health care professionals in the field; the confidential helpline service is open to them as well as parents. Throughout its lifetime ARC has run a well-evaluated program of professional training in order to feed back the insight we gather from the helpline regarding the parent experience of caregivers and the screening process [72]. This collaboration and mutual respect means health care professionals are comfortable referring women and couples to us at all stages of the screening process.

As regards cfDNA NIPT, currently, the highest number of contacts ARC receives come from parents making decisions about whether and where to access the testing in the private sector. The majority have had an NHS screening result that has made them anxious but unwilling to undertake invasive testing because of the miscarriage risk. Discussions on the ARC helpline will include their knowledge of cfDNA NIPT, what it screens for and the testing process. ARC does not endorse any provider or laboratory service, instead we discuss what to look out for and direct them to a page on the ARC website that encourages them to ask requisite questions of a test provider.

Providing access to independent services such as ARC and relevant condition-specific support organizations can both be of benefit to parents considering or undergoing cfDNA NIPT and help health care professionals mitigate against what will inevitably be their more medicalized slant. However, parents should only be signposted to external organizations that can demonstrate that they are nondirective and are quality assured in a meaningful way. In the UK this would include checking that they are officially recognized by the relevant national charity commission. In other countries there will be methods of checking credibility, such as evidence of endorsement by professional bodies. This is especially important when it is so easy for anyone to set up web-based information and support services and some "pregnancy support services" are provided by organizations with a bias against abortion.

CONCLUSION

"I am grateful to have had the chance to have the NIPT. We had then time to consider our decision. We had time to research and we opted for an amnio to confirm and then decided to interrupt the pregnancy. Without the NIPT, I would not have had so much time to research my decision. I would presumably have gone for the amnio and then known 3 days later and had to make a prompt decision rather than been given the weeks that we did have to explore everything.

The fact we had time to consider our options means that in the long run I am confident we made the right choice for us and our family. We will always love our daughter and we live with the choices we made. For me these new tests need to absolutely be about informed consent and must be accompanied with advice and guidance so that parents can make the right choice for them and their children."

Email to ARC June 2016 from parent who opted for cfDNA NIPT.

It is clear cfDNA NIPT is here to stay. Internationally, many public sector health care services have or are planning to implement cfDNA NIPT within their antenatal screening offer and private sector provision continues to expand to meet demand from women. Its benefits, particularly the greater positive predictive value than conventional biochemical/ultrasound screening for aneuploidy, are now well established. For women who want to have as accurate an assessment as possible of their fetus having Down's syndrome, so they can avoid invasive procedures, cfDNA NIPT is attractive and many are prepared to pay for it. However, this simple and "safe" test is a recent addition to the ethically charged terrain of prenatal screening and scrutiny of its implementation will help ensure the benefits outweigh the harms.

There is not universal support for cfDNA NIPT. Some of the more politicized critiques can be countered but some concerns regarding how cfDNA NIPT and prenatal screening in general are presented to women have validity; improvements can be made in supporting a parent-centered process. cfDNA NIPT is a relatively new testing technology, but we know the nature of the decisions expectant parents encounter when confronted with an unexpected result from any test can be stressful and health services need to have requisite individualized care pathways in place to deal with the psychological fallout. Supporting parents when cfDNA NIPT results mean adjusting to an uncertain outlook or painful decisions will make demands on health care professionals at a personal as well as professional level. Meeting the training and support needs of these professionals will be instrumental in helping them do a difficult job to the best of their ability.

This chapter has attempted to demonstrate that cfDNA NIPT has added another dimension to the already complex field of prenatal screening. Most expectant parents who opt into screening do so with the simple hope of being assured their baby is healthy; many will not fully engage with the possibility that this will not be their reality until they receive a positive result. Therefore decisional support needs to be proportionate and stepwise. There is scope in examining further how decision aids might best be developed and utilized and what external resources parents might find useful. Ultimately, more qualitative research and service audit that asks expectant parents to reflect on their experience of the process through prenatal screening and its consequences will be valuable in assessing how far services are meeting their decisional support needs.

REFERENCES

[1] Lau TK, Chan MK, Lo PS, et al. Clinical utility of noninvasive fetal trisomy (NIFTY) test – early experience. J Matern Fetal Neonatal Med 2012;25(10):1856–9.
[2] Boyd PA, DeVigan C, Khoshnood B, et al. Survey of prenatal screening policies in Europe for structural malformations and chromosome anomalies, and their impact on detection and termination rates for neural tube defects and Down's syndrome. BJOG 2008;115:689–96.
[3] American College of Obstetricians and Gynecologists Committee. ACOG practice bulletin no. 77: screening for fetal chromosomal abnormalities. Obstet Gynecol 2007;109:217.
[4] NHS. Fetal Anomaly Screening Programme standards, https://www.gov.uk/government/uploads/system/uploads/attachment_data/file/421650/FASP_Standards_April_2015_final_2_pdf; 2015–2016 [accessed 10.11.07].
[5] Green JM, Hewison J, Bekker HL, Bryant LD, Cuckle HS. Psychosocial aspects of genetic screening of pregnant women and newborns: a systematic review. Health Technol Assess 2004;8(33).
[6] Ahmed S, Bryant LD, Tizro Z, Shickle D. Interpretations of informed choice in antenatal screening: a cross-cultural, Q-methodology study. Soc Sci Med 2012;74(7):997–1004.
[7] Hewison J. Psychological aspects of individualized choice and reproductive autonomy in prenatal screening. Bioethics 2015;29(1):9–18.
[8] Marteau TM, Dormandy E, Michie S. A measure of informed choice. Health Expect 2001;4(2):99.
[9] Van den Berg M, Timmermans DRM, Ten Kate L, Van Vugt JMG, Van der Wal G. Are pregnant women making informed choices about prenatal screening? Genet Med 2005;7:332–8.
[10] Hill M, Fisher J, Chitty LS, Morris S. Women's and health professionals' preferences for prenatal tests for down syndrome: a discrete choice experiment to contrast noninvasive prenatal diagnosis with current invasive tests. Genet Med 2012;14(11):905–13.
[11] Skirton H, Barr O. Antenatal screening and informed choice: a cross-sectional survey of parents and professionals. Midwifery 2010;26(6):596–602.
[12] Farrell RM, Nutter B, Agatisa PK. Patient-centered prenatal counseling: aligning obstetric healthcare professionals with needs of pregnant women. Women Health 2015;55(3):280.

[13] Lawson KL. Expectations of the parenting experience and willingness to consider selective termination for down syndrome. J Reprod Infant Psychol 2006;24:43–59.

[14] Seror V, Ville Y. Women's attitudes to the successive decisions possibly involved in prenatal screening for down syndrome: how consistent with their actual decisions? Prenat Diagn 2010;30:1086–93.

[15] Sheets KB, Best RG, Brasington CK, Will MC. Balanced information about down syndrome: what is essential? Am J Med Genet A 2011;155(6):1246–57.

[16] Farrell RM, Dolgin N, Flocke SA, Winbush V, Mercer MB, Simon C. Risk and uncertainty: shifting decision making for aneuploidy screening to the first trimester of pregnancy. Genet Med 2011;13(5):429–36.

[17] Fisher J. First-trimester screening: dealing with the fall-out. Prenat Diagn 2011;31:46–9.

[18] Bryant LD, Ahmed S, Hewison J, Rodeck C, Whittle M. Conveying information about screening. In: Fetal medicine: basic science and clinical practice. second ed. Edinburgh: Elsevier; 2008.

[19] https://www.nuffieldtrust.org.uk/resource/the-nhs-workforce-in-numbers#2-what-is-the-overall-shortfall-in-clinical-staff-in-the-nhs [accessed 10.11.17].

[20] https://www.rcm.org.uk/news-views-and-analysis/news/%E2%80%98new-nmc-report-shows-the-need-for-more-midwives-says-rcm%E2%80%99 [accessed 10.11.17].

[21] Securing the Workforce Supply – Sonography workforce review. UK Centre for Workforce Intelligence, 2017, March

[22] Ukuhor HO, Hirst J, Closs SJ, Montelpere WJ. A framework for describing the influence of service organisation and delivery on participation in fetal anomaly screening in England. J Pregnancy 2017;2017:1–13. 4975091.

[23] Bramwell R, West H, Salmon P. Health professionals' and service users' interpretation of screening test results: experimental study. BMJ 2006;333(7562):284.

[24] Janvier A, Couture E, Deschenes M, Nadeau S, Barrington K, Lantos J. Health care professionals' attitudes about pregnancy termination for different fetal anomalies. Paediatr Child Health 2012;17(8):86–8.

[25] Farsides B, Williams C, Alderson P. Aiming towards "moral equilibrium": health care professionals' views on working within the morally contested field of antenatal screening. Med Ethics 2004;30:505–9.

[26] Lau JYC, Yi H, Ahmed S. Decision-making for non-invasive prenatal testing for down syndrome: Hong Kong Chinese women's preferences for individual vs relational autonomy. Clin Genet 2016;89(5):550–6.

[27] Vanstone M, Kinsella EA, Nisker J. Information-sharing to promote informed choice in prenatal screening in the spirit of the SOGC clinical practice guideline: a proposal for an alternative model. J Obstet Gynaecol Can 2012;34(3):269–75.

[28] Minear MA, Lewis C, Pradhan S, Chandrasekharan S. Global perspectives on clinical adoption of cfDNA NIPT. Prenat Diagn 2015;35(10):959–67.

[29] UK National Screening Committee. The UK NSC recommendation on fetal anomaly screening in pregnancy, https://legacyscreening.phe.org.uk/fetalanomalies; 2016 [accessed 10.11.17].

[30] Van Opstal D, Srebniak MI. Cytogenetic confirmation of a positive cfDNA NIPT result: evidence-based choice between chorionic villus sampling and amniocentesis depending on chromosome aberration. Expert Rev Mol Diagn 2016;16(5):513–20.

[31] Fisher J, Lohr PA, Lafarge C, Robson SC. Termination for fetal anomaly: are women in England given a choice of method? J Obstet Gynaecol 2014;24:1–5.

[32] Abdel Haleem MAS. Medical ethics in Islam. In: Grubb A, editor. Choices and decisions in healthcare. Chichester, West Sussex: Wiley; 1993. p. 1–20.

[33] Lutgendorf MA, Stoll KA, Knutzen DM, Foglia LM. Noninvasive prenatal testing: limitations and unanswered questions. Genet Med 2013;16(4):281–5.

[34] Chiang H-H, Chao Y-M, Yuh Y-S. Informed choice of pregnant women in prenatal screening tests for Down's syndrome. J Med Ethics 2006;32(5):273–7.

[35] van den Heuvel A, Chitty L, Dormandy E, Newson A, Deans Z, Attwood S, Haynes S, Marteau TM. Will the introduction of non-invasive prenatal diagnostic testing erode informed choices? An experimental study of health care professionals. Patient Educ Couns 2010;78(1):24–8.

[36] Skotko BG. With new prenatal testing, will babies with down syndrome slowly disappear? Arch Dis Child 2009;94:823–6.

[37] Hill M, Barrett A, Choolani M, Lewis C, Fisher J, Chitty LS. Has non-invasive prenatal testing impacted termination of pregnancy and live birth rates of infants with down syndrome? Prenat Diagn 2017;.

[38] https://phescreening.blog.gov.uk/2017/07/18/fetal-anomaly-screening-providers-can-sign-up-now-for-nipt-training/ [accessed 10.10.17].

[39] Lavigne J, Sharr C, Ozonoff A, Prock LA, Baumer N, Brasington C, Cannon S, Crissman B, Davidson E, Florez JC, Kishnani P. National down syndrome patient database: insights from the development of a multi-center registry study. Am J Med Genet A 2015;167(11):2520–6.

[40] Ahmed S, Hewison J, Bryant L. 'Balance' is in the eye of the beholder: providing information to support informed choices in antenatal screening via antenatal screening web resource. Health Expect 2008;10(4):309–20.

[41] http://understandingdownsyndrome.org/ [accessed 10.11.17].

[42] Figueiras M, Price H, Marteau TM. Effects of textual and pictorial information upon perceptions of down syndrome: an analogue study. Psychol Health 1999;14:761–71.

[43] Saxton M. Why members of the disability community oppose prenatal diagnosis and selective abortion. In: - Parens E, Asch A, editors. Prenatal testing and disability rights. Washington: Georgetown University Press; 2000.

[44] Klein DA. Medical disparagement of the disability experience: empirical evidence for the "Expressivist objection" AJOB Prim Res 2011;2(2):8.

[45] http://www.savingdownsyndrome.org/ [accessed 10.11.17].

[46] Shakespeare T. Disability rights and wrongs revisited. Routledge; 2013.

[47] https://www.nipt-biomnis.com/ordering-the-kit/ [accessed 10.11.17].

[48] Lewis C, Choudhury M, Chitty LS. 'Hope for safe prenatal gene tests'. A content analysis of how the UK press media are reporting advances in non-invasive prenatal testing. Prenat Diagn 2015;35(5):420–7.

[49] Skirton H, Goldsmith L, Jackson L, Lewis C, Chitty LS. Non-invasive prenatal testing for aneuploidy: a systematic review of internet advertising to potential users by commercial companies and private health providers. Prenat Diagn 2015;35(12):1167–75.

[50] Nuffield Council on Bioethics. Non-invasive prenatal testing: ethical issues. Report of a Working Group. Nuffield Council on Bioethics; 2017, March.

[51] Wapner RJ, Babiarz JE, Levy B, et al. Expanding the scope of noninvasive prenatal testing: detection of fetal microdeletion syndromes. Am J Obstet Gynecol 2015;212:332.e1–9.

[52] Wong FC, Lo DYM. Prenatal diagnosis innovation: genome sequencing of maternal plasma. Annu Rev Med 2016;67:419–32.

[53] The American College of Obstetricians and Gynecologists. Cell-free DNA screening for fetal aneuploidy. Committee Opinion no. 640. Obstet Gynecol 2015;126:e31–7.

[54] https://www.sequenom.com/tests/reproductive-health/maternit-genome [accessed 10.11.17].

[55] https://decisionaid.ohri.ca/ [accessed 10.11.17].

[56] Stacey D, Légaré F, Lewis K, Barry MJ, Bennett CL, Eden KB, Holmes-Rovner M, Llewellyn-Thomas H, Lyddiatt A, Thomson R, Trevena L. Decision aids for people facing health treatment or screening decisions. Cochrane Database Syst Rev 2017;4:CD001431.

[57] Beulen L, Van Den Berg M, Faas BH, Feenstra I, Hageman M, Van Vugt JM, Bekker MN. The effect of a decision aid on informed decision-making in the era of non-invasive prenatal testing: a randomised controlled trial. Eur J Hum Genet 2016;24(10):1409–16.

[58] Vlemmix F, Warendorf J, Rosman A, Kok M, Mol B, Morris J, Nassar N. Decision aids to improve informed decision-making in pregnancy care: a systematic review. BJOG 2013;120:257–66.

[59] Lépine J, Portocarrero ME, Delanoë A, Robitaille H, Lévesque I, Rousseau F, Wilson BJ, Giguère AM, Légaré F. What factors influence health professionals to use decision aids for down syndrome prenatal screening? BMC Pregnancy Childbirth 2016;16(1):262.

[60] Yee LM, Wolf M, Mullen R, Bergeron AR, Cooper Bailey S, Levine R, Grobman WA. A randomized trial of a prenatal genetic testing interactive computerized information aid. Prenat Diagn 2014;34(6):552–7.

[61] Portocarrero ME, Giguère AM, Lépine J, Garvelink MM, Robitaille H, Delanoë A, Lévesque I, Wilson BJ, Rousseau F, Légaré F. Use of a patient decision aid for prenatal screening for down syndrome: what do pregnant women say? BMC Pregnancy Childbirth 2017;17(1):90.

[62] Delanoë A, Lépine J, Portocarrero ME, Robitaille H, Turcotte S, Lévesque I, Wilson BJ, Giguère AM, Légaré F. Health literacy in pregnant women facing prenatal screening may explain their intention to use a patient decision aid: a short report. BMC Res Notes 2016;9(1):339.

[63] https://geneticsupportfoundation.org/videos [accessed 10.11.17].

[64] Allyse M, Sayres LC, Goodspeed TC, Cho MK. Attitudes towards non-invasive prenatal testing for aneuploidy among US adults of reproductive age. J Perinatol 2014;34(6):429–34.

[65] Oepkes D, Yaron Y, Kozlowski P, Rego de Sousa MJ, Bartha JL, van den Akker ES, Dornan SM, Krampl-Bettelheim E, Schmid M, Wielgos M, Cirigliano V, di Renzo GC, Cameron A, Calda P, Tabor A. Counseling for non-invasive prenatal testing (cfDNA NIPT): what pregnant women may want to know. Ultrasound Obstet Gynecol 2014;44:1–5.

[66] Ngan OMY, et al. Obstetric professionals' perceptions of non-invasive prenatal testing for down syndrome: clinical usefulness compared with existing tests and ethical implications. BMC Pregnancy Childbirth 2017;17:285.

[67] Rantanen E, Hietala M, Kristoffersson U, Nippert I, Schmidtke J, Sequeiros J, Kääriäinen H. What is ideal genetic counselling? A survey of current international guidelines. Eur J Hum Genet 2008;16(4):445–52.

[68] Farrelly E, Cho MK, Erby L, Roter D, Stenzel A, Ormond K. Genetic counseling for prenatal testing: where is the discussion about disability? J Genet Couns 2012;21(6):814–24.

[69] Fisher J, Lafarge C. Women's experience of care when undergoing termination for fetal anomaly in England. J Reprod Infant Psychol 2015;30(5):69–87.

[70] Sullivan A, Kean L, Cryer A. Pregnancy loss, breaking bad news and supporting parents. In: Crier A, Kean L, Sullivan A, editors. Midwives' handbook of antenatal investigations. Elsevier; 2006. p. 31–41 [Chapter 3].

[71] www.arc-uk.org [accessed 10.11.17].

[72] http://www.arc-uk.org/for-professionals [accessed 10.11.17].

CELL-FREE DNA-BASED NONINVASIVE PRENATAL TESTING AND SOCIETY

Carla van El, Lidewij Henneman
VU University Medical Center, Amsterdam, The Netherlands

INTRODUCTION

CELL-FREE DNA-BASED NIPT: A NEW TEST IN AN OLD DEBATE

Discussions on ethical and social issues have accompanied the emergence of prenatal testing and screening since the 1970s. Developments in prenatal diagnosis and screening as well as the ethical and social aspects involved can be traced back in many excellent academic and nonacademic publications. Rather than aiming at an exhaustive overview, in this chapter we will first briefly highlight some key developments underlying current prenatal testing and screening and the debate on ethical and social issues involved, then discuss to what extent cell-free DNA-based noninvasive prenatal testing (cfDNA NIPT) differs from earlier diagnostic and screening tests, and how that influences the debate on social and ethical issues. Finally, we will address the challenges for current policies to organize responsible prenatal testing and screening services with cfDNA NIPT.

PRENATAL DIAGNOSIS AND SCREENING FOR FETAL ANOMALIES

Genetic testing emerged in the 1970s allowing families in which children had been born with severe congenital disorders to make reproductive decisions [1]. Until that time the only option for couples to avoid having another affected child was to refrain from having further children. Via invasive procedures, amniocentesis, and later chorionic villus sampling, couples who had experienced the impact of genetic disorders from close by could find out whether the fetus was affected. For instance, in case of neural tube defects (NTDs) it was shown in the early 1970s that elevated alpha-fetoprotein (AFP) levels could be detected in amniotic fluid of mothers carrying a child with a NTD such as spina bifida [2]. As a result, in many hospitals, amniocentesis was promptly offered to women who had had a child with NTD in an earlier pregnancy or in whose family NTDs had occurred, though this was not considered justifiable for the general obstetric population because of the costs and procedure-related miscarriage risk [3]. In these years invasive testing was also used to identify Down syndrome pregnancies by means of karyotyping, and it was offered to women at higher risk for carrying a fetus with Down syndrome because of their age or obstetric history [4,5].

Noninvasive Prenatal Testing (NIPT). https://doi.org/10.1016/B978-0-12-814189-2.00014-1

Options for screening all pregnant women for NTD and not just high-risk pregnancies emerged when it became clear that elevated AFP levels could also be detected in maternal serum [6]. Serum screening combined with ultrasound aided in detecting anomalies [7], and large-scale studies were performed to assess whether the introduction of a screening program for all pregnant women would be feasible [8,9].

When, in the 1980s, it was shown that low levels of AFP were related to fetal chromosomal abnormalities [10], also the possibility of serum screening for Down syndrome based on AFP and other markers was studied. Test accuracy was relatively low and proved closely related to maternal age [11]. In subsequent years accuracy was improved by adding and refining markers, and including the use of ultrasound to obtain a more accurate gestational age [12]. However, the false-positive rate of serum screening remained problematic, causing many women to undergo unnecessary invasive testing with a small risk of procedure-related miscarriage (i.e., 0.5%–1%) [13]. In the 1990s, in many European countries and the United States, serum screening for Down syndrome became available, but several countries restricted offering the test to women over 36 or 38 years of age who otherwise would be directly eligible for amniocentesis [4]. In this context serum screening was used as a risk selection instrument to avoid unnecessary invasive testing.

SOCIAL AND ETHICAL ISSUES

In the 1970s, patient groups started to share experiences and information on living with a disorder to improve support and care [14] (www.nads.org/about-us/history-of-nads/). The possibility of prenatal genetic testing soon was also discussed and was welcomed by some families, though rejected by others [15]. Parents might opt to terminate a pregnancy in case of a serious disorder, for instance, because they wish to avoid the suffering of the future child or think the burden would be too high for them. In the media and scholarly publications, accounts can be found expressing understanding for parents choosing such new reproductive options. At the same time concerns and fears were voiced about testing for more disorders and becoming available for all pregnant women [16]. Comparisons with eugenic aspirations of the first half of the 20th century were made, and fears of "playing God" were expressed, as if lives of people with disabilities would not be worth living [17]. Other scholars and health care professionals pointed toward positive aspects of being allowed individual choices and denied any connotation with the negative past eugenic practices, arguing that there was no official obligation or pressure to test and abort affected fetuses [18]. Others pointed to more subtle and yet pervasive mechanisms of pressure, implying eugenics to enter again via the "back door" [19]. Parents might feel that they should accept an offer of screening, either by social pressure [20], or by the mere fact that screening is offered and in some countries seemed to have become a routine part of prenatal care.

In the early clinical literature on prenatal testing, prenatal tests to assess whether the unborn child had a congenital genetic or structural abnormality were described as a form of *prevention* [7,21] However, it was argued that reducing the prevalence of a condition through selective abortion could not be subsumed under this heading [17,22]. Rather than "prevention," in the years that followed "reproductive choice" became more standard terminology. For screening programs not the uptake or reduction of live births of a certain disorder, but the degree to which parents were able to make an informed reproductive choice was chosen as the measure of success [23].

Efforts were made to improve the process of autonomous informed decision-making and organize screening in such a way to avoid that prenatal screening would be a routine element of normal prenatal care. In this regard, it is crucial to distinguish between prenatal screening for severe fetal abnormalities, on the one hand, and screening tests that aim at early detection of pregnancy-related problems to lower perinatal and maternal morbidity and mortality, on the other hand. Assessment of risk in pregnancy includes obstetric risk factors (e.g., prior preterm birth delivery), medical conditions in the mother (e.g., anemia, diabetes), substance use (e.g., smoking, alcohol use), prenatal screening for infectious diseases (e.g., hepatitis B, rubella), and maternal—and in some countries noninvasive fetal [24]—blood group typing [25]. In this context the aim of screening is unquestionable, it is preferably offered to and accepted by all pregnant women as it is to prevent health problems in mother and child through timely interventions. In prenatal genetic screening for fetal abnormalities that cannot be cured or substantially alleviated, however, the mother's only current option is to decide whether to accept the child's impairment or to terminate the pregnancy. Consequently, the main argument for offering prenatal genetic screening is to enhance the reproductive autonomy of the pregnant woman [26].

In the discussion on prenatal screening some raised fundamental objections against abortion, while others questioned whether the severity of a disorder, such as Down syndrome, justified screening and selective termination [27]. Some argued that the severity of Down syndrome was exaggerated and that the public and health professionals should be educated more about the disorder [28]. In public and ethical discussions on prenatal screening this became a focal point of attention. Patient groups sought media attention influencing public opinion on having a child with Down syndrome. This was fueled after the death of "baby Doe" in 1982. In this case, treatment for a correctable gastrointestinal birth defect was withheld in a newborn with Down syndrome, and a surgeon began a campaign to prevent such discrimination against children with disabilities [29]. The disability rights movement stressed that people living with disability can also have a meaningful life, and it was contended that the offer of prenatal testing suggests that disability is bad, which could be regarded as discriminatory [30]. In addition, it was stressed that prospective parents should be better informed about living with disabilities to be able to make a truly informed choice [31].

Studies started to focus on the psychological burden and moral implications of having to take a decision on prenatal testing and on experiences of women and parents [32]. Criticism was voiced about genetic screening being another example of unnecessary medicalization of childbirth, and as a way for society to control women's bodies for reproduction [33].

Much of the social and ethical aspects of prenatal testing as described previously have been discussed and documented in Northern American and European countries. Countries that adopted prenatal screening later used existing practices as examples. However, ideas on abortion and the ethical dimensions of prenatal testing for fetal abnormalities vary between and within countries [34,35]. Public perceptions and acceptance of disability vary and may also influence variation in uptake of prenatal screening [36]. In Israel, for example, prenatal screening meets with great public support [37]. In other countries testing would not be an option because of the prohibition on abortion [35,38]. Cultural motives have been shown to influence also nonmedical uses of prenatal screening [39]. Most notably in India and China sex selection against girls has led to an unbalanced gender ratio [40]. Whereas in western countries and bioethical literature a strong emphasis is put on genetic testing as a decision of the individual or the couple, in many parts of the world such decisions may involve the wider family [35]. In addition to culture or religious motives, financial circumstances can influence the uptake of prenatal screening. In some countries women do not have access to testing because they cannot afford to take a test or the country has few facilities for such testing [35,38].

Though the offer of screening for fetal anomalies in some countries may have become more or less routine, having the test is not meant to be. Decisions to have prenatal screening or not should be informed by knowledge on the test and the nature of the disorder screened for, and based on the personal values and the circumstances of the future mother or future parents [41].

CELL-FREE DNA NIPT: THE ULTIMATE PRENATAL TEST?

CELL-FREE DNA NIPT: WHAT IS NEW AND HOW DOES THAT IMPACT SOCIAL AND ETHICAL ISSUES?

Early academic papers on cfDNA NIPT focussed on high-risk pregnancies [42]. Research using cell-free DNA (cfDNA) in maternal plasma showed a high detection rate and a low false-positive rate for fetal aneuploidy [43], and all that was needed was a maternal blood sample. Accurate test results could now be obtained safely since there was no miscarriage risk, unlike other forms of prenatal testing, requiring chorionic villus sampling or amniocentesis. Health care professionals, especially obstetricians having experienced women suffering an iatrogenic miscarriage after invasive prenatal diagnosis, felt a noninvasive test as a great relief [44].

Another positive aspect of cfDNA NIPT was that testing can be performed early in pregnancy, which may have less impact on women because of a less-intense emotional bond between mother and unborn child, and because it allows for safer termination of pregnancy. Because cfDNA NIPT requires only a blood draw, testing is also more accessible in countries with less resources and trained personnel for invasive testing [45].

Whereas initially cfDNA NIPT's accuracy was based on, and viewed to be most appropriate in, high-risk pregnancies [46], both researchers and health care professionals were eager to ascertain whether cfDNA NIPT would have similar accuracy in low-risk pregnancies. While large-scale clinical studies comparing cfDNA NIPT to conventional screening methods were only to confirm high accuracy in the general obstetric population after 2014 [47], a huge interest in cfDNA NIPT became evident both in academia, health care, commercial parties, and the public domain already after its introduction in 2011in China and the United States [45]. The expectation of an accurate and safe test helped fast introduction in prenatal care [44]. Commercial parties started offering cfDNA NIPT irrespective of maternal age and this accelerated the introduction of cfDNA NIPT across the globe [45], despite official statements recommending against its use in low-risk groups because of lack of evidence [46]. The positive test characteristics of the test soon became known by many women who shared this information and their experiences on the Internet, leading to a growing demand from pregnant women and health care professionals.

Content analyses of media reports in the United Kingdom until mid-2014 showed a predominantly positive view about cfDNA NIPT, focusing on the accuracy and safety of testing [48]. The rapidly growing demand put pressure on the health care systems of various countries. In countries where cfDNA NIPT was not available or not allowed to be offered because of legislation, for example, in the Netherlands, this led to "prenatal tourism" to other countries. In Europe, restrictive national legislations and lack of harmonization of health care services may lead to such reproductive tourism as has also been reported for assisted reproductive technologies [49].

The rapid introduction of cfDNA NIPT also raised concerns. One concern was that more women who otherwise would not have chosen for first trimester combined testing (FCT), because of its low accuracy (only a risk estimation) and relatively high risk of being referred for invasive testing, would now opt for screening. Furthermore, earlier testing might lead to trivialization of selective abortions [41]. "Routinization" of prenatal screening for fetal anomalies might be imminent, in the sense that such screening would be seen as a standard and required element of prenatal care and if more women would use prenatal screening an increase in termination of Down syndrome pregnancies might be expected [50,51]. Moreover, a reduction in the number of invasive procedures because of cfDNA NIPT might imply that the experiences of clinicians who perform these procedures would decline, potentially increasing the risk of procedure-associated miscarriages [45].

The involvement of commercial companies also raised concern. Most notably questions on whether adequate information and pretest counselling would be provided, and the high costs of testing were mentioned [45,52]. Another fear concerned the possible use cfDNA NIPT for sex-selection. Parents in countries where officially returning information about fetal sex to parents is not allowed might find ways to use online services or resort to "sex-selection tourism" to other countries [53].

EXPECTATIONS OF CELL-FREE DNA NIPT: (UN)CERTAINTY, THE ULTIMATE DIAGNOSTIC TEST?

Already in the early 1990s the question was raised whether noninvasive methods for prenatal testing will make women's decisions to have prenatal diagnosis more or less complicated [54]. The introduction of cfDNA NIPT in 2011, quickly spreading to countries around the world, created a paradigm shift in prenatal care [55]. In the early days of testing some even suggested that cfDNA NIPT could eventually be used in a one-step diagnostic approach [41]. It was expected that a new test had become available to diagnose, as opposed to predict, the presence of fetal trisomies. Although cognitively this may be less demanding for women [56], this one-step diagnostic procedure was actually seen as problematic, especially when offered as first-line screening test in prenatal screening programs. It was argued that because there would only be one contact moment between the pregnant woman and the health professional, this could potentially undermine autonomous decision-making, and cfDNA NIPT could easily be perceived as "an offer you cannot refuse" [57]. In addition, Hewison [56] argued that non-invasive *diagnosis* could make decision-making more emotionally burdensome, especially when more conditions are involved, by removing a so-called psychological shelter: previously people had been protected from having to make decisions because the rationale for not having a test—avoidance of miscarriage risk—had simplified decision-making for many women.

The earlier presumed one-step diagnostic test procedure appeared to be an unrealistic scenario. For biological reasons—fetal cfDNA being derived from the placenta—confirmatory testing should always be advised to confirm the diagnosis. It is however known from clinical studies that this is not always acted upon. For example, in a large study in the United States in 2013 6.2% of the women with positive (unfavorable) cfDNA NIPT results terminated the pregnancy without confirmation [58]. Since some health professionals still consider cfDNA NIPT a diagnostic test, education of those offering cfDNA NIPT is crucial [59]. To emphasize that cfDNA NIPT is not diagnostic some started to refer to the test as NIPS (noninvasive prenatal *screening*). Counseling women with positive results to wait before taking action until their diagnosis is confirmed is of utmost importance, and it is even more important in the general obstetric population. This is because of the lower positive predictive value (PPV) in low-risk populations given the lower prevalence rates of aneuploidy compared to high-risk groups [60]. The concept of PPV, indicating the probability that a *positive* test result is true positive, is difficult to understand for most women, but also health professionals and scholars have difficulty understanding what it means. In addition, most websites providing information on cfDNA NIPT seem to focus on the test accuracy (detection rate) and not on the possibility of a false-positive result [61]. In Belgium, lack of attention for the PPV of cfDNA NIPT led to a critical debate in 2017. cfDNA NIPT is an almost free test offered by genetics centers and laboratories in Belgium: in 2017 pregnant women covered by the public health insurance only paid ~€ 8. The test was voted "Product of the year" in 2017 by readers of the national newspaper *De Standaard* (www.standaard.be/cnt/dmf20180101_03275535). Academic researchers explained that the sensitivity and specificity communicated in the media are not the same as PPV, and that the test incorrectly had been portrayed as an almost diagnostic test (https://www.demaakbaremens.org/is-hype-rond-nipt-terecht/). To understand implications of testing, the PPV for the population a woman belongs to should be considered [62].

SHIFTING AIMS OF PRENATAL SCREENING
BEING PREPARED FOR THE BIRTH OF A HANDICAPPED CHILD

Prenatal screening for fetal anomalies aims to provide autonomous reproductive choices to the pregnant couple including termination of pregnancy, preparing for having a child with a condition and, in some cases, fetal therapy. Fetal therapy offers an intervention before birth for the purpose of correcting, treating, or diminishing the deleterious effects of a fetal condition. Though the idea may be appealing, fetal therapy is still in its infancy [63] (see Chapter 20). Most parents decide to terminate the pregnancy of a child with a congenital and/or genetic condition after the diagnosis is confirmed but there are differences between countries [5,38,64]. Some current studies on decisions after cfDNA NIPT testing, however, mentioned that parents increasingly choose to use the test results to prepare for a child with a condition and not necessarily for decision-making about termination of pregnancy [65,66]. Thus while the uptake of screening may increase with cfDNA NIPT, either as contingent test or as a first-line test, the impact of the introduction of this new technology on the live birth prevalence of Down syndrome may be less linear than expected.

VISIONS OF FETAL MEDICINE

The results of prenatal testing, including ultrasound, may also be useful for pregnancy and childbirth management and enable the best possible start for the affected child. In this context the use of cfDNA NIPT for "fetal personalized medicine" has been discussed [67]. Recently, studies have started to find out whether administration of substances, such as choline and fluoxetine, to mothers pregnant of a fetus with Down syndrome might be beneficial to brain development and cognitive abilities [68]. If such interventions would be successful, prenatal diagnosis of Down syndrome would have a new rationale besides reproductive choice or preparing for the birth of child with Down syndrome. In such cases the fetus would be regarded as a patient, which would raise new ethical questions. If a treatment would be safe and effective it has been argued future parents might be morally obliged to use prenatal testing to promote the health of their future child [69]. Similarly, new neonatal treatment options for serious disorders might change our views on prenatal screening.

EXPANDING SCOPE

Since the entire fetal genome is represented in the maternal plasma [70,71], it is possible to use cfDNA NIPT to screen and diagnose disorders beyond aneuploidy. Noninvasive prenatal *diagnosis* (NIPD) can increasingly be used for the detection or exclusion of several single-gene de novo, dominant, or recessive conditions because of a known family history (e.g., cystic fibrosis) or ultrasound abnormalities (e.g., achondroplasia) [72].

In fetal aneuploidy screening, when sequencing and analysis are performed in a nontargeted way, and depending on resolution, other chromosome aberrations than the common aneuploidies can be detected [73,74]. This may identify other fetal aberrations, pregnancies at risk due to placental insufficiency, and in rare cases maternal aberrations including signs of maternal cancer [75]. With cfDNA NIPT, an increasing number of test providers include (optional) screening for rare microdeletion syndromes, large duplications, and sex chromosomal aneuploidies. It is expected that cfDNA and cfRNA NIPT will also be used to detect feto-maternal risk factors, for example, for preeclampsia and preterm birth [76,77].

Due to technology-driven improvements in DNA sequence analysis and decreasing costs it is thus very likely that cfDNA NIPT/D will lead to major changes in the landscape of prenatal care. In the long run screening may also become available for many other (non)treatable conditions ranging from congenital lethal disorders, serious nonlethal cognitive and mental disorders, to milder genetic conditions, and for feto-maternal risk factors. In light of these future developments it becomes increasingly important to consider "what to offer" and "how to offer" [77], and who decides what to include, not only from a medical perspective but also from an ethical perspective. Survey research has shown that prenatal screening for mild disorders, gender for nonmedical purposes, and nonmedical traits (e.g., height) generally receives little support [78–80]. Some studies, however, showed that about one-third of both health professionals [79] and pregnant women [81] are positive about fetal screening for late-onset disorders such as predisposition for hereditary cancer. In case of a continuing pregnancy, which will most often be the case, this information might entail an infringement of the child's autonomy, of its "right to an open future" [41,82]. It is therefore widely endorsed that predictive genetic testing for late-onset conditions in minors has to be discouraged [83].

cfDNA NIPT may increasingly be used to detect both untreatable and treatable or actionable conditions and therefore the distinction between parental "autonomy," the traditional goal of prenatal screening, and "prevention," may become blurred [84]. The comparable tension between prenatal ultrasound for pregnancy monitoring and the detection of congenital anomalies has not led to much ethical debate. Dondorp et al. [84] have argued that this is probably because ultrasound is routinely seen as a pregnancy monitoring tool and for future parents an opportunity to see their child. The challenge will be to give pregnant women the amount of information needed for taking a well-informed decision, including the aim(s) of screening, and at the same time avoiding "information overload" [41]. The uncertainty about health implications of certain findings may further complicate counseling. As more conditions are added, more false-positive results and additional findings will be seen leading to an increase in invasive procedures for confirming findings that, because of their low prevalence, have a low PPV [26].

When broadening the scope, one solution could be to group conditions into (optional) categories containing disorders similar in type and severity, and to develop an approach based on "generic" consent [41,85]. It is unclear whether this is feasible and desirable, also because individual variation in preferences and cultural differences of what to include in those packages will remain [56]. Pregnant women themselves seem, in principle, in favor of "having a free choice" of what conditions are included in the test, but also acknowledge that the decision-making process might become too complex and might overburden women [81]. Expanding the scope of prenatal screening by cfDNA NIPT thus demands reflection on the ethical, psychological, social, and legal consequences.

RESPONSIBLE IMPLEMENTATION OF CELL-FREE DNA NIPT
AVAILABILITY AND ACCESS ACROSS HEALTH CARE SYSTEMS

cfDNA NIPT's great potential and advantages have made it a highly requested test offered in many countries, with prices rapidly decreasing [45,52]. Worldwide the availability of cfDNA NIPT varies depending on the legal and health care system and prenatal care services. Tests can be offered as part of well-organized public health programs, by obstetric health professionals as part of their policy to adhere to professional guidelines, or by private providers. Industry has been a major driver in the development and introduction [45,52,86]. In many countries, cfDNA NIPT is almost exclusively available through commercial laboratories [52]. Some of these tests, for example, for fetal sex determination, are offered online on a direct-to-consumer basis outside the traditional health care system, obviating the requirements for informed consent [87]. Gekas et al. [88] stated that "direct marketing to patients and end-users may have facilitated the early introduction of cfDNA screening into clinical practice despite limited evidence-based independent research data supporting this rapid shift." Another concern the authors vowel is the aggressive marketing methods to push its use by women [88]. Test prices of cfDNA NIPT greatly vary between, but also within, countries which creates inequality of access [52]. It has been argued that the financial barriers to access cfDNA NIPT are unfair discrimination and weaken reproductive autonomy [89].

cfDNA NIPT has been integrated in screening programs for Down syndrome in several countries. Implementation in publicly funded health care systems is mainly determined by costs and affordability. In countries where screening is paid for from public funds, in principle all pregnant women have equal access but funding requires justification in the context of total health care spending. Among others due to automation and multiplexing, the costs of cfDNA NIPT have decreased substantially. In 2016, based on an implementation study (RAPID study) involving eight different hospitals across the country in a National Health Service (NHS) setting [66], the UK National Screening Committee recommended to evaluate the introduction of cfDNA NIPT in the existing screening program (https://legacyscreening. phe.org.uk/fetalanomalies). This was approved by the UK government by the end of that year. cfDNA NIPT will become available in 2018 for women with a high-risk screening result. With data from the RAPID study and national data it was calculated that cfDNA NIPT will improve the performance of the program without increasing costs if offered as a contingent screening test at a risk threshold of 1:150 based on first trimester screening for Down, Edward, or Patau syndrome [66]. Other initiatives exploring or advising the adoption of cfDNA NIPT in publicly funded services are seen in New Zealand [90], France (www.has-sante.fr/portail/upload/docs/application/pdf/2017-05/dir42/recommandation_en_sante_ publique__place_des_tests_adn_libre_circulant_dans_le_sang_maternel_dans_le_depistage_ de_la_trisomie.pdf), Italy (www.salute.gov.it/imgs/C_17_pubblicazioni_2438_allegato.pdf), and Germany. The latter country is currently performing a Health Technology Assessment analysis of cfDNA NIPT with a decision expected in 2019 (www.g-ba.de/institution/presse/

pressemitteilungen/668/#n). In Denmark, cfDNA NIPT is being offered as contingent screening test in the publicly funded national prenatal screening program since 2017 [5]. Also in two provinces in Canada (British Columbia and Ontario) [91] and in Switzerland (Avis d'experts No 52 1.1.2018), cfDNA NIPT as a contingent test has been implemented. In Switzerland women with a risk according to the first trimester combination test above 1:1000 have access to cfDNA NIPT.

Although cfDNA NIPT as a contingent or second-line screening test reduces the number of women who need invasive testing, it cannot detect the cases missed by first trimester serum screening or combined test screening (FCT), as the first screening test will always be the "weakest link." For national prenatal screening programs, it would thus make sense to consider cfDNA NIPT as first-line test. In most countries, the costs are considered too high to implement cfDNA NIPT in the public sector.

In the Netherlands, cfDNA NIPT has been implemented as part of the TRIDENT studies. In 2011, the Dutch NIPT Consortium was installed representing all institutions, organizations, and stakeholders involved in cfDNA prenatal testing [44]. In 2014 the Consortium obtained a ministerial license to offer cfDNA NIPT nationwide to women at high risk ($\geq$1:200) for fetal aneuploidy based on the FCT or medical history [92]. In 2017 the Consortium obtained another 3-year license for an evaluative study where cfDNA NIPT is offered as a first-line screening test to all pregnant women irrespective of their risk. Women are still able to choose for FCT first and if needed cfDNA NIPT as a contingent screening test. Women are counselled by trained and certified counsellors, mostly midwives [93]. A first-tier cfDNA NIPT costs women € 175, which is almost similar to the costs for the FCT ($\sim$€ 168). cfDNA NIPT is performed by three Dutch university clinical genetic laboratories using an in-house validated test [92]. Women can choose to have all autosomes analyzed or chromosomes 21, 13, and 18 only. Sex chromosomal abnormalities (and fetal sex) are not analyzed. It is expected that the TRIDENT studies will provide all the results needed for the final decision on the introduction of cfDNA NIPT in the Dutch National prenatal screening program.

Different scenarios of implementing cfDNA-based prenatal screening are possible which, as is stated by Dondorp et al. [26], "should not just be regarded as a matter of screening technology and health economics; the question is also how these […] enable or impede meaningful reproductive choices and how they affect both the balance of benefits and burdens for pregnant women and their partners, and the screening goals and values acceptable to society." Since preferences and opinions among women and health professionals differ between countries, approaches to implementation should be country specific and strategies and needs should be explored before implementation [94]. The Dutch initiative shows that implementing advanced technologies such as cfDNA NIPT benefits from a learning phase involving all stakeholders where pregnant women can be offered cfDNA NIPT, while taking time to evaluate and fine-tune its offer [44].

PUBLIC CONCERNS AND CHALLENGES TO AUTONOMOUS CHOICE

Many stakeholders including pregnant women, health care professionals, and Councils on Bioethics such as the Nuffield Council [53] brought up points to consider while implementing cfDNA NIPT. As described in a previous section of this chapter concerns around its uncritical use or routinization were raised [50,79,95]. Since cfDNA NIPT only entails sampling a tube of blood, it may be hard to distinguish this test from other blood tests during pregnancy, and women may feel unable to refuse an "easy" test. Public concerns about expanding the test "menu," especially when including less serious medical conditions, are that it might lead to further trivialization of testing and termination of pregnancy, and discrimination against people with a disability [96,97].

From the perspective of parents of children with Down syndrome, fears were expressed that cfDNA NIPT might lead to reducing the acceptance and care and facilities for children with a disability such as Down syndrome [98,99]. The effect of such a scenario would be that women would not really have a choice anymore to turn down prenatal screening, resulting in more people deciding to screen. Some feared that this might even lead to an "eradication" of Down syndrome (www.bbc.com/news/magazine-37500189). Parents emphasized the importance of giving balanced and accurate information during prenatal counseling about living with a person with Down syndrome not only at birth but also later in life [98,99].

An important challenge for the implementation of cfDNA NIPT and any prenatal test for fetal anomaly is to safeguard women's autonomy and to ensure informed consent. Informed decision-making for prenatal screening is currently already a challenge. The relative ease of cfDNA NIPT and expanding scope might reinforce this challenge and might lead to a not well thought-through use of the test offer in practice. This calls for continuing education and training of health care professionals, as well as the development of online interactive tools [100], to support the decision-making process in a way that is valued by pregnant women and their partners. It is important to create and maintain space for discussion on prenatal testing to help parents and providers develop their ideas and values on this subject. One of the most important things is that our societies need to be supportive of patients with disabilities and their families. A decision about prenatal testing can only truly be a choice if the care for patients is guaranteed.

CONCLUSIONS

The rapid spread of cfDNA NIPT across the globe underscores the interest pregnant women and clinicians have in a good prenatal test. Stakeholders have focused on access to testing, the provision of balanced information on the conditions tested for and what living with a child with such a condition means, and guaranteeing informed decision-making. Future challenges for such informed decision-making will come with extending the scope of cfDNA based NIPT/D, and a blurring distinction between testing for reproductive choice vs testing for improving general pregnancy outcome. cfDNA NIPT is part of a changing landscape of reproductive medicine as knowledge on the genome and development of techniques to unravel it will improve our capacity for detecting and eventually treating disorders. Much will also depend on test developers and providers, health care professionals, and women and couples engaging in reflection and debate on the most responsible use of prenatal genetic testing.

REFERENCES

[1] Milunsky A, Littlefield JW, Kanfer JN, Kolodny EH, Shih VE, Atkins L. Prenatal genetic diagnosis. N Engl J Med 1970;283(25):1370–81.

[2] Brock DJH, Sutcliffe RG. Alpha-fetoprotein in the antenatal diagnosis of anencephaly and spina bifida. Lancet 1972;300(7770):197–9.

[3] Editorial. Screening for neural tube defects. J R Coll Gen Pract 1980;30:581–2.

[4] Wald NJ, Kennard A, Hackshaw A, McGuire A. Antenatal screening for Down's syndrome. J Med Screen 1997;4(4):181–246.

[5] Lou S, Petersen OB, Jørgensen FS, Lund ICB, Kjærgaard S, Danish Cytogenetic Central Registry Study Group, et al. National screening guidelines and developments in prenatal diagnoses and live births of Down syndrome in the period 1973-2016 in Denmark. Acta Obstet Gynecol Scand 2018;97:195–203.

[6] Wald NJ, Brock DJH, Bonnar J. Prenatal diagnosis of spina bifida and anencephaly by maternal serum-alpha-fetoprotein measurement: a controlled study. Lancet 1974;303(7861):765–7.

[7] Harris R, Read AP. New uncertainties in prenatal screening for neural tube defect. Br Med J 1981; 282(6274):1416–8.

[8] Wald, NJ, Cuckle H, Brock JH, Peto R, Polani PE, Woodford FP. Report of UK collaborative study on alpha-fetoprotein in relation to neural tube defects. Maternal serum-alpha-fetoprotein measurement in antenatal screening for anencephaly and spina bifida in early pregnancy. Lancet 1977;1(8026):1323–32.

[9] Roberts CJ, Elder GH, Laurence KM, Woodhead JS, Hibbard BM, Evans KT, et al. The efficacy of a serum screening service for neural-tube defects: the South Wales experience. Lancet 1983;321(8337):1315–8.

[10] Merkatz IR, Nitowsky HM, Macri JN, Johnson WE. An association between low maternal serum α-fetoprotein and fetal chromosomal abnormalities. Am J Obstet Gynecol 1984;148(7):886–94.

[11] Cuckle H, Wald NJ, Thompson SG. Estimating a woman's risk of having a pregnancy associated with Down's syndrome using her age and serum alpha-fetoprotein level. BJOG 1987;94:387–402.

[12] Wald NJ, Cuckle HS, Densem JW, Kennard A, Smith D. Maternal serum screening for Down's syndrome: the effect of routine ultrasound scan determination of gestational age and adjustment for maternal weight. BJOG 1992;99(2):144–9.

[13] Tabor A, Madsen M, Obel E, Philip J, Bang J, Gaard-Pedersen B. Randomised controlled trial of genetic amniocentesis in 4000 low risk women. Lancet 1986;327(8493):1287–93.

[14] Black RB, Weiss JO. Genetic support groups in the delivery of comprehensive genetic services. Am J Hum Genet 1989;45(4):647–54.

[15] Frets PG. The reproductive decision after genetic counseling. Thesis, Rotterdam: Erasmus University; 1990.

[16] Kolata G. Tests of fetuses rise sharply amid doubts. The New York Times, 22 September, 1987.

[17] van El C, Pieters T, Cornel M. Genetic screening and democracy: lessons from debating genetic screening criteria in the Netherlands. J Community Genet 2012;3(2):79–89.

[18] Paul DB. What was wrong with eugenics? Conflicting narratives and disputed interpretations. Sci & Edu 2014;23(2):259–71.

[19] Duster T. Backdoor to eugenics. New York: Routledge; 2003.

[20] Clarke A. Is non-directive genetic counselling possible? Lancet 1991;338(8773):998–1001.

[21] Stein Z, Susser M, Guterman A. Screening programme for prevention of Down's syndrome. Lancet 1973;301(7798):305–10.

[22] Grunberg F. Screening for chromosomal abnormalities. Letter to the editor. Lancet 1973;1:543.

[23] Marteau TM, Dormandy E, Michie S. A measure of informed choice. Health Expect 2001;4(2):99–108.

[24] Scheffer PG, van der Schoot CE, Page-Christiaens GC, de Haas M. Noninvasive fetal blood group genotyping of rhesus D, c, E and of K in alloimmunised pregnant women: evaluation of a 7-year clinical experience. BJOG 2011;118(11):1340–8.

[25] Kilpatrick SJ, Papile LA. Guidelines for perinatal care. 8th ed. American Academy of Pediatrics and American College of Obstetrics and Gynecology; 2017.

[26] Dondorp W, de Wert G, Bombard Y, Bianchi DW, Bergmann C, Borry P, et al. Non-invasive prenatal testing for aneuploidy and beyond: challenges of responsible innovation in prenatal screening. Eur J Hum Genet 2015;23(11):1438–50.

[27] Cooper M. Ethics for screening for Down's syndrome. BMJ 1991;303(6793):56.

[28] Steele MW. Screening for chromosomal abnormalities. Letter to the editor. Lancet 1973;1:542–3.

[29] White M. The end at the beginning. Ochsner J 2011;11:309–16.

[30] Asch A. Reproductive technology and disability. In: Cohen STN, editor. Reproductive laws for the 1990s contemporary issues in biomedicine, ethics, and society. Humana Press; 1989. pp. 69–124.

[31] Parens E, Asch A. Disability rights critique of prenatal genetic testing: reflections and recommendations. Ment Retard Dev Disabil Res Rev 2003;9(1):40–7.

[32] Rapp R. Testing women, testing the fetus: the social impact of amniocentesis in America. Routledge; 1999.

[33] Oakley A. The captured womb: a history of the medical care of pregnant women. New edition. Oxford: Blackwell Publisher; 1986.

[34] Hewison J, Green JM, Ahmed S, Cuckle HS, Hirst J, Hucknall C, et al. Attitudes to prenatal testing and termination of pregnancy for fetal abnormality: a comparison of white and Pakistani women in the UK. Prenat Diagn 2007;27(5):419–30.

[35] Mozersky J, Ravitsky V, Rapp R, Michie M, Chandrasekharan S, Allyse M. Toward an ethically sensitive implementation of noninvasive prenatal screening in the global context. Hastings Cent Rep 2017;47(2):41–9.

[36] Crombag NM, Vellinga YE, Kluijfhout SA, Bryant LD, Ward PA, Iedema-Kuiper R, et al. Explaining variation in Down's syndrome screening uptake: comparing the Netherlands with England and Denmark using documentary analysis and expert stakeholder interviews. BMC Health Serv Res 2014;14:437.

[37] Raz A. Important to test, important to support: attitudes toward disability rights and prenatal diagnosis among leaders of support groups for genetic disorders in Israel. Soc Sci Med 2004;59(9):1857–66.

[38] Loane M, Morris JK, Addor MC, Arriola L, Budd J, Doray B, et al. Twenty-year trends in the prevalence of Down syndrome and other trisomies in Europe: impact of maternal age and prenatal screening. Eur J Hum Genet 2013;21(1):27–33.

[39] Sleeboom-Faulkner. Frameworks of choice: predictive and genetic testing in Asia. Amsterdam University Press; 2010.

[40] Madan K, Breuning MH. Impact of prenatal technologies on the sex ratio in India: an overview. Genet Med 2014;16(6):425–32.

[41] de Jong A, Dondorp WJ, Frints SG, de Die-Smulders CE, de Wert GM. Advances in prenatal screening: the ethical dimension. Nat Rev Genet 2011;12(9):657–63.

[42] Chiu RW, Akolekar R, Zheng YW, Leung TY, Sun H, Chan KC, et al. Non-invasive prenatal assessment of trisomy 21 by multiplexed maternal plasma DNA sequencing: large scale validity study. BMJ 2011;c7401:342.

[43] Gil MM, Akolekar R, Quezada MS, Bregant B, Nicolaides KH. Analysis of cell-free DNA in maternal blood in screening for aneuploidies: meta-analysis. Fetal Diagn Ther 2014;35(3):156–73.

[44] van Schendel RV, van El CG, Pajkrt E, Henneman L, Cornel MC. Implementing non-invasive prenatal testing for aneuploidy in a national healthcare system: global challenges and national solutions. BMC Health Serv Res 2017;17(1):670.

[45] Allyse M, Minear MA, Berson E, Sridhar S, Rote M, Hung A, et al. Non-invasive prenatal testing: a review of international implementation and challenges. Int J Womens Health 2015;7:113–26.

[46] American College of Obstetricians and Gynecologists Committee on Genetics. Committe Opinion No.545: noninvasive prenatal testing for fetal aneuploidy. Obstet Gynecol 2012;120:1532–4.

[47] Gil MM, Quezada MS, Revello R, Akolekar R, Nicolaides KH. Analysis of cell-free DNA in maternal blood in screening for fetal aneuploidies: updated meta-analysis. Ultrasound Obstet Gynecol 2015;45(3):249–66.

[48] Lewis C, Choudhury M, Chitty LS. 'Hope for safe prenatal gene tests'. A content analysis of how the UK press media are reporting advances in non-invasive prenatal testing. Prenat Diagn 2015;35(5):420–7.

[49] Soini S, Ibarreta D, Anastasiadou V, Ayme S, Braga S, Cornel M, et al. The interface between assisted reproductive technologies and genetics: technical, social, ethical and legal issues. Eur J Hum Genet 2006;14(5):588–645.

[50] Lewis C, Hill M, Silcock C, Daley R, Chitty LS. Non-invasive prenatal testing for trisomy 21: a cross-sectional survey of service users' views and likely uptake. BJOG 2014;121(5):582–94.

[51] Vanstone M, King C, de Vrijer B, Nisker J. Non-invasive prenatal testing: ethics and policy considerations. J Obstet Gynaecol Can 2014;36(6):515–26.

[52] Minear MA, Lewis C, Pradhan S, Chandrasekharan S. Global perspectives on clinical adoption of NIPT. Prenat Diagn 2015;35(10):959–67.

[53] Nuffield Council on Bioethics. Non-invasive prenatal testing: ethical issues. London: Nuffield Council on Bioethics; 2017.

[54] Rothenberg KH, Thomson EJ. Women and prenatal testing: facing the challenges of genetic technology. Columbus: Ohio State University Press; 1994.

[55] Chitty LS, Bianchi DW. Noninvasive prenatal testing: the paradigm is shifting rapidly. Prenat Diagn 2013;33(6):511–3.

[56] Hewison J. Psychological aspects of individualized choice and reproductive autonomy in prenatal screening. Bioethics 2015;29(1):9–18.

[57] Schmitz D, Netzer C, Henn W. An offer you can't refuse? Ethical implications of non-invasive prenatal diagnosis. Nat Rev Genet 2009;10(8):515.

[58] Dar P, Curnow KJ, Gross SJ, Hall MP, Stosic M, Demko Z, et al. Clinical experience and follow-up with large scale single-nucleotide polymorphism-based noninvasive prenatal aneuploidy testing. Am J Obstet Gynecol 2014;211(5): 527. e1–e17.

[59] Brewer J, Demers L, Musci T. Survey of US obstetrician opinions regarding NIPT use in general practice: implementation and barriers. J Matern Fetal Neonatal Med 2017;30(15):1793–6.

[60] Morain S, Greene MF, Mello MM. A new era in noninvasive prenatal testing. N Engl J Med 2013;369(6):499–501.

[61] Mercer MB, Agatisa PK, Farrell RM. What patients are reading about noninvasive prenatal testing: an evaluation of Internet content and implications for patient-centered care. Prenat Diagn 2014;34(10):986–93.

[62] Benn P. Posttest risk calculation following positive noninvasive prenatal screening using cell-free DNA in maternal plasma. Am J Obstet Gynecol 2016;214(6): 676. e1–7.

[63] Moon-Grady AJ, Baschat A, Cass D, Choolani M, Copel JA, Crombleholme TM, et al. Fetal treatment 2017: the evolution of fetal therapy centers—a joint opinion from the International Fetal Medicine and Surgical Society (IFMSS) and the North American Fetal Therapy Network (NAFTNet). Fetal Diagn Ther 2017; 42(4):241–8.

[64] Mansfield C, Hopfer S, Marteau TM. Termination rates after prenatal diagnosis of Down syndrome, spina bifida, anencephaly, and Turner and Klinefelter syndromes: a systematic literature review. Prenat Diagn 1999;19(9):808–12.

[65] van Schendel RV, Page-Christiaens GC, Beulen L, Bilardo CM, de Boer MA, Coumans AB, et al. Trial by Dutch laboratories for evaluation of non-invasive prenatal testing. Part II-women's perspectives. Prenat Diagn 2016;36(12):1091–8.

[66] Chitty LS, Wright D, Hill M, Verhoef TI, Daley R, Lewis C, et al. Uptake, outcomes, and costs of implementing non-invasive prenatal testing for Down's syndrome into NHS maternity care: prospective cohort study in eight diverse maternity units. BMJ 2016;i3426:354.

[67] Bianchi DW. From prenatal genomic diagnosis to fetal personalized medicine: progress and challenges. Nat Med 2012;18:1041–51.

[68] Guedj F, Bianchi DW, Delabar J-M. Prenatal treatment of Down syndrome: a reality? Curr Opin Obstet Gynecol 2014;26(2):92–103.

[69] de Wert G, Dondorp W, Bianchi DW. Fetal therapy for Down syndrome: an ethical exploration. Prenat Diagn 2017;37(3):222–8.

[70] Lo YM, Chan KC, Sun H, Chen EZ, Jiang P, Lun FM, et al. Maternal plasma DNA sequencing reveals the genome-wide genetic and mutational profile of the fetus. Sci Transl Med 2010;2(61). 61ra91.

[71] Kitzman JO, Snyder MW, Ventura M, Lewis AP, Qiu R, Simmons LE, et al. Noninvasive whole-genome sequencing of a human fetus. Sci Transl Med 2012;4(137). 137ra76.

[72] Drury S, Hill M, Chitty LS. Cell-free fetal DNA testing for prenatal diagnosis. In: Makowski GS, editor. Advances in clinical chemistry, vol. 76. Elsevier; 2016. p. 1–35 [chapter one].

[73] Van Opstal D, van Maarle MC, Lichtenbelt K, Weiss MM, Schuring-Blom H, Bhola SL, et al. Origin and clinical relevance of chromosomal aberrations other than the common trisomies detected by genome-wide NIPS: results of the TRIDENT study. Genet Med 2018;20:480–5.

[74] Pertile MD, Halks-Miller M, Flowers N, Barbacioru C, Kinnings SL, Vavrek D, et al. Rare autosomal trisomies, revealed by maternal plasma DNA sequencing, suggest increased risk of feto-placental disease. Sci Transl Med 2017;9(405).

[75] Bianchi DW, Chudova D, Sehnert AJ, Bhatt S, Murray K, Prosen TL, et al. Noninvasive prenatal testing and incidental detection of occult maternal malignancies. JAMA 2015;314(2):162–9.

[76] Wong FCK, Lo YMD. Prenatal diagnosis innovation: genome sequencing of maternal plasma. Annu Rev Med 2016;67(1):419–32.

[77] Tamminga S, van Maarle M, Henneman L, Oudejans CB, Cornel MC, Sistermans EA. Maternal plasma DNA and RNA sequencing for prenatal testing. Adv Clin Chem 2016;74:63–102.

[78] Yotsumoto J, Sekizawa A, Koide K, Purwosunu Y, Ichizuka K, Matsuoka R, et al. Attitudes toward noninvasive prenatal diagnosis among pregnant women and health professionals in Japan. Prenat Diagn 2012;32(7):674–9.

[79] Benn P, Chapman AR, Erickson K, Defrancesco MS, Wilkins-Haug L, Egan JF, et al. Obstetricians and gynecologists' practice and opinions of expanded carrier testing and noninvasive prenatal testing. Prenat Diagn 2014;34(2):145–52.

[80] Tamminga S, van Schendel RV, Rommers W, Bilardo CM, Pajkrt E, Dondorp WJ, et al. Changing to NIPT as a first-tier screening test and future perspectives: opinions of health professionals. Prenat Diagn 2015;35(13):1316–23.

[81] van Schendel RV, Dondorp WJ, Timmermans DR, van Hugte EJ, de Boer A, Pajkrt E, et al. NIPT-based screening for Down syndrome and beyond: what do pregnant women think? Prenat Diagn 2015;35(6):598–604.

[82] Bredenoord AL, de Vries MC, JJM VD. Next-generation sequencing: does the next generation still have a right to an open future? Nat Rev Genet 2013;14(5):306.

[83] European Society of Human Genetics. Genetic testing in asymptomatic minors: recommendations of the European Society of Human Genetics. Eur J Hum Genet 2009;17(6):720–1.

[84] Dondorp WJ, Page-Christiaens GCML, de Wert GMWR. Genomic futures of prenatal screening: ethical reflection. Clin Genet 2016;89(5):531–8.

[85] Elias S, Annas GJ. Generic consent for genetic screening. N Engl J Med 1994;330(22):1611–3.

[86] Agarwal A, Sayres LC, Cho MK, Cook-Deegan R, Chandrasekharan S. Commercial landscape of noninvasive prenatal testing in the United States. Prenat Diagn 2013;33(6):521–31.

[87] Skirton H. Direct to consumer testing in reproductive contexts—should health professionals be concerned? Lif Sci Soc Policy 2015;11:4.

[88] Gekas J, Langlois S, Ravitsky V, Audibert F, van den Berg DG, Haidar H, et al. Non-invasive prenatal testing for fetal chromosome abnormalities: review of clinical and ethical issues. Appl Clin Genet 2016;9:15–26.

[89] Rolfes V, Schmitz D. Unfair discrimination in prenatal aneuploidy screening using cell-free DNA? Eur J Obstet Gynecol Reprod Biol 2016;198(Suppl. C):27–9.

[90] Filoche S, Cram F, Lawton B, Beard A, Stone P. Implementing non-invasive prenatal testing into publicly funded antenatal screening services for Down syndrome and other conditions in Aotearoa New Zealand. BMC Pregnancy Childbirth 2017;17:344.

[91] Audibert F, De Bie I, Johnson J-A, Okun N, Wilson RD, Armour C, et al. No. 348-Joint SOGC-CCMG Guideline: Update on Prenatal Screening for Fetal Aneuploidy, Fetal Anomalies, and Adverse Pregnancy Outcomes. J Obstet Gynaecol Can 2017;39(9):805–17.

[92] Oepkes D, Page-Christiaens GC, Bax CJ, Bekker MN, Bilardo CM, Boon EM, et al. Trial by Dutch laboratories for evaluation of non-invasive prenatal testing. Part I-clinical impact. Prenat Diagn 2016;36 (12):1083–90.

[93] Martin L, Gitsels-van der Wal JT, de Boer MA, Vanstone M, Henneman L. Introduction of non-invasive prenatal testing as a first-tier aneuploidy screening test: a survey among Dutch midwives about their role as counsellors. Midwifery 2018;56:1–8.

[94] Hill M, Johnson J-A, Langlois S, Lee H, Winsor S, Dineley B, et al. Preferences for prenatal tests for Down syndrome: an international comparison of the views of pregnant women and health professionals. Eur J Hum Genet 2015;24:968.

[95] van Schendel RV, Kleinveld JH, Dondorp WJ, Pajkrt E, Timmermans DR, Holtkamp KC, et al. Attitudes of pregnant women and male partners towards non-invasive prenatal testing and widening the scope of prenatal screening. Eur J Hum Genet 2014;22(12):1345–50.

[96] Farrimond HR, Kelly SE. Public viewpoints on new non-invasive prenatal genetic tests. Public Underst Sci 2013;22(6):730–44.

[97] Greely HT. Get ready for the flood of fetal gene screening. Nature 2011;469(7330):289–91.

[98] Kellogg G, Slattery L, Hudgins L, Ormond K. Attitudes of mothers of children with down syndrome towards noninvasive prenatal testing. J Genet Couns 2014;23(5):805–13.

[99] van Schendel RV, Kater-Kuipers A, van Vliet-Lachotzki EH, Dondorp WJ, Cornel MC, Henneman L. What do parents of children with Down syndrome think about non-invasive prenatal testing (NIPT)? J Genet Couns 2017;26(3):522–31.

[100] Kuppermann M, Pena S, Bishop JT, Nakagawa S, Gregorich SE, Sit A, et al. Effect of enhanced information, values clarification, and removal of financial barriers on use of prenatal genetic testing: a randomized clinical trial. JAMA 2014;312(12):1210–7.

ETHICS OF CELL-FREE DNA-BASED PRENATAL TESTING FOR SEX CHROMOSOME ANEUPLOIDIES AND SEX DETERMINATION

15

Wybo Dondorp*, Angus Clarke[†], Guido de Wert*

Department of Health, Ethics & Society, Research Schools CAPHRI & GROW, Maastricht University, Maastricht, The Netherlands* Institute of Cancer and Genetics, Cardiff University, United Kingdom[†]

INTRODUCTION

Cell-free DNA-based prenatal testing (further: cfDNA testing) is a new technology that may both be used as a diagnostic test in the context of prenatal diagnosis of monogenetic disorders and as a second or first tier test in the context of prenatal screening for chromosome abnormalities. cfDNA testing is commonly also known as "noninvasive prenatal testing" (NIPT). When specifically used as a diagnostic test for monogenetic disorders, it is often referred to as "noninvasive prenatal diagnosis" (NIPD), whereas "noninvasive prenatal screening" (NIPS) has been proposed for its use as a screening test [1]. However, in line with other chapters in this volume, we will speak of cfDNA testing.

In the context of screening for common autosomal aneuploidies, cfDNA testing is available in a growing number of countries [2], either as a testing offer made available to patients through individual practitioners or practices in the private sector, or in the context of national or regional prenatal screening programs. While the emphasis in the literature is on how best to use this new technology for improving existing prenatal screening for Down syndrome (trisomy 21) and the other two common autosomal aneuploidies Patau and Edwards Syndrome (trisomy 13 and 18) [3], part of the debate is also about how cfDNA testing may lead to a widening of the scope of prenatal screening beyond these conditions.

While with cfDNA testing technology it may eventually become possible to turn noninvasive prenatal screening into a comprehensive fetal genome scan, more immediate candidate conditions are sex chromosome aneuploidies (SCAs), microdeletions and -duplications and rare autosomal trisomies. In recent years, commercial providers have moved in this direction, optionally offering cfDNA testing for a wider range of chromosomal abnormalities, including for SCAs [4].

Noninvasive Prenatal Testing (NIPT). https://doi.org/10.1016/B978-0-12-814189-2.00015-3

With cfDNA analysis, the sex chromosomes of the fetus can be noninvasively identified much earlier in pregnancy than was previously the case with ultrasound. cfDNA testing allows for easy and safe sex determination already from 7 weeks of gestation. This has opened up possibilities for early fetal sex determination for medical reasons, for example, for women who are known carriers of a sex-linked disorder such as hemophilia or Duchenne muscular dystrophy [5,6], but also as an add-on to prenatal screening for women wanting to know the sex of the fetus for curiosity or other nonmedical reasons [4].

In this chapter we will first give background information on SCAs, briefly summarizing their etiology and diversity as well the benefit of early diagnosis. Next, we will provide information on how the sex chromosomes may be brought to light with cfDNA testing and what is known about the performance of this test for SCAs. In the two subsequent sections we will then discuss the ethical aspects of two main themes of this chapter: prenatal testing and screening for SCAs and the use of cfDNA testing for sex determination, followed by our conclusions.

SEX CHROMOSOME ANEUPLOIDIES

Those born with one of the SCAs have an atypical number of the sex chromosomes, X and Y. A female has two X chromosomes and a male has a single X chromosome, the other sex chromosome in a male being the Y chromosome. The X chromosome is an average-sized chromosome carrying many genes that influence a wide range of developmental processes and physical and nervous system functioning. Some of the genes on this chromosome relate to sexual development and reproductive function but most do not. The Y chromosome is a very small chromosome whose function relates especially to sexual development: the presence of the Y chromosome in the embryo leads to development as a male and is required for male fertility. The genes involved in these functions are passed from father to son and to grandson; they never pass through a female. One can appreciate immediately that a few genes involved in being male are never part of the genetic content of a female whereas many genes, important for a wide range of biological functions, are present in two copies in the female and one copy in the male. Mammals cope with this difference between the sexes in the dosage of many genes by inactivating large parts of one X chromosome in any female cell that has two: X chromosome inactivation (XCI) is the mammalian approach to X chromosome dosage compensation [7]. Different solutions to this problem are found in other classes of the animal kingdom.

A small number of other genes are present on the X and Y chromosomes that are not inherited in a sex-linked fashion. These "pseudo-autosomal" genes are inherited as if they were on the usual type of chromosome (an autosome) because both copies are active in a female and the Y chromosome has an active equivalent, so there is no need for a mechanism of dosage compensation. Changes in the number of the sex chromosomes therefore lead to changes in the number of copies of active genes not involved solely in sex and reproduction. If the process of XCI applied to the whole of the X chromosome, and if the Y chromosome dealt only with male-specific traits, such SCAs would have little if any phenotypic effect. As it is, however, there is a range of consequences associated with this group of conditions although, because XCI affects much of the X chromosome, these effects are much less marked than might be expected for an equivalent block of autosomal chromatin. The sex chromosome trisomies are rather well tolerated and their effects are often rather mild or subtle; Turner syndrome (often 45,X) is mostly well tolerated in the small proportion of affected conceptions that survive the pregnancy. There are other types of SCA with more marked effects but these are much less common and will not be considered further here.

COMMON TYPES OF SCA

There are four relatively common types of SCA, all of which can have consequences although these will often be mild. Combined together, these occur at a birth incidence of approximately 1 in 1000, being somewhat more common in phenotypic males than females. This compares to a birth incidence for trisomy 21 of approximately 1 in 700 live births. One of the genes in the pseudo-autosomal region of the X chromosome is *SHOX*, which promotes skeletal growth. People with only one sex chromosome—who have Turner syndrome (45,X)—are on average shorter than women who have two X chromosomes, while people with any of the three conditions with an additional chromosome (XXX, XXY, and XYY) tend to be taller. Dosage differences at other loci will contribute to the other effects of the SCAs, only some of which can be fully accounted for in a simple "gene dosage–phenotype" relationship.

People with Turner syndrome are usually infertile and may have one or more congenital anomalies (such as coarctation of the aorta or other cardiovascular defects, renal anomalies or cystic hygroma, that develops in utero but usually resolves by the time of birth to leave webbing of the neck). Girls with Turner syndrome may also be fully affected by sex-linked disorders, as if they were male, as they are hemizygous for all X chromosome genes. It should be remembered that Turner syndrome can also be caused by deletions and other rearrangements affecting the X chromosome. People with Klinefelter syndrome (47, XXY) are also usually infertile; people with XYY and XXX may have subfertility. All four of these conditions are associated with a modest drop in mean IQ but most of the individuals involved have an IQ in the normal range. Subtle neurocognitive problems may also be found in Turner syndrome. There is also a modest association with some behavioral problems in the SCAs, with XYY syndrome associated with autistic spectrum disorder. People with four or more X chromosomes have more serious cognitive and behavioral difficulties.

BENEFIT OF EARLY DIAGNOSIS

Early diagnosis is very helpful in Turner and Klinefelter syndromes because it enables the prompt recognition and appropriate management of endocrine problems. In Turner syndrome, expert management is important if growth is to be optimized. As well as growth hormone in childhood, the prescription of estradiol and then also progesterone triggers puberty with its growth spurt and enables pseudomenstrual cycles. In Klinefelter syndrome, prompt treatment with supplementary testosterone permits a normal male puberty with normal male musculature; it may also have helpful behavioral effects.

It is difficult to make objective quality of life assessments of these conditions as there has been a long-standing problem with ascertainment bias. Although some unbiased, population-based cytogenetic studies have been carried out, much clinical experience is skewed to those with more marked difficulties and many affected individuals are probably never diagnosed [8]. However, experience indicates that an early diagnosis is on balance helpful; it enables the optimization of medical and educational support [9,10] and avoids the sudden discovery of a diagnosis as a cause of infertility in adult life [11]. Putting to one side the question of infertility in most of the people with Turner and Klinefelter syndromes, these conditions are often compatible with "normal," happy, and fulfilled lives.

IDENTIFICATION OF SCAs IN PRENATAL TESTING

When not deliberately sought for in an antenatal screening program, SCAs may still be identified antenatally in three circumstances: (i) triggered by ultrasound findings: Turner syndrome may be suspected—and then tested for—in the presence of fetal hydrops, cystic hygroma, or coarctation of the aorta found at fetal ultrasound scan; (ii) as an additional finding of karyotyping or molecular analysis after amniocentesis or chorion villus sampling performed for a different indication; (iii) inadvertently when cfDNA testing is performed, either as part of population screening or for a different specific purpose, when it has been decided to conduct the analysis in such a way that SCAs are identified. SCAs may be sought as part of a package of cfDNA testing conducted primarily to screen for the autosomal trisomies but included among the additional options. One should note that the performance of cfDNA-based tests is often only moderate and sometimes rather poor [17]. Also, a discordance between sex predicted by cfDNA based testing and phenotype of the external genitalia is likely to occur in 1 in 1500–2000 pregnancies [12]. In addition to technical reasons, there can be several biological reasons for such discordances, for instance, a Y-signal picked up by cfDNA testing in a pregnancy of a female fetus may be due to a vanishing twin. Many of those discordant results may also be caused by complex disorders of sexual differentiation [12]. This is further discussed in Chapter 5.

STRATEGIES FOR DETECTING THE SEX CHROMOSOMES IN CELL-FREE DNA TESTING

There are three principal molecular strategies to detect the sex chromosomes in cfDNA testing: (i) PCR amplification of sequences from the Y chromosome, as used in the offer of fetal sexing by cfDNA testing. It is essentially a Yes/No test that aims to detect Y chromosome sequences, often the sex-determining SRY gene. It is good at determining whether or not there is a Y chromosome present but is not so effective at counting how many Y chromosomes there are, and it cannot determine how many X chromosomes are present. The other two methods are based on sequencing either (ii) unselected cfDNA in maternal plasma, as performed in whole genome sequencing, or (iii) cfDNA that has been enriched for sequences (i.e., chromosomes) of interest, often by means of a microarray that contains target sequences from the X and Y chromosomes.

The sequencing methods rely on counting the copies of the sequences of interest, to determine their ratio against sequences of reference chromosomes. Sequencing has to be performed to a sufficient depth: enough copies have to be sequenced that one can be confident of the ratio of the placenta-derived sequences in the maternal plasma.

The key is the calibration of the test on pregnancies of known chromosomal constitution, using the sequences that are to be detected and counted when performing the test in a clinical situation. This is technically simpler but demands more sequencing if one sequences unselected cfDNA—method (ii)—than if one enriches the cfDNA for specific chromosomal regions-method (iii). With the latter method one can achieve statistical significance more readily, and much more cheaply [13,14]. The disadvantage of the enrichment step is that it could introduce a bias in sequence representation unless great care is taken at the design stage and in each analytical step, hence the need for careful calibration.

In the absence of a pregnancy, the woman's plasma will contain roughly equal proportions of DNA from her two X chromosomes but in a pregnancy there will usually be a slight excess of DNA from one of the mother's X chromosomes, along with DNA from either a Y chromosome or from the father's X chromosome. To detect deviations from these two normal scenarios, a 46,XX or a 46,XY fetus, will require quantification of a large number of DNA fragments. Simply examining the ratio of autosomes: X chromosome:Y chromosome will answer some questions. More precise information about the origin of the chromosomal nondisjunction can be obtained if sequences are generated from polymorphic regions where the mother's two X chromosomes differ from one another and from the father's

X chromosome. Given a fetal fraction of 5%–10%, the ratio expected between sequences derived from the two maternal and one paternal X chromosome in different scenarios will depend upon both the chromosome constitution of the fetus and the precise meiotic error underlying the SCA. In 47, XXY, for example, the additional X chromosome sequences may derive from meiosis I or II of the mother or come from the father.

Let us take Klinefelter syndrome as an example and consider one X-specific sequence and one Y-specific sequence. If we assume that the fetal fraction of the cfDNA in maternal plasma is 12%, so that the maternal fraction is 88%, then one has to distinguish between a ratio of 94:6 (expected X:Y copy ratio if the fetus is 46,XY) and 96:4 (expected X:Y ratio if the fetus is 47,XXY). Many copies of these sequences have to be counted for us to distinguish these ratios with confidence, more sequence information being required if the fetal fraction is less.

TEST QUALITY OF CELL-FREE DNA TESTING FOR SCAs

In terms of analytical validity, reliability of laboratory methods for cfDNA prenatal testing is good, and the results are readily reproducible, when the numbers of chromosomes 21, 18, and 13 are calibrated against other autosomes [13,15,16]. In terms of clinical validity, it is a much better test for these classic trisomies than previous testing options that combined biochemical analytes and nuchal translucency measurement. But even with a sensitivity of ~99% and a specificity of ~99.9% for the classic trisomies [17], the test is less than fully accurate. Because the DNA tested represents a combination of maternal and fetal cell-free DNA, with the latter deriving from the placenta, a result signaling a suspected aneuploidy may be generated by other events such as placental mosaicism, a vanishing twin, or a maternal tumor [18]. The impact of this becomes clear if the test is assessed in terms of its predictive value, rather than only its sensitivity and specificity. PPV also takes the low prevalence of the relevant conditions in the target population into account [19]. The positive predictive value (PPV) is the chance that an increased risk result corresponds to the *fetus* being affected. In a screening test for common autosomal trisomies applied to an unselected (population-risk) pregnancies, a positive screening test often corresponds to an 80% or so probability of the fetus being affected, while if the pregnancy is already known through conventional screening to have a somewhat increased chance of an affected fetus, the PPV may be greater than 90% [19,20]. Both these values demonstrate that a diagnostic procedure should be advised after a positive cfDNA test result for autosomal trisomy, because the chance of a false-positive result is of the order of 5%–15% (depending on the detailed circumstances).

There are limited data on cfDNA test performance for SCAs. These data indicate that cfDNA testing is screen "positive" in up to 1.1% of pregnancies [21] but has a lower accuracy for SCA than for trisomy 21 and 18, with an overall PPV of 50%, and a much lower PPV for Turner syndrome (20%–30%) [14,17,22,25–28]. For determining detection rates one has to be informed about the number of affected cases missed, but since most children born with SCAs do not have clinical features [23] and healthy newborns are for obvious reasons not genotyped, detection rates cannot be calculated. Also, one would not necessarily expect cfDNA testing to detect mosaic cases or cases associated with more complex chromosome rearrangements. False-positive SCA findings may, in addition to the causes referred to previously, be attributed to maternal X chromosome aneuploidy and maternal X chromosome copy number variations [24].

ETHICS OF PRENATAL TESTING AND SCREENING FOR SCAs

Although SCAs constitute about 25% of all chromosomal abnormalities detected prenatally, there is much emphasis in the literature on the fact that for many women the diagnosis of a SCA comes as an unforeseen result that leads to difficult decision-making given the relative mildness of the involved phenotypes [29,30]. As said, some will be referred because of an abnormal ultrasound finding indicating potential Turner syndrome, in others cfDNA testing for the "classic trisomies" may lead to incidental or nonincidental findings triggering invasive testing for SCA.

With the exception of nonmosaic Turner syndrome (45,X), internationally reported termination rates for SCA are much lower than for Down syndrome, for which pre-cfDNA testing figures are >90% in many countries [31]. A report of EUROCAT data covering the years 2000–05 gives a termination rate for sex chromosomal trisomies of 36% [32]. A 21-year (1994–2014) retrospective cohort study from Hong Kong found rates of 92% for 45,X; 48% for XXY; and lower percentages for XXX, XYY, and mosaic SCAs, findings that are in line with studies from other parts of the world [33]. Abnormal ultrasound findings (most often in combination with a 45,X diagnosis) were significantly associated with termination decisions. A consistent finding is also that termination rates for SCAs show a downward trend, which is linked with a generally more optimistic counseling than given in the past. This reflects experience with and knowledge about milder phenotypes of prenatally diagnosed cases as compared with postnatal cases where a diagnosis is typically triggered by clinical features [34], as well as a multidisciplinary approach [35]. Main parental concerns seem to focus on abnormal sexual development or infertility rather than on relatively minor cognitive and behavioral problems [29].

SCAs AS SECONDARY FINDINGS AT INVASIVE TESTING: THE OPTION OF TARGETED TESTING

Most SCAs are found as secondary results of invasive testing offered to women at a higher risk of having a child with a common autosomal aneuploidy (trisomies 21, 18, 13). Given the challenges for decision-making and counseling posed by SCAs, it has been suggested that in pregnancies at elevated risk for trisomies 21, 18, or 13 (but without abnormal ultrasound findings), a targeted diagnostic test, fluorescence in situ hybridization (FISH) or quantitative fluorescence polymerase chain reaction (QF-PCR) might be used that would avoid other findings including SCA. This would protect women from being confronted with secondary findings that they may find difficult to handle. The maxim behind this proposal has been summarized as "test what you screen for" [36]. When targeted follow-up testing is feasible, which has become the case with the availability of molecular tests such as FISH or QF-PCR, forgoing this option would amount to using prenatal diagnosis as a platform for adding opportunistic screening, for which a justification would be needed [37]. The prevailing view, however, is that "testing for less" than can be found with karyotyping (or nowadays chromosomal microarray), amounts to making suboptimal use of risky invasive testing, and denies prospective parents relevant information about abnormalities with potential clinical significance that they might consider relevant for reproductive decision-making [38].

An important question is of course what women or couples would want. This question was addressed in a small-scale qualitative interview study among Dutch women who chose to continue the pregnancy after finding themselves unexpectedly confronted with an SCA diagnosis (full and mosaic forms of Turner, Klinefelter, and Triple-X syndromes) after invasive testing for advanced maternal age. Although all respondents reported initial feelings of anguish and fear, they were sufficiently reassured in posttest counseling and expressed that they expected a good quality of life for their child. When asked whether they would want to have testing for the sex chromosomes included when having follow-up invasive testing in a further pregnancy, most couples said they would [29]. These couples found it important to be given the information and the opportunity to make a personal decision on the basis of it. However, there were also couples stating that they would rather not have been told.

The latter finding is as important as the former. Even if most pregnant women or couples do indeed appreciate the benefits of "testing for more" [39], it does not follow that individual women need not be asked about their preferences in this regard [40,41]. Ideally, women having prenatal diagnosis for Down syndrome should be given the right to indicate that they do not want to receive secondary information about milder conditions such as SCAs [43].

PRENATAL SCREENING: THE "AUTONOMY PARADIGM" AND ITS LIMITS

Generally, the potential benefits (or clinical utility) of screening are defined in terms of prevention: the extent to which the screening contributes to a lower burden of disease in the relevant population. But prenatal screening for fetal abnormalities is a special case [42]. In order to avoid the moral pitfalls connected with presenting selective terminations as the intended result of the offer of prenatal screening for conditions such as Down syndrome, screening authorities and committees concur that the practice should be understood as aimed, not at prevention, but at providing pregnant women with meaningful options for autonomous reproductive decision-making [43]. In the ethics literature, this is known as the "autonomy paradigm" of prenatal screening. However important this perspective may be when it comes to responding to the critique of "disability rights" advocates, who argue that the practice of prenatal screening sends a discriminatory message about the worth of people living with the relevant conditions [44], the problem with the autonomy paradigm is that it does not help us answering the increasingly important question of what the scope of prenatal screening should be [45]. Clearly, there is no reason why prenatal screening for fetal abnormalities should be limited to Down, Edwards, and Patau syndrome, as other conditions can be brought to light that are no less serious than those. But when it comes to milder or highly variable conditions, it is not obvious that anything that anyone would consider a reason for selective abortion should be regarded as a justification for offering screening. The reductivist view that in the ethics of prenatal screening "it is all about choice" [46] tends to silence all concerns about possible harms, suggesting that whatever these are, they can simply be dealt with as part of informed consent. However, this is at odds with how the responsibility of those offering screening is understood in the international framework as developed on the basis of the Wilson & Jungner principles. This does not only apply to national or regional screening programs, but also to individual practitioners offering screening to their patients and to commercial companies operating on the "screening market" [44]. Given that inevitably all screening also leads to harms, it is essential only to offer screening when benefits clearly outweigh harms for the persons being tested [47]. When considering prenatal screening for SCAs, this is at least not obvious.

CONCERNS WITH PRENATAL SCREENING FOR SCAs

First, with screening for SCAs, positive test results indicative of SCAs will confront women with even more difficult decision-making than is the case with SCAs identified as a secondary result of prenatal diagnosis. Here, the question is not only whether to continue or to terminate the pregnancy after a confirmed SCA diagnosis, but also whether or not to take the risk of invasive testing in order to exclude or confirm the cfDNA test result. Although the chances that invasive diagnostic procedures will lead to a miscarriage are now believed to be much smaller than previously thought [48], this does not make these procedures entirely safe.

Given the low PPV of a positive screening test for SCAs, this means a probably more than 50% chance of a false alarm. It is true that this is still better than the ~5% PPV of traditional screening for Down syndrome [49], but given the relatively mild phenotype of SCAs, the question remains whether providing (or confronting) women with this choice is a matter of helping or harming them. There are a few papers reporting on women's actual choices in the context of screening for SCAs, based on small numbers [26,50]. These suggest that upon a cfDNA result positive for SCAs most (but far from all) women accept the offer of diagnostic testing. However, the authors of one of these papers also state that several women expressed regret that they had learned this information prenatally [26]. Their experience of being caught in a "screening trap" suggests layers of complexity in the dynamics of women's decision-making that need to be further explored before the routine offer of cfDNA testing for SCAs can be considered to be a responsible policy. In this matter, the specifics of different SCAs will have to be taken into account, for instance, the fact that >99% of nonmosaic Turner fetuses miscarry, and that those who survive often also have abnormalities that are detected by ultrasound [51].

Second, given the much lower PPV of cfDNA testing for SCAs, adding testing for these conditions to prenatal screening for common autosomal trisomies will have the potential of reversing the large reduction in invasive testing that is widely considered to be the main benefit of this new screening test. Again, it is true that the PPV for SCAs is higher than was regarded as acceptable for traditional first trimester prenatal screening for Down syndrome. However, the question remains whether the potential higher loss of wanted pregnancies would be outweighed by the benefits of added screening for SCAs.

BENEFITS FOR THE CHILD

The recommendation by the European and American Societies of Human Genetics not to offer screening for SCAs at this time [43] has met with criticism. Not only was this recommendation ethically troubling because it restricts patient choice, but it would also deny future children the benefit of early diagnosis [52]. The latter argument has also been used to question the policy of offering women the option of targeted testing as to avoid SCA as secondary findings from prenatal diagnosis [53]. Interestingly, the argument that screening for SCAs might benefit the future child reintroduces "prevention" as a (secondary) aim of prenatal screening for fetal abnormalities, though without raising the moral problems of connecting prevention with selective termination. What is meant here is prevention in the sense of improving long-term health outcomes for children born after a timely diagnosis enabled by prenatal screening. It is expected that prevention in this sense will become more important as an aim of prenatal screening with improved options for fetal therapy [54]. Even for conditions such as Down syndrome, a timely prenatal diagnosis may in the future open up options for in utero therapeutic intervention with better long-term results than postnatal treatment [55]. This may in the future lead to possible tensions with the autonomy paradigm, especially with regard to pregnant women's "right not to know" [56,57].

But as long as there is no prenatal or perinatal treatment for SCAs, there is also no benefit for the future child of having a SCA diagnosis during pregnancy rather than early in postnatal life. This means that if an early diagnosis is on balance beneficial, this would be an argument for considering newborn screening for SCAs [58], rather than trying to convince women that they should have prenatal SCA screening or testing in the interest of any affected child that they would allow to be born.

The professional consensus that an early diagnosis of SCA is beneficial specifically pertains to Turner and Klinefelter syndrome as it allows the timely management of endocrine problems. Whether this conclusion also holds for other SCAs is less clear. Given concerns about potential psychosocial harm (effect on self-esteem, parent–child interaction, and stigmatization) as a consequence of being born with a diagnosis that otherwise might have been made only much later or never at all, the balance of possible benefits and harms for the future child depends on the precise SCA.

DIRECT-TO-CONSUMER CELL-FREE DNA TESTING

Direct-to-consumer (DTC) offers of cfDNA testing for fetal abnormalities in average population-risk pregnancies are likely to add screening for SCAs and indeed microdeletion syndromes as a marketing ploy to increase their attractiveness to naïve customers, who will often fail to appreciate whether or not these tests have met test performance criteria and will eventually greatly increase the chance of an inappropriate follow-on invasive test procedure that puts the pregnancy at risk for no clear purpose. Such unvalidated cfDNA testing will also impose costs of follow-up invasive investigations on either the patient, their health insurer, or the national health care system. In the UK, the Nuffield Council of Bioethics has recommended that such additional costs should be met by the cfDNA testing provider from the initial fee for performing the test. This may discourage inappropriate marketing [19].

ETHICS OF FETAL SEX DETERMINATION

cfDNA testing can reliably and noninvasively determine the sex of the fetus from 7 weeks of gestation, bringing this forward by 6 weeks as compared with the earliest reliable use of ultrasound for this purpose [59]. In low-risk pregnancies, fetal sex is normally only identified with routine ultrasound at around 20 weeks. As we will discuss in this section, information about fetal sex can be sought for medical as well as for nonmedical reasons.

FETAL SEX DETERMINATION FOR MEDICAL AND NONMEDICAL REASONS

Early fetal sex determination has the important medical benefit of reducing the need for invasive prenatal diagnosis (amniocentesis or chorion villus sampling) for women who are known carriers of certain sex-linked disorders by 50% [5,6]. For instance, if the woman is a carrier of hemophilia, 1:2

of any boys will have the disorder, whereas 1:2 of any girls will be a healthy carrier. If early noninvasive sex determination reveals a female fetus, invasive testing can be avoided. As cfDNA-based sex determination for this purpose (preselection for invasive testing) is offered to women at a known risk of a child with a genetic disorder, this is a form of noninvasive prenatal diagnosis (NIPD), rather than prenatal screening. In some sex-linked disorders (such as fragile-X syndrome, FRAXA) female carriers may to some extent also be affected; in these cases the benefits of this approach are of course less evident [59].

The clear benefits of early fetal sex determination as a preselection step for invasive testing in the context of sex-linked disorders are expected to be superseded in the near future by further developments in noninvasive mutation detection. Until recently, diagnostic cfDNA testing for monogenetic disorders was clinically available only for paternally inherited dominant disorders. With the introduction of haplotyping technology, a form of indirect mutation detection has become available that will allow the clinical use of such diagnostic cfDNA testing for recessive disorders as well [60].

However, in the private sector, fetal sex determination is often offered as an add-on to cfDNA prenatal screening for fetal abnormalities, thus catering to the wishes of prospective parents who want to know for nonmedical reasons whether their child will be a boy or a girl. In the pre-cfDNA era, those wanting to know the sex of the fetus had to wait until the second trimester when this would be seen on a routine ultrasound scan. As a DTC test, cfDNA-based sex determination has already been available for more than 10 years [61], but its incorporation in a screening offer for fetal abnormalities has led to concerns that this might encourage or even normalize prenatal testing for fetal sex.

For most, having this information will be a matter of simple curiosity: wanting to know about one's future child [62]. The option to find this out so much earlier may be regarded as "a benign diversion for a couple impatient to bond with their baby" but has also raised concerns over the potential use of this information by those considering a sex-selective termination when the fetus is not of the desired sex [19,63–65]. Since many persons regard early abortions as morally less problematic than later ones, the option of getting to know the sex of the fetus so much earlier in pregnancy might lower the threshold for using this information for a sex-selective termination.

SEX SELECTION FOR NONMEDICAL REASONS

In the debate about the ethics of sex selection for nonmedical reasons, it is important to distinguish between sex selection as such, the use of specific sex-selective methods at different reproductive stages (preconception sperm sorting, preimplantation embryo selection, and termination of pregnancy), and the possible impact of the practice in different societal contexts [66]. Those rejecting all forms of sex selection for nonmedical reasons suspect that sexist motives will be at work in most if not all cases. They argue that this renders the practice deeply problematic from a moral point of view even when the method used may in itself be morally innocuous (as in preconception sex selection using insemination with enriched fractions of X- or Y-bearing sperm) and also when the motive of the prospective parents is to complete a family of only boys with a girl or vice versa—so-called family balancing. These authors stress that "sexism" refers not only to the discriminatory belief that one sex is better than the other but also to stereotyping views about gender role behavior. In their view, this also holds for sex selection aimed at having a "balanced family." Although not necessarily based on a view about the supremacy of one sex, this application would still be sexist in the sense of expressing and reinforcing preconceived gender role expectations [67,68]. Opponents of this view argue that although individual requests for sex selection may indeed flow from a sexist parental motivation, it cannot be maintained that this is necessarily the case [69,70]. These commentators do not agree that seeing a difference between what boys and girls might contribute to family life would necessarily stand in the way of valuing one's children for their own sake and allowing them to grow into autonomous individuals.

Even when so much is accepted, it does not follow that the notion of "family balancing" is beyond criticism. For one thing, it sends the false message that there might be something wrong with families of only boys or girls [71]. Moreover, where applied to families without boys, the apparently neutral language of "balancing" may well serve as a justifying cloak for a practice driven by the same ideas about inequality between the sexes that were supposed not to play a role in sex selection for this purpose [72].

For a moral evaluation, much also depends on how great the efforts are that prospective parents are prepared to make in order to see their preferences with regard to the sex of their children fulfilled [73]. If they are willing to pay large sums of money for a burdensome in vitro fertilization procedure followed by preimplantation genetic diagnosis, just in order to be able to select the sex of the embryo [74], the motive behind those investments becomes an issue of moral concern. Indeed, the question emerges whether any resulting children will still be given the freedom to choose their own path in life [75]. Similar concerns apply to prospective parents who would be willing to go as far as terminate an otherwise wanted pregnancy only because the fetus turns out to be of the "wrong" sex. On the other hand, parents with strong stereotypical gender expectations will raise their children accordingly, regardless of whether they have made use of sex-selective procedures.

Those parental expectations and preferences are not just personal but culturally shared. Together with lower fertility rates, the fact that in large parts of the world there is a strong preference for boys over girls has been an important driver of sex selection, mainly in the form of sex-selective abortion of female fetuses [76,77]. As argued by Mary Ann Warren (who coined the term "gendercide" in her seminal book on the issue), individual women who ask for a termination in this context need not themselves harbor any sexist views. For instance, they may be motivated by a "reluctance to bring a female child into a society in which she will be abused and devalued as they themselves have been" [69]. However, on a societal level, the practice of sex-selective abortion clearly reflects and reinforces sexist cultural patterns.

This practice is not only morally condemnable in itself, but it has also led to a serious distortion of the numeric balance between the sexes in several Asian countries, with wider societal consequences in terms of a rise of criminality, social unrest, and human rights violations [78]. The strong preference for sons is based on a complex amalgam of cultural and economic factors, including, for example, the burden that the (outlawed) dowry system in (regions of) India imposes on the bride's family [79].

As in Western countries son preference is now only weak or even absent, there would seem no reason to fear that the wider availability of sex selection technologies might lead to a general distortion of the sex ratio in this part of the world [80,81]. Still, concerns have been raised about sex-selective abortion as (supposedly) practiced in Western countries by members of minorities from countries with a strong preference for sons. This at least is strongly suggested by findings of skewed sex ratios among these groups [82–86]. However, in studies from the UK and Norway, findings from the late 20th century that were taken to point in this direction could not be confirmed in follow-up studies looking at the same groups in later periods [87,88]. A recent series of semiyearly reports on sex ratios in Great Britain issued by the UK Department of Health since 2013 have also found no evidence of sex-selective abortion [89]. Although reassuring, these findings do not rule out that on a smaller scale also in countries such as the UK, pregnancies may still be terminated to avoid the birth of a girl. Moreover, the present figures do not yet reflect a possible impact of the wider availability of early sex determination with modalities of cfDNA testing. Continued monitoring will therefore be needed.

SEX SELECTION AND LEGAL ABORTION

Although a large number of countries have legislation forbidding sex selection for nonmedical reasons, most of these laws only pertain to the use of assisted reproductive technologies [66]. This also holds for the prohibition in article 14 of the Convention on Human Rights and Biomedicine of the Council of Europe, which states that "the use of techniques of medically assisted procreation shall not be allowed for the purpose of choosing a future child's sex, except where serious hereditary sex-related disease is to be avoided." Only some Asian countries, including India and China, have enacted specific legislation also forbidding the use of medical technologies for sex-selective abortion [90]. Regarding this as a legislative lacuna, the Parliamentary Assembly of the Council of Europe has called for additional legal measures [78].

Recent sociological studies show how this call was taken up in the UK (as in other countries, including Australia and the USA) by an unlikely alliance of prolife antiabortionists and liberals, campaigning together against what was framed as "prenatal [sic] violence against women."[91–93] Between 2012 and 2015, a press campaign suggesting widespread misuse of prenatal diagnosis and abortion services accompanied the unsuccessful attempt to legally criminalize sex-selective abortion under the Protection of Children clause of the Serious Crime Act [92]. A similar debate was ignited in the Netherlands in 1997 by a public statement of the Minister of Health in which she expressed her understanding for the plight of women whose marriage or whose life was at stake because of the prospect of delivering a further daughter [94]. Also in this debate, calls for further specifying the scope for a legal termination were heard.

However, as many commentators including the former Dutch Minister have rightly pointed out, restricting legal abortion in the interests of women is a problematic proposition [95,96]. According to the World Health Organization in its report on "preventing gender-biased sex selection," relying on restrictive legislation tends to aggravate the situation: "Discouraging health-care providers from conducting safe abortions for fear of prosecution (…) potentially places women in greater danger than they would otherwise face" [76]. Instead, the report calls for policies aimed at addressing inequalities and awareness raising.

AVOIDING MISUSE OF CELL-FREE DNA TESTING

More specifically with regard to cfDNA testing for fetal sex determination for nonmedical reasons (either as an add-on to prenatal screening or as a separate DTC service), the British Nuffield Council on Bioethics has advised banning this practice, reasoning that "the consequences of an increase in sex selective terminations in the UK are potentially serious, particularly within specific cultural communities, and the medical benefits to a pregnant woman of finding out the sex of her fetus at 9–10 weeks, rather than at later ultrasound scans, are few" [19]. This is part of the more comprehensive recommendation that cfDNA testing should not be used to reveal nonmedical traits of the fetus [19].

A ban on early sex determination for nonmedical reasons interferes with women's reproductive freedom. Whether this would be justified to avoid sex-selective terminations is a matter for debate [97].

However, in the UK and other Western countries, there is a strong feeling among professionals that both in the light of the global dimension of the problem and in view of some indirect evidence that sex-selective terminations do also take place among minority groups, it would send out a wrong signal to leave the possible misuse of cfDNA testing for sex selection unchecked.

There are further reasons why adding fetal sex determination to prenatal screening for fetal abnormalities is not a good idea. First, it may attract women to prenatal screening for reasons of curiosity about the sex of the fetus, without fully considering that the test may also lead to information that is less trivial and may require difficult decision-making about whether or not to continue a wanted pregnancy [98]. This is indeed reminiscent of the challenge of misguided expectations with regard to routine ultrasound. But whereas this cannot easily be avoided with routine ultrasound, cfDNA testing (if not for SCAs) can very well be conducted without fetal sex determination.

Second, it may happen that after cfDNA testing has predicted a male or female fetus, second trimester ultrasound reveals the opposite phenotype [12]. Such discordant findings will turn what may have seemed an innocent offer of early fetal sex determination into a highly distressing experience of uncertainty about the implications for the future child [25]. As discussed, the causes range from false alarm to complex disorders of sexual differentiation. Some of those findings may be welcome in so far as they enable timely reproductive decision-making or alter perinatal management, but others may be emotionally burdensome without offering clear benefits. This is more specifically discussed in Chapter 5.

When done for screening rather than for sex determination, cfDNA testing may still generate information about fetal sex that could be used for sex selection. In the literature, it has been suggested that information about fetal sex should be withheld from women having prenatal screening until later in pregnancy, except when this information is medically relevant. However, if women expressly ask for this information, withholding it will not be legally possible in many jurisdictions. One might of course consider creating a legal title for this, as is already the case in Germany (where it is forbidden to report the sex of the fetus to the pregnant woman in the first 12 weeks), and as has been debated in Sweden [93]. However, it has also been argued that this would amount to a disproportionate infringement of women's right to information [97] and to treating all pregnant women as suspects in advance [75]. A more appropriate strategy would be to perform cfDNA testing without generating information about fetal sex. Generating this information requires an active step (either PCR or presequencing enrichment or bioinformatic analysis) that can be avoided except when screening for SCAs.

CONCLUSIONS

cfDNA-based prenatal screening for SCAs has no established clinical utility and is likely to be unhelpful to the pregnant woman and her health professionals. The performance of cfDNA testing for the SCAs is not good, especially for Turner syndrome, and it is likely to trigger a significant increase in invasive follow-up testing. Given these circumstances, and in line with the recommendations of the European and American Human Genetics Societies, we would suggest that cfDNA testing should not at present be used to screen for the SCAs (as opposed to its use when a SCA has been suspected on other grounds). More insight is needed into the balance of possible benefits and inevitable harms before adding these conditions to the screening package may be considered.

Where cfDNA testing is offered for SCAs, it is crucial that women are adequately informed about the range of possible test results and their reliability, as well as about the phenotypic diversity and variability of SCAs and their impact on the quality of life. This information should all be part of pretest counseling and not only given after a positive SCA result is obtained. Women should also be made aware of the fact that discordant results are regularly, although not frequently, found between the sex genotype reported in cfDNA testing and a later ultrasound assessment of the external genitalia and that this will then require further testing. On the basis of the pretest information, women should be given the option to accept or decline added screening for SCAs.

The use of cfDNA testing for early sex determination may facilitate sex-selective terminations for nonmedical reasons. Although there can be different views about whether all forms of sex selection for nonmedical reasons are inherently sexist, it is difficult to see how a willingness to terminate a pregnancy just because the fetus happens to be a girl would not reflect and reinforce problematic views about male superiority. In order to avoid contributing to this, cfDNA testing should not be used for sex determination (unless for medical reasons) and should be performed in a way that avoids generating information about fetal sex which is possible when SCAs are not screened for.

In order to maintain minimum quality standards, prenatal screening for fetal abnormalities should always be offered through health professionals with the expertise and training to provide the necessary pre- and posttest information and counseling rather than as DTC test. The use of cfDNA testing as a DTC service just for sex determination is problematic not just because it may facilitate sex-selective terminations, but also in view of the risk of sex discordance being found later in pregnancy, a finding for which pregnant women will be ill-prepared and that they may find very difficult to handle.

REFERENCES

[1] Harper JC, Aittomaki K, Borry P, Cornel MC, de Wert G, Dondorp W, et al. Recent developments in genetics and medically assisted reproduction: from research to clinical applications. Eur J Hum Genet 2018;26(1):12–33.

[2] Minear MA, Lewis C, Pradhan S, Chandrasekharan S. Global perspectives on clinical adoption of NIPT. Prenat Diagn 2015;35(10):959–67.

[3] Griffin B, Edwards S, Chitty LS, Lewis C. Clinical, social and ethical issues associated with non-invasive prenatal testing for aneuploidy. J Psychosom Obstet Gynaecol 2018;39(1):11–8.

[4] Deans ZC, Allen S, Jenkins L, Khawaja F, Hastings RJ, Mann K, et al. Recommended practice for laboratory reporting of non-invasive prenatal testing of trisomies 13, 18 and 21: a consensus opinion. Prenat Diagn 2017;37(7):699–704.

[5] Devaney SA, Palomaki GE, Scott JA, Bianchi DW. Noninvasive fetal sex determination using cell-free fetal DNA: a systematic review and meta-analysis. JAMA 2011;306(6):627–36.

[6] Wright CF, Wei Y, Higgins JP, Sagoo GS. Non-invasive prenatal diagnostic test accuracy for fetal sex using cell-free DNA a review and meta-analysis. BMC Res Notes 2012;5:476.

[7] Lyon MF. Gene action in the X-chromosome of the mouse (Mus musculus L.). Nature 1961;190:372–3.

[8] Ratcliffe S. Long-term outcome in children of sex chromosome abnormalities. Arch Dis Child 1999; 80(2):192–5.

[9] Gravholt CH, Andersen NH, Conway GS, Dekkers OM, Geffner ME, Klein KO, et al. Clinical practice guidelines for the care of girls and women with Turner syndrome: proceedings from the 2016 Cincinnati International Turner Syndrome Meeting. Eur J Endocrinol 2017;177(3):G1–G70.

[10] Sutton EJ, McInerney-Leo A, Bondy CA, Gollust SE, King D, Biesecker B. Turner syndrome: four challenges across the lifespan. Am J Med Genet A 2005;139a(2):57–66.

[11] Sutton EJ, Young J, McInerney-Leo A, Bondy CA, Gollust SE, Biesecker BB. Truth-telling and Turner syndrome: the importance of diagnostic disclosure. J Pediatr 2006;148(1):102–7.

[12] Richardson EJ, Scott FP, McLennan AC. Sex discordance identification following non-invasive prenatal testing. Prenat Diagn 2017;37(13):1298–304.

[13] Juneau K, Bogard PE, Huang S, Mohseni M, Wang ET, Ryvkin P, et al. Microarray-based cell-free DNA analysis improves noninvasive prenatal testing. Fetal Diagn Ther 2014;36(4):282–6.

[14] Nicolaides KH, Musci TJ, Struble CA, Syngelaki A, Gil MD. Assessment of fetal sex chromosome aneuploidy using directed cell-free DNA analysis. Fetal Diagn Ther 2014;35(1):1–6.

[15] Chiu RWK, Chan KCA, Gao YA, Lau VYM, Zheng WL, Leung TY, et al. Noninvasive prenatal diagnosis of fetal chromosomal aneuploidy by massively parallel genomic sequencing of DNA in maternal plasma. Proc Natl Acad Sci USA 2008;105(51):20458–63.

[16] Stokowski R, Wang E, White K, Batey A, Jacobsson B, Brar H, et al. Clinical performance of non-invasive prenatal testing (NIPT) using targeted cell-free DNA analysis in maternal plasma with microarrays or next generation sequencing (NGS) is consistent across multiple controlled clinical studies. Prenatal Diag 2015;35 (12):1243–6.

[17] Gil MM, Accurti V, Santacruz B, Plana MN, Nicolaides KH. Analysis of cell-free DNA in maternal blood in screening for aneuploidies: updated meta-analysis. Ultrasound Obstet Gynecol 2017;50(3):302–14.

[18] Bianchi DW, Wilkins-Haug L. Integration of noninvasive DNA testing for aneuploidy into prenatal care: what has happened since the rubber met the road? Clin Chem 2014;60(1):78–87.

[19] Nuffield Council on Bioethics. Non-invasive prenatal testing: ethical issues. London: NCoB; 2017.

[20] Taylor-Phillips S, Freeman K, Geppert J, Agbebiyi A, Uthman OA, Madan J, et al. Accuracy of non-invasive prenatal testing using cell-free DNA for detection of Down, Edwards and Patau syndromes: a systematic review and meta-analysis. BMJ Open 2016;6(1):e010002.

[21] Bevilacqua E, Ordóñez E, Hurtado i, Rueda L, Mazzone E, Cirigliano V, Jani J. Screening for sex chromosome aneuploidy by cell-free DNA testing: patient choice and performance. Fetal Diagn Ther 2017.

[22] Petersen AK, Cheung SW, Smith JL, Bi WM, Ward PA, Peacock S, et al. Positive predictive value estimates for cell-free noninvasive prenatal screening from data of a large referral genetic diagnostic laboratory. Am J Obstet Gynecol 2017;217(6).

[23] Yu B, Lu BY, Zhang B, Zhang XQ, Chen YP, Zhou Q, et al. Overall evaluation of the clinical value of prenatal screening for fetal-free DNA in maternal blood. Medicine (Baltimore) 2017;96(27) e7114.

[24] Wang YL, Chen Y, Tian F, Zhang JG, Song Z, Wu Y, et al. Maternal mosaicism is a significant contributor to discordant sex chromosomal aneuploidies associated with noninvasive prenatal testing. Clin Chem 2014;60 (1):251–9.

[25] Kornman L, Palma-Dias R, Nisbet D, Scott F, Menezes M, da Silva Costa F, et al. Non-invasive prenatal testing for sex chromosome aneuploidy in routine clinical practice. Fetal Diagn Ther 2017.

[26] Reiss RE, Discenza M, Foster J, Dobson L, Wilkins-Haug L. Sex chromosome aneuploidy detection by non-invasive prenatal testing: helpful or hazardous? Prenat Diagn 2017;37(5):515–20.

[27] Scibetta EW, Valderramos SG, Rao RR, Silverman NS, Han CS, Platt LD. Clinical accuracy of abnormal cell-free fetal DNA (cfDNA) results for the sex chromosomes. Prenat Diagn 2017;37(13):1291–7.

[28] Zhang B, Lu BY, Yu B, Zheng FX, Zhou Q, Chen YP, et al. Noninvasive prenatal screening for fetal common sex chromosome aneuploidies from maternal blood. J Int Med Res 2017;45(2):621–30.

[29] Pieters JJ, Kooper AJ, Eggink AJ, Verhaak CM, Otten BJ, Braat DD, et al. Parents' perspectives on the unforeseen finding of a fetal sex chromosomal aneuploidy. Prenat Diagn 2011;31(3):286–92.

[30] Lalatta F, Tint GS. Counseling parents before prenatal diagnosis: do we need to say more about the sex chromosome aneuploidies? Am J Med Genet A 2013;161A(11):2873–9.

[31] Hill M, Barrett A, Choolani M, Lewis C, Fisher J, Chitty LS. Has noninvasive prenatal testing impacted termination of pregnancy and live birth rates of infants with Down syndrome? Prenat Diagn 2017;37(13):1281–90.

[32] Boyd PA, Loane M, Garne E, Khoshnood B, Dolk H, EUROCAT Working Group. Sex chromosome trisomies in Europe: prevalence, prenatal detection and outcome of pregnancy. Eur J Hum Genet 2011; 19(2):231–4.

[33] So PL, Cheng KYY, Cheuk KY, Chiu WK, Mak SL, Mok SL, et al. Parental decisions following prenatal diagnosis of sex chromosome aneuploidy in Hong Kong. J Obstet Gynaecol Res 2017;43(12):1821–9.

[34] Pieters JJ, Kooper AJ, van Kessel AG, Braat DD, Smits AP. Incidental prenatal diagnosis of sex chromosome aneuploidies: health, behavior, and fertility. ISRN Obstet Gynecol 2011;2011:807106.

[35] Gruchy N, Blondeel E, Le Meur N, et al. Pregnancy outcomes in prenatally diagnosed 47,XXX and 47,XYY syndromes: a 30-year French, retrospective, multicentre study. Prenat Diagn 2016;36:523–9.

[36] de Jong A, Dondorp WJ, Timmermans DR, van Lith JM, de Wert GM. Rapid aneuploidy detection or karyotyping? Ethical reflection. Eur J Hum Genet 2011;19(10):1020–5.

[37] Dondorp WJ, Page-Christiaens GCML, de Wert GMWR. Genomic futures of prenatal screening: ethical reflection. Clin Genet 2016;89(5):531–8.

[38] Leung WC, Lau ET, Lau WL, Tang R, Wong SF, Lau TK, et al. Rapid aneuploidy testing (knowing less) versus traditional karyotyping (knowing more) for advanced maternal age: what would be missed, who should decide? Hong Kong Med J 2008;14(1):6–13.

[39] van der Steen SL, Diderich KE, Riedijk SR, Verhagen-Visser J, Govaerts LC, Joosten M, et al. Pregnant couples at increased risk for common aneuploidies choose maximal information from invasive genetic testing. Clin Genet 2015;88(1):25–31.

[40] Wolf SM, Annas GJ, Elias SP-c. Patient autonomy and incidental findings in clinical genomics. Science 2013;340(6136):1049–50.

[41] Hewison J. Psychological aspects of individualized choice and reproductive autonomy in prenatal screening. Bioethics 2015;29(1):9–18.

[42] Juth N, Munthe C. The ethics of screening in health care and medicine: serving society or serving the patient? Dordrecht, Heidelberg, London, New York: Springer; 2012.

[43] Dondorp W, de Wert G, Bombard Y, Bianchi DW, Bergmann C, Borry P, et al. Non-invasive prenatal testing for aneuploidy and beyond: challenges of responsible innovation in prenatal screening. Eur J Hum Genet 2015;23(11):1438–50.

[44] Parens E, Asch A. The disability rights critique of prenatal genetic testing. Reflections and recommendations. Hastings Cent Rep 1999;29(5):S1–22.

[45] Donley G, Hull SC, Berkman BE. Prenatal whole genome sequencing: just because we can, should we? Hastings Cent Rep 2012;42(4):28–40.

[46] McGillivray G, Rosenfeld JA, McKinlay Gardner RJ, Gillam LH. Genetic counselling and ethical issues with chromosome microarray analysis in prenatal testing. Prenat Diagn 2012;32(4):389–95.

[47] Gray JA, Patnick J, Blanks RG. Maximising benefit and minimising harm of screening. BMJ 2008;336 (7642):480–3.

[48] Akolekar R, Beta J, Picciarelli G, Ogilvie C, D'Antonio F. Procedure-related risk of miscarriage following amniocentesis and chorionic villus sampling: a systematic review and meta-analysis. Ultrasound Obstet Gynecol 2015;45(1):16–26.

[49] Gregg AR, Skotko BG, Benkendorf JL, Monaghan KG, Bajaj K, Best RG, et al. Noninvasive prenatal screening for fetal aneuploidy, 2016 update: a position statement of the American College of Medical Genetics and Genomics. Genet Med 2016;18(10):1056–65.

[50] Yao H, Jiang F, Hu H, Gao Y, Zhu Z, Zhang H, et al. Detection of fetal sex chromosome aneuploidy by massively parallel sequencing of maternal plasma DNA: initial experience in a Chinese hospital. Ultrasound Obstet Gynecol 2014;44(1):17–24.

[51] Huang B, Thangavelu M, Bhatt S, JS C, Wang S. Prenatal diagnosis of 45,X and 45,X mosaicism: the need for thorough cytogenetic and clinical evaluations. Prenat Diagn 2002;22(2):105–10.

[52] Benn P, Chapman AR. Ethical and practical challenges in providing noninvasive prenatal testing for chromosome abnormalities: an update. Curr Opin Obstet Gynecol 2016;28(2):119–24.

[53] Pieters JJ. Incidental findings of sex chromosome aneuploidies in routine prenatal diagnostic procedures. [chapter 8]. In: General discussion and future prospects. Thesis: Radboud University Nijmegen; 2013.

[54] Bianchi DW. From prenatal genomic diagnosis to fetal personalized medicine: progress and challenges. Nat Med 2012;18(7):1041–51.

[55] Guedj F, Bianchi DW, Delabar JM. Prenatal treatment of Down syndrome: a reality? Curr Opin Obstet Gynecol 2014;26(2):92–103.

[56] de Wert G, Dondorp W, Bianchi DW. Fetal therapy for Down syndrome: an ethical exploration. Prenat Diagn 2017;37(3):222–8.

[57] Dondorp W, De Wert G. The 'normalization' of prenatal screening: prevention as prenatal beneficence. In: Schmitz D, Clarke A, Dondorp W, editors. The fetus as a patient A contested concept and its normative implications. Abingdon, UK: Routledge (Taylor & Francis); 2018. p. 144–53.

[58] Campos-Acevedo LD, Ibarra-Ramirez M, Lugo-Trampe JD, Zamudio-Osuna MD, Torres-Munoz I, Velasco-Campos MD, et al. Dosage of sex chromosomal genes in blood deposited on filter paper for neonatal screening of sex chromosome aneuploidy. Genet Test Mol Biomarkers 2016;20(12):786–90.

[59] Finning KM, Chitty LS. Non-invasive fetal sex determination: impact on clinical practice. Semin Fetal Neonatal Med 2008;13(2):69–75.

[60] Hu P, Qiao F, Yuan Y, Sun R, Wang Y, Meng L, et al. Noninvasive prenatal diagnosis for X-linked disease by maternal plasma sequencing in a family of Hemophilia B. Taiwan J Obstet Gynecol 2017;56(5):686–90.

[61] Javitt GH. Pink or blue? The need for regulation is black and white. Fertil Steril 2006;86(1):13–5.

[62] Deans Z, Clarke AJ, Newson AJ. For your interest? The ethical acceptability of using non-invasive prenatal testing to test 'purely for information'. Bioethics 2015;29(1):19–25.

[63] Newson AJ. Ethical aspects arising from non-invasive fetal diagnosis. Semin Fetal Neonatal Med 2008; 13(2):103–8.

[64] Hall A, Bostanci A, John S. Ethical, legal and social issues arising from cell-free fetal DNA technologies. Appendix III to the report: cell-free fetal nucleic acids for noninvasive prenatal diagnosis. Cambridge: PHG Foundation; 2009.

[65] Chapman AR, Benn PA. Noninvasive prenatal testing for early sex identification: a few benefits and many concerns. Perspect Biol Med 2013;56(4):530–47.

[66] Dondorp W, De Wert G, Pennings G, Shenfield F, Devroey P, Tarlatzis B, et al. ESHRE task force on ethics and Law 20: sex selection for non-medical reasons. Hum Reprod 2013;28(6):1448–54.

[67] Bayles MD. Reproductive ethics. Englewood Cliffs, NJ: Prentice-Hall Inc.; 1984.

[68] Seavilleklein V, Sherwin S. The myth of the gendered chromosome: sex selection and the social interest. Camb Q Healthc Ethics 2007;16(1):7–19.

[69] Warren MA. Gendercide: the implications of sex-selection. Totowa, NJ: Rowman & Allanheld; 1985.

[70] Dahl E. The presumption in favour of liberty: a comment on the HFEA's public consultation on sex selection. Reprod Biomed Online 2004;8(3):266–7.

[71] Holm S. Like a frog in boiling water: the public, the HFEA and sex selection. Health Care Anal 2004; 12(1):27–39.

[72] Wilkinson S. Sexism, sex selection and 'family balancing'. Med Law Rev 2008;16(3):369–89.

[73] Davis DS. Genetic dilemmas: reproductive technology, parental choices, and children's futures. 2nd ed. Oxford: Oxford University Press; 2010.

[74] Harper JC, Wilton L, Traeger-Synodinos J, Goossens V, Moutou C, SenGupta SB, et al. The ESHRE PGD Consortium: 10 years of data collection. Hum Reprod Update 2012;18(3):234–47.

[75] Health Council of the Netherlands. Sex selection for non-medical reasons. The Hague: Gezondheidsraad; 1995.

[76] World Health Organisation (WHO). Preventing gender-biased sex selection. Geneva: WHO; 2011.

[77] Madan K, Breuning MH. Impact of prenatal technologies on the sex ratio in India: an overview. Genet Med 2014;16(6):425–32.

[78] Council of Europe, Parliamentary Assembly. Prenatal sex selection. In: Resolution 1829. Strasbourg: CoE; 2011.

[79] Hvistendahl M. Unnatural selection. Choosing boys over girls, and the consequences of a world full of men. New York: Public Affairs; 2011.

[80] Dahl E, Beutel M, Brosig B, Grussner S, Stobel-Richter Y, Tinneberg HR, et al. Social sex selection and the balance of the sexes: empirical evidence from Germany, the UK, and the US. J Assist Reprod Genet 2006;23 (7–8):311–8.

[81] Van Balen F. Attitudes towards sex selection in the Western world. Prenat Diagn 2006;26(7):614–8.

[82] Singh N, Pripp AH, Brekke T, Stray-Pedersen B. Different sex ratios of children born to Indian and Pakistani immigrants in Norway. BMC Pregnancy Childbirth 2010;10:40.

[83] Egan JF, Campbell WA, Chapman A, Shamshirsaz AA, Gurram P, Benn PA. Distortions of sex ratios at birth in the United States; evidence for prenatal gender selection. Prenat Diagn 2011;31(6):560–5.

[84] Dubuc S, Coleman D. An increase in the sex ratio of births to India-born mothers in England and Wales: evidence for sex-selective abortion. Popul Dev Rev 2007;33(2):383–400.

[85] Puri S, Adams V, Ivey S, Nachtigall RD. "There is such a thing as too many daughters, but not too many sons": a qualitative study of son preference and fetal sex selection among Indian immigrants in the United States. Soc Sci Med 2011;72(7):1169–76.

[86] Urquia ML, Ray JG, Wanigaratne S, Moineddin R, O'Campo PJ. Variations in male-female infant ratios among births to Canadian- and Indian-born mothers, 1990-2011: a population-based register study. CMAJ Open 2016;4(2):E116–23.

[87] Tonnessen M, Aalandslid V, Skjerpen T. Changing trend? Sex ratios of children born to Indian immigrants in Norway revisited. BMC Pregnancy Childbirth 2013;13:170.

[88] Smith C, Fogarty A. Is the mothers' country of birth associated with the sex of their offspring in England and Wales from 2007 to 2011? BMC Pregnancy Childbirth 2014;14:332.

[89] Department of Health. Gender ratios at birth in Britain, 2011-15. London: DOH; 2017.

[90] Eklund L, Purewal N. The bio-politics of population control and sex-selective abortion in China and India. Fem Psychol 2017;27(1):34–55.

[91] Amery F. Intersectionality as disarticulatory practice: sex-selective abortion and reproductive politics in the United Kingdom. New Polit Sci 2015;37(4):509–24.

[92] Lee E. Constructing abortion as a social problem: "sex selection" and the British abortion debate. Fem Psychol 2017;27(1):15–33.

[93] Purewal N, Eklund L. 'Gendercide', abortion policy, and the disciplining of prenatal sex-selection in neo-liberal Europe. Glob Public Health 2017;1–18.

[94] Saharso S. Sex-selective abortion—gender, culture and Dutch public policy. Ethnicities 2005;5(2):248–65.

[95] Dickens BM, Serour GI, Cook RJ, Qiu RZ. Sex selection: treating different cases differently. Int J Gynecol Obstet 2005;90(2):171–7.

[96] Nie JB. Limits of state intervention in sex-selective abortion: the case of China. Cult Health Sex 2010;12 (2):205–19.

[97] Raphael T. Disclosing the sex of the fetus: a view from the UK. Ultrasound Obstet Gynecol 2002;20 (5):421–4.

[98] Benn PA. Prenatal technologies and the sex ratio. Genet Med 2014;16(6):433–4.

COST-EFFECTIVENESS OF CELL-FREE DNA-BASED NONINVASIVE PRENATAL TESTING: SUMMARY OF EVIDENCE AND CHALLENGES

16

Stephen Morris*, Caroline S. Clarke[†], Emma Hudson*

Department of Applied Health Research, University College London, London, United Kingdom* Research Department of Primary Care & Population Health, University College London, London, United Kingdom[†]

INTRODUCTION

Systematic reviews have demonstrated that noninvasive prenatal testing (NIPT) based on sequencing of cell-free DNA in maternal plasma is a highly effective screening test for Down's syndrome (trisomy 21) [1–4]. It can also be used to screen for Patau syndrome (trisomy 13), Edwards' syndrome (trisomy 18), and sex chromosome aneuploidies, though with less accuracy [1, 2]. For example, a recent meta-analysis showed that the pooled detection rate (DR) of cfDNA testing for Down's syndrome was 99.7% (95% confidence interval (CI) 99.1%–99.9%) and the false positive rate (FPR) was 0.04% (0.03%–0.07%); for Patau's syndrome the pooled DR and FPR were 97.9% (94.9%–99.1%) and 0.04% (0.03%–0.07%), respectively; for Edwards' syndrome they were 99.0% (65.8%–100.0%) and 0.04% (0.02%–0.07%), respectively [3].

In most developed countries, pregnant women are offered some form of prenatal testing, but the delivery of these services varies between countries. "Standard" screening comprises either first-trimester combined (serum parameters and ultrasound) screening, second-trimester serum screening or integrated screening (a two-stage test where risk is estimated after all first- and second-trimester screening tests). Risk is based on the computation of a series of factors such as maternal age, maternal serum markers, and measurement of nuchal translucency with ultrasound. Those who screen positive, with a predictive risk higher than a predetermined value, are offered a definitive invasive diagnostic test, typically amniocentesis or chorionic villus sampling (CVS) [5]. Current evidence suggests that the procedure-related risks of miscarriage for amniocentesis and CVS are 0.11% (95% CI −0.04% to 0.26%) and 0.22% (−0.71% to 1.16%), respectively [6].

Noninvasive Prenatal Testing (NIPT). https://doi.org/10.1016/B978-0-12-814189-2.00016-5

Cell-free DNA-based NIPT has been proposed as an addition to the current prenatal testing pathway to decrease the number of invasive diagnostic tests, and consequently reduce the number of procedure-related miscarriages [7–11].

There are two main ways in which cfDNA NIPT may be introduced into a prenatal screening pathway. The first is as a contingent strategy to lower the false positive rate of testing prior to invasive diagnostic procedures. In this case, all pregnant women are offered standard prenatal screening and those with a risk above a prespecified threshold are offered cfDNA NIPT rather than an invasive test. Invasive diagnostic testing is then offered to those with an abnormal cfDNA NIPT result. The second alternative is to use cfDNA NIPT as the principal screening test—a first-line or universal testing strategy. In this case cfDNA NIPT replaces standard prenatal screening and it is offered to all women instead of the combined testing or serum screening; invasive diagnostic testing is offered to those with abnormal cfDNA NIPT results. Note that in both testing strategies cfDNA NIPT is not a replacement for invasive procedures because it can produce false positive results, as discussed in Chapter 5— a diagnosis can only be made by invasive diagnostic testing [5]. However, while cfDNA NIPT does not completely remove the need for invasive testing, it can potentially reduce the number of women needing it, thereby also lowering the number of procedure-related miscarriages.

Cell-free DNA-based NIPT is now becoming increasingly available worldwide, but before it can be implemented into routine practice, information is required to identify whether and how it fits best in the prenatal testing pathway, based on the predicted costs and benefits. The benefits of NIPT when used as a contingent test or as first-line testing are expected to be different. With contingent testing, the main benefit is expected to be fewer invasive tests and procedure-related miscarriages. With universal testing, in addition to a reduction in false positives and number of invasive tests and procedure-related miscarriages, the number of cases detected is expected to rise compared with standard prenatal screening. The impact of the introduction of cfDNA NIPT on costs depends on several factors, such as the cost of the test and the range of costs included, which will depend on the perspective taken to measure costs.

The aims of this chapter are twofold. First, we consider current evidence on the cost-effectiveness of cell-free DNA-based NIPT for prenatal testing both as a contingent test and as a universal test. Second, we explore challenges in evaluating the cost-effectiveness of cell-free DNA-based NIPT, which need to be borne in mind when considering the extant evidence. In the following section we provide an overview of economic appraisal, introducing the concepts and methodological framework utilized in the rest of this chapter.

OVERVIEW OF ECONOMIC APPRAISAL

Organizing and planning health care services to meet the needs of a population has to be considered in the context of the finite resources available to do this. Decisions are needed about whether or not to implement a health care program (e.g., should cfDNA NIPT for prenatal testing be made available?), in what ways should it be made available (e.g., as a contingent or universal test?), and to whom should it be made available (e.g., to all pregnant women or a subset?). Economic appraisal is a systematic way of analyzing such choices. The term "economic appraisal" describes a set of techniques that weigh up the costs of an action, such as providing cfDNA NIPT, against the benefits that the action may elicit. The term "economic evaluation" is also used. The distinction between the two terms is that appraisal is undertaken before the action is taken, to help in deciding whether or not and how the action is to be done, and evaluation is undertaken after the action, to judge its effects.

The principles of economic appraisal can be applied to all kinds of health programs, from interventions at the individual level (e.g., whether cfDNA NIPT should be offered or what type of surgery should be used to treat a specific type of cancer) to the population level (e.g., whether there ought to be a nationally organized prenatal testing program for Down's syndrome or whether acute stroke services should be centralized).

The methods and principles of economic appraisal are well articulated and described [12, 13]. Several of these methods and principles are particularly relevant when undertaking an economic appraisal of cfDNA NIPT, including scarcity, opportunity, cost, incremental analysis, perspective, cost-effectiveness analysis, cost-effectiveness plane, and cost-effectiveness acceptability.

SCARCITY

Scarcity exists because the claims on health care resources outstrip the resources that are available. This provides the rationale for economic appraisal, because without scarcity there would be no need to make difficult resource allocation decisions. For example, the decision about whether or not to implement a cfDNA NIPT testing strategy would not need to depend on the cost of that strategy, but could focus instead purely on the benefits to pregnant women and families.

OPPORTUNITY COST

The existence of scarcity leads directly to an important economics concept, namely, opportunity cost. If scarce resources are used to produce a particular health care program or service, those resources cannot be used to produce other goods or services. Opportunity costs reflect the benefits that are forgone by not producing those other services. The opportunity cost of using resources in a particular way is defined as the benefits that would have resulted from their best alternative use. When economists refer to costs, they usually mean opportunity costs. This is different from the financial costs—the costs of goods and services in terms of money—though very often financial costs are used to measure opportunity costs. As an example, from the perspective of a national committee responsible for all screening programs in a public health system, the opportunity costs of additional investment in cfDNA NIPT could be the potential health benefits of investing instead in new screening programs for cancer.

INCREMENTAL ANALYSIS

Economic appraisal involves a comparison between two or more alternatives in terms of their costs and consequences. An economic appraisal ought to be comparative, because a health care program can only be thought to be worth doing or not relative to some alternative—even if that alternative is a do-nothing strategy. An incremental analysis is carried out to express the results of an economic appraisal. This is an assessment of the extra costs and benefits of one option compared with another—the additional costs incurred by one health care program are compared with another, in relation to the additional benefits of that health program compared with the other. Note that if the comparator is a do-nothing strategy it is unlikely to have zero costs and consequences, and these should also be accounted for.

To give an illustration, Torgerson and Spencer [14] showed the importance of incremental analysis by analyzing data on the cost-effectiveness of biochemical screening for Down's syndrome as a replacement for screening by maternal age. An estimate had been published showing that introducing biochemical screening in a population of 20,000 pregnant women would cost £413,500 (including costs of screening, invasive testing, and terminations of pregnancy) and might prevent eleven affected births, giving a cost-effectiveness ratio of £37,591 per case identified. Torgerson and Spencer pointed out that for the same number of women existing antenatal screening procedures based on maternal age already cost £79,000 (including invasive testing and terminations of pregnancy) with four affected births potentially prevented. So, the true cost-effectiveness of biochemical screening when compared to preexisting services was $(£413,500 - £79,000)/(11 - 4) = £334,500/7 = £47,786$, a higher cost per extra affected case identified than for the estimate based on average cost alone.

PERSPECTIVE

The perspective of an economic appraisal is important because among other things it determines what constitutes a cost, that is, what costs are included in the appraisal. Perspectives commonly taken in economic appraisals are those of the health service (or individual components within it, e.g., the hospital), the purchaser or payer of health care, and society as a whole. A societal perspective, which is the broadest perspective, might include health services costs, costs borne by patients and families, and costs in other sectors, for example, costs arising due to production losses from time off work. Researchers should be explicit about the viewpoint they adopt so an assessment can be made as to whether the costs included in an economic appraisal are complete and appropriate. The choice of perspective is important for an economic appraisal of cfDNA NIPT, as it affects the costs included. For example, one perspective is that of the prenatal test provider, who might only be interested in the costs of providing standard prenatal screening, cfDNA NIPT, and invasive diagnostic testing. A broader health service perspective might also include health care costs of pregnancy and childbirth, and termination of pregnancy. A societal perspective additionally includes lifetime costs of caring for individuals born with Down's syndrome, such as lost earnings from parents taking time off work.

COST-EFFECTIVENESS ANALYSIS

Cost-effectiveness analysis is one form of economic appraisal; others are cost-benefit analysis, cost-utility analysis, cost-minimization analysis, and cost-consequences analysis. The main form of economic appraisal that is encountered in the context of cfDNA NIPT is cost-effectiveness analysis, so we focus on that here. As with other forms of economic appraisal, cost-effectiveness analysis seeks to evaluate the costs and consequences of different options. One alternative is preferred to another if it provides greater benefit at the same or lower cost, or has lower cost for the same or greater benefit. Once data on the costs and benefits of the options being compared has been calculated, the first decision rule for this kind of appraisal is to reject any intervention dominated by another alternative or combination of alternatives; a dominated alternative has a greater cost with fewer or the same benefits, or lower benefits with the same or greater costs. For example, Garfield et al. [15] calculated that a contingent cfDNA NIPT test strategy incurred a cost of US\$59,228,142 in 100,000 pregnant women to detect 170 cases of Down's syndrome. This compared favorably with standard prenatal screening, which costed US$59,748,721 and detected 148 cases of Down's syndrome. Therefore in this case cfDNA NIPT dominated standard prenatal screening, and the latter strategy was therefore rejected.

The choice between nondominated alternatives is more complex because it leaves open the question which of two alternatives is preferred if one provides greater benefit than the other, but at a higher cost. A cost-effectiveness ratio (CER), defined as costs divided by benefits, can be calculated to inform this. The CER most often used in health economics is called an incremental CER, or ICER, which measures the incremental cost of an activity relative to its best alternative divided by its incremental effect.

In cost-effectiveness analysis, costs are regarded as opportunity costs measured in monetary terms, and benefits are measured in units other than money. For example, in the case of cfDNA NIPT, benefits might be measured in terms of the number of cases of Down's syndrome diagnosed or the number of procedure-related miscarriages. Costs and benefits are therefore measured in different units and are not comparable with each other; directly weighing the value of effects against the value of costs is not possible. For decision-making, the ICER is measured against a ceiling ratio or threshold; to be regarded as cost effective any intervention must have an ICER which is on, or below, this threshold. To illustrate, Morris et al. [16] calculated the cost-effectiveness of cfDNA NIPT as a universal testing strategy compared with standard prenatal testing in a population of 10,000 pregnant women. The costs of the cfDNA NIPT universal strategy and standard screening were estimated to be UK£1,825,000 and UK£279,000, respectively. The two strategies were estimated to produce 16.49 and 13.24 diagnoses of Down's syndrome, respectively. Neither option was dominant. The incremental cost and incremental effect of cfDNA NIPT as a universal testing strategy versus standard prenatal testing strategy were UK£1,546,000 and 3.25 cases, respectively, with an associated ICER of UK£475,692 per extra case detected. By using the aforementioned ceiling ratio it is possible to judge whether or not cfDNA NIPT is cost effective in this case. For example, if the ceiling ratio was UK£500,000 per extra case identified, that is, we are prepared to pay up to UK£500,000 to diagnose an additional case of Down's syndrome, then cfDNA NIPT is good value for money because the ICER is below the threshold. Alternatively, if we are only prepared to pay UK£100,000 per extra case identified then cfDNA NIPT would not represent good value for money. A potential problem with cost-effectiveness analysis, which affects its usefulness for decision-making, is that the ceiling ratio may not be known (see later).

COST-EFFECTIVENESS PLANE

Incremental analyses in economic appraisal are often illustrated using the cost-effectiveness plane (Fig. 1). ICERs are presented graphically as a combination of the costs and the effects of a health intervention, described in the figure as cfDNA NIPT compared to an alternative. Incremental costs are conventionally placed on the north-south axis and incremental effects on the east-west axis. The origin is the point where costs and benefits of both options are equal. Incremental costs and effects can be positive, negative, or zero.

An intervention can be positioned anywhere on this diagram according to its incremental costs and benefits versus the comparator. For example, suppose we were comparing a contingent cfDNA NIPT testing strategy to standard prenatal screening. If cfDNA NIPT lies in the northwest quadrant, such as point A, the costs of cfDNA NIPT are higher than standard prenatal screening, and its benefits are lower. It is therefore dominated by the alternative. Conversely, in the southeast quadrant, at a point

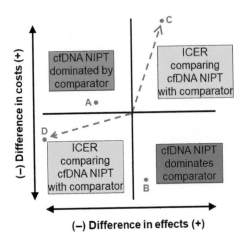

FIG. 1

The cost-effectiveness plane comparing cfDNA NIPT vs comparator.

such as B, costs are lower, and benefits are higher, so cfDNA NIPT dominates standard prenatal screening. In the northeast quadrant, at a point such as C, higher benefits of cfDNA NIPT are gained at a higher cost over standard prenatal screening. So, we can calculate an ICER, the incremental cost per unit of effect gained, measured as the slope of the line from the origin (the point where costs and effects are equal between both strategies) to point C. In the southwest quadrant, at a point such as D, the cfDNA NIPT costs are lower, but also the benefits are lower than the alternative strategy. Again, we can calculate an ICER, although this is now a cost saving per unit of effect lost, which is again measured as the slope of the line from the origin to the point.

COST-EFFECTIVENESS ACCEPTABILITY

The ceiling ratio can also be visualized in the cost-effectiveness plane diagram (Fig. 2). In this diagram, the dotted diagonal line marked Rc represents the ceiling ratio. If cfDNA NIPT lies above the line, it will not be acceptable on cost-effectiveness grounds versus standard prenatal screening. This is either because cfDNA NIPT is dominated by standard prenatal screening, whatever the value of the ceiling ratio, as in point A (or any other point in the northwest quadrant), or its ICER does not satisfy the ceiling ratio, as in points B and C (or any other point above the Rc line in the southwest and northeast quadrants). It should be noted that for points in the southwest quadrant above Rc, the value of the ICER (i.e., the incremental cost per unit of effect gained or the slope of the line from the origin to the point) is smaller than the ceiling ratio, as in point B, while for points in the northeast quadrant above Rc it is larger, as in point C. Below the Rc line cfDNA NIPT is acceptable on cost-effectiveness grounds. This is either because it dominates the alternative, as in point D, or its ICER satisfies the ceiling ratio, as in points E and F. In this case, the ICER is then either larger than the ceiling ratio in the southwest quadrant, as in point E, or smaller than it in the northeast quadrant, as in point F.

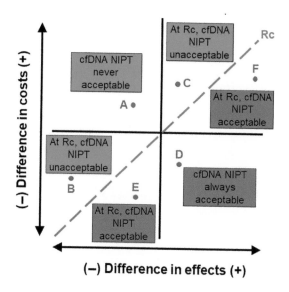

FIG. 2

Cost-effectiveness acceptability of cfDNA NIPT vs comparator.

EVIDENCE ON THE COST-EFFECTIVENESS OF CELL-FREE DNA NIPT
SYSTEMATIC REVIEW: GARCÍA-PÉREZ ET AL. [17]

In this section we review empirical evidence on the cost-effectiveness of cfDNA NIPT for prenatal screening and subsequent diagnosis. We focus on economic studies identified by a recent, up-to-date systematic review of the cost-effectiveness of cfDNA NIPT for trisomies 21, 18, and 13 based on papers published in English or Spanish between January 2006 and April 2017 [17]. The review included universal and contingent strategies, compared with screening strategies that do not include cfDNA NIPT or with no screening. To be included in the review the study had to present the number of cases of trisomy 21, 18, and/or 13 detected for every comparator, and their costs or ICERs.

Fourteen papers related to 12 studies, all published after 2012, were identified and included in the review [15, 16, 18–29] (one study was reported three times [22–24]). The review authors assessed the methodological quality of the economic appraisals and found it to be acceptable in most cases, but noted that "the lack of transparency and details on sources prevented a more accurate assessment of the bias" [17]. The sensitivity, specificity, and false positive rate of the strategies compared were consistent with wider published evidence [17].

Five studies were undertaken in the United States; two in Australia; two in the United Kingdom; and one each in Canada, the Netherlands, and Belgium. All studies were model based, combining data from several sources. All studies compared a prenatal screening strategy (usually first-trimester screening or a combination of first- and second-trimester screening or integrated screening) and some form of screening with cfDNA-based NIPT. Nine studies evaluated cfDNA-based NIPT in a contingent testing strategy; nine evaluated it in a universal testing strategy. Two studies included detection of trisomies

13, 18, and 21 [15, 21]; the remainder focused on trisomy 21. The perspective was not always stated but a health care system perspective was most common (see later). Some form of health care costs were included in every study. Two studies used multiple perspectives, including educational costs and lost productivity costs as well as costs to the health care system [28, 29]. The time horizon was the duration of pregnancy in all studies; three studies included longer time horizons as well (5 years [27], lifetime [28, 29]).

There are several published studies that have assessed the cost-effectiveness of cfDNA NIPT but which were not included in García-Pérez et al.'s [17] systematic review because they did not meet the inclusion criteria [30–34].

NIPT AND PROCEDURE-RELATED MISCARRIAGES

Ten of the 12 studies included procedure-related miscarriages as an outcome. In all of these studies the number of procedure-related miscarriages was lower for cfDNA NIPT, whether in a contingent or a universal testing strategy, compared with the strategy without cfDNA NIPT.

CONTINGENT CELL-FREE DNA NIPT VERSUS STANDARD PRENATAL SCREENING

Nine studies evaluated cfDNA NIPT in a contingent testing strategy compared with the usual screening strategy [15, 16, 18–20, 24–26, 29]. In terms of the incremental costs and benefits, the studies were represented in all four quadrants of the cost-effectiveness plane, with contingent cfDNA NIPT being shown to be more or less costly than standard prenatal screening, and more or less effective at identifying true cases of Down's syndrome, across different studies (Table 1).

cfDNA NIPT was in the northwest quadrant of the cost-effectiveness plane in one Australian study [26]. Under the baseline scenario in which uptake of cfDNA testing was assumed to be the same as uptake of invasive testing prior to the introduction of the new test, there was no difference in Down's syndrome cases detected and an increase in costs. When cfDNA NIPT uptake was assumed to be 100%, it moved to the northeast quadrant.

Three studies reported findings in the southwest quadrant of the cost-effectiveness plane. In all three cases cfDNA detected slightly fewer Down's syndrome cases, had fewer procedure-related miscarriages, and was less costly than standard screening. In an Australian study, Ayres et al. [18] calculated that in a population of 300,000 pregnant women cfDNA testing would detect three fewer Down's syndrome cases, have 96 fewer procedure-related miscarriages, and save approximately Aus$900,000. In a UK study Morris et al. [16] calculated that in a cohort of 10,000 pregnant women cfDNA NIPT would detect around two fewer Down's syndrome cases, lead to around one fewer procedure-related miscarriage, and save approximately UK£35,000. A potential explanation was that adding another layer of screening introduced a further opportunity for people to "drop out" of the screening program. In a study of 129,000 births in Belgium, Neyt et al. [24] calculated that cfDNA testing would detect around one fewer Down's syndrome case, have 42 fewer procedure-related miscarriages, and save approximately €1.6 million. The small reduction in Down's syndrome cases detected in all three studies was probably due to assumptions made concerning the uptake of cfDNA testing and subsequent invasive testing among those with a positive cfDNA test [16].

When cfDNA NIPT was in the northeast quadrant of the cost-effectiveness plane, the incremental cost per case of Down's syndrome diagnosed varied between studies from US$2246 to US$119,539.

Table 1 Cost-Effectiveness of cfDNA NIPT for Down's Syndrome in a Contingent Testing Strategy vs Standard Prenatal Screening			
Difference in costs	Higher	O'Leary [26][a]	Beulen [19] $119,539 Okun [25] $2246[b] O'Leary [26] $58,855[c] Walker [29] $26,699[d]
	Lower	Ayres [18] Morris [16][e] Neyt [24] $3,363,495[f] Lower	Chitty [20][e] Garfield [15] Walker [28][g] Higher
		Difference in Down's syndrome cases diagnosed	

Studies are referred to by the name of the first author. Data derived from García-Pérez et al. [17]. Monetary values in the northeast quadrant are ICERs (incremental cost per additional case of Down's syndrome diagnosed), reported in 2016 US$. Monetary values in the southwest quadrant (where reported) are ICERs reflecting the cost saving per case of Down's syndrome not detected, reported in 2016 US$. A large ICER in the southwest quadrant favors the intervention because it indicates a large saving per extra case missed (see text). A large ICER in the northeast quadrant is unfavorable for the intervention because it indicates that the extra costs per extra case detected are high.
[a]Assuming no change in uptake of standard prenatal screening after the introduction of cfDNA NIPT testing as a contingent test.
[b]Assuming current performance of standard prenatal screening.
[c]Assuming increased uptake of standard prenatal screening after the introduction of cfDNA NIPT testing as a contingent test.
[d]Taking a payer perspective.
[e]Assuming screening risk cutoff ≥1/150.
[f]Three studies reporting the same data. Assuming screening risk cutoff ≥1/300.
[g]Taking a government or societal perspective.

There was variation between studies in terms of the choice of comparator (reflecting usual screening practice within each country), cutoff risk to classify a pregnancy as high risk, cfDNA NIPT performance, perspective, and costs of cfDNA NIPT. All of these factors may have contributed to the variation in results between studies, explaining why they cover all four quadrants of the cost-effectiveness plane. García-Pérez et al. [17] showed that, even among studies within the same country, different results were obtained, suggesting that the extent of the variation was not explained solely by country-level variations in screening practice.

UNIVERSAL CELL-FREE DNA NIPT VERSUS USUAL PRENATAL SCREENING

Nine studies evaluated cfDNA NIPT as a universal screening strategy compared with the usual screening [16, 18, 19, 21, 24, 25, 27–29]. In every study cfDNA NIPT was more effective than the alternative in terms of cases of Down's syndrome diagnosed (Table 2). In the majority of studies (seven), cfDNA NIPT was in the northeast quadrant of the cost-effectiveness plane (more effective, more costly). In this case, the incremental costs per additional case of Down's syndrome diagnosed were larger than for contingent testing, varying between studies from US$174,121 to US$1,031,745.

In four studies cfDNA NIPT was in the southeast quadrant of the cost-effectiveness plane (more effective, less costly), being a dominant strategy compared with the comparator. There are idiosyncrasies with each of these four studies: Fairbrother et al. [21] concluded that universal cfDNA NIPT was cost saving compared with first-trimester screening when the unit cost of NIPT was US$453 or less, but did not report any incremental costs; Song et al. [27] evaluated a combined strategy using cfDNA NIPT

Table 2 Cost-Effectiveness of cfDNA NIPT for Down's Syndrome in a Universal Testing Strategy vs Standard Prenatal Screening

Difference in costs	Higher		Ayres [18] $767,516 Beulen [19] $584,979 Morris [16] $719,413 Neyt [24] $1,031,745[a] Okun [25] $396,759 Walker [28] $174,121–$357,074[b] Walker [29] $210,537–$273,602[c]
	Lower		Fairbrother [21] Song [27][d] Walker [28][e] Walker [29][e]
		Lower	Higher
		Difference in Down's syndrome cases diagnosed	

Studies are referred to by the name of the first author. Data derived from García-Pérez et al. [17]. Monetary values in the northeast quadrant are ICERs (incremental cost per additional case of Down's syndrome diagnosed), reported in 2016 US$. A large ICER in the northeast quadrant is unfavorable for the intervention because it indicates that the extra costs per extra case detected are high (see text).
[a]*Three studies reporting the same data.*
[b]*Taking a government, health sector, or payer perspective.*
[c]*Taking a government or payer perspective.*
[d]*Universal NIPT for women aged >35 years or at high risk due to pregnancy history; contingent NIPT for all others who test positive to standard prenatal screening.*
[e]*Taking a societal perspective.*

as a universal and contingent test in different population subgroups (universal NIPT for women aged >35 years or at high risk due to pregnancy history; contingent NIPT for all others who test positive to standard prenatal screening); Walker et al. [28, 29] showed that universal cfDNA NIPT was a dominant strategy when a societal perspective was taken—when a narrower perspective was taken cfDNA NIPT was more effective and more costly than the comparator (i.e., in the northeast quadrant of the cost-effectiveness plane).

TRISOMIES 13 AND 18

Two studies also reported results for trisomies 13 and 18 [15, 21], with one evaluating contingent cfDNA NIPT [15], and the other universal cfDNA NIPT [21]. Both put NIPT in the southeast quadrant of the cost-effectiveness plane. The limited availability of data means that it is difficult to draw robust conclusions about the cost-effectiveness of cfDNA NIPT for diagnosing trisomies 13 and 18.

SUMMARY OF EVIDENCE

There is considerable variation in the cost-effectiveness assessments of cfDNA NIPT when used in a contingent testing strategy, in that cfDNA NIPT ranges from being dominated by usual screening practice to dominating it. Universal cfDNA NIPT is more effective than usual screening practice, but also more costly, with ICERs reported in excess of US$175,000 per additional case of Down's syndrome diagnosed. In both strategies cfDNA NIPT is associated with fewer procedure-related miscarriages than usual screening. There is little evidence on the cost-effectiveness of cfDNA NIPT for prenatal screening and diagnosis of trisomies 13 and 18. At the end of their review, García-Pérez et al. [17] concluded that "the cost-effectiveness of contingent cfDNA NIPT is uncertain according to several studies, while the universal NIPT is not cost-effective currently."

METHODOLOGICAL CHALLENGES FOR THE ECONOMIC APPRAISAL OF CELL-FREE DNA-BASED NIPT

The economic appraisal of cfDNA NIPT for prenatal screening and diagnosis raises several distinct methodological challenges. In this section we describe and illustrate these challenges and provide suggestions for overcoming them. We use this to identify a future research agenda.

CHOICE OF OUTCOME MEASURE AND THE CEILING RATIO

Outcome measures that are most commonly used in economic appraisals of cfDNA NIPT are the numbers of Down's syndrome cases diagnosed and procedure-related miscarriages. If, using these measures, cfDNA NIPT lies in the northwest or southeast quadrants of the cost-effectiveness plane then it is dominated by or dominates the alternative and the decision whether or not to adopt cfDNA NIPT on cost-effectiveness grounds is straightforward. If, alternatively, cfDNA NIPT is in the southwest or northeast quadrants the decision rule for adopting it, or not, is based on a comparison of the ICER with the ceiling ratio. For example, in the northeast quadrant the ICER could be expressed in terms of the incremental cost per additional case of Down's syndrome diagnosed or the incremental cost per additional procedure-related miscarriage avoided. Unfortunately, this approach is of limited use for decision-making unless the ceiling ratio is known. Put another way, unless the maximum amount of extra money a decision-maker is prepared to pay to diagnose one additional Down's syndrome case or to avoid an extra procedure-related miscarriage is known, it is not possible to judge whether or not NIPT is cost effective. There is little evidence as to what the ceiling ratios are with respect to these outcome measures. The implications of this for decision-making are important. We showed earlier that the minimum incremental cost per additional case of Down's syndrome diagnosed for universal cfDNA NIPT versus standard screening was around US$175,000; however, if the magnitude of the ceiling ratio is not known it is unclear whether or not universal NIPT is cost effective.

One solution is to use an alternative outcome measure for which the ceiling ratio is known. For example, in the UK the recommended outcome for economic evaluations is quality-adjusted life years (QALYs), and the recommended ICER is the incremental cost per QALY gained [35]. One advantage of using QALYs is that in the UK there is a published ceiling ratio that can be used to judge whether or not an intervention is good value for money (£20,000–£30,000 per QALY gained [35]). Unfortunately, measuring QALYs from prenatal testing is challenging. Only one study has reported QALY gains for cfDNA NIPT [32]. The authors assumed that women would have a lower quality of life during their remaining years of life after giving birth to a baby with Down's syndrome, during 2 years after miscarriage or termination and during 1 year after a false positive result. It was difficult to determine the utility score for a child with Down's syndrome and for other unaffected children in the family as well as for the parents. There was also uncertainty about what happens when a pregnant mother decides to terminate an affected pregnancy. She might become pregnant again and have another baby, or not, and that baby might be healthy, or not. Some women might decide not to terminate the pregnancy in the case of an affected fetus and use the information to prepare for the birth of an affected baby. This, as well as the reassurance that a baby does not have the condition, could increase the utility for all parents regardless of whether they have a child with the condition screened for or not. Due to these measurement difficulties, outcomes of economic appraisals for prenatal screening and diagnosis of genetic conditions are not usually reported in terms of QALYs.

Another solution is to identify the ceiling ratio and use that to inform the analysis. Few economic appraisals of cfDNA NIPT have done this. An exception is Chitty et al. [20], who based their judgment on the cost-effective use of cfDNA NIPT (as a contingent test implemented at a Down's syndrome risk threshold of 1/150) on the grounds that it significantly decreased the number of invasive diagnostic tests and procedure-related miscarriages, with no significant effect on costs or number of cases of Down's syndrome cases detected. This judgment was based on feedback from the UK National Screening Committee stating that cfDNA NIPT was required to be at least as effective as standard prenatal screening and no more costly, to be implemented by the health service. This was equivalent to specifying a ceiling ratio equal to £0, that is, incremental costs were zero or negative, with an additional constraint that increment effects were zero or positive.

ACCOUNTING FOR IMPACT ON UPTAKE

Uptake rates of screening and invasive diagnostic testing in conventional prenatal screening strategies are unlikely to be 100%. For example, in a UK-based study Morris et al. assumed Down's syndrome screening uptake of 69% and uptake of invasive diagnostic testing in screen positive cases of 80%–90% based on published data [16]. Uptake of cfDNA-based NIPT is also unlikely to be 100%. Morris et al. assumed that uptake of cfDNA NIPT as a contingent testing strategy was 80%–90%, reflecting the invasive diagnostic test uptake rate; they also assumed that uptake of cfDNA NIPT as a first-line test was 69%, reflecting the uptake rate of standard prenatal screening. While there may have been some logic to the uptake figures selected, the actual values were unknown. When uptake was increased in sensitivity analyses by an arbitrary 10 percentage points, this increased the number of Down's syndrome cases diagnosed, procedure-related miscarriages, and costs. Similar findings were obtained by O'Leary et al. [26], who also ran sensitivity analyses of cfDNA NIPT uptake. The implication is that uptake of cfDNA NIPT is a potentially important variable affecting its cost-effectiveness and ought to be scrutinized in economic appraisals.

It is also possible that introducing cfDNA NIPT into the screening pathway affects the uptake of traditional screening and invasive diagnostic testing. That is, the availability of a highly sensitive and risk-free contingent test as the next line of testing for high-risk pregnancies may increase the uptake of the primary screening, and this could affect overall costs and outcomes of the screening pathway. Few studies have accounted for this possibility. Morris et al. [16] modeled a hypothetical scenario in a sensitivity analysis in which the introduction of cfDNA NIPT as a contingent test increased the uptake of screening by an arbitrary 10 percentage points. Chitty et al. [20], in their evaluation of cfDNA NIPT in a contingent testing strategy, measured the actual uptake of screening and invasive diagnostic testing with and without cfDNA NIPT. They ultimately assumed that screening uptake would not change, but found that uptake of invasive diagnostic testing was higher after an abnormal cfDNA NIPT result than after a positive screening result in the conventional screening pathway.

The implications are that uptake of cfDNA NIPT, and its impact on uptake of screening and invasive diagnostic testing, are potentially important factors affecting cost-effectiveness. Uptake ought to be considered explicitly in economic appraisals of cfDNA NIPT.

CHOICE OF PERSPECTIVE

The choice of perspective determines the costs included and is therefore important for an economic appraisal of cfDNA NIPT. Of the 12 studies identified by García-Pérez et al. [17] the perspective was explicitly stated in 10 studies. The perspectives taken were the health care system (four studies), the health care payer (two studies), the UK National Screening Committee, and the Ministry of Health; two studies reported findings from multiple perspectives (societal, government, payer, health system). Walker et al. [29] justified their choice of multiple perspectives on the basis that although a societal perspective is often recommended, decisions are often based on narrower perspectives. When taking a societal perspective, they included the costs of screening, diagnosis, and termination of pregnancy as well as the lifetime costs associated with trisomies 21, 18, and 13. Lifetime costs were defined as the average difference in direct medical and educational costs between a person with the trisomy and an average person, plus the indirect costs of lost productivity for the person with the condition due to morbidity and mortality. The government perspective included screening, diagnosis, and termination of pregnancy costs plus the direct lifetime medical and education costs. The payer perspective included only costs associated with screening, diagnosis, and termination of pregnancy. The authors showed that universal and contingent cfDNA NIPT dominated usual screening when a societal perspective was taken, but that when the narrower payer perspective was taken cfDNA NIPT was in the northeast quadrant of the cost-effectiveness plane. This difference was mainly due to the fact that each case of Down's syndrome avoided led to a reduction in lifetime direct and indirect costs to society of around US\$1.5 million. The implication is that a broader societal perspective is more likely to show that cfDNA NIPT is cost effective due to the cost savings from avoiding Down's syndrome births.

The most common perspective in the review was the health care system, and as a consequence most studies focused exclusively on direct health care costs. However, there was some uncertainty as to what costs ought to be included. For example, Morris et al. [16] ran analyses including costs of screening, invasive diagnostic testing, miscarriage, and termination of pregnancy. They also ran a supplementary analysis including cost of live birth outcomes, on the basis that this was consistent with the perspective adopted. They reported that these costs were similar between the different strategies, but so large compared with screening and diagnosis costs, that they masked all cost differences between usual screening and cfDNA NIPT.

The main implication is that costs included in any economic appraisal of cfDNA NIPT should be considered carefully and reported transparently. The choice of study perspective can affect whether or not cfDNA NIPT is good value for money. The perspective ought therefore to be explicitly stated in all economic appraisals to allow readers to judge the applicability of the findings to their own context and whether the correct costs have been included. In addition, the perspective itself ought to be justified in terms of its relevance to the specific decision-making context, and the costs included in the economic evaluation ought to be justified with regards to the perspective taken.

TRANSFERABILITY

Care must be taken when making comparisons about the cost-effectiveness of cfDNA NIPT between countries, especially when deciding whether or not it is cost effective in one country based on an economic appraisal conducted in another country. In the case of universal cfDNA NIPT, the results in Table 2 are reasonably consistent in that it falls largely in the northeast quadrant of the cost-effectiveness plane with a sizeable ICER. However, in the case of contingent testing strategy the results are more variable (Table 1), both within and between countries, and researchers and decision-makers should take care when inferring results between countries.

There are several reasons why cost-effectiveness results may not be transferable between countries, including: the comparator may be different (i.e., usual screening tests may vary between countries); in the case of contingent cfDNA NIPT the screening cutoff values for identifying high-risk pregnancies might be different; population characteristics might be different, affecting the underlying risk of trisomies 13, 18, and 21; uptake of screening, cfDNA NIPT, and invasive diagnostic testing may vary; the appropriate perspective might be different; and unit costs (e.g., of screening, invasive diagnostic testing, and cfDNA NIPT) might vary between countries. To illustrate, the costs of the cfDNA NIPT in the 12 studies reviewed previously were (converted into 2016 US$ [36], in increasing order of magnitude): $378, $378, $403–$631, $408–$714, $415, $415, $522, $565, $668, $838, $987, $1200, and $1224. There are several plausible reasons for this variation including differences in cost base years and the type of health system, but since cfDNA NIPT cost is an important driver of cost-effectiveness [17] variations in this cost affect transferability.

A related issue is the variation in cost-effectiveness over time, even within countries, that is, the extent to which cost-effectiveness findings are likely to persist over time. For example, a decline in next generation sequencing costs over time [37] will produce an improvement in the cost-effectiveness of cfDNA NIPT, and previous findings may become outdated.

CONCLUSION

If cfDNA NIPT is to be considered for implementation into the prenatal screening pathway, either as a contingent test or first-line test, then its cost-effectiveness compared with standard screening needs to be assessed. According to existing evidence, there is variation in the cost-effectiveness when used in a contingent testing strategy for screening for Down's syndrome: cfDNA NIPT ranges from being dominated by usual screening to dominating it. Primary cfDNA NIPT for screening for Down's syndrome is more effective than standard screening practice, but also more costly, with high ICERs. In both strategies cfDNA NIPT is associated with lower procedure-related miscarriages compared with usual screening due to the reduction in invasive diagnostic testing. There is little evidence on the cost-effectiveness of cfDNA NIPT for prenatal screening and diagnosis of trisomies 13 and 18.

Several factors need to be considered when considering the cost-effectiveness of cfDNA NIPT, including the outcome measure used and the ceiling ratio for judging cost-effectiveness, the uptake of cfDNA NIPT and its impact on uptake of screening and invasive diagnostic testing, study perspective and associated costs included, and transferability of findings between countries and over time. Further research is needed to evaluate the cost-effectiveness of cfDNA NIPT in countries where it is being considered without existing relevant economic evidence.

REFERENCES

[1] Taylor-Phillips S, Freeman K, Geppert J, Agbebiyi A, Uthman OA, Madan J, et al. Accuracy of non-invasive prenatal testing using cell-free DNA for detection of down, Edwards and Patau syndromes: a systematic review and meta-analysis. BMJ Open 2016;6:e010002.

[2] Mackie FL, Hemming K, Allen S, Morris RK, Kilby MD. The accuracy of cell-free fetal DNA-based non-invasive prenatal testing in singleton pregnancies: a systematic review and bivariate meta-analysis. Br J Obstet Gynaecol 2017;124:32–46.

[3] Gil MM, Accurti V, Santacruz B, Plana MN, Nicolaides KH. Analysis of cell-free DNA in maternal blood in screening for fetal aneuploidies: updated meta-analysis. Ultrasound Obstet Gynecol 2017;50(3):302–14.

[4] Iwarsson E, Jacobsson B, Dagerhamn J, Davidson T, Bernabé E, Heibert Arnlind M. Analysis of cell-free fetal DNA in maternal blood for detection of trisomy 21, 18 and 13 in a general pregnant population and in a high-risk population—a systematic review and meta-analysis. Acta Obstet Gynecol Scand 2017;96:7–18.

[5] Dondorp W, de Wert G, Bombard Y, Bianchi DW, Bergmann C, Borry P, et al. European Society of Human Genetics, American Society of Human Genetics. Non-invasive prenatal testing for aneuploidy and beyond: challenges of responsible innovation in prenatal screening. Eur J Hum Genet 2015;23:1438–50.

[6] Akolekar R, Beta J, Picciarelli G, Ogilvie C, D'Antonio F. Procedure-related risk of miscarriage following amniocentesis and chorionic villus sampling: a systematic review and meta-analysis. Ultrasound Obstet Gynecol 2015;45:16–26.

[7] Allyse M, Minear MA, Berson E, Sridhar S, Rote M, Hung A, et al. Non-invasive prenatal testing: a review of international implementation and challenges. Int J Women's Health 2015;7:113–26.

[8] Khalifeh A, Weiner S, Berghella V, Donnenfeld A. Trends in invasive prenatal diagnosis: effect of sequential screening and noninvasive prenatal testing. Fetal Diagn Ther 2016;39:292–6.

[9] Gregg AR, Stotko BG, Benkendorf JL, Monaghan KG, Bajaj K, Best RG, et al. on behalf of the ACMG Noninvasive Prenatal Screening Work Group. Noninvasive prenatal screening for fetal aneuploidy, 2016 update: a position statement of the American College of Medical Genetics and Genomics. Genet Med 2016;18 (10):1056–65.

[10] Benn P, Borrell A, Chiu RWK, Cuckle H, Dugoff L, Faas B, et al. Position statement from the Chromosome Abnormality Screening Committee on behalf of the Board of the International Society for Prenatal Diagnosis. Prenat Diagn 2015;35:725–34.

[11] Salomon LJ, Alfirevic Z, Audibert F, Kagan KO, Paladini D, Yeo G, et al. on behalf of the ISUOG Clinical Standards Committee. ISUOG updated consensus statement on the impact of cfDNA aneuploidy testing on screening policies and prenatal ultrasound practice. Ultrasound Obstet Gynecol 2017;49:815–6.

[12] Morris S, Devlin N, Parkin D, Spencer A. Economic analysis in health care. 2nd ed. Chichester: Wiley; 2012.

[13] Drummond M, Sculpher M, Claxton G, Stoddard M, Torrance GW. Methods for the economic evaluation of health care programmes. 3rd ed. Oxford: Oxford University Press; 2015.

[14] Torgerson DJ, Spencer A. Marginal costs and benefits. Br Med J 1996;312:35.

[15] Garfield SS, Armstrong SO. Clinical and cost consequences of incorporating a novel non-invasive prenatal test into the diagnostic pathway for fetal trisomies. J Manag Care Med 2012;15:34–41.

[16] Morris S, Karlsen S, Chung N, Hill M, Chitty LS. Model-based analysis of costs and outcomes of non-invasive prenatal testing for Down's syndrome using cell free fetal DNA in the UK National Health Service. PLoS ONE 2014;9(4):e93559.

[17] García-Pérez L, Linertová R, Álvarez-de-la-Rosa M, Bayón JC, Imaz-Iglesia I, Ferrer-Rodríguez J, et al. Cost-effectiveness of cell-free DNA in maternal blood testing for prenatal detection of trisomy 21, 18 and 13: a systematic review. Eur J Health Econ 2017. https://doi.org/10.1007/s10198-017-0946-y.

[18] Ayres AC, Whitty JA, Ellwood DA. A cost-effectiveness analysis comparing different strategies to implement noninvasive prenatal testing into a down syndrome screening program. Aust N Z J Obstet Gynaecol 2014;54:412–7.

[19] Beulen L, Grutters JP, Faas BH, Feenstra I, van Vugt JM, Bekker MN. The consequences of implementing non-invasive prenatal testing in Dutch national health care: a cost-effectiveness analysis. Eur J Obstet Gynecol Reprod Biol 2014;182:53–61.

[20] Chitty LS, Wright D, Hill M, Verhoef TI, Daley R, Lewis C, et al. Uptake, outcomes, and costs of implementing non-invasive prenatal testing for Down's syndrome into NHS maternity care: Prospective cohort study in eight diverse maternity units. Br Med J 2016;354:i3426.

[21] Fairbrother G, Burigo J, Sharon T, Song K. Prenatal screening for fetal aneuploidies with cell-free DNA in the general pregnancy population: a cost-effectiveness analysis. J Matern Fetal Neonatal Med 2016;29:1160–4.

[22] Gyselaers W, Hulstaert F, Neyt M. Contingent non-invasive prenatal testing: an opportunity to improve non-genetic aspects of fetal aneuploidy screening. Prenat Diagn 2015;35:1347–52.

[23] Hulstaert F, Neyt M, Gyselaers W. The non-invasive prenatal test (NIPT) for trisomy 21—health economic aspects. Health Technology Assessment (HTA) Brussels, Belgian Health Care Knowledge Centre (KCE); 2014. KCE Reports 222, D/2014/10.273/36.

[24] Neyt M, Hulstaert F, Gyselaers W. Introducing the non-invasive prenatal test for trisomy 21 in Belgium: a cost-consequences analysis. BMJ Open 2014;4:e005922.

[25] Okun N, Teitelbaum M, Huang T, Dewa CS, Hoch JS. The price of performance: a cost and performance analysis of the implementation of cell-free fetal DNA testing for Down syndrome in Ontario, Canada. Prenat Diagn 2014;34:350–6.

[26] O'Leary P, Maxwell S, Murch A, Hendrie D. Prenatal screening for Down syndrome in Australia: costs and benefits of current and novel screening strategies. Aust N Z J Obstet Gynaecol 2013;53:425–33.

[27] Song K, Musci TJ, Caughey AB. Clinical utility and cost of non-invasive prenatal testing with cfDNA analysis in high-risk women based on a US population. J Matern Fetal Neonatal Med 2013;26:1180–5.

[28] Walker BS, Jackson BR, LaGrave D, Ashwood ER, Schmidt RL. A cost-effectiveness analysis of cell free DNA as a replacement for serum screening for down syndrome. Prenat Diagn 2015;35:440–6.

[29] Walker BS, Nelson RE, Jackson BR, Grenache DG, Ashwood ER, Schmidt RL. A cost-effectiveness analysis of first trimester non-invasive prenatal screening for fetal trisomies in the United States. PLoS ONE 2015;10 (7):e0131402.

[30] Palomaki GE, Kloza EM, Lambert-Messerlian GM, Haddow JE, Neveux LM, Ehrich M, et al. DNA sequencing of maternal plasma to detect down syndrome: an international clinical validation study. Genet Med 2011;13:913–2020.

[31] Wald NJ, Bestwick JP. Incorporating DNA sequencing into current prenatal screening practice for Down's syndrome. PLoS ONE 2013;8:e58732.

[32] Ohno M, Caughey AB. The role of non-invasive prenatal testing as a diagnostic versus a screening tool – a cost-effectiveness analysis. Prenat Diagn 2013;33:630–5.

[33] Cuckle H, Benn P, Pergament E. Maternal cfDNA screening for down syndrome – a cost sensitivity analysis. Prenat Diagn 2013;33:636–42.

[34] Benn P, Curnow KJ, Chapman S, Michalpoulos SN, Hornberger J, Rabinowitz M. An economic analysis of cell-free DNA non-invasive prenatal testing in the US general pregnancy population. PLoS ONE 2015;10(7): e0132313.

[35] National Institute for Health and Care Excellence (NICE). Guide to the methods of technology appraisal 2013, PMG9. NICE; 2013.

[36] Campbell and Cochrane Economics Methods Group (CCEMG). Evidence for Policy and Practice Information and Coordinating Centre (EPPI-Centre). Cost Converter v.1.5, CCEMG and EPPI-Centre; 2016.http://eppi.ioe.ac.uk/costconversion.

[37] National Human Genome Research Institute (NHGRI). The cost of sequencing a human genome, NHGRI; 2016.https://www.genome.gov/sequencingcosts/.

GLOSSARY

Ceiling ratio the cost-effectiveness threshold; the level of the cost-effectiveness ratio that an intervention must meet for it to be regarded as cost effective.

Cost-effectiveness analysis a form of economic appraisal or evaluation that seeks to assess the costs and consequences of different options; costs are opportunity costs measured in monetary terms, benefits are measured in units other than money.

Cost-effectiveness plane a diagram showing the costs and the effects of an activity compared to some alternative.

Cost-effectiveness ratio (CER) the cost per unit of effect.

Dominate an activity is said to dominate another if its beneficial effects are higher and costs are lower; an activity is said to be dominated by another if its beneficial effects are lower and costs are higher.

Economic appraisal a set of techniques that weigh up the costs of an action against the benefits that it may provide; an economic appraisal is undertaken before an action is taken, to help in deciding whether or not and how the action is to be done.

Incremental analysis a type of analysis used to express the results of an economic appraisal, which is an assessment of the extra costs of an activity compared with an alternative, and the extra benefits.

Incremental cost-effectiveness ratio (ICER) a type of cost-effectiveness ratio that measures the incremental (or extra) cost of an activity relative to its best alternative divided by its incremental (or extra) effect.

Opportunity cost the opportunity cost of using resources in a particular way is defined as the benefits that would have resulted from their best alternative use; the concept underpinning economic appraisal.

Perspective also sometimes referred to as the viewpoint of an economic appraisal; determines among other things what constitutes a cost, that is, what costs are included in the appraisal; perspectives taken in economic appraisals include the health service, the payer and society.

Quality-adjusted life year (QALY) a generic health measure including both quality of life and years of life lived; used as an outcome measure in economic appraisals; one QALY equates to 1 year lived in full health.

Scarcity the notion that the claims on health care resources outstrip the resources that are available; the rationale for economic appraisal.

THE FUTURE

EXOME SEQUENCING IN THE EVALUATION OF THE FETUS WITH STRUCTURAL ANOMALIES

17

Elizabeth Quinlan-Jones[*,†], **Mark D. Kilby**[†,‡]

Department of Clinical Genetics, Birmingham Women's & Children's NHS Foundation Trust, Birmingham, United Kingdom * *Centre for Women's & Newborn Health, Institute of Metabolism & Systems Research, University of Birmingham, Birmingham Health Partners, Birmingham, United Kingdom*[†] *West Midlands Fetal Medicine Centre, Birmingham Women's and Children's NHS Foundation Trust, Birmingham, United Kingdom*[‡]

INTRODUCTION

INCIDENCE OF FETAL STRUCTURAL ANOMALIES

Ultrasound detects structural anomalies in the fetus in up to 3% of pregnancies [1] and those are associated with a significant global burden of disease [2]. Pregnancy outcome is variable depending on the number and type of abnormalities, and the underlying genetic etiology [3]. After identification of a fetal structural anomaly, genetic testing is available for parents. Recent advances in molecular genetics enable increasingly detailed genetic testing leading to an accurate prenatal diagnosis in more cases [4]. This prenatal information on the genetic causes of fetal anomalies is relevant for prognosis for the affected fetus and recurrence risk for subsequent pregnancies [3]. Such prenatal genetic diagnosis is of significant value to parents to inform decisions on continuation or termination and for their future reproductive planning, and for caregivers for optimal perinatal management [5]. Within these decisions are strong moral and ethical considerations.

PRENATAL GENETIC TESTING METHODS

In the United Kingdom (UK), prenatal genetic diagnosis is available for individuals with an increased risk of aneuploidy based on history or routine antenatal screening results, or following a diagnosis of fetal ultrasound anomaly. Such testing involves increasingly routine quantitative-fluorescence polymerase chain reaction (QF-PCR) to identify the most common aneuploidies, followed by chromosomal microarray (CMA) to detect copy number variations (CNVs) and microscopic insertions/deletions (indels) in cases with normal QF-PCR [6,7]. Indels are small (typically <5bases) alterations in DNA sequence and differ from larger CNVs classified as microdeletions and microduplications. Some genetic laboratories continue to use fluorescence in situ hybridization and G-banded (conventional) karyotyping to determine structural chromosomal rearrangements. Targeted exome capture (TEC) also called clinical exome sequencing of specific exonic regions of interest, and whole exome sequencing

(WES) of all exonic regions, are used to detect single nucleotide variants (SNVs) associated with various monogenic disorders, but these modalities have limited potential to identify CNVs [8]. TEC enables filtering against a prespecified gene list or panel with known disease association such as Noonan syndrome. WES, on the other hand, allows for all exonic regions to be filtered according to an expanded list (typically thousands) of potentially relevant genes to look for alterations that may be associated with multiple genetic disorders. The exome encompasses only 1.5% of the genome but harbors 85% of the variants that cause single-gene disorders. Whole genome sequencing (WGS), in theory more powerful than WES, is also beginning to be used, as tools for interpretation improve and data sources are developed. It permits genome-wide (entire exonic and intronic) filtering against a comprehensive list of potentially relevant genes to identify all types of disease-causing mutations. More detailed information relating to the various prenatal genetic testing methods can be found in Chapters 2 and 3.

NEXT-GENERATION SEQUENCING APPLICATIONS

Next-generation Sequencing (NGS) applications (TEC, WES, and WGS) are broadening the scope of prenatal diagnosis to identify the genetic etiology of sporadic and inherited disease [9] and are changing current practice [10]. Sequencing analysis of trio DNA (fetus and both parents) aids the assignment of pathogenicity and improves timeliness of interpretation. Fetal or placental DNA is currently obtained by amniocentesis or chorionic villus sampling, [11] but in time testing will be possible on placental cell-free DNA (cfDNA) in maternal blood [12]. WES identifies SNVs and small indels and captures the regions of the genome that encode proteins [9]. As a technique, it is useful for the diagnosis of known genetic disease and for the discovery of novel disorder genes [13]. It is increasingly being used to diagnose rare Mendelian conditions in fetuses with a single major anomaly or anomalies in multiple organ systems, when standard tests results are normal [14].

The use of WES in prenatal diagnosis enables more accurate prospective risk assessment, focused genetic counseling, and personalized pregnancy care. For future pregnancies reproductive genetic counseling, preventative-assisted reproduction approaches, such as preimplantation genetic diagnosis, or invasive or noninvasive prenatal diagnosis will help the family to avoid recurrence [15]. Prenatal NGS can present challenges around the interpretation of results. Not all genomic alterations have been linked to a phenotype and the significance of some findings may be uncertain. Challenges of all genetic tests, but of CMA, WES, and WGS in particular, are "incidental findings" (ICFs), "variants of uncertain significance" (VUS) and susceptibility loci. The prenatal detection of these types of findings may have significant emotional effects for parents and their relatives and further complicate prenatal decision-making.

CELL-FREE DNA-BASED NONINVASIVE PRENATAL TESTING AND NONINVASIVE PRENATAL DIAGNOSIS

The development of massively parallel sequencing has enabled extremely accurate detection of common chromosomal aneuploidies, i.e., trisomy 13, 18, and 21 in cfDNA within maternal plasma [16]. Cell-free DNA-based NIPT (cfDNA NIPT) is widely available for this purpose, but the technology is also increasingly being used to identify indels and various single-gene alterations [17] for Noninvasive

Prenatal Diagnosis (NIPD). Uptake of cfDNA NIPT for aneuploidy has increased considerably due to the improved accuracy of the technology when compared with conventional screening methods, leading to significant reductions in the number of invasive diagnostic procedures alongside a concomitant decrease in the procedure-related pregnancy loss rate [18]. The analysis process is technically challenging, however, and false-positive results are inevitable [17] because cfDNA originates from the cytotrophoblast and may be subject to confined placental mosaicism. Confirmatory testing on amniotic fluid cells or cytotrophoblast and mesenchymal culture of chorionic villi is required when cfDNA-based prenatal testing indicates an aneuploidy or a CNV. NIPD for single-gene disorders is available for parents with recurrence risks—a concept explored further in Chapter 9. The potential for diagnostic sequencing analysis using cfDNA samples is currently being explored in a research capacity as part of the PAGE (prenatal assessment of genomes and exomes) Study [19] and final results are pending. As NGS technology improves and the cost of sequencing falls, it is likely that analysis of cfDNA will play an increasingly important role not only in prenatal screening but also in prenatal diagnosis in cases of ultrasound-detected congenital anomalies [17].

NGS IN PERINATAL LOSS

Various sequencing approaches have recently been used to investigate perinatal loss [20–23]. Armes et al. performed trio WGS in 16 cases of fetal, perinatal, and early neonatal death where postmortem information was available [20]. In all cases conventional cytogenetic and single nucleotide polymorphism (SNP) array analysis was normal, and in some cases targeted gene testing was undertaken and reported as negative. A likely genetic diagnosis was made in 2 cases. In the first case compound heterozygous variants (a paternally inherited splice-site variant, and a maternally inherited missense variant) in RYR1 were detected and assessed as pathogenic and causative of RYR1-associated congenital myopathy. This was in keeping with the phenotype of arthrogryposis multiplex congenita with fetal akinesia deformation sequence and pulmonary hypoplasia. The second case involved a heterozygous, de novo, splice-site variant in COL2A1 associated with Kniest dysplasia, also in keeping with the phenotype of respiratory failure secondary to Pierre Robin sequence and skeletal dysplasia (Kniest type). These data demonstrate that WGS is useful for genetic diagnosis but the alterations described could also have been detected with WES, a likely more cost-effective approach with a potentially quicker turnaround time. Shebab et al. similarly used WGS to identify the genetic etiology of recurrent male intrauterine fetal death (~19) in a large multigenerational pedigree [21]. Sequence analysis of 5 healthy obligatory carrier females and an unaffected male offspring demonstrated an X-linked frame-shift mutation in FOXP3 associated with IPEX syndrome in all of the tested females and absent in the tested male. In DNA extracted from paraffin embedded tissue derived from an intrauterine demised male fetus in this family the same FOXP3 variant was confirmed. This finding was consistent with the prenatal phenotype of hydrops fetalis, and although in this case the variant was identified on WGS, it could have been detected on WES as well. A WES approach was employed by Shamseldin et al. in a cohort of 44 families with a history of intrauterine fetal death, nonimmune hydrops fetalis, or congenital malformation resulting in termination of pregnancy [22]. All cases were reported as having normal chromosome analysis although SNP array analysis was not performed. Where fetal DNA was available solo WES was carried out, otherwise duo WES on both parents was carried out to identify shared heterozygous variants under an autosomal recessive model of inheritance. The authors report that pathogenic or likely pathogenic variants were identified in 22 families (50%). Given that pathogenic findings in the fetus could not be confirmed as having arisen de novo (as trio analysis was not performed), and as it was not possible to directly confirm candidate variants in the fetus after duo-exome analysis in the parents, it is not possible to accurately determine the diagnostic yield of WES in this cohort. In addition, postmortem information was not available in any case and thus the lack of a genotype–phenotype correlation is a limitation of this study. Yates et al. carried out WES in 84 deceased fetuses with ultrasound diagnosed anomalies: 29 cases as solo fetal analysis, 4 maternal–fetal duos, 45 fetal–parent trios, and 6 fetal–parent plus sibling quads [23]. Diagnostic yield in this cohort was 20%, 24% for trio cases, and 14% for fetus-only cases. There was postmortem information available for some cases to inform on genotype–phenotype correlation, and for other cases the fetal phenotype was deduced from prenatal ultrasound information. Research to evaluate WES in cases of perinatal mortality as part of the PAGE Study [19] is currently underway. A retrospective cohort of ~100 fetus/parental trios with normal QF-PCR/CMA/NIPT results will undergo WES to investigate the potential diagnostic yield of this approach. DNA extracted from postmortem fetal tissue will be sequenced alongside parental DNA extracted from blood or saliva samples. Preliminary data indicates that WES has the potential to determine the underlying genetic etiology in approximately 30% of perinatal mortality cases above standard cytogenetic methods. The results of this research are expected toward the end of 2018.

NGS IN CRITICALLY ILL NEWBORNS

In critically ill newborns most commonly NGS panel sequencing (also called clinical exome sequencing or TEC) is used. This technique focuses on specific diseases or phenotypes to identify disease-causing gene variants. The technique enables simultaneous analysis of hundreds of genes with comprehensive coverage for defined phenotypes such as cardiac defects, skeletal dysplasias, mitochondrial disorders, and Noonan syndrome. Disease-focused analysis is generally less expensive than WES or WGS and can be used alongside other technologies such as CMA, to identify alterations, e.g., CNVs that are more challenging to detect with some NGS methods [8]. The turnaround time for panel analysis can be prolonged depending on the need for sample send-away to specialist national/international laboratories, delaying the time to diagnosis with potential implications for clinical management. Studies are currently underway to compare cost and time to diagnosis between traditional diagnostic pathways and "whole exome or whole genome sequencing first." Turnaround times for TEC and WES are becoming shorter: a mean turnaround time of 13 days was recently reported [24]. These techniques will therefore likely become standard practice for clinical diagnosis of neonates with rare unknown genetic disorders [15]. A few papers have recently been published [24–26]. Meng et al. performed clinical exome sequencing on 278 unrelated infants as proband-only (~176) or as trio analysis (~102) with a median turnaround time of 13 days and reported an overall diagnostic yield of 36.7%. Obtaining a genetic diagnosis in this cohort enabled improved medical management of the affected neonate in 52% of cases, leading the authors to suggest that clinical exome sequencing should be considered as a first-line test for neonates with suspected monogenic disease.

WES as first-tier diagnostics in children with congenital or early onset disorders showed potential to achieve high diagnostic yields [26]. Use of trio WES in early diagnostic workup will potentially enable reductions in the use of financial resources and costs related to days of admission and facilitate individualized clinical management.

PRENATAL WES: THE POTENTIAL
EXISTING EVIDENCE

WES in patients with suspected Mendelian disease has a diagnostic yield in the order of 25% [27,28] thus it is likely that WES will also be valuable in prenatal diagnosis [29]. Research on the use of genetic sequencing for prenatal diagnosis is limited [5] but feasibility has been demonstrated in small case series [4,29,30]. Pangalos et al. performed WES on fetal DNA targeting 758 genes associated with fetal malformation in 11 ongoing pregnancies all with various anomalies identified on prenatal ultrasound examination [4]. A definitive or highly likely diagnosis was made in 3 (27%) cases: Citrullinemia, Noonan Syndrome, and PROKR2-related Kallmann syndrome. Trio WES analysis in 30 nonaneuploid fetuses and neonates with diverse ultrasound-detected structural anomalies was reported by Carss et al. and showed a diagnostic yield of 10% [29]. All findings were de novo mutations with low recurrence risks in FGFR3, COL1A1, and OFD1. Drury et al. also used a prenatal WES approach in 24 euploid fetuses with abnormal ultrasound findings. They sequenced fetal DNA only in the first 14 cases and performed trio analysis in the 10 subsequent cases [30]. A genetic diagnosis was made in 5 cases (5%): Milroy disease, hypophosphatasia, achondrogenesis type 2, Freeman–Sheldon syndrome, and Baraitser–Winter syndrome.

More recently, the potential for trio WES to increase genetic diagnosis in structurally abnormal fetuses, with normal QF-PCR and CMA analysis, has been reported in two large prospective prenatal cohorts [31,32]. Wapner et al. have reported on 166 parent–fetus trios where WES was carried out following ultrasound detection of varying fetal structural anomalies. This resulted in a genetic diagnosis in 13 fetuses (7.8%) [31]. Fetuses with multiple anomalies affecting different anatomical systems had a higher diagnosis rate (14.9%) than those with isolated anomalies involving a single anatomical system (5%). McMullan et al. have similarly performed trio WES in 406 structurally abnormal fetuses as part of the PAGE Study [19]. They demonstrated an overall diagnostic yield of 6.2% [32]. Here again the highest yield (16%) was reached in the group of fetuses with multisystem disorders. These studies have demonstrated that the application of WES can substantially improve prenatal genetic diagnosis in fetuses with congenital abnormalities identified on ultrasound assessment.

Successful implementation of WES for prenatal diagnosis will require health economic assessment, and clinical utility will among others depend upon the development of comprehensive and rapid analytical and interpretation pipelines [29]. Meaningful results need to be available within a timeframe that allows for timely informing the parents and give the opportunity for altering pregnancy management. Also, a robust system for relating the genetic results to a potential syndrome or disease, the so-called variant calling procedure, needs to be in place [29]. The challenge of prenatal WES will be the integration of sequencing analysis into prenatal diagnostics as part of a responsible and ethical framework for clinical practice [3].

PRENATAL ASSESSMENT OF GENOMES AND EXOMES STUDY

The PAGE consortium project funded by the Department of Health, Wellcome Trust, Health Innovation Challenge Fund, is currently recruiting parent/fetus trios across the UK to investigate the use of WES as a diagnostic tool in cases of structural anomaly identified on prenatal ultrasound scan [19]. The study will analyze ~1000 trio whole exomes with three primary objectives:

- Elucidate the relative contribution of different forms of genetic variation to prenatal structural anomalies
- Design cost-effective genome sequencing assays for improved prenatal diagnosis of structural anomalies
- Catalyze the adoption, by the National Health Service (NHS), of prenatal diagnostic sequencing through translation of acquired knowledge, rigorous health economic assessment, and establishment of an ethical social science framework for clinical implementation

As a secondary objective, the consortium will also explore the feasibility of targeted sequencing of cfDNA from maternal plasma in a selected series of positive control study samples to promote rapid translation of the technology to noninvasive prenatal diagnostics. Prospective recruitment of cases is currently ongoing with findings expected toward the end of 2018.

PRENATAL WES: THE CLINICAL UTILITY

In the context of rare disease, the time taken to establish a genetic diagnosis can be extremely protracted. This process, sometimes referred to as a "diagnostic odyssey," is attributed to the heterogeneity of genetic disorders, the diversity of disease manifestations, and a lack of clinical genetic knowledge among some healthcare professionals [33]. In the absence of an accurate diagnosis, it is not always possible for clinicians to develop an appropriate plan of care for the individual concerned; hence opportunities for early intervention and effective treatment may be missed. The availability of comprehensive prenatal diagnostic assessment, including genomic sequence analysis, can help to avoid this delay in postnatal diagnosis and negate the need for prolonged admissions and serial testing and thus reduce associated costs [11]. Prenatal diagnosis of a lethal or severe genetic condition enables parents to make informed choices around termination of pregnancy, pregnancy and delivery management, and immediate clinical management of the neonate [11]. Genetic diagnosis can also enable parents to choose reproductive genetic technologies such as preimplantation genetic diagnosis, or targeted prenatal diagnosis, in a subsequent pregnancy [11]. Finally, genomic-driven prenatal diagnosis has the potential to improve perinatal outcome through fetal treatment, and immediate and long-term clinical management of the neonate as demonstrated by de Koning [34] and Brison [35]. Fetal therapy is further explored in Chapter 20. In cases where in utero fetal therapy is an option, WES may help select cases that are most likely to benefit from intervention, based on the prognostic value of prenatal genomic analysis [11]. Finally, the ability to perform prenatal WES in fetuses affected with various congenital anomalies is leading to increased understanding of early human development and the multiple prenatal phenotypes, both lethal and nonlethal that may be associated with a given mutation. WES has already, and will likely continue to enable the identification of new genes pivotal for normal human development, and also provide a means to further classify known causative genes associated with fetal structural malformation.

CASE EXAMPLE

CASE STUDY: PIK3CA-RELATED OVERGROWTH SPECTRUM (PROS) DISORDER

Background: G1P0, age 24 years, no relevant family history, fit and well, referred for fetal medicine opinion at 22 weeks gestation. Ultrasound and MRI findings: 22/40 USS—macrocephaly and associated asymmetry of the cerebral hemispheres, localized lymphangiomata of the lower abdomen, 23/40 MRI—extensive subcutaneous swelling in both fetal flanks and marked enlargement of the left cerebral hemisphere with advanced sulcation, 24,28,32,34 week USS—similar appearances. Normal QF-PCR, chromosomal microarray, and maternal hypermethylation studies for Beckwith–Wiedemann syndrome

Presumptive diagnosis: PROS disorder

Targeted genetic testing: Mutation screen of seven genes associated with fetal overgrowth syndromes on cultured amniocytes identified a pathogenic de novo missense variant (c.1633G > A p.Glu545Lys) in PIK3CA estimated to be present at a level of 36% mosaicism confirmed on Sanger sequencing.

Outcome: Livebirth by LSCS at 36 + 6 weeks. Girl with a birthweight of 3334 g, born in good condition with no resuscitation required. Asymmetrical discoloration and swelling on the skin of the left upper abdomen noted alongside mildly splayed toes bilaterally. Cranial and abdominal USS confirmed left-sided mild hemimegalencephaly and marked multiloculated and heavily septated areas of fluid within the abdomen

Key point: Prenatal ultrasound phenotype enabled prospective diagnosis of PROS disorder which was then confirmed with targeted genetic sequencing of relevant genes associated with fetal overgrowth disorders. This informed the parents and facilitated multidisciplinary perinatal management

Citation: Quinlan-Jones E, Williams D, Bell C, et al. Prenatal detection of PIK3CA-related overgrowth spectrum in cultured amniocytes using long-range PCR and next-generation sequencing. Pediatr Dev Pathol 2017;20(1):54–57.

PRENATAL WES: THE CHALLENGES
TECHNICAL ISSUES

From a technical perspective prenatal WES presents various challenges. Currently, invasive testing by CVS or amniocentesis is required to obtain a suitable sample for DNA extraction. This limits the quantity of available material. Maternal cell contamination (MCC) affects the purity of DNA [8] and is estimated to occur in approximately 0.3%–0.5% of amniotic fluid samples [36,37], and in 1%–2.5% of CVS specimens [38]. Discarding of the first few milliliters of amniotic fluid is useful to reduce the presence of maternal cells. MCC is also decreased through cell culture by selectively enhancing fetal cell growth [39]. The prenatal clinical pipeline from sample retrieval to reportable results is currently lengthy (approximately 9–12 months) and relevant findings cannot be used to inform decision-making and management of the index pregnancy. Turnaround times can be reduced if variants are filtered according to a defined group of genes related to a specific phenotype, [8] but adopting a targeted approach can miss potentially relevant genes and prevent diagnosis [40]. Trio analysis can expedite interpretation if a de novo variant is detected, or if parental samples are needed to confirm the phase of heterozygous alterations in a single gene [8]. It is also feasible to avoid the need for confirmation with Sanger sequencing with good quality metrics for high confidence true positive variants and adequate sample mix-up controls [8]. Improved genome library preparation protocols amenable to small DNA amounts that utilize direct tissue as a starting material can remove the requirement for lengthy culture times [8]. Furthermore, it is not possible to cover all genomic regions equally, and some regions, e.g., those with high GC content, pseudogenes, and repetitive sequences, may not be covered at all. The accuracy of WES is determined by the depth of coverage, which if too low (generally <30-fold), will reduce the mutational detection rate. And finally, there is variability between sequencing platforms.

IMPLICATIONS FOR COUNSELING

Prospective parents are entitled to pretest counseling that includes the benefits and limitations of WES for genetic diagnosis, the types of findings that may be revealed, the inclusion or exclusion of findings unrelated to the primary indication and variants of uncertain significance (VUS) in the results disclosure, and the strategies for sample and data storage and reanalysis [41]. Given the amount and complexity of information trained healthcare professionals should be responsible for this counseling as well as for returning results [42]. Prenatal testing may not necessarily be the best option in all circumstances, thus genetic counseling that assists parents to make informed choices [3,43] is essential to ensure that parents have realistic expectations [14]. The aim of post-test counseling is to inform on the significance of the results in terms of diagnosis, prognosis, and management options, and to facilitate the decisions parents can make. It is also geared toward providing information on risk of recurrence and potential ways to avoid recurrence [11,44]. Depending on the results it may also be necessary to organize carrier testing in family members.

MANAGING FINDINGS

Accurate interpretation and classification of WES findings are complicated and time-consuming aspects of molecular prenatal genetic diagnosis [5]. In the prenatal setting, WES result interpretation requires a multidisciplinary team approach involving clinical scientists and geneticists, genetic counselors, and fetal medicine specialists with expert knowledge in prenatal dysmorphology [11]. Currently, there is no existing comprehensive fetal variant database, or international registry of fetal phenotypes and associated variants, to assist genetic scientists/clinicians with the interpretation of sequence findings. There is a need for a formal curation of prenatally relevant variants given the increasing use of NGS methods in the prenatal space [11]. UK practice guidance for the evaluation of pathogenicity and reporting of sequence variants in clinical molecular genetics have been disseminated by the Association of Clinical Genetic Science [45]. They recommend use of a five-class system which they consider to be essential for the standardization of report wording and follow-up studies. According to these guidelines Sequence findings are classified as follows:

Class 1—clearly not pathogenic, common polymorphism, not reported
Class 2—unlikely to be pathogenic, diagnosis not confirmed molecularly, not reported
Class 3—VUS, uncertain pathogenicity, does not confirm or exclude diagnosis, local team to determine whether to report
Class 4—likely to be pathogenic, consistent with diagnosis, reported
Class 5—clearly pathogenic, predicted to be pathogenic, result confirms diagnosis, reported.

Joint American College of Medical Genetics and Genomics (ACMG) and Association for Molecular Pathology standards and guidance for the interpretation of sequence variants [46] have been formally adopted for use in UK laboratories to assist clinical scientists with classification of sequence variants identified prenatally.

Variants of uncertain significance

WES may lead to the identification of VUS; these are variants with unclear pathogenicity. Their frequency is currently unknown. VUS pose complex counseling challenges prenatally as well as postnatally [5]. When the multidisciplinary team decides to disclose the VUS to the patient, they may experience confusion and anxiety, and their decisions relating to the pregnancy become more complex especially if the information is unexpected. Parental anxiety may be further compounded by the need to initiate additional testing and follow-up. This impacts on the healthcare system as well [14]. Appropriate pre- and post-test counseling that addresses the various implications of identifying VUS is crucial to reduce this negative emotional impact on parents who choose to have WES for prenatal diagnosis. Although parents are generally keen to have access to as much information as possible about the genetics of their fetus, the majority feel cautious about receiving findings that have an uncertain meaning [47]. One of the options for parents is to receive VUS at a later time, or not at all, depending on the judgment of a multidisciplinary team. As our understanding of prenatal genomics improves the likelihood and burden of VUS will decrease [14]. The final clinical utility of WES and WGS depends on our knowledge and understanding of the role of genetic variants in disease.

Secondary findings

Secondary (or additional) findings are genetic alterations considered to be medically actionable but unrelated to the primary indication for testing (e.g., BRCA1 indicating increased risk for breast cancer). Currently in the UK there is no existing guidance that recommends the prospective identification of sequence variants of secondary significance in known disease-associated genes when testing is performed for a prenatal indication. Guidance issued by the ACMG relating to the use of genetic sequencing performed postnatally recommends that diagnostic laboratories actively look for and report on pathogenic variants in 59 genes predominantly related to cancer predisposition and cardiac disease [48]. Patients recruited to the UK 100,000 Genome Project [49] have been given the option to receive information on additional findings. Primary results are currently being returned to participants in this study, and in time secondary findings will also be fed back to participants who "opted in" to receiving this information. Insight into how this process is managed, and the implications for the families involved, will hopefully inform future best practice guidance relating to the management of secondary findings in the context of prenatal diagnostics.

Incidental findings

ICFs are unexpected discoveries that are unrelated to the primary indication for testing and not medically actionable (e.g., childhood disorders). Currently, ICFs are unlikely to be reported, although guidance for the management of ICFs identified as a result of prenatal testing is not yet available. The use of trio prenatal sequencing approaches also has the potential to reveal both nonpaternity and parental consanguinity, adding a further dimension on the complexity of prenatal WES.

ETHICAL AND SOCIETAL CONSIDERATIONS FOR PRENATAL WES
ETHICAL EVIDENCE AND DEBATE

The introduction of new genomic technologies for prenatal diagnostics presents various ethical challenges [31]. Existing evidence on the ethics of prenatal genomics is limited; however, the available literature has recently been reviewed [50]. Horn et al. identified five areas of concern: valid consent, management and feedback of information, responsibilities of health professionals, priority setting and resources, and duties toward the future child [50]. Facilitating valid consent requires parents to have sufficient understanding of the implications of genetic diagnosis prior to deciding whether or not to have prenatal testing [50]. This is not always straightforward due to the complexity of information to be conveyed and difficulties with the interpretation of certain types of results. Opinion differs as to the type and amount of information parents require and the degree of understanding necessary for valid consent. There is concern that providing too much information during the consent process may overload parents and lead to frustration, [51] thus it is argued that health professionals should provide sufficient, albeit layered information, to avoid this risk and to prevent misinformed consent [52].

As to reporting results, questions relating to what, when, and how results are fed back to families are the key areas for discussion. The return of VUS and ICFs has been discussed previously. Parents may choose to terminate pregnancies based on genomic information; therefore ethical dilemmas relating to the return of genomic findings are amplified in the prenatal setting [53]. Targeted analysis based on the primary indication for prenatal testing offers an alternative approach to avoid secondary diagnoses and ICFs [53]. Parents who continue pregnancies after the identification of disease causing genetic findings may treat their children differently once they are born, presenting a significant psychological burden for the individuals concerned [54]. On the other hand, a lack of clarity about eventual genetic causes of a given congenital anomaly in the child may be equally burdensome for some parents. It is also conceivable that parents with continuing pregnancies might feel better able to cope postnatally, if they have had more time to prepare for the birth of their baby given the genomic information available to them prenatally. Thus the pros and cons of each scenario should be considered in terms of the needs of the individual parents involved [11].

The duty of health professionals to recontact parents when previously uncertain findings are subsequently determined as pathogenic, due to increased knowledge over time, is another pressing ethical issue. The duty to provide families with new genomic information has inevitable resource implications. Publicly funded healthcare systems such as the NHS will have to decide how to prioritize limited resources. Finally, prenatal disclosure of certain genetic information may be beneficial to the child, but in certain situations it can also deny children the autonomy to decide for themselves whether they wish to be informed or not. As such the prenatal situation does not differ from the situation where newborns and young children are tested [50,55]. It has been suggested that parents should be able to choose whether to receive information relating to their child's predisposition to delayed-onset disorders, irrespective of their medical actionability and that parental authority to make health-related decisions they believe to be in the best interests of their children, should be endorsed as a family-centered principle, different from the child-centered principle of a right to an open future [56].

RECOMMENDATIONS AND GUIDELINES

Guidance published by the ACMG states that WES should be considered when targeted testing for a specific phenotype fails to identify a cause for multiple fetal anomalies that are suggestive of a genetic condition [57]. Similar guidelines have been disseminated by the American College of Obstetrics and Gynaecology (ACOG) who recommend that due to the complexity of analysis and interpretation of results, WES should only be offered after consultation with a clinical geneticist [58]. In the UK, workshops convened by the Joint UK Committee on Genomics in Medicine are currently underway do develop "New genomic technologies in pregnancy" guidelines the outcome of which are awaited [11]. With regard to the return of secondary findings in the prenatal context there is no existing guidance. It is anticipated that UK recommendations for prenatal sequencing in fetuses with congenital structural anomalies will be developed in the near future following completion of the PAGE Study [19].

Recently the International Society of Prenatal Diagnosis (ISPD), the Society of Maternal Fetal Medicine (SMFM) and the Perinatal Quality Foundation (PGF) have published points requiring consideration for practitioners and laboratories offering WES, targeted analysis with clinical panels, and WGS [59]. The joint position statement lists the following indications for offering prenatal diagnostic sequencing:

- fetus with single major or multiple organ system anomalies where standard genetic testing including CMA is uninformative and the anomaly pattern strongly suggests genetic etiology
- personal history of prior undiagnosed fetus or child affected with major single or multiple anomalies suggestive of genetic etiology and a recurrence of similar anomalies where standard genetic testing is uninformative
- families with history of recurrent stillbirth without a genetic diagnosis with recurrence of similar pattern of anomalies in current pregnancy

As to the return of results the statement emphasizes that the range of possible results has to be discussed with the future parents prior to the test by a knowledgeable genetic professional and that result reporting is best focused on pathogenic and likely pathogenic variants that are relevant to the fetal phenotype. It is recognized that some laboratories may report VUS in strong candidate genes in some situations, as well as pathogenic or likely pathogenic variants that are secondary findings, and carrier status that might have implications for future pregnancies.

COSTS

The cost-effectiveness of prenatal WES is not yet known although this is currently being evaluated as part of the PAGE Study collaboration [19]. To be useful and to aid parental decision-making WES results are needed within the timeframe of pregnancy, thus trio (fetus and biparental) analysis with a rapid turnaround time is required. The costs associated with NGS technologies have decreased dramatically over the last years. However, costs involved with variant interpretation will likely remain the same for some time due to the complexity of the process and the resources required [60]. Also, the availability of specialist genetic counseling required to support the use of WES prenatally, and in general, is limited and additional resources will be needed to satisfy growing demand. If genetic sequence information is to be used as a lifelong resource, the costs involved with long-term data storage and reinterpretation are also considerable [11,61]. Finally, the additional costs involved with the identification of ICFs, i.e., confirmatory testing, carrier screening, and targeted treatment all require consideration in relation to the overall cost-effectiveness of WES undertaken prenatally. The cost of prenatal WES must of cause be evaluated in relation to the costs saved by performing sequence analysis postnatally. If a diagnosis is made prior to birth, there is the potential to avoid the myriad of unnecessary clinical investigations required to enable a diagnosis in the newborn, leading to significant cost savings in the wider context.

PATIENT AND PROVIDER VIEWS

Currently there is limited evidence on views of parents who have experience with prenatal WES. Survey data involving 186 expectant parents in the United States demonstrated that 83% thought prenatal WES should be offered [47]. Fifteen women who experienced fetal demise described that they had high hopes and expectations that WES would provide a genetic explanation for their pregnancy loss despite knowing that testing would only enable a diagnosis in 1 in 3 (30%) cases [62]. Qualitative interview data involving parents who underwent prenatal WES following the diagnosis of a fetal structural anomaly on routine ultrasound examination has been undertaken [63]. This research aimed to elucidate parental experiences of prenatal sequencing, understand more about what influenced their decision-making to have testing, and elicit their thoughts around how they perceived WES to be of potential benefit. The study found that parents wanted as much information as possible and appreciated information being repeated and provided in various formats. Some parents struggled with clinical uncertainty relating to the cause and prognosis following a fetal anomaly diagnosis, and found it difficult to balance the risks of invasive testing, i.e., miscarriage against their need for more definitive information. Parents trusted their clinicians and valued their support with decisions in pregnancy. Testing was sometimes pursued as a means for reassurance that their baby was "normal" rather than to confirm an underlying genetic problem. Parents were also motivated to undergo WES for personal and altruistic reasons but disliked waiting times for results and were uncertain about what findings would be returned.

As to provider views around WES in the prenatal context, focus group data involving clinicians, scientists, genetic counselors, and patient representatives have been published [42]. As a group, representatives from patient groups/charities with experience of prenatal genetic diagnosis felt that parents want access to all information possible and would like their results reinterpreted over time. Conversely, clinical professionals did not feel that VUS or ICFs should be fed back to parents and believed that variant interpretation should be performed at the point of testing only. Survey data involving 1114 members of the ACOG around the ethics of prenatal sequencing and the use of this technology revealed concerns around overtreatment, elevated care costs, and increased parental anxiety [64]. Concerns were also raised regarding reporting nonmedical information. The importance of specialist training for clinicians alongside support from genetic counselors was highlighted as essential to effectively counsel parents considering prenatal WES.

WHOLE GENOME SEQUENCING
WGS IN CHILDREN AND ADOLESCENTS

Lionel et al. carried out WGS on 103 individuals $\leq$18 of age with clinical phenotypes suggestive of an underlying genetic disorder, and compared the diagnostic yield and coverage of WGS to conventional testing methods including TEC, CMA, and supportive clinical investigations for specific disease [25]. The authors report that WGS identified diagnostic variants in 41% of the cases, of which 17 were new diagnoses including structural and nonexonic sequence alterations not detectable on WES. Variants included deep intronic SNVs ($\sim$2), SNVs in noncoding RNA ($\sim$5), and mitochondrial DNA ($\sim$2) all of which would have escaped exonic capture alone. This indicates a diagnostic yield of approximately 8.7% above WES in cases with normal chromosomes and negative CMA findings.

WGS IN PREGNANCY

The clinical effectiveness of WGS in cases of ultrasound diagnosed fetal structural anomaly will be evaluated as part of a selected cohort of prenatal cases within the PAGE Study [19]. It is anticipated that ~100 parent/fetus trios, with normal exomes but potentially enriched for intronic variants, will have WGS performed with the aim to identify alterations that are causative of the fetal phenotype in each case. Once published, data from this cohort (the largest known to date), with accurate genotype–phenotype correlation, will provide valuable information and inform future research around the use of prenatal WGS.

NONINVASIVE PRENATAL WES

The potential for WES on cfDNA retrieved from maternal plasma is being explored but currently not clinically available. In a proof-of-principle study Lo et al. performed deep sequencing (65-fold coverage) on cfDNA from maternal plasma and demonstrated proportional representation of the entire fetal genome [65]. Using parental SNP genotypes (from SNP array data) to distinguish fetal vs maternal sequencing reads, it was possible to determine that cfDNA fragments are slightly smaller (143 bp) than maternal cfDNA fragments (166 bp). Lo et al. showed that construction of a genome-wide fetal genetic map was possible demonstrating that noninvasive sequencing of the fetal genome is technically achievable [65]. Subsequently, Fan et al. [66] and Kitzman et al. [12] independently published data to show that ultradeep sequencing of cfDNA in maternal plasma can be used to determine the fetal genome sequence. Fan et al. [66] used a shotgun approach to sequence the fetal genome (of two pregnant women) by molecular counting of the parental haplotypes in maternal plasma to >99% accuracy. They were able to identify fetal inheritance of a maternal 2.85-Mb deletion including a critical region associated with DiGeorge syndrome and were also able to detect fetal paternally inherited or de novo germline mutations through the application of exome capture. Kitzman et al. [12] similarly used shotgun sequencing of cell-free biparental genomic DNA to >30-fold coverage but combined this with deep sequencing (78-fold coverage) of cfDNA from maternal plasma. They likewise reported a >99% genotype accuracy at maternal heterozygous sites when predicting the fetal genotype and were also able to detect de novo mutations and recombination switch breakpoints using a Hidden Markov model. Chan et al. [67] have developed a second-generation approach for noninvasive prenatal sequencing demonstrating the feasibility of detecting de novo fetal mutations on a genome-wide level with a sensitivity of 85% and a positive predictive value of 74% without the use of maternal haplotypes [67]. The feasibility of prenatal sequencing to single base pair resolution using noninvasive samples has therefore been demonstrated in the research setting, but there are significant limitations to the clinical translation of this technology. Currently the cost to sequence the fetal genome noninvasively to sufficient depth and analyze relevant findings is prohibitively expensive. Indeed, Chan et al. [67] acknowledge that the potential to clinically implement their approach is limited by the costs of sequencing maternal plasma DNA to a required depth of 195X [67]. Until these costs are significantly reduced, noninvasive prenatal testing in the clinical context will likely be focused on the detection of chromosomal aneuploidies and CNVs, specific sex-linked and single-gene disorders, and bespoke analysis for known familial mutations.

CONCLUSIONS

The use of prenatal WES in the fetus with ultrasound diagnosed structural anomalies has the potential to enable a genetic diagnosis in approximately 8% more cases than standard genetic tests [4,29–32]. This percentage is higher in fetuses with multiorgan anomalies than in fetuses with single organ anomalies. It is technically possible to employ both "targeted" and "genome-wide" prenatal exome capture. Prenatal sequencing is currently limited to DNA obtained by invasive approaches but development of WES using cfDNA from maternal blood is a focus of current research. The results can inform perinatal management and parental decision-making. Whereas these technologies come with great opportunities for parents confronted with concerns in their pregnancies, healthcare providers, ethicists, philosophers, and society will have to take on their duty in assuring appropriate delivery of these opportunities.

ACKNOWLEDGMENTS

With thanks to Elizabeth Young (Clinical Scientist, West Midlands Regional Genetics Laboratory) for her input in the section on cfDNA-based (noninvasive) prenatal testing.

FUNDING SOURCES

Elizabeth Quinlan-Jones is funded through the Department of Health, Wellcome Trust, Health Innovation Challenge Fund (Award Number HICF-R7-396) as a research midwife for the PAGE and PAGE2 research studies.

REFERENCES

[1] Springett A. Congenital anomaly statistics. In: Morris J, editor. Congenital anomaly statistics: England and Wales. London: British Isles Network of Congenital Anomaly Registers; 2010.

[2] Boyle B, Addor MC, Arriola C, et al. Estimating global burden of disease due to congenital anomaly: an analysis of European data. Arch Dis Child Fetal Neonatal Ed 2018;103:F22–8.

[3] Hillman SC, Williams D, Carss KJ, et al. Prenatal exome sequencing for fetuses with structural abnormalities: the next step. Ultrasound Obstet Gynaecol 2015;45:4–9.

[4] Pangalos C, Hagnefelt B, Lilakos K. First applications of a targeted exome sequencing approach in fetuses with ultrasound abnormalities reveals an important fraction of cases with associated gene defects. Peer J 2016;4:e41955.

[5] Van den Veyver IB, Eng CM. Genome-wide sequencing for prenatal detection of fetal single-gene disorders. Cold Spring Harb Perspect Med 2015;5:a023077.

[6] Hillman SC, McMullan DJ, Hall G, et al. Use of prenatal chromosomal microarray: prospective cohort study and systematic review and meta-analysis. Ultrasound Obstet Gynaecol 2013;41(6):610–20. Jun.

[7] Robson SC, Chitty LS, Ambler G, et al. Evaluation of array comparative genomic hybridisation in prenatal diagnosis of fetal anomalies: a multicentre cohort assessment of patient, health professional and commissioner preferences for array comparative genomic hybridisation. Southampton, UK: NIHR Journals Library; 2017.

[8] Abou Tayoun AN, Spinner NB, Rehm H, et al. Prenatal DNA sequencing: clinical, counselling, and diagnostic laboratory considerations. Prenat Diagn 2017;37:1–7.

[9] Babkina N, Graham JM. New genetic testing in prenatal diagnosis. Semin Fetal Neonatal Med 2014;19:214–9.

[10] Rabbani B, Tekin M, Mahdieh N. The promise of whole exome sequencing in medical genetics. J Hum Genet 2014;59:5–15.

[11] Best S, Wou K, Vora N, et al. Promises, pitfalls and practicalities of prenatal whole exome sequencing. Prenat Diagn 2018;38:10–9.

[12] Kitzman JO, Snyder MW, Murray M, et al. Non-invasive whole-genome sequencing of a human fetus. Sci Transl Med 2012;(4) 137ra176.

[13] Sawyer SL, Hartley T, Dyment DA, et al. Utility of whole-exome sequencing for those near the end of the diagnostic odyssey: time to address gaps in care. Clin Genet 2016;89:275–84.

[14] Volk A, Conboy E, Wical B, et al. Whole-exome sequencing in the clinic: lessons from six consecutive cases from the clinician's perspective. Mol Syndromol 2015;6:23–31.

[15] Peters DG, Svetlana A, Yatsenko MD, et al. Recent advances of genomic testing in perinatal medicine. Semin Perinatol 2015;39(1):44–54. Feb.

[16] Mackie FL, Hemming K, Allen S, et al. The accuracy of cell-free fetal DNA-based non-invasive prenatal testing in singleton pregnancies: a systematic review and bivariate meta-analysis. Br J Obstet Gynaecol 2017;124(1):32–46. Jan.

[17] Wong KCK, Lo YMD. Prenatal diagnosis innovation: genome sequencing of maternal plasma. Annu Rev Med 2016;67:419–32.

[18] Warsof SL, Larion S, Abuhamad AZ. Overview of the impact of noninvasive prenatal testing on diagnostic procedures. Prenat Diagn 2015;35:972–9.

[19] Wellcome Trust Sanger Institute. Prenatal Assessment of Genomes and Exomes (PAGE) Study, http://www.sanger.ac.uk/science/collaboration/prenatal-assessment-genomes-and-exomes-page. Accessed 9 January 2018.

[20] Armes JE, Williams M, Price G, et al. Application of whole genome sequencing technology in the investigation of genetic causes of fetal, perinatal and early infant death. Paediat Dev Pathol 2018;21:54–67.

[21] Shebab O, Tester DJ, Ackerman NC, et al. Whole genome sequencing identifies etiology of recurrent male interuterine fetal death. Prenat Diagn 2017;37(10):1040–5.

[22] Shamseldin HE, Kurdi W, Almusafri F, et al. Molecular autopsy in maternal-fetal medicine. Genet Med 2018;20:420–7.

[23] Yates CL, Monaghan KG, Copenheaver D, et al. Whole-exome sequencing on deceased fetuses with ultrasound anomalies: expanding our knowledge of genetic disease during fetal development. Genet Med 2017;19(10):1171–8.

[24] Meng L, Pammi M, Saronwala A, et al. Use of exome sequencing for infants in intensive care units. Ascertainment of severe single-gene disorders and effect on medical management. JAMA Paediatr 2017;171(12). e173438.

[25] Lionel AC, Costain G, Monfared N, et al. Improved diagnostic yield compared with targeted gene sequencing panels suggests a role for whole-genome sequencing as a first-tier genetic test. Genet Med 2018;20:435–43.

[26] Lemke JR. High-throughput sequencing as first-tier diagnostics in congenital and early-onset disorders. JAMA Paediatr 2017;171(9):833–5.

[27] Yang Y, Muzny DM, Reid JG, et al. Clinical whole-exome sequencing for the diagnosis of Mendelian disorders. N Engl J Med 2013;369:1502–11.

[28] Yang Y, Muzny DM, Xia F, et al. Molecular findings among patients referred for clinical whole-exome sequencing. JAMA 2014;312(18):1870–9. November 12.

[29] Carss KJ, Hillman SC, Parthiban V, et al. Exome sequencing improves genetic diagnosis of structural fetal abnormalities revealed by ultrasound. Hum Mol Genet 2014;23(12):3269–77.

[30] Drury S, Williams H, Trump N, et al. Exome sequencing for prenatal diagnosis of fetuses with sonographic abnormalities. Prenat Diagn 2015;35:1010–7.

[31] Wapner RPS, Brennan K, Bier L, et al. Whole exome sequencing in the evaluation of fetal structural anomalies: a prospective study of sequential patients. Am J Obstet Gynaecol 2017;216:S5–6. http://www.ajog.org/article/S0002-9378(16)30988-7/fulltext.

[32] McMullan DJ, Eberhardt R, Rinck G, et al. In: Exome sequencing of 406 parental/fetal trios with structural abnormalities revealed by ultrasound in the UK Prenatal Assessment of Genomes and Exomes (PAGE) project. Annual Meeting. Copenhagen, Denmark: European Society of Human Genetics; 2017.

[33] Lohmann K, Klein C. Next generation sequencing and the future of genetic diagnosis. Neurotherapeutics 2014;11:699–707.

[34] de Koning TJ, Klomp LWJ, van Oppen ACC, et al. Prenatal and early postnatal treatment in 3-phosphoglycerate-dehydrogenase deficiency. Lancet 2004;364:2221–2.

[35] Brison N, van den Bogaert K, Dehaspe L, et al. Accuracy and clinical value of maternal incidental findings during noninvasive prenatal testing for fetal aneuploidies. Genet Med 2016;19(3):306–13.

[36] Steed HL, Tomkins DJ, Wilson DR, et al. Maternal cell contamination of amniotic fluid samples obtained by open needle versus trocar technique of amniocentesis. J Obstet Gynaecol Can 2002;24(3):233–6.

[37] Stojikovic-Mikic T, Mann K, Docherty Z, et al. Maternal cell contamination of prenatal samples assessed by QF-PCR genotyping. Prenat Diagn 2005;25(1):79–83.

[38] Steinberg S, Katsanis S, Moser A, et al. Biochemical analysis of cultured chorionic villi for the prenatal diagnosis of peroxsomal disorders: biochemical thresholds and molecular sensitivity for maternal cell contamination detection. J Med Genet 2005;42(1):38–44.

[39] Taglauer ES, Wilkins-Haug L, Bianchi DW. Review: cell free fetal DNA in the maternal circulation as an indication of placental health and disease. Placenta 2014;35(Suppl):S64–8.

[40] Stals KL, Wakeling M, Baptista J, et al. Diagnosis of lethal or prenatal-onset autosomal recessive disorders by parental exome sequencing. Prenat Diagn 2018;38:33–43.

[41] Norton ME, Britton DR. Changing indications for invasive testing in an era of improved screening. Semin Perinatol 2016;40:56–66.

[42] Quinlan-Jones E, Kilby MD, Greenfield SG, et al. Prenatal whole exome sequencing: the views of clinicians, scientists, genetic counsellors and patient representatives. Prenat Diagn 2016;36:935–41.

[43] Sijmons RH, Langen IMV, Sijmons JG. A clinical perspective on ethical issues in genetic testing. Account Res 2016;18(3):148–62.

[44] Filges I, Friedman JM. Exome sequencing for gene discovery in lethal fetal disorders—harnessing the value of extreme phenotypes. Prenat Diagn 2014;35:1005–9.

[45] Wallis Y, Payne S, McAnulty C, et al. Practice guidelines for the evaluation of pathogenicity and the reporting of sequence variants in clinical molecular genetics. ACGS/VGKL; 2013.

[46] Richards S, Aziz N, Bale S, et al. Standards and guidelines for the interpretation of sequence variants: a joint consensus recommendation of the American College of Medical Genetics and Genomics and the Association for Molecular Pathology. Genet Med 2015 May;17(5):405–24.

[47] Kalynchuk EJ, Althouse A, Parker LS, et al. Prenatal whole exome sequencing: parental attitudes. Prenat Diagn 2015;35:1030–6.

[48] Kalia SS, Adelman K, Bale SJ, et al. Recommendations for reporting of secondary findings in clinical exome and genome sequencing, 2016 update (ACMG SF v2.0): a policy statement of the American College of Medical Genetics and Genomics. Genet Med 2017;19:249–55.

[49] The 100,000 Genomes Project Protocol V3 Genomics England. 2017. https://www.genomicsengland.co.uk/the-100000-genomes-project/

[50] Horn R, Parker M. Opening Pandora's Box?: ethical issues in prenatal whole genome and exome sequencing. Prenat Diagn 2017;37:1–6.

[51] De Jong A, Dondorp WJ, Frints SG, et al. Advances in prenatal screening: the ethical dimension. Nat Rev Genet 2011;12(9):657–63.

[52] Elias S, Annas GJ. Generic consent for genetic screening. N Engl J Med 1994;330(22):1611–3.

[53] Mackie FL, Carss KJ, Hillman SC, et al. Exome sequencing in fetuses with structural malformations. J Clin Med 2014;3:747–62.

[54] Botkin JR, Belmont JW, Berg JS, et al. Point to consider: ethical, legal and psychosocial implications of genetic testing in children and adolescents. Am J Hum Genet 2015;97:6–21.

[55] Yurkiewicz IR, Korf BR, Lehmann LS. Prenatal whole genome sequencing—is the quest to know a fetus's future ethical? N Engl J Med 2004;370:195–7.

[56] Bredenoord AL, de Vries MC, van Delden JJM. Next generation sequencing: does the next generation still have a right to an open future? Nat Rev Genet 2016;14(5):306.

[57] ACMG Board of Directors. Points to consider for informed consent for genome/exome sequencing. Genet Med 2013;15:748–9.

[58] ACOG Committee Opinion no 682. Microarrays and next-generation sequencing technology: the use of advanced genetic diagnostic tools. Obstet Gynaecol 2016;128:1462–3.

[59] International Society of Prenatal Diagnosis, the Society of Maternal and Fetal Medicine, the Perinatal Quality Foundation. Joint Position Statement from the International Society for Prenatal Diagnosis (ISPD), the Society for Maternal Fetal Medicine (SMFM), and the Perinatal Quality Foundation (PQF) on the use of genome-wide sequencing for fetal diagnosis. Prenat Diagn 2018;38:6–9.

[60] Beckmann JS. Can we afford to sequence every new born baby's genome? Hum Mutat 2015;36(3):283–6.

[61] Sboner A, Mu XJ, Greenbaum D, et al. The real cost of sequencing: higher that you think! Genome Biol 2011;12:125.

[62] Vora NL, Brandt A, Strande N, et al. Prenatal exome sequencing in anomalous fetuses: new opportunities and challenges. Genet Med 2017;19:1207–16.

[63] Quinlan-Jones E, Hillman SC, Kilby MD, et al. Parental experiences of prenatal whole exome sequencing (WES) in cases of ultrasound diagnosed fetal structural anomaly. Prenat Diagn 2017;37:1225–31.

[64] Bayefsky MJ, White A, Wakim P, et al. Views of American OB/GYN on the ethics of prenatal whole-genome sequencing. Prenat Diagn 2016;36:1250–6.

[65] Lo YM, Chan KC, Sun H, et al. Maternal plasma DNA sequencing reveals the genome-wide genetic and mutational profile of the fetus. Sci Transl Med 2010;(2):61ra91.

[66] Fan HC, Gu W, Wang J, et al. Non-invasive prenatal measurement of the fetal genome. Nature 2012;487:320–4.

[67] Chan KCA, Jiang P, Sun J, et al. Second generation noninvasive fetal genome analysis reveals do novo mutations, single-base parental inheritance, and preferred DNA ends. PNAS 2016;113(50):E8159–68. Dec 13.

CELL-BASED NONINVASIVE PRENATAL TESTING: A PROMISING PATH FOR PRENATAL DIAGNOSIS

18

Liesbeth Vossaert, Arthur L. Beaudet

Department of Molecular and Human Genetics, Baylor College of Medicine, Houston, TX, United States

INTRODUCTION: THE CHANGING LANDSCAPE OF PRENATAL DIAGNOSIS

In recent years, increased interest in noninvasive prenatal testing (NIPT) methods led to the commercialization and clinical implementation of NIPT based on the analysis of "fetal" cell-free DNA (cfDNA) isolated from maternal blood samples; this has caused a drastic transformation of the field of prenatal screening and diagnosis. Since the initial report on "fetal" cfDNA being present in maternal plasma by Dennis Lo and colleagues in 1997 [1], the clinical potential of this finding has been elaborately described and validated, with a substantial decrease in amniocentesis and chorionic villus sampling (CVS) diagnostic procedures as a consequence [2].

However, one inherent disadvantage is the overwhelming amount of maternal cfDNA that is inevitably analyzed together with the "fetal" fraction. The "fetal" cfDNA only comprises 5%–15% of the total plasma cfDNA pool. cfDNA-based NIPT can thus currently only be recommended to be offered as a screening test for trisomy 21, 13, and 18, and any positive result should be confirmed by diagnostic methods. For these common autosomal aneuploidies, cfDNA-based NIPT has an undeniably higher positive predictive value (PPV) than serum marker analysis using a combination of 3, 4, or 5 markers (alpha-fetoprotein [AFP], human chorionic gonadotropin [hCG], and unconjugated estriol [uE3] for a triple screen, with the addition of dimeric inhibin A [DIA] for a quad screen and hyperglycosylated hCG as well for a penta screen). The detection of other autosomal and sex chromosome aneuploidies, and subchromosomal copy number abnormalities is less adequate, and multiple laboratories have reported high false-positive rates and low PPVs [3,4]. cfDNA-based prenatal testing is thus not recommended for these indications by the leading professional organizations [5,6], although multiple commercial providers do offer such testing.

About 6000 children per year are born in the United States with Down syndrome. Trisomy 21 is the most common genetic disability; however, it makes up only about 5% of all genetic disabilities. Other autosomal and sex chromosome aneuploidies and microdeletions and duplications on their own are significantly rarer but adding them all together leads to a much higher incidence than Down syndrome.

Noninvasive Prenatal Testing (NIPT). https://doi.org/10.1016/B978-0-12-814189-2.00018-9

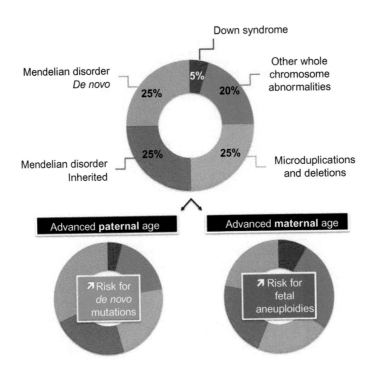

FIG. 1

Considering that about 20 pregnancies in 1000 are affected by a genetic disability, the distribution of the nature of those genetic abnormalities is illustrated here: about half concern copy number abnormalities, while the other half are single gene disorders. The frequencies of these are dependent on parental age.

Additionally, a whole range of both inherited and de novo Mendelian disorders can occur, adding yet another large population of affected pregnancies deserving adequate prenatal testing and diagnosis. Taking this into consideration, it is thus clear that current cfDNA-based prenatal testing only provides few pieces of the total landscape (Fig. 1).

Additionally, the interfering maternal cfDNA makes the incidental discovery of maternal findings unavoidable, which is one of the main causes for false-positive cfDNA-based test results. Furthermore, sometimes no results can be generated in case of low fetal fraction, due to fetal or maternal characteristics such as a very small placenta or a high maternal body mass index (BMI) [7]. Also, samples from multiple pregnancies pose challenges for cfDNA-based NIPT.

Prenatal testing starting from fetal cells, on the other hand, provides a better alternative. The trafficking of intact fetal or trophoblast cells into the maternal circulation is a natural phenomenon during pregnancy, and these cells provide a DNA source of purely fetal or placental origin without any maternal contamination. The isolation and analysis of multiple single fetal or trophoblast cells gives multiple results per sample, which contributes to a high confidence test result and quality. This would also enable a more nuanced detection of potential placental mosaicism. Specifically determining the individual genetic profile of each analyzed cell makes it possible to obtain a result for every fetus within a multiple pregnancy.

FETAL CELLS IN THE MATERNAL CIRCULATION

Fetal cells can be detected in the mother's circulation already very early on; Ariga et al. report the detection of fetal cells already at 4 weeks' gestational age (GA) [8]. Altogether, there are four different groups of fetal cells that can be distinguished: (i) trophoblastic cells, (ii) fetal nucleated red blood cells (fnRBC), (iii) lymphocytes, and (iv) stem cells and progenitor cells. The life span of these cell types is determining their use in prenatal testing: fetal lymphocytes and stem/progenitor cells may persist in the maternal circulation for even years after delivery [9], making them unsuitable for prenatal testing/diagnosis. The reasons behind their persistence are multifold and they are likely to have several beneficial roles such as promoting maternal tolerance against the fetus, but are contrastingly also involved in the development of maternal autoimmune diseases—the immunological implications of pregnancy-related microchimerism are reviewed elsewhere [10].

Trophoblasts and fnRBCs, on the other hand, disappear after delivery and thus only reflect the pregnancy ongoing at the time of testing, without confounding of previous pregnancies. Fetal cytotrophoblasts from anchoring villi invade the maternal endometrium and migrate to remodel the spiral arteries into larger vessels [11]. The migration of these extravillous trophoblasts is a prerequisite for normal fetal and placental development, and occurs in two waves (one at 8–10 weeks and one at 16–18 weeks GA) [9,11]. In contrast to the multinucleate syncytiotrophoblasts, cytotrophoblasts are mononucleate and an attractive target for prenatal testing. However, these cells are technically derived from the placenta instead of the fetus itself, and as a consequence concerns have been raised that they are susceptible to confined placental mosaicism. Fetal trophoblastic cells can be obtained both from maternal blood and through cervical sampling [12,13].

Also, fnRBCs escape into the mother's blood stream and are considered to be "truly fetal" instead of placenta derived. In the first trimester, these fnRBCs consist of both primitive and definitive erythroblasts, which differ both in morphology and in location where they are produced [14]. The definitive fnRBCs are produced in the fetal liver and bone marrow, but before 6 weeks' GA erythropoiesis still takes place in the yolk sac, where thus the primitive cells are formed. Precautions need to be taken for the isolation and analysis of fnRBCs, as the simultaneously present maternal immature RBCs might be isolated along with them: the portion of fnRBCs is estimated to be 30% of the total nRBC population, although this may vary between patients and across gestation [15]. It has been suggested that fnRBCs only contribute a very small fraction of the total fetal cell population [16,17].

Independent of which cell type is focused on, the biggest obstacle to overcome remains the very low number of fetal cells available. The publications by Hamada et al. [18] and Krabchi et al. [19] describe some of the most laborious efforts to quantify the total number of fetal cells. Both groups used a similar approach of only minimally manipulating blood samples collected from male pregnancies, prior to an elaborate microscopic analysis. In short, after maternal RBC lysis and fixation, the obtained nucleated cell suspensions were spread on microscopy slides for fluorescent in situ hybridization (FISH) specifically for the X and Y chromosome. No specific enrichment was done, nor were the microscopy slides analyzed for any characteristics specific for a certain fetal cell type. By simply counting the cells with an XY profile, the total number of fetal cells was obtained. Hamada et al. [18] reported a range of 1 fetal cell in 144,000 maternal to 1 in 4000, while Krabchi et al. [19] found 2–6 fetal cells per mL of maternal blood for a normal pregnancy. The latter group confirmed all fetal cells by use of reverse FISH, in which the same X and Y probes were used as the first step but now with reversed color.

Different factors might influence the number of fetal cells such as GA, fetal genetic abnormalities, maternal and placental health, physical activity of the mother before the blood draw, and normal interindividual biological variation. It is important to keep in mind that differences in blood collection conditions, sample size, targeted fetal cell type, isolation protocol with or without enrichment, detection method, and so on, might substantially contribute to the variability between study results.

Multiple studies have been reported on the association of the number of fetal cells with GA, with varying results. The study by Hamada et al. [18] reports a significant increase in total fetal cells from <1/100,000 in the first trimester to about 1/10,000 at term. Ariga and colleagues [8] collected samples from 4 weeks' GA until 39 days after delivery, and noticed a temporary peak in fetal cells around 12 weeks and another more steady increase after 20 weeks. The samples taken after delivery showed a quick clearing of the fetal cell population after birth. Other groups have focused on one specific fetal cell type. Bianchi and colleagues did not detect any fnRBCs after 16 weeks [20], while Takabayashi et al. described that fnRBCs could be retrieved at earliest at 8 weeks but no more after 24 weeks GA [21]. Rodríguez de Alba et al. found a significant increase in fnRBCs from first to second trimester [22]. Shulman and colleagues found no significant association between the number of fnRBCs recovered and the GA [15], and neither did Lim et al., although they noted a trend of decreasing numbers for both fnRBCs and trophoblast cells with increasing GA [23]. For trophoblast cells retrieved from the cervix, no correlation with GA was seen in a large series of samples collected between 5 and 20 weeks' gestation [24]. The variability in results makes it difficult to draw a definitive conclusion on the potential association between GA and number of fetal cells. The total fetal cell population seems to increase over the course of a pregnancy, but fnRBCs might only contribute to this increase until about 20–25 weeks' GA. The number of trophoblastic cells appears to be rather decreasing over time, although no significant correlation was found so far. Most importantly, all above-mentioned studies show that it should be feasible to obtain a sufficient amount of fetal cells noninvasively in the first or early second trimester, which is the time frame aimed for in prenatal genetic diagnosis (Table 1).

Also, fetal genetic abnormalities are known to affect the number of fetal cells. Bianchi and colleagues reported a sixfold increase in fetal cells in blood of women carrying a male trisomy 21 fetus, with an increase from a mean of 1.2 cells/mL of blood to 6.9 cells/mL, as measured by Y-PCR quantitation [25]. Also Krabchi et al. described that the total number of nucleated fetal cells was 2–5 times higher in aneuploid compared to euploid pregnancies around the same GA [26,27].

Fetal-maternal cell trafficking and even more so the release of placental cfDNA might be substantially altered in other pregnancy-related pathologies as well, as was reviewed by Hahn and colleagues [28]. They, for example, refer to one study reporting a 37-fold elevation of fetal trophoblasts in uterine vein blood of preeclamptic women, and also the number of erythroblasts went up drastically. An interesting observation was that the number of circulating fetal cells was most often already elevated before the onset of any clinical symptoms. Therefore both Hahn et al. and van Wijk et al. proposed to use fetal cells and fetal cfDNA as a tool to investigate abnormal placentation [28,29]. This trend was not confirmed for endocervical trophoblast sampling, as Fritz et al. described a nonsignificant decrease in numbers of cells in samples collected from women with preeclampsia or a fetus with IUGR [24].

Contrary to fetal cfDNA-based testing [7], cell-based NIPT is not impacted by maternal BMI, neither for trophoblasts collected from maternal blood [30] (Vossaert et al., unpublished observation), nor for cells obtained via cervical sampling [24].

Another factor to consider is the potential influence of an invasive procedure. Christensen and colleagues compared the yield in fnRBCs before and after CVS, which clearly causes fetal bleeding into

Table 1 Different Factors Influence the Yield of Fetal Cells

Author	Fetal Cell Type	Isolation and Enrichment	Gestational Age	Pregnancy Conditions	Number of Samples	Conclusion
Association with GA						
Hamada [18]	Total pop.	No enrichment/ XY FISH	7–40 wks	Not specified	50	Fetal cells detected from 9 wks; significant increase with ↗ GA
Ariga [8]	Total pop.	No enrichment/ Y-PCR	4 wks to 39 days after delivery	Uncomplicated pregnancies	25	Fetal cells detected for every GA; transient peak at 12 wks, more steady increase after 24 wks with peak at parturition
Bianchi [20]	fnRBC	DGC followed by FACS (CD71)	11–20 wks	Uncomplicated pregnancies	25	No fnRBCs found after 16 wks
Takabayashi [21]	fnRBC	DGC	4–40 wks	Not specified	60	fnRBCs detected between 8 and 24 wks
Rodríguez de Alba [22]	fnRBC	MACS (CD71)	10–20 wks	Normal and cases with fetal abnormalities	146	fnRBCs detected at 10–20 wks; significant increase with ↗ GA
Shulman [15]	fnRBC	DGC followed by CFS	7–25 wks	Normal pregnancies	225	fnRBCs detected at 7–25 wks; no association with GA
Lim [23]	fnRBC +tropho	Mab340 and CD71	9–35 wks	Normal pregnancies	41	Fetal cells detected at 9–35 wks; nonsignificant decrease of both cell types with ↗ GA
Fritz [24]	Tropho	Cervical sampling/ HLA-G	5–20 wks	Normal vs pregnancies with complications	224	Trophoblasts isolated at 5–20 wks; no association with GA
Association with fetal genetic abnormalities						
Bianchi [25]	Total pop.	No enrichment/ Y-PCR	11–32 wks	Normal and cases with fetal abnormalities	199 normal vs 31 fetal aneuploidy	Significant increase for T21 and 47, XY,+inv(dup)15 compared to normal pregnancies, but

Continued

Table 1 Different Factors Influence the Yield of Fetal Cells—cont'd

Author	Fetal Cell Type	Isolation and Enrichment	Gestational Age	Pregnancy Conditions	Number of Samples	Conclusion
Krabchi [26]	Total pop.	No enrichment/ XY FISH	1–4 wks after amnio	T21	16 T21	Significant increase compared to normal pregnancies
Krabchi [27]	Total pop.	No enrichment/ XY FISH	17–22 wks (+one sample at 27 wks)	Fetal abnormalities	T18: 7 T13: 1 69,XXX: 2 47,XXX: 2 47,XXY: 1 47,XYY: 1 47,XY,r(22),+r (22): 1	nonsignificant increase for T18, T13, and 47,XXY T18: nonsignificant increase compared to normal pregnancies. Insufficient samples were collected for the other aneuploidies, but the absolute numbers of fetal cells/ sample are higher
Association with other pregnancy-related pathologies						
			Review of multiple studies, with varying parameters			
Hahn [28]	fnRBC and Tropho					Substantial increase in PE, eclampsia, and IUGR for both cell types
Fritz [24]	Tropho	Cervical sampling/ HLA-G	5–20 wks	Uncomplicated vs complicated	75 Uncomplicated term vs 20 PE and/ or IUGR and 18 EPL	Significant decrease in EPL Nonsignificant decrease in PE and IUGR

Abbreviations used: CD, cluster of differentiation; CFS, charge flow separation; DGC, density gradient centrifugation; FACS, fluorescence-activated cell sorting; FISH, fluorescent in situ hybridization; fnRBCs, fetal nucleated red blood cells; GA, gestational age; HLA-G, human leukocyte antigen-G; IUGR, intrauterine growth retardation; Mab340, monoclonal antibody 340; MACS, magnetic-activated cell sorting; PE, preeclampsia; T21/13/18, trisomy 21, 13, or 18; Total pop., total population of fetal cells; Tropho, fetal trophoblastic cells; wks, weeks of gestational age; Y-PCR, PCR specifically for the Y chromosome.

the maternal circulation as they obtained many fnRBCs post-CVS but practically none prior to the procedure [16]. This thus emphasizes the importance of working with preprocedure samples during the process of cell-based NIPT protocol optimization.

EVOLUTION OF CELL-BASED NIPT
EARLY REPORTS: A CENTURY OF RESEARCH

Fetal cells in the maternal circulation have been studied for decades. Schmorl wrote one of the first reports on fetal-maternal cell trafficking in 1893, describing the presence of trophoblasts in the maternal pulmonary capillaries in women who died of eclampsia [31,32]. He had found large thrombi in maternal lung tissues, which partially consisted of large multinucleated cells, most likely originating from the placental villi and invading the lungs during eclampsia. These thrombi were not present in the lungs of women who had died shortly after giving birth but not due to eclampsia, leading Schmorl to conclude that the number of fetal cells exchanged in noneclamptic women were probably much lower. Several studies from the 1950s and 1960s describe the presence of fetal cells in the maternal circulation, although often these findings were based on samples collected after delivery [33–37]. In 1969, the usefulness of circulating fetal cells for prenatal genetic testing was proposed for the first time by Walknowska and colleagues [38], who reported on the presence of male cells in lymphocyte culture preparations from maternal blood collected from 14 to 37 weeks of gestation. Fetal cells have been studied ever since, and the evidence gathered over the years has only corroborated their diagnostic potential.

As mentioned previously, the two fetal cell types suitable for prenatal diagnosis are trophoblastic cells and fnRBCs, and various protocols have been published for the specific isolation of each cell type, with the early focus mainly on fnRBCs. The earliest studies relied solely on the detection of an XY profile among the vast pool of maternal cells, independent of any cell type-specific characteristics. One of the first reports including an extra selection step was published in 1979 by Herzenberg and colleagues [39,40]. They incorporated fluorescence-activated cell sorting (FACS) based on paternal human leukocyte antigens (HLA) absent in the mother. The selected fetal candidates were consequently confirmed via microscopic detection of the Y chromosome.

Dozens of other reports followed over the years, further consolidating the existence of circulating fetal cells. Generally, the basic protocol for enrichment included removing the excess of maternal cells, a specific enrichment step for fetal cells and finally applying an adequate detection method.

The initial blood-processing step consisted of bulk separation or more selective density gradient centrifugation (DGC), selective lysis of maternal RBCs, or a combination of aforementioned techniques.

For bulk separation, the blood sample is centrifuged in a specialized tube to separate the maternal RBCs and plasma with a layer of nucleated cells, the so-called "buffy coat" in between. This buffy coat includes the fetal and trophoblast cells [16,41]. This method was further refined by adding a specific density component, such as Ficoll or Percoll [42]. Ficoll is a hydrophilic polysaccharide (polymerized sucrose) solution, which is osmotically inert but yet toxic for cells. Percoll solution contains colloidal silica particles that are coated with polyvinylpyrrolidone, which makes them nontoxic for cells. They serve as a means to better mark the interface between the RBC layer and the plasma and make it easier to isolate the layer of nucleated cells including the fetal and trophoblast cells. Several studies have used and compared these reagents, and generally conclude that adding a higher density component results in a higher fetal cell yield [21,43,44].

The selective lysis of maternal but not fetal RBCs had already been suggested in 1976 by Boyer et al. [45]. This method was based on the fact that fetal cells only have carbonic anhydrase II while maternal cells have both anhydrase I and II activity. After adding the appropriate substrate (potassium bicarbonate and ammonium chloride), the ammonium bicarbonate that is formed subsequently will attract water into the cells, until they eventually burst. Given the higher enzyme concentration in maternal cells and with the help of acetazolamide, a selective blocker for carbonic anhydrase II, maternal cells are selectively lysed over fetal cells [41,45].

Additional depletion was often included by means of magnetic-activated cell sorting (MACS) with specific maternal white blood cell (WBC) markers. Surface antigens such as cluster of differentiation (CD) markers are good targets for this purpose: CD45, also known as leukocyte common antigen, and CD14, or monocyte differentiation antigen, are widely used to deplete maternal WBCs [42].

The confirmation methods applied after cell isolation were however not ideal: XY FISH can only be used for male fetuses and HLA genotyping is patient specific. Extensive efforts were put into finding specific markers for fnRBCs and trophoblastic cells, to allow for specific fetal cell selection by FACS or more often MACS, and optimized confirmation by immunostaining (Table 2).

CD71 is the most extensively tested fnRBC selection marker, although studies showed it is insufficient on its own for adequate cell isolation [46,47]. Bianchi described the use of anti-CD71 or anti-CD36 antibodies in combination with anti-glycophorin A (GPA) [48]. When analyzing the samples by flow cytometry, they noticed two different groups of CD71+/GPA+ cells, on one hand, and a considerably larger group of CD36+/GPA+ cells, on the other hand. The addition of GPA to the other antibodies significantly brought up the fetal gender prediction accuracy: from 57% correct with CD71 alone or 88% with CD36 alone, up to 100% correct prediction when the combination was used. These markers also were popular targets for fnRBC selection in multiple other studies (e.g., Refs. [42,43,49,50]), as was reviewed by Jackson [54]. Additional strategies included targeting antigen i [44], the use of specific antibody HAE9 [44], or fnRBC selection based on CD34 [51], although these did not prove to be as useful as GPA.

In the meantime the suitability of cytoplasmic fetal (gamma) and embryonic (epsilon and zeta) hemoglobins (Hb) was investigated as well, both for cell sorting [52] and detection/identification of fetal cells [42,50]. Fetal gamma hemoglobin (HbF) is significantly higher expressed in fnRBCs than in adult nRBCs [52], but the maternal HbF expression levels in blood can be upregulated under certain conditions, which hinders its use as a fetal cell marker [50]. Choolani et al. studied the levels of epsilon Hb by making mixtures of fetal blood and blood of women who had never been pregnant and by studying samples of women carrying a male fetus, collected at 7–14 weeks' gestation [53]. They described a linear decrease in epsilon Hb+ cells with increasing GA, to an almost insignificant level at 14 weeks. Christensen and colleagues found that staining for epsilon Hb showed a higher fetal specificity than staining for HbF in samples collected post-CVS at 9–14 weeks' GA, but did not find any epsilon+ cells pre-CVS at that same time frame [16]. Given that the expression of zeta Hb is lost even earlier than epsilon Hb [54], both epsilon and zeta Hb are not useful as fetal cell markers.

Most reports showed a rather low fnRBC recovery ranging from 0 to 20 cells per blood sample [42,61]. In contrast, Wachtel and colleagues reported a far higher recovery of fnRBCs by applying

Table 2 Markers and Characteristics for Fetal Cell Selection and Identification

Fetal Cell Type		Marker or Characteristic		References
fnRBCs	Surface markers	CD71	Transferrin receptor, or cluster of differentiation 71, expressed during different stages of erythrocyte maturation	[42,46–50]
		CD36	Thrombospondin receptor, expressed on erythrocytes during early differentiation	[48]
		GPA	Glycophorin A, a RBC-specific marker	[43,48,49]
		Antigen i	Only expressed on fetal and newborns erythrocytes	[44]
		HAE9	Specific antibody to cell surface antigen of human nucleated erythroid cells	[44]
		CD34	Hematopoietic progenitor cell antigen, expressed on progenitor cells in very early developmental stages	[51]
	Cytoplasmic markers	HbF	Fetal gamma hemoglobin is the predominant cytoplasmic protein in fnRBC after 7 wks GA	[16,42,50,52]
		Epsilon Hb	Embryonic epsilon hemoglobin, more specific for fetal cells than HbF but only detected at very early gestation	[16,53]
		Zeta Hb	Embryonic zeta hemoglobin, expression decreases even earlier than epsilon Hb	[50,54]
	Charge flow separation		Separation based on the difference in surface charge density of different cell types	[55,56]
Tropho	Surface markers	FD0161G	Antibody specific for extravillous trophoblastic cells	[57]
		hPL	Human placental lactogen or human chorionic somatomammotropin, a placental hormone	[58]
		HLA-G	Human leukocyte antigen-G, a nonclassical major histocompatibility complex class I antigen	[29]
		Mab340	Monoclonal antibody 340, specifically for cytotrophoblasts	[59]
	Isolation based on size (ISET)		The slightly larger trophoblasts are caught on a filter, while the smaller maternal WBCs flow through	[60]

Abbreviations used: fnRBCs, *fetal nucleated red blood cells;* GA, *gestational age;* Tropho, *fetal trophoblastic cells;* WBC, *white blood cell;* wks, *weeks of gestational age.*

charge flow separation (CFS), even claiming a recovery of about 2000 fnRBCs per sample as confirmed by XY FISH [55,56]. This method uses specific surface charge densities of each cell type to separate the different blood cell types into individual compartments.

Also for trophoblasts multiple studies were done in the search for more specific markers. Mueller et al. developed an antibody, FD0161G, that is specific to extravillous trophoblasts [57], and the same group found later that also human placental lactogen can be used as a target in MACS-based positive selection [58]. van Wijk and colleagues adopted the marker HLA-G into their workflow, which is a nonclassical major histocompatibility complex class I antigen present on extravillous trophoblasts, where the classical HLA-A and HLA-B are not expressed [29]. They only recovered few cells, but as their laborious workflow resulted in about 17 microscopy slides per sample, they only analyzed one slide per sample; the amount of fetal cells reported is thus not a definitive number. Durrant et al. [59] developed yet another antibody specifically for cytotrophoblasts, Mab 340, but this antibody cross-reacted with syncytiotrophoblasts as well.

Vona and colleagues took a slightly different approach and set out to select for trophoblastic cells based on their size [60]. In the isolation by size of epithelial tumor/trophoblastic cells (ISET) protocol, diluted blood is filtered through a polycarbonate filter with 8-μm pores, onto which larger cells including the trophoblasts are retained. After hematoxylin and eosin staining or immunostaining (cytokeratin+, placental alkaline phosphatase+, CD45−), the cells are removed from the filter by laser capture microdissection (LCM) to undergo further analysis. They found 0.5–3.5 trophoblastic cells/mL of maternal blood.

As illustrated here, a myriad of variably successful protocols for blood processing and fetal cell selection has been published so far. It can be appreciated that less manipulation of the sample is better for preserving intact fetal cells. None of the described markers has 100% fetal cell specificity; CD71 whether or not in combination with GPA, and HLA-G seemed among the better ones for, respectively, fnRBC and trophoblast selection.

The question has also been raised whether it would be possible to culture the fetal cells after isolation in order to increase the yield. Theoretically, CD34+ progenitor cells are still immature and have potential for growth and differentiation. Wachtel and colleagues set out to prepare RBC progenitor cultures, as they achieved a >99% viability of the nucleated cell suspension obtained with their CFS method [55]. After CFS enrichment of the blood samples, maternal RBC lysis and further lymphocyte depletion was performed. Thereafter the surviving cells were cultured in methylcellulose. The investigators succeeded in culturing several tens of clones in all samples, each consisting of about 100–200 cells. However, during validation, only a minority of the clones turned out to be Y-positive in male pregnancies, showing that the majority of the formed clones were of maternal origin. Also other attempts were unsuccessful [51].

Another interesting finding was the presence of apoptotic fetal cells in maternal plasma. When studying cells isolated from the plasma fraction after Percoll DGC, van Wijk et al. found between 1 fetal in 500 maternal cells to 1 in 2000 [62]. These cells had both morphological features suggesting apoptosis and a positive terminal deoxynucleotidyl transferase-mediated dUTP nick end labeling (TUNEL) staining, indicative of ongoing apoptosis. Nevertheless, despite these cells being in an apoptotic state, XY FISH and PCR analysis still yielded successful results. Simultaneously, Sekizawa reported on the incidence of apoptosis among fnRBCs isolated from maternal blood samples [63]. After DGC and subsequent fnRBC flow cytometric sorting based on HbF, the fetal cells obtained were assessed microscopically for potential positive TUNEL staining. Fourth 3% of the fnRBCs had undergone apoptotic change. This phenomenon can explain some of the lower numbers sometimes reported, and undoubtedly provides a source for fetal cfDNA (Fig. 2).

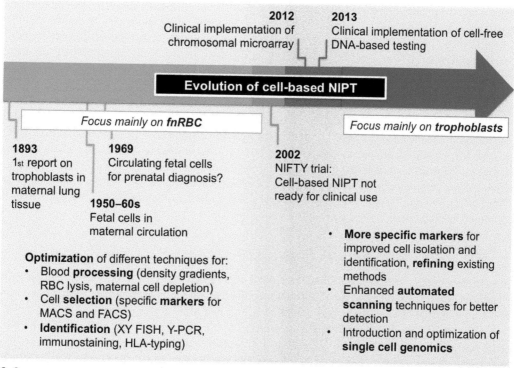

FIG. 2

Fetal cells have been a subject of interest for over a century. Substantial progress had been made by the early 2000s, but the NIFTY trial made it clear that cell-based NIPT was not yet ready then for clinical implementation. Nevertheless, the field of cell-based NIPT benefited substantially from the more recent developments and achievements for single cell genomics and improved isolation and detection methods.

EARLY 2000s: DISAPPOINTMENT AFTER THE NIFTY TRIAL, COINCIDING WITH UPCOMING FETAL CELL-FREE DNA TEST DEVELOPMENT

Above-mentioned studies mostly only included a small number of samples. In 2002, the results of the National Institute of Child Health and Human Development Fetal Cell Isolation Study (NIFTY) trial were published, a study focusing on the isolation and diagnostic use of fnRBCs, with over 2700 samples [64]. Samples were collected at multiple sites from 1995 to 1999 and were processed by DGC (1.077 or 1.119 g/mL), whether or not followed by an additional MACS or FACS depletion step (based on CD14, CD15, and/or CD45). Subsequent positive selection for fetal cells was done with MACS (CD71+) or FACS (HbF+/CD45−). All samples were analyzed with XY FISH, and occasionally also probes for chromosomes 21, 13, and 18 were included for aneuploidy detection. Unfortunately, the disappointing conclusion of the study was that fetal cell analysis was far from ready for clinical implementation, even though it had an aneuploidy detection rate of 74% comparable to single serum marker prenatal screening. Only in 41% of the cases carrying a male fetus at least one XY cell was found, and the false-positive rate for gender detection was 11%. More technological advances would be needed to turn the enrichment process and analysis into a streamlined routine protocol yielding consistent results. Jackson summarized the predominant opinion on cell-based NIPT at that time in his review written after an NICHD conference: it was generally acknowledged that both fetal cells and fetal cfDNA are present in detectable numbers or copies in the maternal blood. However, there was no agreement on which of both paths was the best one to follow in the future [54]. The shortcomings of the cell-based NIPT methods were clear, and needed to be seriously improved, but the idea was not given up on.

The enthusiasm for cell-based NIPT was partially dimmed when cfDNA testing was developed and clinically implemented around 2013 [1,65]. This profoundly changed the field of prenatal genetic testing. Also chromosomal microarray had just been introduced in diagnostic labs in 2012 and was a great improvement for invasive prenatal diagnosis compared to karyotyping [66]. It was expected that, as a consequence, more women would opt for CVS or amniocentesis as results for copy number variant (CNV) aberrations could now be reported. However, a considerable decrease in the number of invasive procedures has been seen over the years [2].

YET CELL-BASED NIPT EFFORTS CARRY ON

The interest in fetal cells and research efforts to make cell-based NIPT work persisted, as publications by Hamada et al. [18] and Krabchi et al. [19] had reported very hopeful numbers of about 2–6 fetal cells per mL of maternal blood. Studies like the NIFTY trial had made it clear that much more work was needed on both the enrichment and the detection procedure, and its throughput. Given the disappointing results for fnRBCs so far, the focus shifted toward trophoblastic cells.

The Danish group led by Kølvraa (the group later transformed to the company FCMB ApS, and eventually to Arcedi Biotech) studied the expression patterns of fetal trophoblastic cells in more detail. In their 2013 publication [67], Hatt and colleagues proposed a combination of mesenchymal and endothelial markers for trophoblast isolation and detection. Comparing CD34, CD105, CD141, and CD146 for MACS enrichment, CD105 turned out to give by far the highest yield, followed by CD141. A subsequent transcriptome microarray analysis of the CD105 + population pointed out a series of 39 overexpressed fetal markers, of which 13 were expressed in extravillous trophoblasts, thus confirming the specificity of CD105. Simultaneously, they also validated the use of a cocktail of anticytokeratin (CK) antibodies for the detection of fetal trophoblasts. Cytoplasmic CK is a typical endothelial marker and is also expressed in trophoblastic cells in a specific vesicular pattern. In a follow-up study [68], using CD105 and CD141 positive selection combined with CK staining followed by FISH, they could retrieve a range of 0–18 cells per sample, but noted as well that some cells were lost during the FISH procedure. They since filed patents in the United States and Europe.

Zimmermann described the finding of novel unique monoclonal antibodies that could specifically target fetal erythroblasts, although they were not tested on actual peripheral blood samples from pregnant women [69]. They undertook an elaborate screening of 690 monoclonal antibodies obtained after immunizing mice with fetal erythroid cell membranes. They evaluated the antibodies' reactivity with several fetal and adult samples (both blood and tissue), which led them to the selection of two of those antibodies that showed the desired fnRBC specificity. The surface antigens targeted by those antibodies were different from known markers as GPA, CD36, and CD71.

Emad and Drouin took a deeper look into the influence that DGC might have on cell recovery: they found that DGC with Histopaque (1.119 g/mL) introduced a loss of about 60%–70% of fetal cells [70], suggesting that other approaches might be more suitable. They also invested a lot of time in improving automated scanning methods for the detection of fetal cells to replace manual scanning, as the latter is both very time consuming and laborious, and might underestimate the real number of fetal cells [71,72].

SINGLE CELL GENOMICS

Besides the developments specifically for fetal cells, the field of single cell genomics in general got a tremendous boost over recent years, in part also promoted by increased interest in circulating tumor cell (CTC) analysis. A review by Macaulay and Voet from 2014 summarizes the advances in single cell genomics and transcriptomics, which were also beneficial to the fetal cell field [73]. Initial fetal cell enrichment studies focused on getting a maximum purity of the obtained fetal cells, but thanks to improvements in (automated) microscopic detection and techniques for isolating cells at a single cell level, reaching 100% purity is not essential. One prerequisite to enable single cell analysis is that the individual cell needs to undergo whole genome amplification (WGA) to obtain a sufficient amount of DNA to allow for further analysis, such as SNP-based analysis. A single cell only contains about 6 pg of DNA. This can be amplified by WGA to the μg range. There are various WGA technologies on the market, each having advantages and disadvantages depending on the downstream application [74,75]. In general, WGA methods are divided into: (i) purely PCR-based amplification, (ii) multiple displacement amplification (MDA)-based methods, (iii) hybrid methods that combine those two, and (iv) other methods that do not fall under one of the previous categories.

Pure PCR methods, such as degenerate oligonucleotide primed-PCR (DOP-PCR, e.g., GenomePlex by Sigma-Aldrich) and primer extension preamplification-PCR (PEP-PCR), are older methods based on a standard PCR reaction with reduced specificity to enable a general DNA amplification. They use a lower fidelity DNA polymerase, resulting in a higher error rate, which can lead to allelic dropout and thus to an incorrect detection of the cell's genetic signature. This is important when performing SNP- and short tandem repeat (STR)-based analysis on single fetal cells.

MDA methods (e.g., REPLI-g by Qiagen or GenomiPhi by GE), on the other hand, make use of the higher fidelity φ29 polymerase which also has $3'$–$5'$ proofreading activity, resulting in less erroneous nucleotides being incorporated during the amplification process. This results in better coverage than with the methods based on PCR but has a higher nonuniformity across the genome.

To overcome the issues described previously, hybrid methods such as PicoPLEX by Rubicon/TakaraBio or multiple annealing and looping-based amplification (MALBAC) [76] by Yikon have been developed. They deal with potential amplification and representation bias by making the process semilinear: starting off with a limited isothermal amplification followed by a PCR amplification, to avoid the immediate exponential amplification of potential errors that were included in the first steps. In general, hybrid methods are better for CNV analysis, whereas MDA methods score better for single nucleotide polymorphism (SNP)-based analysis. An ideal WGA method would work well for both applications.

Other types of WGA have been developed as well, for example, Ampli1 by Silicon Biosystems. This so-called ligation-based method uses restriction enzymes to create well-defined fragments to be amplified by sequence-specific primers. Most recently, the linear amplification by transposon insertion (LIANTI) method was published [77], making use of random fragmentation of the cell's genomic DNA by transposition. During this fragmentation a T7 promoter is included that serves as a starting point for subsequent linear in vitro transcription amplification. Next, the resulting RNA fragments undergo reverse transcription and second strand synthesis, leaving LIANTI amplicons ready for further library preparation and sequencing.

One drawback of most fetal cell isolation protocols is that a fixation and permeabilization step is included, which can have a detrimental effect on the DNA quality and WGA performance. It thus has to be evaluated for each fetal cell enrichment protocol what the best downstream approach is: for example, MDA kits are in general less well suited for fixed cells, where hybrid methods and Ampli1 might work better. One such comparison for genome-wide CNV analysis by array comparative genomic hybridization (aCGH) was described by Normand et al. [78]: they compared three different WGA kits for their performance with both fixed and unfixed cells, showing that a 2.5-Mb DiGeorge deletion was detected consistently in all samples, but differences were seen when smaller CNVs were included.

WHERE WE ARE TODAY
SAMPLE COLLECTION

There are three sources for obtaining fetal DNA: (i) fetal cfDNA from maternal plasma, (ii) fetal cells from maternal blood, or (iii) fetal cells collected through (endo)cervical sampling.

Literature describes the use of various blood collection tubes, whether it be commercially available EDTA [79,80], Cell-Free DNA BCT Streck [79,81], acid-citrate-dextrose [15,46], heparinized [27,82,83], or customized RareCyte collection tubes [84]. The fact that circulating fetal cells are obtained with all collection tubes in a comparable efficiency suggests that the tube type may not be the most critical feature of the process, as long as it is compatible with the subsequent blood-processing protocol applied. The reported working volumes of maternal blood range from 5 mL to almost 40 mL, which is also a factor to take into account for clinical implementation, as the pregnant woman might object to larger volumes. Other questions that need to be addressed are how long the blood can be stored before processing and what the optimal storage conditions are. Most protocols start the blood processing as soon as possible after collection, but it is known that adequate results can still be obtained after keeping the sample overnight or even several days at 4°C or room temperature [79,84]. The fact that prolonged storage does not pose a problem is important for cases where a diagnostic facility is not immediately available at the site of blood collection, so that the sample can be transported safely to the appointed diagnostic lab.

Several techniques can be applied for the endocervical collection of trophoblasts, such as an endocervical canal lavage [85], or a transcervical smear taken with a cytobrush. Concerns may be raised about whether this is truly noninvasive and thus safe. The Paterlini group reported cervical trophoblasts being isolated with a cytological brush that is rotated at the external os, similarly to a routine PAP test [12]. The Armant laboratory at Wayne State University developed a method for trophoblast retrieval and isolation from the cervix (TRIC) whereby 2 cm of a cytobrush is inserted and rotated in the endocervical canal to collect cervical mucus [13], thus a more invasive approach than a regular PAP smear. The question about the safety of the procedure is not so straightforward to investigate. The simultaneous collection of tough cervical mucus can pose a problem for fetal cell identification: studies investigating the use of mucolytics in the postcollection processing do not report superior results compared to when no mucolytic is used [85]. A drawback is that more specialized personnel is needed to collect these samples, compared to a simple blood collection. Also, the inconvenience and discomfort for the patient need to be taken into account—hence, cell-based analysis starting from a maternal blood sample has received the most attention so far.

CURRENT METHODOLOGIES FOR FETAL CELL ENRICHMENT AND ISOLATION

Following we summarize the studies published over the last few years, illustrating the improvements and currently used methodologies for cell-based NIPT, mainly focused on trophoblast isolation from maternal blood.

Our group at Baylor College of Medicine collaborated with both RareCyte Inc. and Arcedi Biotech for the development of two fetal trophoblast isolation protocols [79,84]. For the RareCyte protocol [84], the first phase comprises a DGC step using a rubber float of a specific density to separate the layer of nucleated cells from plasma and maternal RBCs. In combination with WBC depletion using a CD45/CD36 RosetteSep depletion cocktail added prior to DGC, 75%–99% of maternal WBCs are depleted (Vossaert et al., unpublished observation). By adding a series of higher density fluids and centrifugation steps, the WBC band is then lifted above the float into a smaller collection tube. The float serves as a more easily workable solution compared to Ficoll gradients and also has the advantage of being nontoxic for the cells. The resulting cell suspension is stained with a specific antibody cocktail for cytoplasmic CK and CD45 to prepare the sample for microscopic analysis. Fetal cell candidates are first selected by automated scanning on the RareCyte CyteScanner/CytePicker device, based on a positive DAPI and CK staining, and a negative CD45 staining. Subsequently these fetal cell candidates are manually validated, for which also the nuclear morphology of the cells is taken into account (Fig. 3). All putative trophoblasts are picked as single cells with a 40-μm needle and deposited into a PCR tube, to undergo WGA (mainly PicoPlex). As such, an average fetal cell yield of 0.74 cells/mL of maternal blood was obtained without depletion and 0.36 cells/mL when RosetteSep was included [84]. In a follow-up study on a larger sample cohort including more samples collected at later GA, a range of 0–1.58 cells/mL was found, sometimes sticking together in a cluster of two, three, or more trophoblastic cells (Vossaert et al., unpublished observation).

Arcedi Biotech uses maternal RBC lysis as a first step [79]. The resulting suspension then undergoes MACS, to positively select for fetal cells based on their CD105 and CD141 expression, which then are stained with a cocktail of anti-CK antibodies. Also, here fetal cell candidates are selected after automated scanning and manual validation on a MetaSystems scanner, mainly based on their CK staining pattern. Individual cells are picked on the CytePicker or the ALS CellCelector. Their reported average fetal cell yield is at currently 0.43 trophoblasts/mL.

Specific cell enrichment can also take place in a microfluidics setting: several companies have developed microfluidics platforms allowing for the dielectrophoretic (e.g., Silicon Biosystems [86]) or immunomagnetic (e.g., Cynvenio [87], Fluxion [88]) isolation of rare cells, initially for CTC isolation. With the implementation of the appropriate capturing antibodies, these platforms can also serve for the isolation of fetal cells. A group in California led by Hsian-Rong Tseng has developed so-called Nano-Velcro microchips [82], which are coated with anti-EpCAM antibodies to specifically capture fetal trophoblasts. The cell suspension obtained after initial dilution and DGC (Histopaque, 1.119 g/mL density) of the blood is injected into one of those microfluidics chips at a flow rate of 1 mL/h. Captured cells are subsequently fixed and stained with Hoechst and antibodies directed against CK-7, HLA-G, and CD45. Fetal cell candidates are isolated by LCM, and thereafter analyzed by FISH, aCGH, and/or

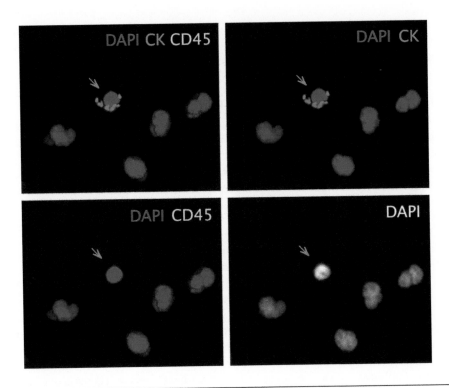

FIG. 3

A single fetal trophoblastic cell in a pool of maternal cells. The fetal trophoblast has a specific positive cytokeratin staining in a vesicular pattern (*green*) but is negative for CD45 (*yellow*). When considering the nuclear morphology of the fetal cell (DAPI only), it is clear that the fetal nucleus is rounder and does not have a lobed structure, compared to the surrounding maternal WBCs.

STR. A recovery of 1.5–3 cells/mL for 6 healthy pregnancies was reached, and a recovery of 2.5–7.5 cells/mL for pregnancies in which a fetal genetic abnormality was diagnosed.

Chen and colleagues recently published their method for fetal cell isolation, independent of fetal cell type and based on a double-negative selection strategy [83]. After RBC lysis, the samples undergo depletion by CD45-MACS. All CD45 − cells are subsequently stained with Hoechst and propidium iodide (PI), and sorted via flow cytometry. CD45 −/PI −/Hoechst + cells are individually picked afterward using a specific single cell isolation glass capillary pipette. They isolated two and five confirmed male fetal cells from two 10-mL blood samples. A limitation of this study is thus that only two samples were included so far, but these preliminary results and their data on spike-in cord blood samples seem promising.

Guanciali Franchi et al. described a method for fnRBC and CD34+ cell isolation and analysis, as part of an improved screening before referring pregnant women for invasive procedures [80]. In short, after blood collection, the blood is kept overnight in nonphysiological conditions (4°C, slightly acidic pH, and raised osmolarity) [89], and separated the next day by DGC (Biocoll, 1.082 g/mL density) with the help of a device specifically designed for this purpose. The cell layer containing the fetal cells is subsequently isolated, washed, and depleted of monocytes and platelets by leaving the sample overnight in a culture flask at 37°C and 5% CO_2 so that these contaminants can adhere to the flask. The supernatant is used the following day for FISH analysis for chromosomes 21 and 18. The authors reported a yield of several hundred cells per maternal sample this way [90].

The group at Wayne State University also uses specific immunomagnetic isolation of trophoblasts but starts from endocervical samples [91,92]. After fixation and washing of the cervical specimens (collected as early as 5 weeks' gestation), the cell suspension is incubated overnight at 4°C with anti-HLA-G-coupled nano-sized magnetic beads. After removal of the unbound cells using a DynaMag-Spin Magnet, the HLA-G-positive trophoblasts are placed on microscopy slides either to undergo FISH, or to undergo DNA extraction for further analysis. The authors reported a yield of approximately 500-1500 trophoblasts per sample isolated by TRIC. The fetal origin was confirmed by the trophoblast-specific expression of β-hCG for 95%–100% of all cells and the presence of an XY profile for >99% of the cells in male pregnancies [91].

Also the group led by Paterlini-Bréchot has started working with cervical samples, which are subsequently processed with the ISET technology [12]. In a cohort of 21 samples collected between 8 and 21 weeks GA, they were able to recover 2–12 fetal cells/sample. They are still working on improving the isolation and sequencing analysis of rare cells retrieved from blood, but more in the field of CTCs [93].

DOWNSTREAM ANALYSIS

Downstream analysis of single fetal cells is twofold: the fetal origin of each analyzed cell needs to be confirmed, and the cell needs to be assessed for fetal genetic abnormalities. As described earlier, WGA is performed to obtain a sufficient amount of input material for downstream analysis for aneuploidy detection, subchromosomal CNV analysis, and/or single gene mutation detection.

For the detection of common autosomal and sex chromosome aneuploidies in fetal cells, karyotyping and FISH have been extensively used [26,27,64], and are still valuable today: Guanciali Franchi et al. recently reported 100% specificity and 100% sensitivity for the 39 normal and 7 abnormal (4 T21, 2 T18 and 1 triploidy) cases that were screened by fetal cell FISH analysis [80]. FISH has its limitations however, in that it is not very useful for the genome-wide detection of aneuploidy and smaller subchromosomal events. Chromosomal microarray, clinically introduced in 2012, led to a huge improvement in prenatal diagnostic accuracy [66]. The Arcedi Biotech group reported on five cases for which isolated trophoblastic cells were analyzed by aCGH after PicoPlex WGA [81]. One T21 case, one case with a partial T21 (12.4 Mb duplication), two cases with high-grade mosaicism for T2 and T13, respectively, and one case in which an unbalanced translocation (31.4 Mb terminal deletion of 4p and 30.1 Mb terminal duplication of 8p) diagnosed by invasive testing and confirmed by cell-based NIPT in a blinded study. All blood samples were collected prior to invasive testing. Of note, they worked with a pool of 2–7 trophoblasts per analysis instead of single cells, and in general do not perform any separate genotyping assay to confirm the fetal origin of cells.

Also the NanoVelcro group has used microarray for CNV analysis and described a series of nine samples from pregnancies in which the fetus had been shown to be affected by invasive testing, with respectively T21 (four cases), T18, T13, tetrasomy X, both T18 and disomy Y, and a 17.2 Mb 9p24.3p22.2 deletion combined with a 11.7 Mb 14q32.13q32.33 duplication [82]. All samples were taken prior to the invasive procedure. All fetal genders and genetic abnormalities were detected correctly, concordant with the results obtained after invasive procedure. They applied STR analysis for the confirmation of fetal origin for the analyzed cells.

Our lab so far has focused on copy number abnormalities as well, initially by aCGH analysis, but now by single cell, low-pass, whole genome sequencing. In 2016, we reported on several samples, taken before the invasive procedure, with aneuploidies confirmed by cell-based NIPT, all in concordance with the clinical data [84]. In one of the cases the fetus carried a microdeletion of 15q of only 2.7 Mb in size, showing that also small subchromosomal events can be detected reliably. The fetal origin of the cells analyzed was confirmed by STR analysis or by an in-house developed SNP-based genotyping sequencing assay. The latter uses multiple amplicons containing informative SNPs, distributed across the genome, to compare the fetal cell's content to the maternal, and if available paternal genomic profile. In the meantime, we collected further data showing the detection of abnormalities of only 1 Mb in size (Vossaert et al., unpublished observation). Cell-based NIPT has thus become suitable for the detection of both aneuploidies and a whole range of microdeletions and duplications.

Also for the detection of single gene mutations, cell-based testing has the advantage over cfDNA-based testing in that it is not hindered by the excess of simultaneously present maternal alleles [94]. One of the first reports demonstrating the feasibility of diagnosing Mendelian disorders with cell-based NIPT was written by Camaschella et al. [95]: they developed a PCR protocol for the detection of paternally inherited Lepore–Boston disease (specifically for the δ-β globin fusion gene) with DNA extracted from buffy coats obtained from maternal blood samples collected between 8 and 10 weeks' GA. They described three cases including two affected and one unaffected fetus, and the results were concordant with the clinical data. Also Cheung described the successful diagnosis of two hemoglobinopathy cases after CD71 enrichment [50].

The Paterlini group published studies on the diagnosis of spinal muscular atrophy (SMA) and cystic fibrosis (CF) [96]. By applying STR genotyping of the isolated circulating trophoblasts after ISET enrichment, the investigators could make a correct diagnosis of the normal, carrier, or affected status in all samples (blinded study of 63 pregnancies, of which 31 at risk for SMA and 32 at risk for CF). Later on, they also applied their workflow for samples obtained through cervical collection, with equal success [12]. Also the Armant laboratory developed a comprehensive SNP-profiling protocol for cervical trophoblastic cells and applied it for the detection of single gene mutations [92].

One point of consideration for cell-based NIPT is whether one wants to pool several fetal cells per sample and analyze them together, or whether it is better to analyze multiple cells per sample individually. In favor of the latter is that by analyzing and genotyping the cells separately, certainty about their fetal origin can be obtained, while in fetal cell pools, contamination with an occasional maternal cell might skew the analysis. Additionally, individual cell analysis allows for a more accurate detection of confined placental mosaicism. Confined placental mosaicism is well documented for aneuploidies but likely is less frequent for de novo CNVs or point mutations. Analyzing fetal cell pools, on the other hand, would be less time consuming and cheaper, and could smooth out single cell "noise" in sequencing plots, yielding a higher resolution. This can however also be achieved by pooling sequencing data of individual cells. If somehow a method could be developed with which the number of isolated fetal cells would be a multifold of what is obtained currently, then those fetal cells could be analyzed in bulk.

POTENTIAL AS A CLINICAL TEST

According to a recent meta-analysis, the risk of fetal loss after amniocentesis and CVS is about 0.11% and 0.22%, respectively [97], which is much lower than previously reported. Nevertheless, it would be convenient if in the future cell-based NIPT could replace invasive testing, on the condition that the same test accuracy and confidence can be reached. So far cell-based NIPT is not yet offered as a stand-alone diagnostic test. Guanciali Franchi et al. recently described their experience in adding fetal cell analysis to conventional screening to be able to make a better selection of women who need invasive prenatal diagnosis [80]. They thus already clinically implemented cell-based NIPT but at this stage as a screening step for high-risk pregnancies, rather than truly diagnostic.

Cell-based NIPT is currently sufficiently sensitive to detect copy number abnormalities down to the 1–2 Mb size range (Vossaert et al., in preparation) and is consistently concordant with clinical data [81,84]. More and larger, blinded studies including both high- and low-risk pregnancies will be necessary to establish cell-based testing PPVs and NPVs, but for screening purposes a better performance is expected as compared to cfDNA-based testing [3]. Also, the work on cell-based NIPT for single gene mutations is promising.

What about the practical side of implementing cell-based NIPT as a new test? Considering this is a noninvasive test starting from a blood sample, this procedure does not pose any greater risk to the patient than any other blood collection procedure. A peripheral blood sample is easy to collect and can be drawn by any phlebotomist. It has been shown that prolonged storage of samples is not harming their quality and subsequent cell analysis.

Another factor to take into account is the turnaround time (TAT). cfDNA-based testing currently has a TAT of 2–5 days, depending on which test is chosen. Assuming that the fetal cell isolation and WGA can be accomplished in about 2–3 days, this would be the only increase in TAT for cell-based NIPT, as the downstream process is similar to the current cfDNA-based methods.

The total cost for the current cfDNA-based tests range from $500 to $900 [98]. It is estimated that the initial cost for cell-based NIPT will exceed the cost of cfDNA-based testing, due to a more elaborate blood processing, the fetal cell enrichment, and WGA. Once a cell-based NIPT test would be launched successfully and the demand increases, reagent costs can be reduced, although the labor costs would still be higher than for cfDNA-based testing. The costs for WGA, library preparation, and sequencing increase when per sample multiple cells are analyzed individually, but are partially compensated by the fact that a much lower sequencing depth is needed. Additionally, as the field of NGS itself is so rapidly evolving, actual sequencing costs will also decrease further over time. Regardless, one could argue that the added diagnostic value is worth the additional cost by any means if it enables a more informed decision about the ongoing pregnancy. So all of these factors considered, it is amply clear that cell-based NIPT has the potential of becoming a routine clinical diagnostic test.

CONCLUSIONS AND FUTURE PERSPECTIVE

After decades of trial and error and varying successes, it seems that cell-based NIPT has reached the final stages before clinical implementation. Given its potential for both adequate CNV analysis and single gene disorder detection, it would bring a huge improvement to the field of prenatal diagnosis.

Current efforts aim at gathering data on larger sample cohorts, so that the enrichment protocols and single cell analysis techniques can be fine-tuned in terms of performance and scalability toward an actual clinical launch. Fetal trophoblast-based NIPT could initially be launched as a screening test, to be confirmed by CVS or amniocentesis. Especially for aneuploidy detection and CNV screening, it is anticipated to reach a higher accuracy than conventional serum screening and cfDNA-based testing. Once a good PPV and NPV are established, it could become a diagnostic test. Cell-based NIPT would surely be adequate for the definitive diagnosis of single gene mutations, on the condition that a sufficient number of fetal cells can be isolated from each sample. For aneuploidy and CNV analysis, the concern about mosaicism remains when trophoblastic cells are used. More elaborate sample sizes are necessary to determine how many single cells exactly need to be analyzed per sample for a correct diagnosis. It is to be expected that analyzing multiple single cells per sample will increase the test performance. The use of fnRBCs can remove this concern. Nevertheless, considering the methods currently available, the first commercially launched cell-based NIPT will most likely be fetal trophoblast based, starting from a maternal blood sample rather than a cervical swab.

REFERENCES

[1] Lo YM, Corbetta N, Chamberlain PF, Rai V, Sargent IL, Redman CW, et al. Presence of fetal DNA in maternal plasma and serum. Lancet (London, England) 1997;350(9076):485–7.

[2] Williams J, Rad S, Beauchamp S, Ratousi D, Subramaniam V, Farivar S, et al. Utilization of noninvasive prenatal testing: impact on referrals for diagnostic testing. Am J Obstet Gynecol 2015; 213(1):102.e1–6.

[3] Petersen AK, Cheung SW, Smith JL, Bi W, Ward PA, Peacock S, et al. Positive predictive value estimates for cell-free noninvasive prenatal screening from data of a large referral genetic diagnostic laboratory. Am J Obstet Gynecol 2017;217:691.e1–6.

[4] Hartwig TS, Ambye L, Sørensen S, Jørgensen FS. Discordant non-invasive prenatal testing (NIPT)—a systematic review. Prenat Diagn 2017;37(6):527–39.

[5] American College of Obstetricians and Gynecologists. ACOG committee opinion. [Internet]. American College of Obstetricians and Gynecologists. Available from: https://www.acog.org/Resources-And-Publications/Committee-Opinions/Committee-on-Genetics/Cell-free-DNA-Screening-for-Fetal-Aneuploidy.

[6] Practice Bulletin No. 163 Summary: screening for fetal aneuploidy. Obstet Gynecol 2016;127(5):979–81.

[7] Zhou Y, Zhu Z, Gao Y, Yuan Y, Guo Y, Zhou L, et al. Effects of maternal and fetal characteristics on cell-free fetal DNA fraction in maternal plasma. Reprod Sci 2015;22(11):1429–35.

[8] Ariga H, Ohto H, Busch MP, Imamura S, Watson R, Reed W, et al. Kinetics of fetal cellular and cell-free DNA in the maternal circulation during and after pregnancy: implications for noninvasive prenatal diagnosis. Transfusion 2001;41(12):1524–30.

[9] Klonisch T, Drouin R. Fetal–maternal exchange of multipotent stem/progenitor cells: microchimerism in diagnosis and disease. Trends Mol Med 2009;15(11):510–8.

[10] Kinder JM, Stelzer IA, Arck PC, Way SS. Immunological implications of pregnancy-induced microchimerism. Nat Rev Immunol 2017;17(8):483–94.

[11] Lyall F, Bulmer JN, Duffie E, Cousins F, Theriault A, Robson SC. Human trophoblast invasion and spiral artery transformation: the role of PECAM-1 in normal pregnancy, preeclampsia, and fetal growth restriction. Am J Pathol 2001;158(5):1713–21.

[12] Pfeifer I, Benachi A, Saker A, Bonnefont JP, Mouawia H, Broncy L, et al. Cervical trophoblasts for non-invasive single-cell genotyping and prenatal diagnosis. Placenta 2016;37:56–60.

[13] Imudia AN, Suzuki Y, Kilburn BA, Yelian FD, Diamond MP, Romero R, et al. Retrieval of trophoblast cells from the cervical canal for prediction of abnormal pregnancy: a pilot study. Hum Reprod 2009;24(9):2086–92.

[14] Choolani M, O'Donoghue K, Talbert D, Kumar S, Roberts I, Letsky E, et al. Characterization of first trimester fetal erythroblasts for non-invasive prenatal diagnosis. Mol Hum Reprod 2003;9(4):227–35.

[15] Shulman LP, Phillips OP, Tolley E, Sammons D, Wachtel SS. Frequency of nucleated red blood cells in maternal blood during the different gestational ages. Hum Genet 1998;103(6):723–6.

[16] Christensen B, Kølvraa S, Lykke-Hansen L, Lörch T, Gohel D, Smidt-Jensen S, et al. Studies on the isolation and identification of fetal nucleated red blood cells in the circulation of pregnant women before and after chorion villus sampling. Fetal Diagn Ther 2003;18(5):376–84.

[17] Kølvraa S, Christensen B, Lykke-Hansen L, Philip J. The fetal erythroblast is not the optimal target for non-invasive prenatal diagnosis: preliminary results. J Histochem Cytochem 2005;53(3):331–6.

[18] Hamada H, Arinami T, Kubo T, Hamaguchi H, Iwasaki H. Fetal nucleated cells in maternal peripheral blood: frequency and relationship to gestational age. Hum Genet 1993;91(5):427–32.

[19] Krabchi K, Gros-Louis F, Yan J, Bronsard M, Massé J, Forest JC, et al. Quantification of all fetal nucleated cells in maternal blood between the 18th and 22nd weeks of pregnancy using molecular cytogenetic techniques. Clin Genet 2001;60(2):145–50.

[20] Bianchi DW, Stewart JE, Garber MF, Lucotte G, Flint AF. Possible effect of gestational age on the detection of fetal nucleated erythrocytes in maternal blood. Prenat Diagn 1991;11(8):523–8.

[21] Takabayashi H, Kuwabara S, Ukita T, Ikawa K, Yamafuji K, Igarashi T. Development of non-invasive fetal DNA diagnosis from maternal blood. Prenat Diagn 1995;15(1):74–7.

[22] Rodríguez de Alba M, Palomino P, González-González C, Lorda-Sanchez I, Ibañez MA, Sanz R, et al. Prenatal diagnosis on fetal cells from maternal blood: practical comparative evaluation of the first and second trimesters. Prenat Diagn 2001;21(3):165–70.

[23] Lim TH, Tan AS, Goh VH. Relationship between gestational age and frequency of fetal trophoblasts and nucleated erythrocytes in maternal peripheral blood. Prenat Diagn 2001;21(1):14–21.

[24] Fritz R, Kohan-Ghadr HR, Sacher A, Bolnick AD, Kilburn BA, Bolnick JM, et al. Trophoblast retrieval and isolation from the cervix (TRIC) is unaffected by early gestational age or maternal obesity. Prenat Diagn 2015;35(12):1218–22.

[25] Bianchi DW, Williams JM, Sullivan LM, Hanson FW, Klinger KW, Shuber AP. PCR quantitation of fetal cells in maternal blood in normal and aneuploid pregnancies. Presented in part at the annual meeting of the Society for Pediatric Research, Washington, DC, May 7, 1996, and at the 46th meeting of the American Society of Human Genetics, San Francisco, October 31, 1996. Am J Hum Genet 1997;61(4):822–9.

[26] Krabchi K, Gadji M, Samassekou O, Grégoire M-C, Forest J-C, Drouin R. Quantification of fetal nucleated cells in maternal blood of pregnant women with a male trisomy 21 fetus using molecular cytogenetic techniques. Prenat Diagn 2006;26(1):28–34.

[27] Krabchi K, Gadji M, Forest J-C, Drouin R. Quantification of all fetal nucleated cells in maternal blood in different cases of aneuploidies. Clin Genet 2006;69(2):145–54.

[28] Hahn S, Huppertz B, Holzgreve W. Fetal cells and cell free fetal nucleic acids in maternal blood: new tools to study abnormal placentation? Placenta 2005;26(7):515–26.

[29] van Wijk IJ, Griffioen S, Tjoa ML, Mulders MA, van Vugt JM, Loke YW, et al. HLA-G expression in trophoblast cells circulating in maternal peripheral blood during early pregnancy. Am J Obstet Gynecol 2001;184(5):991–7.

[30] Schlütter JM, Kirkegaard I, Petersen OB, Larsen N, Christensen B, Hougaard DM, et al. Fetal gender and several cytokines are associated with the number of fetal cells in maternal blood—an observational study. PLoS One 2014;9(9):e106934.

[31] Schmorl G. Pathologisch-anatomische Untersuchungen über Puerperal-Eklampsie. Leipzig: Verlag FCW Vogel; 1893.

[32] Lapaire O, Holzgreve W, Oosterwijk JC, Brinkhaus R, Bianchi DW. Georg Schmorl on trophoblasts in the maternal circulation. Placenta 2007;28(1):1–5.

[33] Chown B. Anaemia from bleeding of the fetus into the mother's circulation. Lancet (London, England) 1954;266(6824):1213–5.

[34] Zipursky A, Hull A, White FD, Israels LG. Foetal erythrocytes in the maternal circulation. Lancet (London, England) 1959;1(7070):451–2.

[35] Gordon H, Bhoyroo SK. A study of foetal erythrocytes in the maternal circulation during the antenatal period. J Obstet Gynaecol Br Commonw 1966;73(4):571–4.

[36] Weiner W, Child RM, Garvie JM, Peek WH. Foetal cells in the maternal circulation during pregnancy. Br Med J 1958;2(5099):770–1.

[37] Douglas GW, Thomas L, Carr M, Cullen NM, Morris R. Trophoblast in the circulating blood during pregnancy. Am J Obstet Gynecol 1959;78:960–73.

[38] Walknowska J, Conte FA, Grumbach MM. Practical and theoretical implications of fetal-maternal lymphocyte transfer. Lancet (London, England) 1969;1(7606):1119–22.

[39] Herzenberg LA, Bianchi DW, Schröder J, Cann HM, Iverson GM. Fetal cells in the blood of pregnant women: detection and enrichment by fluorescence-activated cell sorting. Proc Natl Acad Sci USA 1979;76 (3):1453–5.

[40] Iverson GM, Bianchi DW, Cann HM, Herzenberg LA. Detection and isolation of fetal cells from maternal blood using the fluorescence-activated cell sorter (FACS). Prenat Diagn 1981;1(1):61–73.

[41] de Graaf IM, Jakobs ME, Leschot NJ, Ravkin I, Goldbard S, Hoovers JM. Enrichment, identification and analysis of fetal cells from maternal blood: evaluation of a prenatal diagnosis system. Prenat Diagn 1999;19(7):648–52.

[42] Reading JP, Huffman JL, Wu JC, Palmer FT, Harton GL, Sisson ME, et al. Nucleated erythrocytes in maternal blood: quantity and quality of fetal cells in enriched populations. Hum Reprod 1995;10(9):2510–5.

[43] Smits G, Holzgreve W, Hahn S. An examination of different Percoll density gradients and magnetic activated cell sorting (MACS) for the enrichment of fetal erythroblasts from maternal blood. Arch Gynecol Obstet 2000;263(4):160–3.

[44] Troeger C, Holzgreve W, Hahn S. A comparison of different density gradients and antibodies for enrichment of fetal erythroblasts by MACS. Prenat Diagn 1999;19(6):521–6.

[45] Boyer SH, Noyes AN, Boyer ML. Enrichment of erythrocytes of fetal origin from adult-fetal blood mixtures via selective hemolysis of adult blood cells: an aid to antenatal diagnosis of hemoglobinopathies. Blood 1976;47(6):883–97.

[46] Wachtel S, Elias S, Price J, Wachtel G, Phillips O, Shulman L, et al. Fetal cells in the maternal circulation: isolation by multiparameter flow cytometry and confirmation by polymerase chain reaction. Hum Reprod 1991;6(10):1466–9.

[47] Bianchi DW, Flint AF, Pizzimenti MF, Knoll JH, Latt SA. Isolation of fetal DNA from nucleated erythrocytes in maternal blood. Proc Natl Acad Sci USA 1990;87(9):3279–83.

[48] Bianchi DW, Zickwolf GK, Yih MC, Flint AF, Geifman OH, Erikson MS, et al. Erythroid-specific antibodies enhance detection of fetal nucleated erythrocytes in maternal blood. Prenat Diagn 1993;13(4):293–300.

[49] Price JO, Elias S, Wachtel SS, Klinger K, Dockter M, Tharapel A, et al. Prenatal diagnosis with fetal cells isolated from maternal blood by multiparameter flow cytometry. Am J Obstet Gynecol 1991;165 (6 Pt 1):1731–7.

[50] Cheung MC, Goldberg JD, Kan YW. Prenatal diagnosis of sickle cell anaemia and thalassaemia by analysis of fetal cells in maternal blood. Nat Genet 1996;14(3):264–8.

[51] Little MT, Langlois S, Wilson RD, Lansdorp PM. Frequency of fetal cells in sorted subpopulations of nucleated erythroid and CD34+ hematopoietic progenitor cells from maternal peripheral blood. Blood 1997;89 (7):2347–58.

[52] Zheng YL, Demaria M, Zhen D, Vadnais TJ, Bianchi DW. Flow sorting of fetal erythroblasts using intracytoplasmic anti-fetal haemoglobin: preliminary observations on maternal samples. Prenat Diagn 1995;15 (10):897–905.

[53] Choolani M, O'Donnell H, Campagnoli C, Kumar S, Roberts I, Bennett PR, et al. Simultaneous fetal cell identification and diagnosis by epsilon-globin chain immunophenotyping and chromosomal fluorescence in situ hybridization. Blood 2001;98(3):554–7.

[54] Jackson L. Fetal cells and DNA in maternal blood. Prenat Diagn 2003;23(10):837–46.

[55] Wachtel SS, Sammons D, Manley M, Wachtel G, Twitty G, Utermohlen J, et al. Fetal cells in maternal blood: recovery by charge flow separation. Hum Genet 1996;98(2):162–6.

[56] Wachtel SS, Sammons D, Twitty G, Utermohlen J, Tolley E, Phillips O, et al. Charge flow separation: quantification of nucleated red blood cells in maternal blood during pregnancy. Prenat Diagn 1998;18(5):455–63.

[57] Mueller UW, Hawes CS, Jones WR. Identification of extra-villous trophoblast cells in human decidua using an apparently unique murine monoclonal antibody to trophoblast. Histochem J 1987;19(5):288–96.

[58] Latham SE, Suskin HA, Petropoulos A, Hawes CS, Jones WR, Kalionis B. A monoclonal antibody to human placental lactogen hormone facilitates isolation of fetal cells from maternal blood in a model system. Prenat Diagn 1996;16(9):813–21.

[59] Durrant L, McDowall K, Holmes R, Liu D. Non-invasive prenatal diagnosis by isolation of both trophoblasts and fetal nucleated red blood cells from the peripheral blood of pregnant women. Br J Obstet Gynaecol 1996;103(3):219–22.

[60] Vona G, Béroud C, Benachi A, Quenette A, Bonnefont JP, Romana S, et al. Enrichment, immunomorphological, and genetic characterization of fetal cells circulating in maternal blood. Am J Pathol 2002;160 (1):51–8.

[61] Bianchi DW, Shuber AP, DeMaria MA, Fougner AC, Klinger KW. Fetal cells in maternal blood: determination of purity and yield by quantitative polymerase chain reaction. Am J Obstet Gynecol 1994;171 (4):922–6.

[62] van Wijk IJ, de Hoon AC, Jurhawan R, Tjoa ML, Griffioen S, Mulders MA, et al. Detection of apoptotic fetal cells in plasma of pregnant women. Clin Chem 2000;46(5):729–31.

[63] Sekizawa A, Samura O, Zhen DK, Falco V, Farina A, Bianchi DW. Apoptosis in fetal nucleated erythrocytes circulating in maternal blood. Prenat Diagn 2000;20(11):886–9.

[64] Bianchi DW, Simpson JL, Jackson LG, Elias S, Holzgreve W, Evans MI, et al. Fetal gender and aneuploidy detection using fetal cells in maternal blood: analysis of NIFTY I data. Prenat Diagn 2002;22(7):609–15.

[65] Lo YM, Tein MS, Lau TK, Haines CJ, Leung TN, Poon PM, et al. Quantitative analysis of fetal DNA in maternal plasma and serum: implications for noninvasive prenatal diagnosis. Am J Hum Genet 1998;62 (4):768–75.

[66] Wapner RJ, Martin CL, Levy B, Ballif BC, Eng CM, Zachary JM, et al. Chromosomal microarray versus karyotyping for prenatal diagnosis. N Engl J Med 2012;367(23):2175–84.

[67] Hatt L, Brinch M, Singh R, Møller K, Lauridsen RH, Uldbjerg N, et al. Characterization of fetal cells from the maternal circulation by microarray gene expression analysis—could the extravillous trophoblasts be a target for future cell-based non-invasive prenatal diagnosis? Fetal Diagn Ther 2014;35(3):218–27.

[68] Hatt L, Brinch M, Singh R, Møller K, Lauridsen RH, Schlütter JM, et al. A new marker set that identifies fetal cells in maternal circulation with high specificity. Prenat Diagn 2014;34(11):1066–72.

[69] Zimmermann S, Hollmann C, Stachelhaus SA. Unique monoclonal antibodies specifically bind surface structures on human fetal erythroid blood cells. Exp Cell Res 2013;319(17):2700–7.

[70] Emad A, Drouin R. Evaluation of the impact of density gradient centrifugation on fetal cell loss during enrichment from maternal peripheral blood. Prenat Diagn 2014;34(9):878–85.

[71] Emad A, Ayub S, Samassékou O, Grégoire M-C, Gadji M, Ntwari A, et al. Efficiency of manual scanning in recovering rare cellular events identified by fluorescence in situ hybridization: simulation of the detection of fetal cells in maternal blood. J Biomed Biotechnol 2012;2012:610856.

[72] Emad A, Bouchard EF, Lamoureux J, Ouellet A, Dutta A, Klingbeil U, et al. Validation of automatic scanning of microscope slides in recovering rare cellular events: application for detection of fetal cells in maternal blood. Prenat Diagn 2014;34(6):538–46.

[73] Macaulay IC, Voet T. Single cell genomics: advances and future perspectives. Maizels N, editor. PLoS Genet 2014;10(1):e1004126.

[74] Gawad C, Koh W, Quake SR. Single-cell genome sequencing: current state of the science. Nat Rev Genet 2016;17(3):175–88.

[75] Huang L, Ma F, Chapman A, Lu S, Xie XS. Single-cell whole-genome amplification and sequencing: methodology and applications. Annu Rev Genomics Hum Genet 2015;16(1):79–102.

[76] Zong C, Lu S, Chapman AR, Xie XS. Genome-wide detection of single-nucleotide and copy-number variations of a single human cell. Science 2012;338(6114):1622–6.

[77] Chen C, Xing D, Tan L, Li H, Zhou G, Huang L, et al. Single-cell whole-genome analyses by Linear Amplification via Transposon Insertion (LIANTI). Science 2017;356(6334):189–94.

[78] Normand E, Qdaisat S, Bi W, Shaw C, Van den Veyver I, Beaudet A, et al. Comparison of three whole genome amplification methods for detection of genomic aberrations in single cells. Prenat Diagn 2016;36 (9):823–30.

[79] Kølvraa S, Singh R, Normand EA, Qdaisat S, van den Veyver IB, Jackson L, et al. Genome-wide copy number analysis on DNA from fetal cells isolated from the blood of pregnant women. Prenat Diagn 2016;36 (12):1127–34.

[80] Guanciali Franchi P, Palka C, Morizio E, Sabbatinelli G, Alfonsi M, Fantasia D, et al. Sequential combined test, second trimester maternal serum markers, and circulating fetal cells to select women for invasive prenatal diagnosis. Schmidt EE, editor. PLoS One 2017;12(12):e0189235.

[81] Vestergaard EM, Singh R, Schelde P, Hatt L, Ravn K, Christensen R, et al. On the road to replacing invasive testing with cell-based NIPT: five clinical cases with aneuploidies, microduplication, unbalanced structural rearrangement or mosaicism. Prenat Diagn 2017;37(11):1120–4.

[82] Hou S, Chen J-F, Song M, Zhu Y, Jan YJ, Chen SH, et al. Imprinted NanoVelcro microchips for isolation and characterization of circulating fetal trophoblasts: toward noninvasive prenatal diagnostics. ACS Nano 2017;11(8):8167–77.

[83] Chen F, Liu P, Gu Y, Zhu Z, Nanisetti A, Lan Z, et al. Isolation and whole genome sequencing of fetal cells from maternal blood towards the ultimate non-invasive prenatal testing. Prenat Diagn 2017;37 (13):1311–21.

[84] Breman AM, Chow JC, U'Ren L, Normand EA, Qdaisat S, Zhao L, et al. Evidence for feasibility of fetal trophoblastic cell-based noninvasive prenatal testing. Prenat Diagn 2016;36(11):1009–19.

[85] Imudia AN, Kumar S, Diamond MP, DeCherney AH, Armant DR. Transcervical retrieval of fetal cells in the practice of modern medicine: a review of the current literature and future direction. Fertil Steril 2010;93 (6):1725–30.

[86] Fabbri F, Carloni S, Zoli W, Ulivi P, Gallerani G, Fici P, et al. Detection and recovery of circulating colon cancer cells using a dielectrophoresis-based device: KRAS mutation status in pure CTCs. Cancer Lett 2013;335(1):225–31.

[87] Winer-Jones JP, Vahidi B, Arquilevich N, Fang C, Ferguson S, Harkins D, et al. Circulating tumor cells: clinically relevant molecular access based on a novel CTC flow cell. Kyprianou N, editor. PLoS One 2014;9(1):e86717.

[88] Harb W, Fan A, Tran T, Danila DC, Keys D, Schwartz M, et al. Mutational analysis of circulating tumor cells using a novel microfluidic collection device and qPCR assay. Transl Oncol 2013;6(5):528–38.

[89] Sitar G, Brambati B, Baldi M, Montanari L, Vincitorio M, Tului L, et al. The use of non-physiological conditions to isolate fetal cells from maternal blood. Exp Cell Res 2005;302(2):153–61.

[90] Calabrese G, Baldi M, Fantasia D, Sessa MT, Kalantar M, Holzhauer C, et al. Detection of chromosomal aneuploidies in fetal cells isolated from maternal blood using single-chromosome dual-probe FISH analysis. Clin Genet 2012;82(2):131–9.

[91] Bolnick JM, Kilburn BA, Bajpayee S, Reddy N, Jeelani R, Crone B, et al. Trophoblast retrieval and isolation from the cervix (TRIC) for noninvasive prenatal screening at 5 to 20 weeks of gestation. Fertil Steril 2014;102 (1):135–142.e6.

[92] Jain CV, Kadam L, Dijk M, Van Kilburn BA, Hartman C, Mazzorana V, et al. Fetal genome profiling at 5 weeks of gestation after noninvasive isolation of trophoblast cells from the endocervical canal. Sci Transl Med 2016;8(November). 363re4.

[93] Laget S, Broncy L, Hormigos K, Dhingra DM, BenMohamed F, Capiod T, et al. Technical insights into highly sensitive isolation and molecular characterization of fixed and live circulating tumor cells for early detection of tumor invasion. Katoh M, editor. PLoS One 2017;12(1):e0169427.

[94] Hayward J, Chitty LS. Beyond screening for chromosomal abnormalities: advances in non-invasive diagnosis of single gene disorders and fetal exome sequencing. Semin Fetal Neonatal Med 2018;23(2):94–101.

[95] Camaschella C, Alfarano A, Gottardi E, Travi M, Primignani P, Caligaris Cappio F, et al. Prenatal diagnosis of fetal hemoglobin Lepore-Boston disease on maternal peripheral blood. Blood 1990;75(11):2102–6.

[96] Mouawia H, Saker A, Jais J-P, Benachi A, Bussières L, Lacour B, et al. Circulating trophoblastic cells provide genetic diagnosis in 63 fetuses at risk for cystic fibrosis or spinal muscular atrophy. Reprod Biomed Online 2012;25(5):508–20.

[97] Akolekar R, Beta J, Picciarelli G, Ogilvie C, D'Antonio F. Procedure-related risk of miscarriage following amniocentesis and chorionic villus sampling: a systematic review and meta-analysis. Ultrasound Obstet Gynecol 2015;45(1):16–26.

[98] Benn P, Curnow KJ, Chapman S, Michalopoulos SN, Hornberger J, Rabinowitz M. An economic analysis of cell-free DNA non-invasive prenatal testing in the US general pregnancy population. Veitia RA, editor. PLoS One 2015;10(7):e0132313.

MATERNAL CIRCULATING NUCLEIC ACIDS AS MARKERS OF PLACENTAL HEALTH

Francesca Gaccioli*,†, D. Stephen Charnock-Jones*,†, Gordon C.S. Smith*,†

Department of Obstetrics and Gynaecology, University of Cambridge, NIHR Cambridge Comprehensive Biomedical Research Centre, Cambridge, United Kingdom* Centre for Trophoblast Research (CTR), Department of Physiology, Development and Neuroscience, University of Cambridge, Cambridge, United Kingdom†

INTRODUCTION

Adverse pregnancy outcomes are a major cause of the global burden of disease. Many of these outcomes, such as fetal growth restriction (FGR), preeclampsia, and placental abruption, are related to placental dysfunction. These complications can have profound effects on the mother and child. Preeclampsia is one of the leading causes of maternal death globally [1]. Placental abruption is also associated with hemorrhage [2], which is another major cause of death. The mother may experience morbidity related to interventions performed in the fetal interest, such as induction of labor or cesarean section. Placental dysfunction can have multiple adverse effects on the fetus. It can lead to acute and chronic fetal hypoxia resulting in severe short-term complications, such as perinatal death or brain damage through asphyxia [3,4]. It is also recognized that antepartum and intrapartum events are major determinants of long-term adverse neurodevelopmental outcome for the baby. Moreover, placental dysfunction and the management of certain conditions can lead directly to preterm birth [5], which is a major determinant of mortality and long-term morbidity in the offspring. There is a large body of evidence that a suboptimal intrauterine environment may predispose to a range of diseases in adult life, such as ischemic heart disease and stroke [6]. Finally, placenta-related complications of pregnancy are a marker for later risk of cardiovascular disease in the mother [7], indicating possible targeted interventions to reduce cardiovascular risk to women experiencing these complications of pregnancy.

Despite these issues, screening and intervention to prevent placenta-related complications of pregnancy remain relatively unsophisticated. The current primary approach to screening for FGR in the United Kingdom and United States is serial measurement of the external size of the uterus (the "symphyseal-fundal height" [SFH]). This reflects the fact that more advanced approaches, such as universal ultrasonic fetal biometry, have not been shown to improve outcomes in meta-analyses of randomized controlled trials [8]. This failure may, in turn, be explained by the fact that purely imaging-based methods do not perform well as a screening test. An ultrasonic estimated fetal weight (EFW) is a crude estimator of fetal size, commonly with errors of >15% [9]. Moreover, relative fetal smallness for gestational age (SGA) is a proxy for the condition of interest, namely, FGR.

Noninvasive Prenatal Testing (NIPT). https://doi.org/10.1016/B978-0-12-814189-2.00019-0

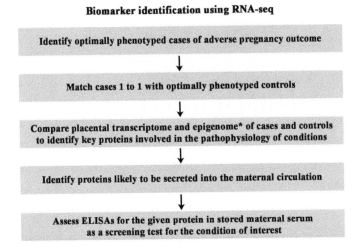

FIG. 1

Current model for biomarker identification based on placental RNA-seq analysis. ELISA: enzyme-linked immunosorbent assay.

We and others have proposed that measurement of circulating markers of placental function may result in improved clinical risk assessment and, consequently, their incorporation into future trials of screening and intervention may result in better clinical outcomes. The current model for approaching this (Fig. 1) uses analysis of placental RNA to identify differentially expressed transcripts. Assays of the encoded proteins are then evaluated as a screening test. However, with the advent of the ability to analyze DNA and RNA derived from the placenta in the maternal circulation, there is now the potential for identification and assessment of placental biomarkers directly in the maternal blood without the need to study the placenta first. In the present review, we outline the major conditions which could be screened for, the existing evidence for the role of the placenta in their pathogenesis, and the existing evidence for the effectiveness of this approach in the diagnostic and screening pathways.

ADVERSE PREGNANCY OUTCOME DUE TO PLACENTAL DYSFUNCTION

PREECLAMPSIA

Preeclampsia is the manifestation of high blood pressure and multisystem disease (typically renal dysfunction resulting in proteinuria) in pregnancy. The disease can either occur de novo or be "superimposed" on preexisting hypertension and/or renal disease. Maternal risk factors for preeclampsia include nulliparity and obesity. Interestingly, the risk of the disease is lower in smokers. Preeclampsia is a major determinant of maternal and perinatal morbidity and mortality. The disease is commonly associated with FGR, but the two conditions can also exist in isolation. Management of the disease involves monitoring and treatment of its symptoms (e.g., antihypertensives) but effective treatment is, currently, only achieved by delivery.

FETAL GROWTH RESTRICTION

FGR is a theoretical concept of failure of the fetus to achieve its genetically determined growth potential. The condition has no gold standard. Frequently the babies are small for gestational age, that is, <10th percentile for sex and week of gestation. However, some cases of FGR may not be SGA and many cases of SGA are not FGR. Corroborative indicators of FGR include the presence of ultrasonic markers (such as abnormal uteroplacental blood flow indices or slowing down of fetal growth), acquired complications of pregnancy (including preeclampsia and preterm birth), and morbidity due to chronic fetal hypoxia. FGR is one of the major causes of stillbirth.

PLACENTAL ABRUPTION

Placental abruption is premature separation of the placenta. Placental detachment from the uterine wall normally occurs following delivery of the baby. Abruption is partial or complete detachment when the baby is still in utero. As the placenta is the site of gaseous exchange, abruption leads to fetal asphyxia with the degree of asphyxia related to the proportion of the placenta which has detached. Typically, the separation extends to the edge of the placenta resulting in blood loss vaginally. The abruption can also be "concealed," that is, the area of detachment is entirely within the placental perimeter and the condition is not associated with vaginal bleeding. The process of abruption can lead to a maternal coagulopathy and disseminated intravascular coagulation. Abruption is a recognized cause of maternal death and accounts for about 10% of stillbirths.

THE PLACENTAL TRANSCRIPTOME IN ADVERSE PREGNANCY OUTCOME

The placenta ensures proper fetal growth and development via its endocrine functions, which sustain maternal adaptation to pregnancy and mobilize resources for the fetus, and by mediating feto-maternal exchange of nutrients, oxygen, and waste products. As the interface between the mother and the fetus, the placenta has also a protective role, representing a barrier for pathogens and toxic substances. Due to its role in orchestrating so many fundamental processes during pregnancy, it is not surprising that alterations in placental development, structure, and function are closely associated with poor pregnancy outcome. Both preeclampsia and FGR have a complex etiology, involving various maternal and fetal responses, and placental alterations. This complexity leads to a wide heterogeneity in the case population that, in turn, makes biomarker discovery particularly challenging. Analysis of the placental transcriptome offers a valuable approach to investigate multifactorial placenta-related complications such as preeclampsia and FGR, as it allows to study several potential components at the same time.

There have been many more studies of the placental transcriptome in preeclampsia than in FGR. Meta-analyses of expression microarray studies using placental biopsies from women with preeclampsia, compared with controls, have demonstrated altered placental levels of messenger RNAs (mRNAs) coding for pro- and antiangiogenic factors, such as fms-like tyrosine kinase (*FLT1*), endoglin (*ENG*), vascular endothelial growth factor A (*VEGFA*), placenta growth factor (*PlGF*) and mir-126, and upregulation of transcripts indicating tissue hypoxia, including hypoxia-inducible factor 2α (*HIF2α*) and mir-210 [10–12]. Other transcripts frequently identified are involved in cellular growth, metabolism, and endocrine function: pregnancy-associated plasma protein A (*PAPP-A*), insulin-like growth factor binding protein 1 (*IGFBP1*), HtrA serine peptidase 1 (*HTRA1*), prolactin (*PRL*), leptin (*LEP*), corticotropin releasing hormone (*CRH*), inhibin subunits (*INHA* and *INHBA*), and follistatin-like 3 (*FSTL3*). Several of these genes are also implicated in defective cytotrophoblast invasion of the maternal decidua and impaired spiral artery remodeling, hallmarks of the preeclamptic placenta and frequently observed in other placenta-related complications (Table 1).

Understanding the placental pathological processes associated with pregnancy diseases might offer a chance for intervention and improved outcome. But the placental undergoes profound changes during gestation [13] and it is only available for analysis during pregnancy via invasive procedures, such as

Table 1 mRNAs and miRNAs With Altered Expression in the Third Trimester Preeclamptic Placenta

	Gene Name	**Expression in Preeclampsia**	**Number of Publications**
Gene symbol			
LEP	Leptin	Upregulated	12
FLT1	fms-like tyrosine kinase 1	Upregulated	11
INHBA	Inhibin β A subunit	Upregulated	9
ENG	Endoglin	Upregulated	8
INHA	Inhibin α subunit	Upregulated	6
SIGLEC6	Sialic acid binding Ig-like lectin 6	Upregulated	6
BCL6	B-cell CLL/lymphoma 6	Upregulated	5
CGB	Chorionic gonadotropin β subunit	Upregulated	5
HTRA1	HtrA serine peptidase 1	Upregulated	5
FSTL3	Follistatin-like 3	Upregulated	4
miRNA name			
hsa-miR-210		Upregulated	3
hsa-miR-1		Downregulated	2
hsa-miR-139-5p		Downregulated	2
hsa-miR-150		Downregulated	2
hsa-miR-181a		Upregulated	2
hsa-miR-542-3p		Downregulated	2
hsa-miR-625		Downregulated	2
hsa-miR-638		Upregulated	2

This meta-analysis included 33 eligible gene expression array studies performed on third trimester placental tissues, comprising 30 mRNA datasets and 4 miRNA datasets. The table summarizes the top 10 mRNAs and all miRNAs with consistent results among studies.
Table based on data in Kleinrouweler CE, van Uitert M, Moerland PD, Ris-Stalpers C, van der Post JA, Afink GB. Differentially expressed genes in the pre-eclamptic placenta: a systematic review and meta-analysis. PLoS One 2013;8(7):e68991.

chorionic villus sampling. Nevertheless, these tissues offer the chance to study placental gene expression early in gestation. Similarly, to what observed in the term preeclamptic placenta, altered levels of mRNAs encoding regulators of angiogenesis, trophoblast invasion, oxidative stress, and inflammation were measured in first trimester biopsies collected from women who subsequently developed preeclampsia [14–16]. Therefore studies conducted on first trimester biopsies support the idea that placental dysfunction leading to complications in the second half of pregnancy often has its origins early in gestation [17]. As these types of samples are not easily accessible, attention has focused on circulating maternal biochemical markers of placental insufficiency to address early detection of pregnancy diseases, but several reports describe the low predictive value of the majority of the candidate proteins analyzed [18,19]. Hence, in more recent years circulating nucleic acids have being investigated, with the hope that the more sensitive techniques available for the detection of these molecules (targeted PCR-based methodologies and, lately, unbiased massively parallel sequencing (MPS)) would lead to screening tests with improved sensitivity and accuracy. Initial studies focused on transcripts highly expressed in the placenta or involved in the etiology of preeclampsia and FGR, such as those mentioned previously.

PLACENTAL ORIGIN OF MATERNAL CIRCULATING NUCLEIC ACIDS

Placental nucleic acids are released into the maternal bloodstream bound to specific proteins or packaged in extracellular vesicles (EVs). EVs are primarily categorized based on their size, but they also differ by their cargo and how they are released into the blood stream (Fig. 2). EVs of placental origin include exosomes (with a size of 30–100 nm), microvesicles (0.1–1 μm), and apoptotic bodies (1–5 μm) [20]. In addition, syncytial nuclear aggregates of syncytiotrophoblast cells (20–500 μm) also shed from the placental surface. Routine methods for vesicle isolation are based on their size or density and include differential centrifugation, density gradient ultracentrifugation, microfiltration, and size exclusion chromatography. In order to purify specific vesicles subtypes, immunocapture with antibodies directed against specific surface antigens such as alkaline phosphatase have been used to enrich for EVs originating from the syncytiotrophoblast layer. The choice of what type of EVs to isolate depends on the biological question, as it is now becoming evident that these vesicles not only facilitate removal of cellular waste material, but they also play a role in intercellular signaling and mediate feto-maternal communication. This function is obviously influenced by the EVs content, which includes immunomodulatory proteins (e.g., antiinflammatory cytokines and syncytin-1, regulating maternal immune reaction), vasoactive proteins (e.g., FLT1, ENG and tissue factor, influencing the maternal vascular system), lipids (e.g., cholesterol and sphingomyelins, modulating coagulation, and hemostasis), and nucleic acids (reviewed by Tong and Chamley [21]). Due to their role in intercellular and feto-maternal communication, the release of EVs is the result of physiological processes and the normal turnover of placental syncytiotrophoblast cells, and their concentration increases with gestational age [22]. In case of placenta-related complications, which are associated with abnormal cellular death and shedding of cellular debris, an increased level of these vesicles is observed. For example, higher maternal plasma concentrations of placenta-derived exosomes were measured throughout gestation in pregnancies with gestational diabetes mellitus (GDM) [23] and in women who subsequently developed preeclampsia as compared to control pregnancies [24]. Therefore the overall abundance of circulating EVs offers a first layer of evidence that altered levels of circulating nucleic acids could be associated with pregnancy complications, and examples of specific feto-placental DNA, RNA, and microRNA (miRNA) molecules are provided later.

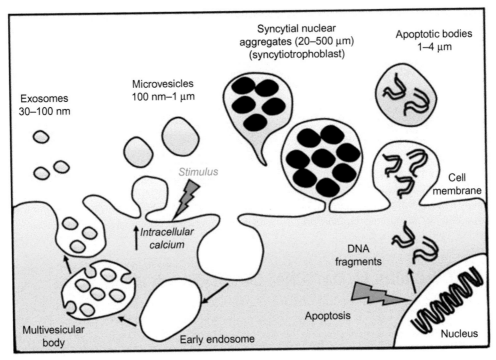

FIG. 2

Extracellular vesicles released by the placenta.

Image from Tannetta D, Dragovic R, Alyahyaei Z, Southcombe J. Extracellular vesicles and reproduction-promotion of successful pregnancy. Cell Mol Immunol 2014;11(6):548–563.

CIRCULATING NUCLEIC ACIDS AND ADVERSE PREGNANCY OUTCOME
CELL-FREE DNA

After the initial discovery of cell-free "fetal"—although more accurately "placental" DNA (cfpDNA) in maternal plasma and serum [25] by Prof. Dennis Lo's team in 1997, it has been established that cfpDNA fragments represent approximately 3%–10% of the total cell-free DNA (cfDNA) in maternal circulation in pregnancy [26,27]. cfpDNA molecules can reliably be measured from 5 to 6 weeks of gestation [28]. Their levels increase with gestational age, paralleling the increase in placental size, and there is rapid clearance from the maternal circulation after delivery [29]. Initially cfpDNA measurements were performed by quantitative PCR, aimed at the detection of Y-chromosome genes for fetal sex determination, or Rhesus D (*RHD*) specific sequences for fetal *RHD* genotyping in RhD⁻ mothers. However, the real breakthrough in cfpDNA-based noninvasive prenatal testing (NIPT) has been the development of massively parallel sequencing (MPS) techniques. These allow accurate counting of fetal DNA fragments in maternal blood and hence the detection of fetal aneuploidies. This application has been a strong incentive to improve cfpDNA-based prenatal testing methods based on placental circulating nucleic acids. Sequencing maternal plasma cfDNA has been particularly successful in Down syndrome and trisomy 18 screening [30,31]. cfpDNA NIPT is now an attractive alternative to chorionic villus sampling and amniocentesis—procedures disliked by pregnant women and associated with increased risk pregnancy loss [32].

It is possible that there is an association between cfpDNA levels and pregnancies complicated by preeclampsia and FGR. The rationale is that the placental insufficiency underlying these pathological conditions might lead to an increased release of necrotic and apoptotic fragments containing cfpDNA from the syncytiotrophoblast layer of the placenta into the maternal circulation [33,34]. In several studies elevated levels of cfpDNA were detected prior to the clinical symptoms of preeclampsia and FGR, fitting the hypothesis that this phenomenon might have a causal role in the onset of the disease [35,36] Reduced clearance from the maternal circulation might also contribute to higher concentrations of circulating cfpDNA in women with preeclampsia [37]. Studies in mice addressing this issue have been inconclusive [38,39].

Although several studies demonstrated higher maternal cfpDNA levels in pregnancies complicated by preeclampsia and FGR compared to normal pregnancies [28,40–44], more recent data are not always consistent with these initial findings [45–49]. A recent systematic review and meta-analysis reported that the effectiveness of cfpDNA as a biomarker for preeclampsia is promising only when measurements are performed in the second trimester and for cases with early onset of the disease [50]. These issues led the authors to conclude that quantifying total cfpDNA concentrations does not seem to be as useful for preeclampsia screening purposes as other biochemical markers, such as the sFLT1/PlGF ratio [51,52]. Nevertheless, the authors recognized the heterogeneity and the fragmentary nature of the data presented in the studies included in their review. Contradictory results might also derive from the different DNA markers used, and this reflects once more the need for standardized protocols in studies aiming at identifying potential disease biomarkers.

miRNAs IN MATERNAL PLASMA

A comprehensive placental miRNA profile was first reported by Liang and colleagues [53]. The villous trophoblast layer is the main source of placental miRNAs, which are expressed in these cells in a spatiotemporal manner throughout pregnancy [54]. Placental miRNAs are released in the maternal bloodstream bound to proteins and high-density lipoproteins or packaged in EVs (i.e., exosomes, microvesicles, and apoptotic bodies), leading to increased stability of the circulating miRNAs. An increase of these vesicles is detected in maternal blood from complicated pregnancies. Moreover, exosomes isolated from the maternal circulation from GDM pregnancies promote migration and release of proinflammatory cytokines from endothelial cells [22,23]. This suggests that EVs and their cargo (miRNAs, proteins, growth factors, etc.) not only regulate the function of adjacent cells, but also contribute to feto-maternal communication by targeting distant tissues. Consistent with this, trophoblast-derived exosomes confer viral resistance to nonplacental recipient cells, such as human primary endothelial cells which would represent an important line of defense against viral infections [55]. More recently, Cronqvist and colleagues demonstrated that placental EVs are internalized by primary human endothelial cells. Treatment of cells with placental EVs derived from pregnancies complicated by preeclampsia led to significant downregulation of several transcripts, including *FLT1* [56]. This is of particular interest as the upregulation of FLT1 mRNA, which encodes a vascular endothelial growth factor receptor, was identified as one of the key changes in the preeclamptic placenta and this was paralleled by elevated levels of sFLT1 protein in the maternal circulation [57]. Furthermore, inhibition of the FLT1 signaling pathway sensitizes endothelial cells to inflammatory mediators and may contribute to the widespread endothelial dysfunction in PE [58].

Studies of the relationship between placental or circulating miRNAs and pregnancy complications have focused on candidates with high or specific expression in the placenta and were grouped into three main clusters (Fig. 3) [59].

Micro RNAs located on chromosome 14 miRNA cluster (C14MC) are highly expressed in embryonic and placental tissues in the first trimester of pregnancy. This cluster includes 52 miRNA genes expressed from the maternally inherited chromosome. The 58 mature miRNAs coded by 46 genes on the chromosome 19 miRNA cluster (C19MC) are mainly expressed from the paternally inherited allele. C19MC miRNAs are primate specific and their placental levels increase from the first trimester to the end of pregnancy. The third cluster, containing miR-371-373, is also located on chromosome 19 and includes three miRNAs conserved in mammals. This cluster is mainly expressed in the placenta and it is upregulated in the third trimester. Placental expression of miRNAs of the C19MC cluster has been a particular focus of study in pregnancy complications and altered expression of some of these was demonstrated in placental tissues from pregnancies with small infants (BW < 10th centile) with or without preeclampsia and in plasma from preeclamptic mothers [60,61].

Other miRNAs of interest were selected for their possible role in the development and progression of placental insufficiency. For example, placental miR-210 (located on chromosome 11) is upregulated in pregnancies with preeclampsia with or without SGA [62–66]. This consistency might reflect the fact that mir-210 is part of several pathways implicated in the onset of preeclampsia. Firstly, mir-210 expression is controlled by hypoxia-inducible factors HIF1α and HIF2α, master regulators of cellular response to hypoxia. Secondly, miR-210 induction causes abnormal trophoblast invasion and vascular remodeling, possibly via downregulation of ephrin-A3 (EFNA3) and homeobox-A9 (HOXA9) [63]. Moreover, elevated placental mir-210 in patients with severe preeclampsia is associated with mitochondrial dysfunction and higher reactive oxygen species (ROS) production [67]. Finally, altered placental mir-210 levels in pregnancy complications are mirrored by higher maternal circulating concentrations throughout pregnancy [65,68,69].

MATERNAL PLASMA mRNAs AS LIQUID BIOPSY OF THE PLACENTAL TRANSCRIPTOME

Measurement of circulating feto-placental mRNAs is appealing as changes in the levels of these transcripts may reflect the placental pathology underlying the disease, and therefore provide useful information on causes and possible interventions. Nevertheless, the detection of mRNAs in the maternal circulation is more challenging than the detection of cfDNA and miRNAs, probably because these molecules are less stable. Mimicking the strategy adopted for the analysis of cfDNA, the first attempts measured male-specific transcripts, such as Y-chromosome-specific zinc finger protein (ZFY) mRNA, with PCR-based techniques [70].

Interest subsequently focused on highly expressed placental transcripts, which might provide a snapshot of placental transcriptome alterations associated with pregnancy complications. The pioneering work of Prof. Dennis Lo and collaborators focused on maternal circulating mRNAs rapidly cleared after delivery (and therefore of probable feto-placental origin) in order to assess their levels in pregnancies with preeclampsia and small fetuses. With this approach, they demonstrated that plasma placental growth hormone (*PGH* or *GH2*) mRNA concentrations correlated with birthweight (BW) and

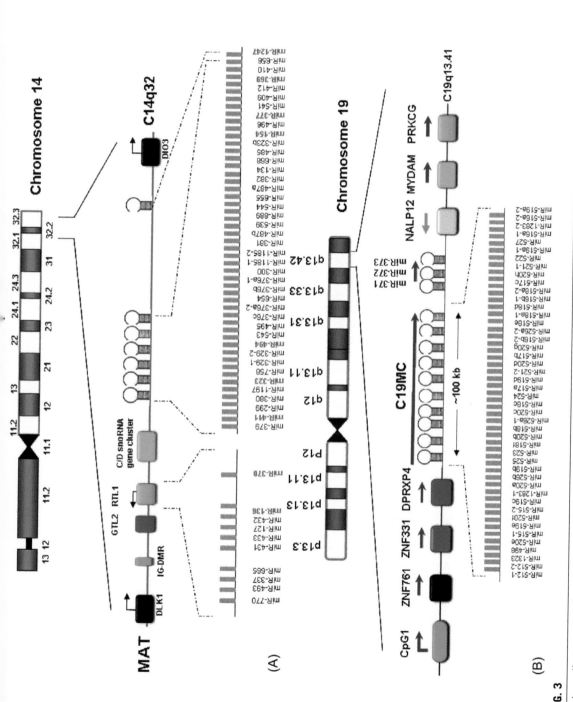

FIG. 3

Schematic representation of miRNA clusters highly expressed in embryonic and placental tissues. (A) Chromosome 14 miRNA cluster (C14MC). (B) Chromosome 19 miRNA cluster (C19MC) and miR-371-373 cluster.

Image from Morales-Prieto DM, Ospina-Prieto S, Chaiwangyen W, Schoenleben M, Markert UR. Pregnancy-associated miRNA-clusters. J Reprod Immunol 2013;97(1):51–61 (reprinted with permission from Elsevier).

fetal biometric measurements, but they were not useful in discriminating SGA (small for gestational age, i.e., BW < 10th centile) from normal pregnancies. On the other hand, maternal plasma concentration of ADAM metallopeptidase domain 12 (*ADAM12*) mRNA was increased in pregnancies with preeclampsia and SGA compared to a control group [71] and *CRH* was increased in women with preeclampsia [72]. Other transcripts, abundantly expressed in the placenta or associated with the onset of the disease, such as placenta-specific 3 and 4 (*PLAC3* and *PLAC4*), *PAPP-A*, and Syncytin (*ERVWE1*), showed high circulating levels in mothers with preeclampsia [73,74].

Transcripts, that can be measured in the maternal circulation early in gestation, have been studied in order to assess their potential as predictors for pregnancy diseases. mRNAs coding for proteins involved in the etiology of preeclampsia, such as pro- and antiangiogenic factors, are of particular interest because altered levels in the placenta could result in altered circulating levels. At 15–20 weeks of pregnancy plasma concentrations of *FLT1*, *VEGFA*, *ENG*, tissue-type plasminogen activator (*PLAT*), plasminogen activator inhibitor-1 (*SERPINE1*), *PLAC1*, and selectin P (*SELP*) mRNAs were higher in 62 preeclamptic women as compared to 310 women with uncomplicated pregnancies [75]. At a 5% false positive rate, these individual circulating mRNAs had detection rates between 18% and 58%. This increased to 84% (95% confidence intervals [CI] 72%–92%) in a multivariable model combining all markers. In the same population, a similar analysis was performed on transcripts extracted from the cellular component of maternal peripheral blood samples [76]. mRNA levels of *FLT1*, *ENG*, *PLAC1*, and *SELP* were significantly higher, while *PlGF* and heme oxygenase-1 (*HO-1*) levels were lower in the preeclampsia group compared to the controls. The univariate analysis showed *FLT1* and *ENG* to be the markers with the highest sensitivity. The best multivariate model was obtained by combining *FLT1*, *ENG*, *PlGF*, and parity with a detection rate of 66%, at a 10% false positive rate [76].

Circulating transcripts coding for angiogenic regulators were also investigated in pregnancies with FGR. Maternal plasma FLT1 mRNA, measured throughout gestation, was higher in 11 cases of severe FGR without preeclampsia as compared to 88 control pregnancies [77]. At a false positive rate of 10%, the detection rate was 50% with measurements performed in the third trimester and 80% with combined measurements from all three trimesters. The group led by Prof. Stephen Tong investigated the levels of placental and circulating transcripts associated with FGR and fetal hypoxic status. In 20 women with severe preterm FGR—BW < 10th centile, delivery before 34 weeks of gestation, and signs of uteroplacental insufficiency—higher mRNA concentrations of *PGH*, insulin-like growth factors 1 and 2 (*IGF1* and *IGF2*), IGF receptor and IGF binding protein (*IGF1R* and *IGFBP2*), and lower mRNA levels of *ADAM12* were measured at 30 weeks of gestation in maternal blood and placental tissue as compared with women with uncomplicated pregnancies [78]. Placental tissues for the control group were collected from women who did not have preeclampsia, altered Doppler flow velocimetry measurements or chorioamnionitis, and delivered preterm of a newborn with an appropriate weight for gestational age. Transcripts coded by hypoxia-regulated genes, including hypoxia-inducible factors *HIF1α* and *HIF2α*, adrenomedullin (*ADM*), and lactate dehydrogenase A (*LDHA*), were described to be higher in maternal blood of mothers belonging to a FGR cohort as compared to controls [79]. In these studies, the authors used whole maternal blood and it is important to be aware that the placental transcriptome signal could have been masked by the abundant transcripts originating from maternal leukocytes. Maternal leukocyte function is altered in pregnancy complications [80,81] and can potentially lead to abnormal cellular mRNA profiles.

Recently, high-throughput RNA sequencing (RNA-seq) methodologies have been used for detecting pregnancy-associated transcripts in maternal plasma [82,83]. To our knowledge cell-free RNA-seq (cfRNA-seq) has not yet been used to investigate altered circulating RNA levels in placenta-related complications. The initial studies using this technique calculated that the fetal fraction of transcripts in maternal plasma increases from ~4% in early pregnancy to ~11% in late pregnancy [82]. In a recent report aiming at studying bacterial and viral infections during pregnancy, analysis of maternal plasma by cfRNA-seq showed increasing abundance of placenta-specific mRNAs, such as *PSG1* (pregnancy-specific beta-1-glycoprotein), *PLAC4*, and *PAPP-A*, throughout pregnancy and a decrease after delivery [84]. Transcripts involved in the immune response and inflammatory processes were also detected at higher concentrations in maternal plasma in the third trimester and after delivery.

PROSPECTS FOR CLINICAL APPLICATION
DIAGNOSIS

One of the major applications of biomarkers at present is in confirming or ruling out a diagnosis. Within Obstetrics, the exemplar for this is the ratio of soluble fms-like tyrosine kinase 1 (sFLT1) to PlGF. Gene expression array studies of the placenta in preeclampsia identified increased *FLT1* mRNA as a feature of the disease. This was associated with release of the soluble form of FLT1 protein, sFLT1, from the placenta. This protein binds pro-angiogenic growth factors, including PlGF. Multiple studies reported elevated sFLT1 and reduced free PlGF in maternal serum in women with preeclampsia. In 2016, a large multicenter clinical study reported that the sFLT1/PlGF ratio provided clinically useful prediction of risk in managing women with suspected preeclampsia [52]. A ratio of <38 had a negative predictive value of >99% for clinical manifestation with preeclampsia in the following 7 days. The test has now been recommended for routine clinical use by the UK's National Institute for Health and Care Excellence (NICE). The main value of this test is that it potentially prevents unnecessary hospital admission. If women have a very low risk of manifesting disease they can be sent home with arrangements made for follow-up at an appropriate interval. For the healthcare system, this means that the purchase of a diagnostic test (typically costing < $100) can then prevent an unnecessary hospital admission (typically costing $1000s). It is also much better for the individual woman given that hospital admission may lead to stress, anxiety, and logistic problems.

The previous example is for reassurance through a test with a high negative predictive value. An alternative approach is that a test may support the diagnosis, and this requires a high positive predictive value. This could improve care for women at high risks of disease by in-hospital monitoring, reduce the risk of severe maternal complications and multisystem failure, and/or indicate delivery since the only effective treatment for preeclampsia is delivery. However, it could also be argued that the negative predictive value of the test is equally important in the context of diagnosis, as a false negative result might reduce the level of care in a woman who is having the disease.

Besides preeclampsia, there is a place for other clinical applications for molecular diagnostic tests such as when FGR is suspected at ultrasound. Reliable additional information could help target resources to the women at highest risk and reduce disease complications by interventions based on the test results.

SCREENING

The previous section dealt with women who are experiencing symptoms or signs of disease and testing is used to rule out or rule in the condition. An extension of this is screening. In this case, tests are applied to all women, irrespective of the presence or absence of symptoms or signs of the disease. The aim of a screening test is to identify women at high risk of the given disease prior to its clinical onset or at an early stage in the disease process. The rationale for screening is predicated on an assumption that this knowledge could indicate interventions with a net beneficial effect. There is a large body of evidence for tests that indicate women at an increased risk of complications. What is lacking, however, is evidence that interventions following an abnormal screening test result in net benefit. Moreover, even if clinically beneficial, such programs would also be required to be cost effective. Hence, a screening program may fail because either the test does not provide strong prediction and/or because the intervention does not (sufficiently) mitigate the risk [85].

The statistical methods for assessing screening are complex and are reviewed elsewhere. In brief, the tests can be summarized at the level of the population or the level of the individual. In the former approach, the sensitivity and specificity are key: the former is the proportion of cases in a population correctly identified as high risk and the latter is the proportion of healthy women correctly classified as low risk. For the individual woman, the key statistics are the positive predictive value and the negative predictive value: the former is the absolute risk of disease in women who are screen positive and the latter is the chance of remaining nondiseased in the women who are screen negative. The positive predictive value is determined by the woman's prior risk for the disease and the proportional increase in risk associated with a positive result (called the positive likelihood ratio). A positive test may therefore predict disease weakly in an individual woman with low a priori risk and/or when the test performs poorly.

Currently, in many placenta-related disorders of pregnancy, the sole intervention is delivery. One of the major determinants of the risk of an adverse outcome for the infant is the gestational age at birth. There is a continuum of declining risk of adverse outcomes with advancing gestational age until ~40 weeks, and this is nicely illustrated with one of the long-term risks, namely, requirement for special educational support at school (Fig. 4).

Therefore when trying to establish a new screening program to prevent placenta-related complications of pregnancy, an initial approach of screening at or near term should be considered given that the primary intervention, i.e., delivery, is less likely to cause harm [86]. When considering the possible adverse effects of the intervention, the positive predictive value (PPV) is key, as this determines the ratio of true positives to false positives. For example, if a test has a PPV of 20%, this means that for every true positive there will be four false positives. The likelihood for net benefit (i.e., the benefit achieved for true positives) or net harm (i.e., the harm caused to false positives) depends on the PPV. The experience of implementing universal ultrasound screening for FGR in France was that there was no clear benefit and quite strong evidence for harm associated with screening due to the large number of false positives [87]. When a test is used to support the diagnosis in women where there is clinical suspicion, as described previously, a false negative result has great potential to cause harm; in the situation of screening low-risk women however, false negatives could be considered less important, as these women would continue the same care pathway as before the test. In contrast, a false positive screening result leads to a low-risk woman being exposed to unnecessary interventions, and has to be avoided.

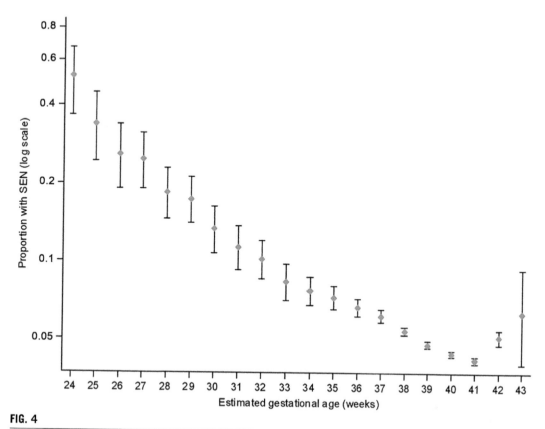

FIG. 4

Relationship between week of gestational age at delivery and the risk of special educational needs (SEN) in childhood.

Graph from MacKay DF, Smith GC, Dobbie R, Pell JP. Gestational age at delivery and special educational need: retrospective cohort study of 407,503 schoolchildren. PLoS Med 2010;7(6):e1000289.

The importance of gestational age in determining short- and long-term complications in the context of screening reflects the fact that the main disease-modifying intervention for complications such as preeclampsia and FGR is delivery. However, looking to the future, the availability of disease-modifying drug therapy raises the possibility of coupling screening to preventative treatment. Low-dose aspirin has been studied for many years as a method for reducing the risk of preeclampsia in women with a prior history of the disease. More recently, a trial has shown that screening the general population for the risk of preeclampsia and treating screen positive women with aspirin from the first trimester reduced the risk of preterm preeclampsia [88]. However, the study screened >25,000 women and the effect of the program was to prevent 20–25 cases of preterm preeclampsia. Further work may be required prior to clinical implementation of this approach, for example, a health economic analysis. Moreover, among multiparous women without a history of preeclampsia, treatment only prevented ~3 cases of preterm preeclampsia and the number needed to screen to prevent each case will be in the region of 5000. This suggests that screening could be more cost effective if targeted at nulliparous women or those with a prior history of disease. Nevertheless, the study does provide a rationale for trying to improve first trimester prediction of preeclampsia and, potentially, other placenta-related complications of pregnancy. Moreover, studies are ongoing to try and identify other novel therapies which could either prevent or treat both preeclampsia and FGR [89].

CONCLUSIONS

Technological developments in the extraction and sequencing of nucleic acids from the maternal plasma have revolutionized screening for aneuploidy and fetal anomalies. Placenta-related complications of pregnancy are associated with changes in the placental transcriptome and epigenome. Since placenta-derived RNA and DNA are present in maternal plasma, there is potential for developing nucleic acid-based diagnostic and screening tests for conditions such as preeclampsia and FGR. However, these conditions are far more complex and less well understood than, for example, autosomal trisomies, and considerable work will be required to turn this promise into clinically useful tests. This is particularly true in the case of screening tests, as development of clinically effective programs will require balancing risks and benefits of interventions such as early delivery and other disease-modifying interventions.

REFERENCES

[1] Ghulmiyyah L, Sibai B. Maternal mortality from preeclampsia/eclampsia. Semin Perinatol 2012;36(1):56–9.
[2] Downes KL, Grantz KL, Shenassa ED. Maternal, labor, delivery, and perinatal outcomes associated with placental abruption: a systematic review. Am J Perinatol 2017;34(10):935–57.
[3] Miller SL, Huppi PS, Mallard C. The consequences of fetal growth restriction on brain structure and neurodevelopmental outcome. J Physiol 2016;594(4):807–23.
[4] Rainaldi MA, Perlman JM. Pathophysiology of birth asphyxia. Clin Perinatol 2016;43(3):409–22.
[5] Goldenberg RL, Culhane JF, Iams JD, Romero R. Epidemiology and causes of preterm birth. Lancet 2008;371(9606):75–84.
[6] Barker DJ. Adult consequences of fetal growth restriction. Clin Obstet Gynecol 2006;49(2):270–83.
[7] Bellamy L, Casas JP, Hingorani AD, Williams DJ. Pre-eclampsia and risk of cardiovascular disease and cancer in later life: systematic review and meta-analysis. BMJ 2007;335(7627):974.
[8] Bricker L, Medley N, Pratt JJ. Routine ultrasound in late pregnancy (after 24 weeks' gestation). Cochrane Database Syst Rev 2015;6:CD001451.
[9] Smith GC, Smith MF, McNay MB, Fleming JE. The relation between fetal abdominal circumference and birthweight: findings in 3512 pregnancies. Br J Obstet Gynaecol 1997;104(2):186–90.
[10] Kleinrouweler CE, van Uitert M, Moerland PD, Ris-Stalpers C, van der Post JA, Afink GB. Differentially expressed genes in the pre-eclamptic placenta: a systematic review and meta-analysis. PLoS One 2013;8(7), e68991.
[11] Vaiman D, Calicchio R, Miralles F. Landscape of transcriptional deregulations in the preeclamptic placenta. PLoS One 2013;8(6),e65498.
[12] Brew O, Sullivan MH, Woodman A. Comparison of normal and pre-eclamptic placental gene expression: a systematic review with meta-analysis. PLoS One 2016;11(8),e0161504.
[13] Tsang JCH, Vong JSL, Ji L, Poon LCY, Jiang P, Lui KO, et al. Integrative single-cell and cell-free plasma RNA transcriptomics elucidates placental cellular dynamics. Proc Natl Acad Sci USA 2017;114(37). E7786–E7795.

[14] Farina A, Sekizawa A, De Sanctis P, Purwosunu Y, Okai T, Cha DH, et al. Gene expression in chorionic villous samples at 11 weeks' gestation from women destined to develop preeclampsia. Prenat Diagn 2008;28(10):956–61.

[15] Farina A, Morano D, Arcelli D, De Sanctis P, Sekizawa A, Purwosunu Y, et al. Gene expression in chorionic villous samples at 11 weeks of gestation in women who develop preeclampsia later in pregnancy: implications for screening. Prenat Diagn 2009;29(11):1038–44.

[16] Founds SA, Conley YP, Lyons-Weiler JF, Jeyabalan A, Hogge WA, Conrad KP. Altered global gene expression in first trimester placentas of women destined to develop preeclampsia. Placenta 2009;30(1):15–24.

[17] Smith GC. First-trimester determination of complications of late pregnancy. JAMA 2010;303(6):561–2.

[18] Conde-Agudelo A, Papageorghiou AT, Kennedy SH, Villar J. Novel biomarkers for predicting intrauterine growth restriction: a systematic review and meta-analysis. BJOG 2013;120(6):681–94.

[19] Conde-Agudelo A, Bird S, Kennedy SH, Villar J, Papageorghiou AT. First- and second-trimester tests to predict stillbirth in unselected pregnant women: a systematic review and meta-analysis. BJOG 2015;122 (1):41–55.

[20] Tannetta D, Dragovic R, Alyahyaei Z, Southcombe J. Extracellular vesicles and reproduction-promotion of successful pregnancy. Cell Mol Immunol 2014;11(6):548–63.

[21] Tong M, Chamley LW. Placental extracellular vesicles and feto-maternal communication. Cold Spring Harb Perspect Med 2015;5(3),a023028.

[22] Salomon C, Torres MJ, Kobayashi M, Scholz-Romero K, Sobrevia L, Dobierzewska A, et al. A gestational profile of placental exosomes in maternal plasma and their effects on endothelial cell migration. PLoS One 2014;9(6),e98667.

[23] Salomon C, Scholz-Romero K, Sarker S, Sweeney E, Kobayashi M, Correa P, et al. Gestational diabetes mellitus is associated with changes in the concentration and bioactivity of placenta-derived exosomes in maternal circulation across gestation. Diabetes 2016;65(3):598–609.

[24] Salomon C, Guanzon D, Scholz-Romero K, Longo S, Correa P, Illanes SE, et al. Placental exosomes as early biomarker of preeclampsia: potential role of exosomal microRNAs across gestation. J Clin Endocrinol Metab 2017;102(9):3182–94.

[25] Lo YM, Corbetta N, Chamberlain PF, Rai V, Sargent IL, Redman CW, et al. Presence of fetal DNA in maternal plasma and serum. Lancet 1997;350(9076):485–7.

[26] Lo YM, Tein MS, Lau TK, Haines CJ, Leung TN, Poon PM, et al. Quantitative analysis of fetal DNA in maternal plasma and serum: implications for noninvasive prenatal diagnosis. Am J Hum Genet 1998;62 (4):768–75.

[27] Lun FM, Chiu RW, Chan KC, Leung TY, Lau TK, Lo YM. Microfluidics digital PCR reveals a higher than expected fraction of fetal DNA in maternal plasma. Clin Chem 2008;54(10):1664–72.

[28] Illanes S, Parra M, Serra R, Pino K, Figueroa-Diesel H, Romero C, et al. Increased free fetal DNA levels in early pregnancy plasma of women who subsequently develop preeclampsia and intrauterine growth restriction. Prenat Diagn 2009;29(12):1118–22.

[29] Lo YM, Zhang J, Leung TN, Lau TK, Chang AM, Hjelm NM. Rapid clearance of fetal DNA from maternal plasma. Am J Hum Genet 1999;64(1):218–24.

[30] Chiu RW, Akolekar R, Zheng YW, Leung TY, Sun H, Chan KC, et al. Non-invasive prenatal assessment of trisomy 21 by multiplexed maternal plasma DNA sequencing: large scale validity study. BMJ 2011;342:c7401.

[31] Bianchi DW, Parker RL, Wentworth J, Madankumar R, Saffer C, Das AF, et al. DNA sequencing versus standard prenatal aneuploidy screening. N Engl J Med 2014;370(9):799–808.

[32] Alfirevic Z, Navaratnam K, Mujezinovic F. Amniocentesis and chorionic villus sampling for prenatal diagnosis. Cochrane Database Syst Rev 2017;9:CD003252.

[33] Hung TH, Skepper JN, Charnock-Jones DS, Burton GJ. Hypoxia-reoxygenation: a potent inducer of apoptotic changes in the human placenta and possible etiological factor in preeclampsia. Circ Res 2002;90 (12):1274–81.

[34] Levine RJ, Qian C, Leshane ES, Yu KF, England LJ, Schisterman EF, et al. Two-stage elevation of cell-free fetal DNA in maternal sera before onset of preeclampsia. Am J Obstet Gynecol 2004;190(3):707–13.

[35] Kim MJ, Kim SY, Park SY, Ahn HK, Chung JH, Ryu HM. Association of fetal-derived hypermethylated RASSF1A concentration in placenta-mediated pregnancy complications. Placenta 2013;34(1):57–61.

[36] Hromadnikova I, Zejskova L, Kotlabova K, Jancuskova T, Doucha J, Dlouha K, et al. Quantification of extracellular DNA using hypermethylated RASSF1A, SRY, and GLO sequences—evaluation of diagnostic possibilities for predicting placental insufficiency. DNA Cell Biol 2010;29(6):295–301.

[37] Lau TW, Leung TN, Chan LY, Lau TK, Chan KC, Tam WH, et al. Fetal DNA clearance from maternal plasma is impaired in preeclampsia. Clin Chem 2002;48(12):2141–6.

[38] Scharfe-Nugent A, Corr SC, Carpenter SB, Keogh L, Doyle B, Martin C, et al. TLR9 provokes inflammation in response to fetal DNA: mechanism for fetal loss in preterm birth and preeclampsia. J Immunol 2012;188 (11):5706–12.

[39] Conka J, Konecna B, Laukova L, Vlkova B, Celec P. Fetal DNA does not induce preeclampsia-like symptoms when delivered in late pregnancy in the mouse. Placenta 2017;52:100–5.

[40] Cotter AM, Martin CM, O'Leary JJ, Daly SF. Increased fetal DNA in the maternal circulation in early pregnancy is associated with an increased risk of preeclampsia. Am J Obstet Gynecol 2004;191(2): 515–20.

[41] Zhong XY, Holzgreve W, Hahn S. The levels of circulatory cell free fetal DNA in maternal plasma are elevated prior to the onset of preeclampsia. Hypertens Pregnancy 2002;21(1):77–83.

[42] Cotter AM, Martin CM, O'Leary JJ, Daly SF. Increased fetal RhD gene in the maternal circulation in early pregnancy is associated with an increased risk of pre-eclampsia. BJOG 2005;112(5):584–7.

[43] Salvianti F, Inversetti A, Smid M, Valsecchi L, Candiani M, Pazzagli M, et al. Prospective evaluation of RASSF1A cell-free DNA as a biomarker of pre-eclampsia. Placenta 2015;36(9):996–1001.

[44] Martin A, Krishna I, Badell M, Samuel A. Can the quantity of cell-free fetal DNA predict preeclampsia: a systematic review. Prenat Diagn 2014;34(7):685–91.

[45] Rolnik DL, O'Gorman N, Fiolna M, van den Boom D, Nicolaides KH, Poon LC. Maternal plasma cell-free DNA in the prediction of pre-eclampsia. Ultrasound Obstet Gynecol 2015;45(1):106–11.

[46] Poon LC, Musci T, Song K, Syngelaki A, Nicolaides KH. Maternal plasma cell-free fetal and maternal DNA at 11–13 weeks' gestation: relation to fetal and maternal characteristics and pregnancy outcomes. Fetal Diagn Ther 2013;33(4):215–23.

[47] Crowley A, Martin C, Fitzpatrick P, Sheils O, O'Herlihy C, O'Leary JJ, et al. Free fetal DNA is not increased before 20 weeks in intrauterine growth restriction or pre-eclampsia. Prenat Diagn 2007;27(2):174–9.

[48] Stein W, Muller S, Gutensohn K, Emons G, Legler T. Cell-free fetal DNA and adverse outcome in low risk pregnancies. Eur J Obstet Gynecol Reprod Biol 2013;166(1):10–3.

[49] Thurik FF, Lamain-de Ruiter M, Javadi A, Kwee A, Woortmeijer H, Page-Christiaens GC, et al. Absolute first trimester cell-free DNA levels and their associations with adverse pregnancy outcomes. Prenat Diagn 2016;36(12):1104–11.

[50] Contro E, Bernabini D, Farina A. Cell-free fetal DNA for the prediction of pre-eclampsia at the first and second trimesters: a systematic review and meta-analysis. Mol Diagn Ther 2017;21(2):125–35.

[51] Sovio U, Gaccioli F, Cook E, Hund M, Charnock-Jones DS, Smith GC. Prediction of preeclampsia using the soluble fms-like tyrosine kinase 1 to placental growth factor ratio: a prospective cohort study of unselected nulliparous women. Hypertension 2017;69(4):731–8.

[52] Zeisler H, Llurba E, Chantraine F, Vatish M, Staff AC, Sennstrom M, et al. Predictive value of the sFlt-1: PlGF ratio in women with suspected preeclampsia. N Engl J Med 2016;374(1):13–22.

[53] Liang Y, Ridzon D, Wong L, Chen C. Characterization of microRNA expression profiles in normal human tissues. BMC Genomics 2007;8:166.

[54] Morales-Prieto DM, Chaiwangyen W, Ospina-Prieto S, Schneider U, Herrmann J, Gruhn B, et al. MicroRNA expression profiles of trophoblastic cells. Placenta 2012;33(9):725–34.

[55] Delorme-Axford E, Donker RB, Mouillet JF, Chu T, Bayer A, Ouyang Y, et al. Human placental trophoblasts confer viral resistance to recipient cells. Proc Natl Acad Sci USA 2013;110(29):12048–53.

[56] Cronqvist T, Tannetta D, Morgelin M, Belting M, Sargent I, Familari M, et al. Syncytiotrophoblast derived extracellular vesicles transfer functional placental miRNAs to primary human endothelial cells. Sci Rep 2017;7(1):4558.

[57] Maynard SE, Min JY, Merchan J, Lim KH, Li J, Mondal S, et al. Excess placental soluble fms-like tyrosine kinase 1 (sFlt1) may contribute to endothelial dysfunction, hypertension, and proteinuria in preeclampsia. J Clin Invest 2003;111(5):649–58.

[58] Cindrova-Davies T, Sanders DA, Burton GJ, Charnock-Jones DS. Soluble FLT1 sensitizes endothelial cells to inflammatory cytokines by antagonizing VEGF receptor-mediated signalling. Cardiovasc Res 2011;89 (3):671–9.

[59] Morales-Prieto DM, Ospina-Prieto S, Chaiwangyen W, Schoenleben M, Markert UR. Pregnancy-associated miRNA-clusters. J Reprod Immunol 2013;97(1):51–61.

[60] Higashijima A, Miura K, Mishima H, Kinoshita A, Jo O, Abe S, et al. Characterization of placenta-specific microRNAs in fetal growth restriction pregnancy. Prenat Diagn 2013;33(3):214–22.

[61] Hromadnikova I, Kotlabova K, Ondrackova M, Kestlerova A, Novotna V, Hympanova L, et al. Circulating C19MC microRNAs in preeclampsia, gestational hypertension, and fetal growth restriction. Mediators Inflamm 2013;2013:186041.

[62] Pineles BL, Romero R, Montenegro D, Tarca AL, Han YM, Kim YM, et al. Distinct subsets of microRNAs are expressed differentially in the human placentas of patients with preeclampsia. Am J Obstet Gynecol 2007;196(3). 261.e1–261.e6.

[63] Zhang Y, Fei M, Xue G, Zhou Q, Jia Y, Li L, et al. Elevated levels of hypoxia-inducible microRNA-210 in pre-eclampsia: new insights into molecular mechanisms for the disease. J Cell Mol Med 2012;16(2):249–59.

[64] Ishibashi O, Ohkuchi A, Ali MM, Kurashina R, Luo SS, Ishikawa T, et al. Hydroxysteroid (17-beta) dehydrogenase 1 is dysregulated by miR-210 and miR-518c that are aberrantly expressed in preeclamptic placentas: a novel marker for predicting preeclampsia. Hypertension 2012;59(2):265–73.

[65] Xu P, Zhao Y, Liu M, Wang Y, Wang H, Li YX, et al. Variations of microRNAs in human placentas and plasma from preeclamptic pregnancy. Hypertension 2014;63(6):1276–84.

[66] Zhu XM, Han T, Sargent IL, Yin GW, Yao YQ. Differential expression profile of microRNAs in human placentas from preeclamptic pregnancies vs normal pregnancies. Am J Obstet Gynecol 2009;200(6). 661. e1–661.e7.

[67] Muralimanoharan S, Maloyan A, Mele J, Guo C, Myatt LG, Myatt L. MIR-210 modulates mitochondrial respiration in placenta with preeclampsia. Placenta 2012;33(10):816–23.

[68] Munaut C, Tebache L, Blacher S, Noel A, Nisolle M, Chantraine F. Dysregulated circulating miRNAs in preeclampsia. Biomed Rep 2016;5(6):686–92.

[69] Murphy MS, Casselman RC, Tayade C, Smith GN. Differential expression of plasma microRNA in pre-eclamptic patients at delivery and 1 year postpartum. Am J Obstet Gynecol 2015;213(3). 367.e1–367.e9.

[70] Poon LL, Leung TN, Lau TK, Lo YM. Presence of fetal RNA in maternal plasma. Clin Chem 2000;46 (11):1832–4.

[71] Pang WW, Tsui MH, Sahota D, Leung TY, Lau TK, Lo YM, et al. A strategy for identifying circulating placental RNA markers for fetal growth assessment. Prenat Diagn 2009;29(5):495–504.

[72] Ng EK, Leung TN, Tsui NB, Lau TK, Panesar NS, Chiu RW, et al. The concentration of circulating corticotropin-releasing hormone mRNA in maternal plasma is increased in preeclampsia. Clin Chem 2003;49(5):727–31.

[73] Paiva P, Whitehead C, Saglam B, Palmer K, Tong S. Measurement of mRNA transcripts of very high placental expression in maternal blood as biomarkers of preeclampsia. J Clin Endocrinol Metab 2011;96(11): E1807–15.

[74] Kodama M, Miyoshi H, Fujito N, Samura O, Kudo Y. Plasma mRNA concentrations of placenta-specific 1 (PLAC1) and pregnancy associated plasma protein A (PAPP-A) are higher in early-onset than late-onset pre-eclampsia. J Obstet Gynaecol Res 2011;37(4):313–8.

[75] Purwosunu Y, Sekizawa A, Okazaki S, Farina A, Wibowo N, Nakamura M, et al. Prediction of preeclampsia by analysis of cell-free messenger RNA in maternal plasma. Am J Obstet Gynecol 2009;200(4). 386.e1–386.e7.

[76] Sekizawa A, Purwosunu Y, Farina A, Shimizu H, Nakamura M, Wibowo N, et al. Prediction of pre-eclampsia by an analysis of placenta-derived cellular mRNA in the blood of pregnant women at 15-20 weeks of gestation. BJOG 2010;117(5):557–64.

[77] Takenaka S, Ventura W, Sterrantino AF, Kawashima A, Koide K, Hori K, et al. Prediction of fetal growth restriction by analyzing the messenger RNAs of angiogenic factor in the plasma of pregnant women. Reprod Sci 2015;22(6):743–9.

[78] Whitehead CL, Walker SP, Mendis S, Lappas M, Tong S. Quantifying mRNA coding growth genes in the maternal circulation to detect fetal growth restriction. Am J Obstet Gynecol 2013;209(2). 133.e1–133.e9.

[79] Whitehead C, Teh WT, Walker SP, Leung C, Mendis S, Larmour L, et al. Quantifying circulating hypoxia-induced RNA transcripts in maternal blood to determine in utero fetal hypoxic status. BMC Med 2013;11:256.

[80] Redman CW, Sacks GP, Sargent IL. Preeclampsia: an excessive maternal inflammatory response to pregnancy. Am J Obstet Gynecol 1999;180(2 Pt 1):499–506.

[81] Gervasi MT, Chaiworapongsa T, Pacora P, Naccasha N, Yoon BH, Maymon E, et al. Phenotypic and metabolic characteristics of monocytes and granulocytes in preeclampsia. Am J Obstet Gynecol 2001;185 (4):792–7.

[82] Tsui NB, Jiang P, Wong YF, Leung TY, Chan KC, Chiu RW, et al. Maternal plasma RNA sequencing for genome-wide transcriptomic profiling and identification of pregnancy-associated transcripts. Clin Chem 2014;60(7):954–62.

[83] Koh W, Pan W, Gawad C, Fan HC, Kerchner GA, Wyss-Coray T, et al. Noninvasive in vivo monitoring of tissue-specific global gene expression in humans. Proc Natl Acad Sci USA 2014;111(20):7361–6.

[84] Pan W, Ngo TTM, Camunas-Soler J, Song CX, Kowarsky M, Blumenfeld YJ, et al. Simultaneously monitoring immune response and microbial infections during pregnancy through plasma cfRNA sequencing. Clin Chem 2017;63(11):1695–704.

[85] Iriye BK, Gregory KD, Saade GR, Grobman WA, Brown HL. Quality measures in high-risk pregnancies: Executive Summary of a Cooperative Workshop of the Society for Maternal-Fetal Medicine, National Institute of Child Health and Human Development, and the American College of Obstetricians and Gynecologists. Am J Obstet Gynecol 2017;217(4):B2–B25.

[86] Smith GC. Researching new methods of screening for adverse pregnancy outcome: lessons from pre-eclampsia. PLoS Med 2012;9(7),e1001274.

[87] Monier I, Blondel B, Ego A, Kaminiski M, Goffinet F, Zeitlin J. Poor effectiveness of antenatal detection of fetal growth restriction and consequences for obstetric management and neonatal outcomes: a French national study. BJOG 2015;122(4):518–27.

[88] Rolnik DL, Wright D, Poon LC, O'Gorman N, Syngelaki A, de Paco Matallana C, et al. Aspirin versus placebo in pregnancies at high risk for preterm preeclampsia. N Engl J Med 2017;377(7):613–22.

[89] Smith GCS. The STRIDER trial: one step forward, one step back. Lancet Child Adolesc Health 2018;2 (2):80–1.

GLOSSARY

Fetal growth restriction (FGR) Condition in which the fetus is unable to achieve its genetically determined potential size.

Small for gestational age (SGA) baby An infant born with a birthweight less than the 10th centile for gestational age.

Preeclampsia Systemic syndrome characterized by gestational hypertension and proteinuria, arising de novo after 20 weeks of gestation in a previously normotensive woman.

Placental abruption Premature separation of the placenta from the uterine wall.

Stillbirth A baby born with no signs of life at or after 28 weeks of gestation (definition recommended by the World Health Organization for international comparison).

PRENATAL TREATMENT OF GENETIC DISEASES IN THE UNBORN

E.J.T. Verweij*, D. Oepkes†

Department of Obstetrics & Gynaecology, Erasmus MC, Rotterdam, The Netherlands* Department of Obstetrics, Leiden University Medical Center, Leiden, The Netherlands†

INTRODUCTION: BENEFITS OF PRENATAL GENETIC DIAGNOSIS BEYOND THE OPTION OF TERMINATION

For several decades, prenatal genetic testing was available to pregnant women in three different situations. Traditional karyotyping was offered to a large group of women at increased risk for trisomy 21, based on population screening using maternal age, serum marker testing, ultrasound or a combination of these. Chorionic villus sampling (CVS) with DNA analysis was offered to couples at high risk for a fetus with dominant or recessive genetic condition. The third situation was karyotyping after the diagnosis of fetal structural anomalies on ultrasound examination. In the majority of cases in all three groups, an abnormal result led to the request by the parents to terminate the pregnancy. In case of potentially treatable fetal structural anomalies, an abnormal karyotype was considered a contraindication to fetal intervention.

Many people believed that with the completion of the Human Genome Project, in the year 2003, we would enter an era of diagnosing, and even curing, many or most genetic diseases. Although since then, enormous progress has been made in technology, leading to both much faster and much cheaper sequencing, clinical application is still mostly limited to testing predisposition for genetic diseases, including certain types of cancer. Major advances have been made in genetic testing methods, and most new modalities are now available for prenatal diagnosis and screening as well. Although a small role for traditional karyotyping remains, testing for fetal diseases is now mostly done using QF-PCR for rapid aneuploidy evaluation, and chromosomal microarray for detailed analysis. Whole exome sequencing is currently introduced in perinatal clinical practice as well, further refining our diagnostic capabilities. Hudecova published a method for noninvasive prenatal diagnosis of inherited single-gene disorders using droplet digital PCR from circulating cell-free DNA in maternal plasma [1]. Noninvasive genetic diagnostic testing, safe and therefore more acceptable to offer to a broad group of pregnant women, is entering the clinic.

The development of treatment options often lag behind the advances in diagnostic testing. This is especially true for fetal diseases, where both high definition ultrasound and high-resolution chromosome analysis enable early and accurate detection of many pathologic conditions. If at all possible, fetal diseases are preferably treated after birth, when direct access to the patient is possible, and risks for the pregnant woman are absent. In a number of fetal conditions however, progressive disease may lead to demise before birth, or to irreversible damage and handicaps at the time of birth. In such diseases, prenatal intervention can be considered to save the child's life or to improve the prognosis for its future health. Any prenatal intervention has the potential to harm the mother, or to inadvertently harm the fetus in case of complications or failure. In particular, the possible harm to the mother makes fetal therapy a challenging subspecialty. In addition, unlike in most other areas of medicine, termination of the pregnancy may be an alternative to prenatal or postnatal treatment. Thus when considering fetal intervention, potential benefits to the fetus also need to be balanced against potential harm to the fetus and to risks to the pregnant woman. If an attempt to save the life of a fetus, or to prevent irreversible damage, leads to the birth of a severely handicapped child, either because of complications of failure of the prenatal treatment, the child and its family may suffer more than with expectant management. The risk of "trading mortality for morbidity" is a valid argument against some forms of fetal therapy. We will explore this aspect in more depth in the part on ethics.

The "classic" forms of fetal therapy are directed at acquired, and not genetic, fetal diseases, such as Rh-alloimmunization or twin-twin transfusion syndrome. In these conditions, the fetus is essentially healthy, only threatened to die from a "hostile environment." Successful prenatal intervention in these conditions often leads to the birth of healthy children with a normal long-term outcome. In genetic diseases, the underlying cause, a DNA mutation, is in most cases present in every cell of the body. Until recently, treatment options were limited to reducing the consequences of the abnormal organ functions or alleviating symptoms. Early versions of gene- and stem-cell therapy were aimed mostly at *addition* of genes or cells with normal function to fetal organs, and/or inducing tolerance for safer postnatal transplantation. Repairing the DNA itself, gene *correction*, is now on the horizon, using DNA cutting techniques such as Cas9, zinc-finger, or TALE nucleases. In this chapter, we will focus on currently or likely soon available prenatal treatments for fetal genetic diseases.

SYMPTOMATIC TREATMENT

Inherited or de novo mutations affecting erythropoiesis, red cell morphology, or the structure/function of hemoglobin may all lead to fetal anemia, which as a final common pathway can cause fetal heart failure, hydrops, and death. Irrespective of the underlying cause, fetal anemia can effectively and quite safely be treated by fetal blood transfusion, a method developed already in the early 1960s. This purely symptomatic treatment, often given multiple times during pregnancy, can be lifesaving. The diagnosis of genetic hematologic diseases in a fetus is made either by testing because of a family history or by workup after the incidental finding of fetal hydrops. To illustrate the complex choices and consequences doctors and parents can be confronted with, the example of fetal treatment of homozygous alpha-thalassemia will be explored further.

Alpha-thalassemia major or Hb Bart's disease is the most common inherited disorder of hemoglobin synthesis in Asia. It is caused by a deletion of all four "alpha genes" on chromosome 16. The gene mutation (a deletion) frequency in Southeast Asia is around 4.5%, resulting in a high prevalence of the homozygous mutation [2]. Traditionally, this is considered a lethal disease due to deletion of all alpha-globin genes. In early fetal life, embryonic hemoglobin types can deliver sufficient oxygen to the tissues, however later on, hemoglobin consisting of four γ chains (Bart's) becomes the dominant fetal hemoglobin. Hemoglobin Bart's has is a very high oxygen affinity, compromising oxygen delivery to the tissues. Screening programs identify carriers, and, in at risk couples, noninvasive fetal testing can be performed by ultrasonography aiming at finding early markers such as cardiomegaly, or by invasive testing and genetic analysis of chorionic villi. Cell-free DNA testing from maternal plasma for thalassemia is being developed [3]. In the vast majority of cases, detection of fetal homozygous alpha-thalassemia within such screening programs will lead to termination of pregnancy.

Some early examples of fetal (intrauterine) intravascular blood transfusion for homozygous alpha-thalassemia were cases where the cause of fetal anemic hydrops was unknown, and diagnostic fetal blood sampling was combined with blood transfusion, to gain time for the diagnostic workup [4]. Then when the diagnosis was made, parents requested continuation of the transfusion therapy, both prenatally and postnatally, to keep their children alive. Technically, this has been a relatively safe and, in fetal therapy centers, common procedure for fetal anemia treatment already for decades. Most often however, it is performed for red cell alloimmunization and parvovirus infection, two acquired fetal diseases carrying an excellent prognosis when treated timely. The main dilemma in alpha-thalassemia is the quality of life of the surviving children. As will be discussed in more detail later in this chapter, doctors feel a responsibility for the burden of the future child. The general view of fetal medicine specialists around the world is that homozygous alpha-thalassemia should still be regarded as a lethal disease, for which termination of pregnancy is justified.

A literature review identified 15 children with homozygous alpha-thalassemia surviving after fetal transfusions [2]. In 5/15, hypospadia was present. Long-term neurodevelopmental outcome, with follow-up between 3 months and 7 years, revealed four children had mild and one had marked psychomotor impairment. Since insufficient tissue oxygenation is already present early in fetal life, fetal transfusion therapy, which at the earliest can be performed from 16 to 18 weeks' gestation, may not be able to prevent neurologic damage [5]. After birth all children required blood transfusions at 3–4-week intervals, and 9/15 were treated with bone marrow/stem cell transplantation. The morbidity related to iron overload and side effects of medication accompanying transplantation is considerable and significantly worse than in β-thalassemia.

This example of a technically feasible symptomatic treatment of fetuses with a serious gene mutation illustrates the main dilemma; the fetal symptomatic treatment may save the child's life, but it is no cure, and the survivors remain patients needing lifelong treatment, with far from normal lives.

SUPPLEMENTATION THERAPY

Genetic diseases generally described as inborn errors of metabolism often originate from (single) gene mutations leading to absent or abnormal enzyme, hormone, or protein function. The mode of inheritance is often autosomal recessive. In particular in metabolic diseases where one specific essential substance is lacking, replacement therapy may solve the clinical problem. In others, preventing the individual to take food containing substances the body cannot process normally, leading to dangerous levels of a toxic product, "nutrition management" may be the solution. The best-known example is phenylalanine hydroxylase deficiency, leading to phenylketonuria (PKU). During pregnancy, the unaffected (heterozygous) carrier mother may provide essential substances via the placenta, and in this way reduce the risk of harm for the fetus with this metabolic disorder. After birth however, completely asymptomatic but affected (homozygous) neonates may soon suffer from their metabolic disorder.

Likewise pregnant women with PKU need strict dietary control at least 1 month prior to conception and during the whole of pregnancy, as well as monitoring of phenylalanine levels, to avoid the severe and irreversible teratogenic effects of hyperphenylalaninemia [6]. Most of the rare diseases screened for with the neonatal dried blood spot test by the "heel prick" fall in this category. In some genetic mutations leading to errors of metabolism, the abnormal function of an enzyme or hormone may lead to irreversible damage already before birth. In such cases, prenatal intervention may improve the outcome and prognosis. Some have argued that several of the diseases screened for by the heel prick should actually be tested for already during pregnancy, to enable earlier intervention [7]. Examples of prenatal fetal treatment of metabolic disorders such as methylmalonic acidemia and multiple carboxylase deficiency have already been given in the 1970s [8] and 1980s [9]. With our increasing ability to screen for fetal genetic alterations during pregnancy, it is expected that a much larger number of "inborn errors of metabolism" will be detected, followed by more research into suitable treatments. We speculate that very few parents (and Institutional Review Boards) are willing to go the uncertain route of experimental fetal treatment. The options and dilemmas will be illustrated by some examples.

Methylmalonic acidemia (MMA) is due to deficient synthesis of 5′-deoxyadenosylcobalamin, which can lead to fetal growth restriction and dilated cardiomyopathy. Children suffer from vomiting, failure to thrive, acidosis, and mental retardation. In 1975, Ampola et al. reported prenatal treatment by vitamin B12 in an attempt to diminish the accumulation of methylmalonic acid to a pregnant woman who lost a previous child to MMA [8]. Dose and response were monitored using maternal urinary excretion of methylmalonic acid. Postnatally, the child was treated with dietary measures. At 19 months, the child was developing normally. In the following decades, a few more apparently successful cases have been published. In 2016, the long-term follow-up was published of a girl with Cobalamin C deficiency (leading to MMA as well as hyperhomocysteinemia) treated in utero by giving the pregnant mother a weekly dose of hydroxycobalamin in 2005, and her untreated older sister [10]. At the age of 11 years the girl had a normal IQ of 103, attended secondary school, and only mild ophthalmic findings. The affected older sister, with the same genotype and similar residual enzymatic activity, received postnatal treatment only and was severely intellectually disabled, cannot walk or talk, shows aggressive behavior, and has severe ophthalmic symptoms. Due to the rarity of the disease and the many genetic variants, it remains difficult to establish optimal dosage and to predict clinical outcome, especially on the long term.

Multiple carboxylase synthetase deficiency presents in the newborn with severe metabolic acidosis. It is treated with lifelong biotin therapy. However, the fetus may already suffer from growth restriction as well as ventriculomegaly [11]. The first case report of prenatal treatment, by maternal oral biotin supplementation, was published in 1982 [9]. Dosage is still a problem, as illustrated by the latest published case where the investigators used 10 mg biotin daily based on a few previous reports. Fetal growth accelerated after biotin treatment, but the child was born with significant lactic acidemia and metabolic acidosis [11].

3-Phosphoglycerate-dehydrogenase deficiency is an amino acid (serine) synthesis disorder, associated with microcephaly, severe psychomotor retardation, and intractable seizures. In 2004, de Koning et al. from Utrecht, The Netherlands published a case of prenatal maternal administration of supplements of L-serine, three doses of 5 g L-serine (190 mg/kg) per day, starting at 26 weeks when the growth of the head circumference was decelerating [12]. The child had low levels of serine at birth, but was born with a normal head size, and at the age of 2 developed normally, in sharp contrast with its two affected older siblings.

FETAL GENETIC THYROID DISEASE

Fetal goitrous hypothyroidism is usually secondary to abnormal maternal iodine intake or autoimmune thyroid disorders. However, the fetus itself may have a genetic mutation causing defects in the thyroglobulin synthesis, leading to primary fetal hypothyroidism. Mutations likely causing congenital hypothyroidism have been described in genes coding for enzymes involved in regulating thyroid functions, such as thyroid oxidase 2. Unless there is a family history, this may go unnoticed before birth. When a fetal goiter develops, complications such as polyhydramnios (due to impaired swallowing) and malposition may occur. Postnatally, the goiter may cause airway problems. Ultrasound can easily detect the thyroid enlargement, and amniocentesis or fetal blood sampling can confirm hypothyroidism. Fetal therapy for goiter was already described in 1980 [13]. The largest single center series of prenatal treatment of fetal goitrous hypothyroidism in euthyroid mothers is from France [14]. The authors described 12 cases where thyroxin was given intraamniotic, with good clinical response defined as reduced size of the goiter. However, most fetuses remained hypothyroid, with sequelae such as jaundice, open posterior fontanel, hypotonia, and delayed bone age. A literature review published in 2017 found 32 in utero treated cases, with evidence of benefit by preventing mechanical difficulties due to the goiter in 28 [15]. The authors confirmed inability of reaching a euthyroid state at birth. In 2017, Vasudevan et al. published a case with fetal goiter and polyhydramnios, in which intraamniotic T3 and T4 was given. The fetus died at 31 weeks. The therapy may not be risk free. Although in a few cases with untreated siblings, the IQ of the prenatally treated children was slightly higher, long-term outcome data are lacking to suggest benefit on neurodevelopmental outcome. Therefore intraamniotic thyroxine injections may have a place, if success of prenatal intervention is defined as size reduction of a large goiter, enabling the fetus to swallow and the neonate to breath at birth. Parents and clinicians should, however, be aware that the fetus likely has suffered from prolonged hypothyroidism, and irreversible sequelae can already be present at the start of fetal treatment.

In conclusion, fetal (and maternal) metabolic disorders may have irreversible effects on fetal health. This justifies attempts to treat the condition before birth. Studies to find the optimal dose of medication needed to reach appropriate fetal tissue levels are difficult to perform, given the rarity and variability of each disorder. Most published cases had a previous severely affected sibling, with better outcome in the prenatally treated pregnancy. Even with the same mutation, clinical presentation may vary between sibs. Improvement may be not only due to the prenatal medication, but also to other aspects of improved care for the second affected child in a family, both pre- and postnatally. Randomized or otherwise properly controlled studies in this field are virtually impossible to perform.

CAUSAL THERAPY I: IN UTERO STEM CELL TRANSPLANTATION

In addition to the rationale of prenatally treating a disease that may cause death or irreversible damage already before birth, there is the potential advantage of the window of fetal immunologic immaturity enabling transplantation or injection of unmatched donor cells. Stem cell transplantation in children and adults is useful in many diseases, but one of the difficulties is that careful matching and immunosuppression are needed to reduce the risk of rejection and graft-versus-host disease (GVHD). Transplantation early in fetal life may be safe and effective without immunosuppression, and even if engraftment is very low, there may be induction of donor-specific tolerance, enabling safer postnatal transplantations from the same donor. The fetus is so to say an immunologically "naïve" target. Proof for this concept has been given since the 1950s [16]. The use of less immunogenic fetal (liver) stem cells (from abortion material) as donor material was already proposed in the 1980s [17]. Also, the fetus is still small, and only requires a small transplant. It has relatively "empty" bone marrow with space for new transplanted cells. Before birth migration of stem cells to bone marrow and other target organs is a natural phenomenon. The transplanted donor stem cells may "piggyback" on this system, but still need to compete with the (defective) host cells in the target organs.

Although for decades, hopes were high that this type of fetal treatment would be successful in the approach of many genetic diseases, in vitro and animal research have not yet resulted in established clinical applications. One major barrier became apparent, when Peranteau et al. demonstrated that even the young fetus has a functioning immune system [18]. Some examples of human trials of fetal stem cell therapy are discussed as follows.

IMMUNODEFICIENCY DISORDERS

Inherited severe immunodeficiency disorders are ideal targets for stem cell transplantation: rejection and GVHD are no issue; the absence of immune reaction is the disease itself.

In 1988, Touraine et al. from Lyon, France did a human proof-of-principle experiment: they injected fetal liver and thymus cells into the umbilical cord of a 30-week fetus, diagnosed with bare lymphocyte syndrome, a combined immunodeficiency disorder with absent expression of HLA antigens [19]. At 1 month of age, 10% of the newborn's lymphocytes expressed HLA class I. The child received multiple fetal stem cell transplantations and was still in isolation at 7 months of age.

In 1995, Flake et al. published the first successful in utero stem cell transplantation, in a case with X-linked recessive severe combined immunodeficiency (SCID) [20]. The parents' first son had died from the disease. In the next pregnancy, CVS revealed again an affected son. This fetus received three stem cell transplantations between 16 and 19 weeks' gestation. The cells were harvested from his fathers' bone marrow. There was a high level of donor cell engraftment, and normal growth and development at 11 months of age.

GENETIC HEMATOLOGIC DISORDERS

Hemoglobinopathies, in particular sickle cell anemia and thalassemia, were among the first diseases thought to be candidates for in utero stem cell transplantation.

A 2017 review article counted a total of 26 reported cases of intrauterine stem cell transplantations for hematopoietic disorders in human fetuses [21]. Only in the three cases of immune deficiency disorders, where the lack of fetal immunity causes selective advantage for donor cells, there was clinically significant donor cell chimerism [19,20] None of the other attempts, for various conditions including Rh disease, leukodystrophy, Chediak-Higashi, thalassemias, Niemann-Pick type A, and Hurler syndrome, were deemed successful, success being defined as sustained donor cell chimerism sufficiently high to ameliorate the disease phenotype. Most transplantations, often intraperitoneally, were done after 14 weeks' gestation, when fetal immunity has already developed, and fetal blood sampling for evaluation of donor cell expression was, with the current knowledge, done too early. Of the 26 fetuses, 4 died unintentionally before birth, and 2 were terminated because of lack of successful engraftment. Current knowledge from animal work suggests that more time may be needed before engraftment can be established, and that intravascular injection (instead of, e.g., intraperitoneally, which was done in some early human cases) is the most successful route of administration. In none of the reported cases overt GVHD occurred. The disappointing engraftment was in part found to be due to fetal immunity, which in contrast to earlier views, is present already early in fetal life. To overcome this barrier, autologous or maternal/paternal transplantation could be used. Another potential improvement could be the use of (fetal donor) mesenchymal stroma cells.

OSTEOGENESIS IMPERFECTA

Osteogenesis imperfecta (OI) is a group of incurable genetic disorders caused by >1350 different dominant and >150 recessive mutations in collagen genes. Seven types have been described. Mutated COL1A1 and COL1A2 are the genes found in most individuals with OI. These mutated genes cause an abnormal production of type I collagen. The abnormal collagen mainly leads to abnormal "brittle bones," resulting in fractures. Other clinical sequelae are deformities, growth problems, protruding eyes, and deafness. Some forms have the typical "blue sclerae." Type I is mild, type II is lethal, types III and IV are not lethal but often severe, with fractures already occurring in utero. These can be seen on ultrasound. Postnatal treatment is symptomatic (mostly by bisphosphonates) and aims at preventing fractures.

In 2002, Westgren, Le Blanc and colleagues from the Karolinska Institutet in Stockholm, Sweden performed the first prenatal transplantation of human fetal liver mesenchymal stem cells with a successful outcome in an immunocompetent fetus suffering from OI type III [22]. Fetal mesenchymal stem cells from a fetal liver were injected in the umbilical vein at 31 weeks. There was long-term donor cell engraftment. Until the age of 8, the patient only suffered one fracture and one vertebral compression fracture per year. At the age of 8 years, the girl received another stem cell transplant from the same donor. Over the following 2 years, she did not acquire any fractures and her linear growth and mobility improved. A child with an identical mutation that did not receive stem cell transplantation exhibited a very severe phenotype of OI and died at the age of 5 months. In recent years, two other fetuses with OI were treated with mesenchymal stem cells with assistance of the same group of investigators. Outcomes thus far are promising.

FETAL TRANSPLANTATION AND IMMUNOLOGY SOCIETY

In 2015, a consensus statement was published on in utero stem cell transplantation [23]. An international organization was founded, FeTIS (Fetal Transplantation and Immunology Society, www.fetaltherapies.org), which will develop a registry of patients treated and outcomes. The international expert community agreed that in utero transplantation is a viable option for selective congenital disorders. Beside its success in SCID and similar diseases, the advantage of in utero stem cell transplantation is the development of donor-specific tolerance. A chimerism of 1%–2% may be sufficient to reach this. A few other diseases may also benefit from low-level engraftment. According to FeTIS, the main target diseases now suitable for human phase I trials are sickle cell disease, thalassemia, and osteogenesis imperfecta.

Two human in utero stem cell trials are currently (2018) ongoing. The first is the BOOSTB4 trial, an EU Horizon 2020 program supported international multicenter phase I/II study (www.boostb4.eu). The investigators, led by the Karolinska group, aim to perform 15 prenatal and 15 postnatal-only fetal mesenchymal stem cell transplantations for OI type III and severe type IV.

The other study is a single center phase I clinical trial combining maternal stem cell transplantation with intrauterine blood transfusion in alpha-thalassemia major, led by MacKenzie, University of California, San Francisco (www.fetaltherapies.org).

CAUSAL THERAPY II: GENE THERAPY

In their consensus statement the above-mentioned society FeTIS, only stated the following on gene therapy: *"Treatment of the fetal patient using gene therapy and gene-modified cells have great future potential and should be fields of active investigation."*

The true causal treatment of diseases caused by gene mutations is correcting the gene itself. With the completion of the Human Genome Project, many hoped for rapid advances in the field of gene therapy. Now, 15 or more years later, hope still exists but progress has been disappointing.

The general idea is to insert DNA with the correct gene into the target cells. When stem cells are used as targets, they can, after injection, proliferate and give rise to permanent (correct) gene expression. Experiments were and mostly are done using viral vectors (adenovirus, recombinant adeno-associated virus, retro/lentivirus). Postnatal gene therapy was hampered by the quick emergence of antibodies or T-cells against the viral vector and sometimes against the transgene product. This was one reason to move to in utero gene transfer to the fetus, with its (perceived) immature immune system [24]. Animal experiments showed development of tolerance to the viral vectors, allowing repeated postnatal transplantations. Other arguments to target the fetus as recipient of gene therapy is based on the same rationale as for stem cell treatment: irreversible damage may already occur in utero. Injection of the vectors can be done directly in the target organ, or into the amniotic fluid or trachea, with absorption via the lungs. Hemophilia, cystic fibrosis, muscular dystrophy, and Wilson's disease are the genetic disorders that have been studied most, with the use of animal models.

The complexity and hazards of gene therapy were well summarized by the group from Anna David in London [25]. In short, the main concerns are acute toxicity and immunogenicity of the vector and transgene product, and the possibility for germ line transmission and insertional mutagenesis/oncogenesis. Although germ line alteration is often mentioned as a concern, the true risk is extremely low [25]. Still, if these risks cannot be excluded, they must be balanced against benefits of the treatment. Oncogenesis obviously is a scary potential complication, first thought to be only hypothetical. In 2003, a case from France was published of acute lymphoblastic leukemia likely caused by insertional mutagenesis after ex vivo, retrovirally mediated transfer of the γc gene into CD34+ cells in four patients with SCID [26]. This case halted human experiments for many years. Although in the laboratory, a lot of progress has been made, prenatal gene therapy is still in the animal study phase.

CAUSAL THERAPY III: GENOME EDITING TECHNOLOGIES

When a double strand DNA breaks, nature has two pathways to repair the defect; nonhomologous end-joining and homology-directed repair, depending on the phase of the cell cycle. For many years, it has been known that endonucleases such as zinc-finger nucleases (ZFNs) and transcription-activator-like effector nucleases (TALENs) can be engineered to modify human cells at single nucleotide resolution. In 2013, a relatively simple, effective, and cheap method was published, the CRISPR (Clustered

Regularly Interspaced Short Palindromic Repeats)-Cas9 (CRISPR-associated protein 9) system, which can precisely cut DNA and repair it again with one of the two mentioned methods, in that way modifying mutations [27,28]. This gene-editing approach is considered revolutionary, since it is able to permanently cure a genetic defect at its origin, in contrast to the supplementation or competition mechanisms with which stem cell and "traditional" gene therapy work [29]. Since then, many proof-of-principle and even preclinical (in animals) studies have shown the great potential of gene editing. Treatment of human patients with genetically engineered (donor) cells has been published, for example, for HIV, leukemia, and lung cancer. Examples of correcting genetic diseases (using ex vivo gene-edited human stem cells transplanted in mice) are Duchenne muscular dystrophy, sickle cell disease, and β-thalassemia [30]. Even more spectacular is the concept of using gene editing manipulation of germ line cells or embryos, to transgenerationally correct or avoid single-gene disorders. A Chinese group already published such an experiment using (nonviable) human embryos [31]. In the past 3 years, many debates and position papers discussed the risks, uncertainties (such as mosaicism and off-target effects), and ways to responsibly control such experiments. Views range from proposing a complete ban on germ line/embryo gene editing to promoting more research in this promising field. Safety and efficacy of this type of gene therapy need to be assessed before clinical application in human monogenic diseases. The main issues still needing a lot of research are potential off-target effects, delivery, immunogenicity, and longevity of its benefits in vivo.

SPECIAL CASE: PRENATAL THERAPY IN DOWN SYNDROME

Since the 1970s, many hundreds of studies have been performed on prenatal diagnosis and screening for the most common form of chromosomal conditions in newborns: trisomy 21 or Down syndrome (DS). Nowadays, in most developed countries, routine prenatal care includes the offer to the pregnant woman to undergo screening for trisomy 21. The majority of women are screened, with increasingly accurate techniques. Of those found to actually carry a fetus with trisomy 21, the vast majority request termination of pregnancy.

With increasingly accurate screening, there is a trend toward more women continuing the pregnancy while knowing their fetus has trisomy 21. Some of these fetuses have clinical symptoms that appear life threatening, such as hydrothorax with secondary hydrops, or prenatal leukemia. Most textbooks and guidelines for fetal therapy list "normal chromosomes" as one of the prerequisites to consider fetal therapy. The concept likely is that even when fetal treatment is successful, the child will have a moderate-to-severe and untreatable mental handicap, for which "dying from natural causes" is perceived to be the preferred course. Some parents however, and this group seems to be growing, think otherwise. A similar discussion happened in the 1980s, when pediatric surgeons debated the (ethical) issues of surgery for gastrointestinal tract or cardiac anomalies in neonates with Down syndrome. This is now done almost routinely. In some institutions, even children with trisomy 18 and 13, conditions known not to be compatible with long survival if not mosaic, are operated for their cardiac defect [32].

We know of several cases, some in our own institution, where well-informed parents insisted on, and were granted, thoraco-amniotic shunt treatment for hydrothorax in their trisomy 21 fetus, with survival of the child and gratitude of the parents. A review published in 2017 described 6 cases in detail, the first one from 1988, with survival of 4/6 children [33]. A German study suggested that fetuses with hydrothorax and trisomy 21, treated with thorax shunting, actually had a *better* prognosis than euploid fetuses [34]. In 2017, the Fetal Therapy Special Interest Group of the ISPD (International Society for Prenatal Diagnosis) formed a committee to issue a statement on fetal therapy in aneuploid fetuses.

A second interesting issue, still remote from human clinical experiments, is the prenatal treatment of fetuses with trisomy 21 aimed at improving their health and cognitive abilities. Part of the interest in treatment of Down syndrome lies in the fact that most of these individuals develop an early form or Alzheimer's disease. In the past, the brain dysfunction in Down syndrome was attributed to overexpression of genes on chromosome 21, however, as usual, reality is far more complex. There may be valuable lessons to be learned from studies in Down syndrome for other patients with Alzheimer's disease [35]. In the past few years, using mouse models of Down syndrome (T(risomy)16 mouse, Ts65Dn and Ts1Cje), researchers have attempted to improve cognitive function using pharmacotherapy, with promising results [36,37]. The limitations of using rodents for this type of research are numerous, reason for research groups to develop trisomy 21 human-induced pluripotent stem cells for further studies [38]. Since the introduction of safe and accurate trisomy 21 screening in pregnancy with cell-free DNA in maternal plasma, an increasing proportion of women now know that their fetus has trisomy 21, and decide to continue the pregnancy. This group may have an interest in prenatal treatment to improve their child's long-term health and cognition. In the second half of pregnancy, brain growth and development are already reduced, at least in part due to reduced neurogenesis and abnormal apoptosis [39]. It is quite possible that many of the brain abnormalities found in Down syndrome might be treatable prenatally, while this is futile in adulthood [40]. Some studies in pregnant Ts65Dn mice showed improved and long-lasting cognitive function in their Ts65Dnpups, after intraperitoneal injection of neurotrophic factors [41]. The group of Bianchi used antioxidants in pregnant mice to reverse the Down syndrome transcriptome, showing improved exploratory behavior in their Ts1Cje pups [42].

Another medication that has been explored with success in early pregnancy in the Ts65Dn mousemodel is fluoxetine (Prozac) [43].

A growth inducer newly identified of neural stem cells (NSCs) was found to rescue proliferative deficits in Ts65Dn-derived neurospheres and human NSCs derived from individuals with DS. The oral administration of this compound, named ALGERNON (altered generation of neurons), restored NSC proliferation in murine models of DS and increased the number of newborn neurons. ALGERNON also rescued aberrant cortical formation in DS mouse embryos and prevented the development of abnormal behaviors in DS offspring [44].

Dr. Robyn Horsager-Boehrer and colleagues of the University of Texas Southwestern Medical Center are performing a pilot human clinical study in pregnant women with trisomy 21 fetuses, comparing maternal administration of fluoxetine to a placebo control. The children will continue treatment or placebo up until 2 years of age.

There is an ongoing debate on the ethical and social aspects of trying to enhance intellectual abilities in Down syndrome, whether prenatally or postnatally. The various arguments are well summarized by de Wert and colleagues [45]. As an example, the likely increased autonomy for individuals with Down syndrome who have, after treatment, a higher IQ seems a worthwhile goal for these individuals, their families, and society. However, the potential change in personality that may come with it may affect the person, and his or her interaction with others, not necessarily leading to a happier life. A person with an IQ of 40–50 may live a "care-free" life, unaware of limitations, while at an IQ of 70–80, there may be full awareness of having a syndrome with many consequences.

Brain dysfunction treatment can still be considered symptomatic therapy. True causal treatment for aneuploidy for long seemed impossible, but appears on the horizon now. Jiang et al., in 2013, showed feasibility of a causal treatment of Down syndrome, by inactivation or silencing of the additional chromosome 21, using gene editing [46]. These investigators used the naturally occurring silencing gene on the X chromosome (XIST; X-Inactivation Specific Transcript). This XIST gene was inserted into a copy of chromosome 21 in induced pluripotent stem cells from Down syndrome patients. When this gene was activated by doxycycline, nearly all genes on the modified extra chromosome 21 were effectively silenced. However, the extra material remains present in the nucleus, which may still give spatial problems. Complete removal of the excess chromosome would likely be a more effective strategy.

Other methods of what is now called "chromosome therapy" are promotion of euploidy in an aneuploid culture, and forced loss and replacement of a chromosome [47]. Many safety issues still need to be resolved, such as the need to remove or silence one of the two identical chromosomes, otherwise a uniparental disomy is created with potentially detrimental effects.

COUNSELING AND ETHICAL ISSUES FOR FETAL GENETIC THERAPY

There seems to be little doubt that in the near future, following improved diagnostic techniques, genetic disease will be treated before birth. The unique feature of fetal therapy is not that the mother needs to decide for her fetus and future child (many other examples of patients unable to decide for themselves exist), but that there is the accepted alternative, at least in many countries, to elect for termination of pregnancy. The other option, to choose for expectant management and if possible, postnatal treatment is rarely controversial, and should perhaps be regarded as standard of care.

In the many examples of prenatally diagnosed genetic diseases where standard care means lifelong severe handicaps, the parents-to-be have to weigh this prospect, both for their future child itself, as well as for themselves and their family, against the also often lifelong burden of possible feelings of doubt and guilt associated with termination of pregnancy. The third option, fetal therapy, would be an easy and logical one if this intervention would be 100% safe and effective, leading to curing the disease completely in all cases, without risks for the fetus and without side effects to the mother. This, however, is rarely if ever the case.

In counseling parents on the option fetal therapy, as an additional choice next to the traditional options, expectant management and termination of pregnancy, a major challenge is to make sure parents understand all possible scenarios, their likelihood and their long-term consequences. Any new treatment, in particular presented by enthusiastic innovative clinician-researchers, bears the risk of being presented or perceived in an overoptimistic way. Obviously, there is hope that the new treatment will lead to a better prognosis, however, by definition there is still little experience, long-term follow-up studies are lacking and both parents and caregivers do not like to emphasize the (often distinct) possibility that the outcome may actually be worse than with standard care. One way to reduce the risk of nonobjective counseling is, in particular in trials, to have the counseling performed by well-informed but independent counselors, supported by properly designed written or online decision aids. Although patients entered in trials generally do better than those not participating (irrespective of the trial arm they are in), it is important to minimize the risk of "therapeutic misconception," emphasizing that a trial by definition means that there is insufficient evidence for the studied treatment.

In their comprehensive review on the scientific and ethical issues of gene editing, the European Society of Human Genetics [48] presented three crucial areas that they thought needed the most attention at this stage in the responsible development and use of gene editing technologies, particularly for uses that directly or indirectly affect humans:

1. Conducting careful scientific research to build an evidence base
2. Conducting ethical, legal and social issues (ELSI) research
3. Conducting meaningful stakeholder engagement, education, and dialogue (SEED).

Especially the last part, involving stakeholders including patient representatives and the general public, seems to gain popularity and acceptance. Since these processes are often time consuming (and thus expensive), they may not be able to keep up with the speed of technical advances. The target group of the treatment options, in particular families carrying genetic diseases, may have difficulties waiting for prolonged comprehensive assessments of all safety issues if there is hope for relief of their burden. Still, scientists and especially the eager, innovative clinicians and their professional societies need to guard their responsibility for careful introduction of these therapies. Preferably, in our view, without the need for legal or governmental control, although this may in many countries be unavoidable.

In the undoubtedly very promising field of gene editing, the main and most complex challenge is how to deal with germ line editing. Although curing a disease not only in the actual patient, but at the same time in his or her offspring seems a great prospect, politicians, lawyers, and the lay public are scared by terms such as "genetically modified children." de Wert and colleagues carefully unravel all legal and ethical aspects in their background document to the ESHRE statement on germ line editing [49].

At this point in time, the debate concentrates on whether or not, and if so what type of germ line editing can be allowed for science to progress. Even early stage preimplantation embryo gene editing will likely result in mosaic embryos and off-target effects that will be hard to detect. To reduce the mosaic risk, gene editing of the sperm cell or oocyte, or their precursors, or even stem cells that differentiate into sperm or oocyte, seems attractive. de Wert et al. state that "whereas most countries currently prohibit germline modification, many of the concepts used in relevant legal documents are ill-defined and ambiguous, including the distinction between research and clinical applications and basic definitions" [49].

We agree with the conclusion from these authors, who represent the European Society of Human Genetics (ESHG) and the European Society for Human Reproduction and Embryobiology (ESHRE), when they state that "from an ethical point of view, scientists and clinicians must respect the legal and regulatory framework in their country. They also have an important responsibility to help society understand and debate the full range of possible implications of the new technologies, and to contribute to regulations that are adapted to the dynamics of the field while taking account of ethical considerations and societal concerns."

REFERENCES

[1] Hudecova I. Digital PCR analysis of circulating nucleic acids. Clin Biochem 2015;48:948–56.
[2] Jatavan P, Chattipakorn N, Tongsong T. Fetal hemoglobin Bart's hydrops fetalis: pathophysiology, prenatal diagnosis and possibility of intrauterine treatment. J Matern Fetal Neonatal Med 2018;31(7):946–57.
[3] Wang W, Yuan Y, Zheng H, Wang Y, Zeng D, Yang Y, et al. A pilot study of noninvasive prenatal diagnosis of alpha- and beta-thalassemia with target capture sequencing of cell-free fetal DNA in maternal blood. Genet Test Mol Biomarkers 2017;21(7):433–9.

[4] Leung WC, Oepkes D, Seaward G, Ryan G. Serial sonographic findings of four fetuses with homozygous alpha-thalassemia-1 from 21 weeks onwards. Ultrasound Obstet Gynecol 2002;19(1):56–9.

[5] Chmait RH, Baskin JL, Carson S, Randolph LM, Hamilton A. Treatment of alpha(0)-thalassemia (–(SEA)/– (SEA)) via serial fetal and post-natal transfusions: can early fetal intervention improve outcomes? Hematology 2015;20(4):217–22.

[6] Levy H, Ghavami M. Maternal phenylketonuria: a metabolic teratogen. Teratology 1996;53:176–84.

[7] Best S, Wou K, Vora N, et al. Promises, pitfalls and practicalities of prenatal whole exome sequencing. Prenat Diagn 2017;37:1–10.

[8] Ampola MG, Mahoney MJ, Nakamura E, Tanaka K. Prenatal therapy of a patient with vitamin-B12-responsive methylmalonic acidemia. N Engl J Med 1975;293(7):313–7.

[9] Packman S, Cowan MJ, Golbus MS, Caswell NM, Sweetman L, Burri BJ, et al. Prenatal treatment of biotin responsive multiple carboxylase deficiency. Lancet 1982;1(8287):1435–8.

[10] Trefz FK, Scheible D, Frauendienst-Egger G, Huemer M, Suomala T, Fowler B, et al. Successful intrauterine treatment of a patient with cobalamin C defect. Mol Genet Metab Rep 2016;6:55–9.

[11] Yokoi K, Ito T, Maeda Y, Nakajima Y, Kurono Y, Sugiyama N, et al. A case of holocarboxylase synthetase deficiency with insufficient response to prenatal biotin therapy. Brain Dev 2009;31(10):775–8.

[12] de Koning TJ, Klomp LW, van Oppen AC, Beemer FA, Dorland L, van den Berg I, Berger R. Prenatal and early postnatal treatment in 3-phosphoglycerate-dehydrogenase deficiency. Lancet 2004;364(9452):2221–2.

[13] Weiner S, Scharf JI, Bolognese RJ, Librizzi RJ. Antenatal diagnosis and treatment of a fetal goiter. J Reprod Med 1980;24:39–42.

[14] Ribault V, Castanet M, Bertrand AM, Guibourdenche J, Vuillard E, Luton D, et al. French Fetal Goiter Study Group. Experience with intraamniotic thyroxine treatment in nonimmune fetal goitrous hypothyroidism in 12 cases. J Clin Endocrinol Metab 2009;94(10):3731–9.

[15] Mastrolia SA, Mandola A, Mazor M, Hershkovitz R, Mesner O, Beer-Weisel R, et al. Antenatal diagnosis and treatment of hypothyroid fetal goiter in an euthyroid mother: a case report and review of literature. J Matern Fetal Neonatal Med 2015;28(18):2214–20.

[16] Billingham RE, Brent L, Medawar PB. Actively acquired tolerance of foreign cells. Nature 1953;172:603–6.

[17] Flake AW, Harrison MR, Adzick NS, Zanjani ED. Transplantation of fetal hematopoietic stem cells in utero: the creation of hematopoietic chimeras. Science 1986;233:776–8.

[18] Peranteau WH, Endo M, Adibe OO, Flake AW. Evidence for an immune barrier after in utero hematopoietic-cell transplantation. Blood 2007;109:1331–3.

[19] Touraine JL, Raudrant D, Royo C, Rebaud A, Roncarolo MG, Souillet G, et al. In-utero transplantation of stem cells in bare lymphocyte syndrome. Lancet 1989;1:1382.

[20] Flake A, Puck J, Almeida-Porada G, et al. Successful treatment of X-linked severe combined immunodeficiency (X-SCID) by the in utero transplantation of CD34 enriched paternal bone marrow. Blood 1995;86:12Sa.

[21] Witt R, MacKenzie TC, Peranteau WH. Fetal stem cell and gene therapy. Semin Fetal Neonatal Med 2017;22(6):410–4.

[22] Le Blanc K, Gotherstrom C, Ringden O, Hassan M, McMahon R, Horwitz E, et al. Fetal mesenchymal stem-cell engraftment in bone after in utero transplantation in a patient with severe osteogenesis imperfecta. Transplantation 2005;79(11):1607–14.

[23] MacKenzie TC, David AL, Flake AW, Almeida-Porada G. Consensus statement from the first international conference for in utero stem cell transplantation and gene therapy. Front Pharmacol 2015;6:15.

[24] Davey MG, Riley JS, Andrews A, Tyminski A, Limberis M, Pogoriler JE, et al. Induction of immune tolerance to foreign protein via adeno-associated viral vector gene transfer in mid-gestation fetal sheep. PLoS One 2017;12:e0171132.

[25] Coutelle C, Themis M, Waddington S, Gregory L, Nivsarkar M, Buckley S, et al. The hopes and fears of in utero gene therapy for genetic disease—a review. Placenta 2003;24(Suppl. B):S114–21.

[26] Hacein-Bey-Abina S, von Kalle C, Schmidt M, Le Deist F, Wulffraat N, McIntyre E, et al. A serious adverse event after successful gene therapy for X-linked severe combined immunodeficiency. N Engl J Med 2003;348:255–6.

[27] Cong L, Ran FA, Cox D, Lin S, Barretto R, Habib N, et al. Multiplex genome engineering using CRISPR/Cas systems. Science 2013;339(6121):819–23.

[28] Mali P, Yang L, Esvelt KM, Aach J, Guell M, DiCarlo JE, et al. RNA-guided human genome engineering via Cas9. Science 2013;339(6121):823–6.

[29] Maeder ML, Gersbach CA. Genome-editing technologies for gene and cell therapy. Mol Ther 2016;24:430–46.

[30] Dever DP, Bak RO, Reinisch A, Camarena J, Washington G, Nicolas CE, et al. CRISPR/Cas9 β-globin gene targeting in human haematopoietic stem cells. Nature 2016;539(7629):384–9.

[31] Liang P, Xu Y, Zhang X, et al. CRISPR/Cas9-mediated gene editing in human tripronuclear zygotes. Protein Cell 2015;6:363–72.

[32] Kosiv KA, Gossett JM, Bai S, Collins 2nd RT. Congenital heart surgery on in-hospital mortality in trisomy 13 and 18. Pediatrics 2017;140(5):e20170772.

[33] Cao L, Du Y, Wang L. Fetal pleural effusion and Down syndrome. Intractable Rare Dis Res 2017;6:158–62.

[34] Mallmann MR, Graham V, Rösing B, Gottschalk I, Müller A, Gembruch U, et al. Thoracoamniotic shunting for fetal hydrothorax: predictors of intrauterine course and postnatal outcome. Fetal Diagn Ther 2017;41:58–65.

[35] Head E, Helman AM, Powell D, Schmitt FA. Down syndrome, beta-amyloid and neuroimaging. Free Radic Biol Med 2018;114:102–9.

[36] Gardiner KJ. Pharmacological approaches to improving cognitive function in Down syndrome: current status and considerations. Drug Des Devel Ther 2014;9:103–25.

[37] Guedj F, Pennings JL, Massingham LJ, Wick HC, Siegel AE, Tantravahi U, et al. An integrated human/murine transcriptome and pathway approach to identify prenatal treatments for Down syndrome. Sci Rep 2016;6:32353.

[38] Lee YM, Zampieri BL, Scott-McKean JJ, Johnson MW, Costa ACS. Generation of integration-free induced pluripotent stem cells from urine-derived cells isolated from individuals with Down syndrome. Stem Cells Transl Med 2017;6:1465–76.

[39] Guidi S, Bonasoni P, Ceccarelli C, Santini D, Gualtieri F, Ciani E, et al. Neurogenesis impairment and increased cell death reduce total neuron number in the hippocampal region of fetuses with Down syndrome. Brain Pathol 2008;18:180–97.

[40] Guedj F, Bianchi DW. Noninvasive prenatal testing creates an opportunity for antenatal treatment of Down syndrome. Prenat Diagn 2013;33:614–8.

[41] Incerti M, Horowitz K, Roberson R, Abebe D, Toso L, Caballero M, et al. Prenatal treatment prevents learning deficit in Down syndrome model. PLoS One 2012;7:e50724.

[42] Guedj F, Hines D, Foley JC, et al. Translating the transcriptome to develop antenatal treatments for fetuses with Down syndrome. Reprod Sci 2013;20(Suppl):64A.

[43] Guidi S, Stagni F, Bianchi P, Ciani E, Giacomini A, De Franceschi M, et al. Prenatal pharmacotherapy rescues brain development in a Down's syndrome mouse model. Brain 2014;137:380–401.

[44] Nakano-Kobayashi A, Awaya T, Kii I, Sumida Y, Okuno Y, Yoshida S, et al. Prenatal neurogenesis induction therapy normalizes brain structure and function in Down syndrome mice. Proc Natl Acad Sci USA 2017;114:10268–73.

[45] de Wert G, Dondorp W, Bianchi DW. Fetal therapy for Down syndrome: an ethical exploration. Prenat Diagn 2017;37:222–8.

[46] Jiang J, Jing Y, Cost GJ, Chiang JC, Kolpa HJ, Cotton AM, et al. Translating dosage compensation to trisomy 21. Nature 2013;500:296–300.

[47] Plona K, Kim T, Halloran K, Wynshaw-Boris A. Chromosome therapy: potential strategies for the correction of severe chromosome aberrations. Am J Med Genet C Semin Med Genet 2016;172:422–30.

[48] Howard HC, van El CG, Forzano F, Radojkovic D, Rial-Sebbag E, de Wert G, et al. One small edit for humans, one giant edit for humankind? Points and questions to consider for a responsible way forward for gene editing in humans. Eur J Hum Genet 2018;26(1):1–11.

[49] de Wert G, Heindryckx B, Pennings G, Clarke A, Eichenlaub-Ritter U, van El CG, et al. Responsible innovation in human germline gene editing: background document to the recommendations of ESHG and ESHRE. Eur J Hum Genet 2018;26(4):450–70.

Index

Note: Page numbers followed by *f* indicate figures, *t* indicate tables, *b* indicate boxes, and *ge* indicate glossary terms.